STATISTICS

IN A NUTSHELL

Second Edition

Sarah Boslaugh

O'REILLY®

Beijing · Cambridge · Farnham · Köln · Sebastopol · Tokyo

Statistics in a Nutshell, Second Edition
by Sarah Boslaugh

Published by O'Reilly Media, Inc., 1005 Gravenstein Highway North, Sebastopol, CA 95472.

O'Reilly books may be purchased for educational, business, or sales promotional use. Online editions are also available for most titles (*http://my.safaribooksonline.com*). For more information, contact our corporate/institutional sales department: 800-998-9938 or *corporate@oreilly.com*.

Editor: Mary Treseler	**Indexer:** Potomac Indexing, LLC
Production Editor: Kristen Borg	**Cover Designer:** Randy Comer
Copyeditor: nSight, Inc.	**Interior Designer:** David Futato
Proofreader: Katie DePasquale	**Illustrator:** Rebecca Demarest

July 2008:	First Edition.
November 2012:	Second Edition.

Revision History for the Second Edition:
    2012-10-31   First release

See *http://oreilly.com/catalog/errata.csp?isbn=9781449316822* for release details.

ISBN: 978-1-449-31682-2

[LSI]

1351704985

Table of Contents

Preface

The first edition of *Statistics in a Nutshell* was a resounding success, but all books can be improved, and I'm grateful to have the opportunity to revise this one. My basic approach to the material hasn't changed: this is much more a book for people who want to think about and understand statistics than it is a book showing you how to use a particular computing package or delving into the mathematical theory behind statistics formulas. This book is also a little different from many titles in the O'Reilly Nutshell series—it's really somewhere between a handbook for people who already know statistics and an introductory textbook for people learning statistics for the first time.

Despite the continued infiltration of statistics into many realms of life, one thing hasn't changed: telling people I work as a statistician is still the best way to derail a promising conversation at a party. For some reason, this seems to prompt people to tell me about how much they hated the required statistics class they needed for their college major or to prompt them to quote that old chestnut popularized by Mark Twain that there are three kinds of lies: lies, damned lies, and statistics.

Personally, I find statistics fascinating, and I love working in this field. I like teaching statistics as well, and I like to believe that I communicate this enthusiasm to others. It's often an uphill battle, however; many people seem to believe that statistics is no more than a set of tricks and manipulations whose purpose is to twist reality to mislead other people. Others take the opposite view, believing that statistics is a collection of magical procedures that will do their thinking for them.

OK, Just What Is Statistics?

Before you jump into the technical details of learning and using statistics, step back for a minute and consider what can be meant by the word "statistics." Don't worry if you don't understand all the vocabulary immediately; it will become clear over the course of reading this book.

When people speak of statistics, they usually mean one or more of the following:

1. Numerical data, such as the unemployment rate, the number of persons who die annually from bee stings, or the population of New York City in 2006 as compared to 1906.

2. Numbers used to describe samples of data as opposed to parameters (numbers used to describe populations). For instance, an advertising firm might be interested in the average age of people who subscribe to *Sports Illustrated*. To answer this question, it could draw a random sample of subscribers, calculate the mean of that sample (a statistic), and use that as an estimate of the mean of the entire population of subscribers (a parameter).

3. Particular procedures used to analyze data, and the results of those procedures, such as the *t* statistic or the chi-square statistic.

4. A field of study that develops and uses mathematical procedures to describe data and make decisions regarding it.

The type of statistics referred to in definition number 1 is not the primary concern of this book. If you simply want to find the latest figures on unemployment, health, or any of the myriad other topics on which governments and other organizations regularly release statistical data, your best bet is to consult a reference librarian or subject matter expert. If, however, you want to know how to interpret those figures (to understand why the mean is often misleading as a statement of average value, for instance, or the difference between crude and standardized mortality rates), *Statistics in a Nutshell* can definitely help you.

The concepts included in definition number 2 will be discussed in Chapter 3, which introduces inferential statistics, but these concepts also permeate the entire book. It is partly a question of vocabulary (*statistics* are numbers that describe *samples*, whereas *parameters* are numbers that describe *populations*) but underscores a fundamental point about the practice of statistics. The concept of using information gained from studying a sample to make statements about a population is the basis of inferential statistics, and inferential statistics is the primary focus of this book (as it is of most books about statistics).

Definition number 3 is also fundamental to most chapters of this book. The process of learning statistics is to some extent the process of learning particular statistical procedures, including how to calculate and interpret them, how to choose the appropriate statistic for a given situation, and so on. In fact, many new students of statistics subscribe primarily to this definition; learning statistics to them means learning to execute a set of statistical procedures. This is not so much an invalid approach to statistics as it is incomplete; learning to execute statistical procedures is a necessary part of the practice of statistics, but it is far from being the entire story. What's more, since computer software has made it increasingly easy for anyone, regardless of mathematical background, to produce statistical analyses, the need to understand and interpret statistics has far outstripped the need to learn how to do the calculations themselves.

Definition number 4 is nearest to my heart because I chose statistics as my professional field. If you are a secondary or post-secondary student, you are probably aware of this definition of statistics because many universities and colleges today either have a separate department of statistics or include statistics as a field of specialization within the department of mathematics. Statistics is increasingly taught in high school as well, and in the United States, enrollment in advanced placement (AP) statistics classes is increasing rapidly.

Statistics is not only a specialist subject at the university level. Many university departments require students to take one or more statistics courses alongside subjects in their major. In addition, it's worth knowing that many important techniques in modern statistics have been developed by people who learned and used statistics as part of their work in another field. Stephen Raudenbush, a pioneer in the development of hierarchical linear modeling, studied Policy Analysis and Evaluation Research at Harvard, and Edward Tufte, perhaps the world's leading expert on statistical graphics, began his career as a political scientist: he wrote his PhD dissertation at Yale on the American Civil Rights movement.

Because the use of statistics in many professions and at all levels from management to line workers is increasing, acquiring a basic knowledge of statistics has become a necessity for many people who have been out of school for years. Such individuals are often ill served by textbooks aimed at introductory college courses, which are too specialized, too focused on calculation, and too expensive.

Finally, statistics cannot be left to the statisticians because it's also a necessity to take part in modern civic life, in particular to understand much of what you read in the newspaper and hear on the television and radio. A working knowledge of statistics is the best check against the proliferation of misleading or outright false numerical claims (whether by politicians, advertisers, or social reformers), which seem to occupy an ever-increasing portion of our daily news diet. There's a reason that Darryl Huff's 1954 classic *How to Lie with Statistics* remains in print: statistics are easy to misuse, the common techniques of statistical distortion have been around for decades, and the best defense against those who would lie with statistics is to educate yourself so you can spot the lies and stop the liars in their tracks.

The Focus of This Book

There are so many statistics books already on the market that you might well wonder why I feel the need to add another to the pile. The primary reason is that I haven't found any statistics books that answer the needs I have addressed in *Statistics in a Nutshell*. In fact, if I may wax poetic for a moment, the situation is, to paraphrase the plight of Coleridge's Ancient Mariner, "books, books, everywhere, nor any with which to learn." The issues I have tried to address with this book are the following:

- The need for a book that focuses on using and understanding statistics in a research or applications context, not as a discrete set of mathematical techniques but as part of the process of reasoning with numbers.

- The need to integrate discussion of issues such as measurement and data management into an introductory statistics text.

- The need for a statistics book that isn't focused on a particular subject area. Elementary statistics is largely the same across subjects (a *t*-test is pretty much the same whether the data comes from medicine, finance, or criminal justice), so there's no need for a proliferation of texts presenting the same information with a slightly different spin.

- The need for an introductory statistics book that is compact, inexpensive, and easy for beginners to understand without being condescending or overly simplistic.

So who is the intended audience of *Statistics in a Nutshell*? I see three groups whose needs it particularly addresses:

- Students taking introductory statistics classes in high schools, colleges, and universities

- Adults who need to learn statistics as part of their current jobs or to be eligible for a promotion

- People who are interested in learning about statistics out of intellectual curiosity

My focus throughout *Statistics in a Nutshell* is not on particular techniques, although many are taught within this work, but on statistical reasoning. You might say that the focus in this book is less on *doing statistics* and more on *thinking statistically*. What does that mean? Several things are necessary to be able in the process of thinking with numbers. More particularly, I focus on thinking about data and using statistics to aid in that process. Most chapters include some practice exercises, but these are meant to provide an opportunity to review the material presented and think about the important concepts covered in the chapter; they are not meant to be mindless calculation.

All the material in *Statistics in a Nutshell* has been revised, and most of the chapters beefed up with new examples and exercises. In particular, more examples working with proportions have been added, as have additional examples using real data sets, from sources such as the United Nations Human Development Project and the Behavioral Risk Factor Surveillance System; both data sets are available for free download from the Internet, so students can experiment with them as well as replicate the analyses in this book. One new chapter has been added to this edition: Chapter 19. I added this chapter because of my observation that, particularly for people learning statistics for vocational reasons, the ability to communicate statistical information is at least as important as the ability to perform statistical computations. Several new appendixes have also been added, mainly to make the book more self-sufficient and user-friendly. These include probability tables for the most common distributions, a bibliography of online sources of information, and a glossary and table of statistical notation.

Statistics in the Age of Information

It's become fashionable to say that we're living in the Age of Information, when so many facts are collected and disseminated that no one could possibly keep up with them. This is one cliché based in truth; as a society, we are drowning in data, and the problem seems likely to increase. There are both positive and negative sides to this circumstance. On the positive side, wide access to computing technology and electronic means of data storage and dissemination have made information easier to access, so researchers have less need to travel to a particular library or archive to peruse printed copies of records.

However, data has no meaning in and of itself. It has to be organized and interpreted by human beings before it becomes meaningful, so participating fully in the Information Age requires becoming fluent in understanding data, including the ways it is collected, analyzed, and interpreted. And because the same data can often be interpreted in many ways to support radically different conclusions, even people who don't engage in statistical work themselves need to understand how statistics work and how to identify invalid claims and arguments based on the misuse of data.

Organization of This Book

Statistics in a Nutshell is organized in three parts: introductory material (Chapters 1–4) that lays the necessary foundation for the chapters that follow; inferential statistical techniques (Chapters 5–13), specialized techniques used in different professional fields (Chapters 14–16); and ancillary topics that are often part of the statistician's job, even if they are not strictly statistical (Chapters 17–20).

Here's a more detailed breakdown of the chapters:

Chapter 1, Basic Concepts of Measurement
    Discusses foundational issues for statistics, including levels of measurement, operationalization, proxy measurement, random and systematic error, reliability and validity, and types of bias.

Chapter 2, Probability
    Introduces the basics of probability, including trials, events, independence, mutual exclusivity, the addition and multiplication laws, combinations and permutations, conditional probability, and Bayes' theorem.

Chapter 3, Inferential Statistics
    Introduces some basic concepts of inferential statistics, including probability distributions, independent and dependent variables, populations and samples, common types of sampling, the central limit theorem, hypothesis testing, Type I and Type II errors, confidence intervals and *p*-values, and data transformation.

Chapter 4, Descriptive Statistics and Graphic Displays
    Introduces common measures of central tendency and dispersion, including mean, median, mode, range, interquartile range, variance, and standard deviation, and discusses outliers. Some of the most commonly used graphical techniques for presenting statistical information are also covered in this chapter,

including frequency tables, bar charts, pie charts, Pareto charts, stem and leaf plots, boxplots, histograms, scatterplots, and line graphs.

Chapter 5, Categorical Data

Reviews the concepts of categorical and interval data and introduces the $R \times C$ table. Statistics covered in this chapter include the chi-squared tests for independence, equality of proportions, and goodness of fit, Fisher's exact test, McNemar's test, large-sample tests for proportions, and measures of association for categorical and ordinal data.

Chapter 6, The t-Test

Discusses the *t*-distribution and the theory and use of the one-sample *t*-test, the two independent samples *t*-test, the repeated measures *t*-test, and the unequal variance *t*-test.

Chapter 7, The Pearson Correlation Coefficient

Introduces the concept of association with graphics displaying different strengths of association between two variables and discusses the Pearson Correlation Coefficient and the Coefficient of Determination.

Chapter 8, Introduction to Regression and ANOVA

Relates linear regression and ANOVA to the concept of the General Linear Model and discusses assumptions made when using these designs. Simple (bivariate) regression, one-way ANOVA, and post hoc testing are discussed and demonstrated.

Chapter 9, Factorial ANOVA and ANCOVA

Discusses more-complex ANOVA designs, including two-way and three-way ANOVA and ANCOVA, and presents the topic of interaction.

Chapter 10, Multiple Linear Regression

Extends the multiple regression model to include multiple predictors. Topics covered include relationships among predictor variables, standardized and unstandardized coefficients, dummy variables, methods of model building, and violations of assumptions of linear regression, including nonlinearity, autocorrelation, and heteroscedasticity.

Chapter 11, Logistic, Multinomial, and Polynomial Regression

Expands the technique of regression to data with binary outcomes (logistic regression), categorical outcomes (multinomial regression), and nonlinear models (polynomial regression) and discusses the problem of overfitting a model.

Chapter 12, Factor Analysis, Cluster Analysis, and Discriminant Function Analysis

Demonstrates three advanced statistical procedures, factor analysis, cluster analysis, and discriminant function analysis, and discusses the types of problems for which each technique might be useful.

Chapter 13, Nonparametric Statistics

Discusses when to use nonparametric rather than parametric statistics and presents nonparametric statistics for between-subjects and within-subjects designs, including the Wilcoxon Rank Sum and Mann-Whitney U tests, the sign test, the median test, the Kruskal-Wallis H test, the Wilcoxon signed rank test, and the Friedman test.

Chapter 14, Business and Quality Improvement Statistics

Demonstrates statistical procedures commonly used in business and quality improvement contexts. Analytical and statistical procedures covered include index numbers; time series; the minimax, maximax, and maximin decision criteria; decision making under risk; decision trees; and control charts.

Chapter 15, Medical and Epidemiological Statistics

Introduces concepts and demonstrates statistical procedures particularly relevant to medicine and epidemiology. Concepts and statistics presented include the definition and use of ratios, proportions, and rates; measures of prevalence and incidence; crude and standardized rates; direct and indirect standardization; measures of risk; confounding; the simple and Mantel-Haenszel odds ratio; and precision, power, and sample-size calculations.

Chapter 16, Educational and Psychological Statistics

Introduces concepts and statistical procedures commonly used in the fields of education and psychology. Subjects covered include percentiles; standardized scores; methods of test construction; classical test theory; the reliability of a composite test; measures of internal consistency, including coefficient alpha; and procedures for item analysis. An overview of item response theory is also provided.

Chapter 17, Data Management

Discusses practical issues in data management, including codebooks, the unit of analysis, procedures to troubleshoot an existing file, methods for storing data electronically, string and numeric data, and missing data.

Chapter 18, Research Design

Discusses observational and experimental studies, common elements of good research designs, the steps involved in data collection, types of validity, and methods to limit or eliminate the influence of bias.

Chapter 19, Communicating with Statistics

Covers general issues about communicating statistical information to different audiences and then provides more detail about writing for a professional journal, for the general public, and for the workplace.

Chapter 20, Critiquing Statistics Presented by Others

Offers guidelines for reviewing the use of statistics, including a checklist of questions to ask of any statistical presentation and examples of when legitimate statistical procedures may be manipulated to support questionable conclusions.

Six appendixes cover topics that are a necessary background to the material covered in the main text and provide references to supplemental reading:

Appendix A, Review of Basic Mathematics

Provides a self-test and review of basic arithmetic and algebra for people whose memory of their last math course is fast receding on the distant horizon. Topics covered include the laws of arithmetic, exponents, roots and logs, methods to solve equations and systems of equations, fractions, factorials, permutations, and combinations.

Appendix B, Introduction to Statistical Packages

Provides an introduction to some of the most common computer programs used for statistical applications, demonstrates basic analyses in each program, and discusses their relative strengths and weaknesses. Programs covered include Minitab, SPSS, SAS, and R; the use of Microsoft Excel (not a statistical package) for statistical analysis is also discussed.

Appendix C, References

An annotated bibliography organized by chapters that includes published works and websites cited in the text and others that are good starting points for people researching a particular topic.

Appendix D, Probability Tables for Common Distributions

Includes tables for the most commonly used statistical distributions—normal, *t*, binomial, and chi-square—as well as directions for using the tables. Even in the age of the computer and the Internet, it's worth knowing how to read a distribution table, and it's convenient to have the tables available in printed form.

Appendix E, Online Resources

A bibliography of some of the best sites on the Internet for people who are learning, using, or teaching statistics. This appendix is organized into general resources, glossaries, probability tables, online calculators, and online textbooks.

Appendix F, Glossary of Statistical Terms

Includes a table of the Greek alphabet (the bane of many a beginning statistician), a table of statistical notation, and a brief glossary of the major statistical terms used in this book.

This book is a tool that can be adapted according to the background and needs of individual readers. Some of the chapters cover subjects that are often skipped in introductory statistics books but that I think are important; these include data management, writing about statistics, and reading statistical articles written by others. These chapters also serve as useful references for people who suddenly find themselves placed in charge of managing the data for a project or who have been appointed, more or less out of the blue, to create a statistical presentation about their team's work. Neither scenario, unfortunately, is particularly uncommon.

Classification of what is elementary and what is advanced depends on an individual's background and purposes. I designed *Statistics in a Nutshell* to answer the needs of many types of users. For this reason, there's no perfect way to organize the material to meet everyone's needs, which brings us to an important point: there's no reason you should feel the need to read the chapters in the order they are presented here. Statistics presents many chicken-and-egg dilemmas. For instance, you can't design experiments without knowing what statistics are available to you, but you can't understand how statistics are used without knowing something about research design. Similarly, it might seem logical that someone assigned to manage data should already have experience in statistical analysis, but I've advised many research assistants and project managers who are put in charge of large data sets before they've completed a single course in statistics. So use the chapters in the way that best facilitates your specific purposes, and don't be shy about skipping around and focusing on whatever meets your particular needs.

Not all the material in this book will be relevant to everyone; this is most obviously the case with Chapters 14–16, which are written with particular subject areas in mind (business and quality improvement, medicine and epidemiology, and education and psychology, respectively). However, it's wise to keep an open mind regarding what statistics you need to know. You might currently believe that you will never need to conduct a nonparametric test or a logistic regression analysis, but you never know what will come in handy in the future. It's also a mistake to compartmentalize too much by subject field; because statistical techniques are ultimately about numbers rather than content, techniques developed in one field often prove to be useful in another. For instance, control charts (covered in Chapter 14) were developed in a manufacturing context but are now used in many fields from medicine to education, whereas the odds ratio (covered in Chapter 15) was developed in epidemiology but is now applied to all sorts of data.

Conventions Used in This Book

The following typographical conventions are used in this book:

Plaintext

> Indicates menu titles, menu options, menu buttons, and keyboard accelerators (such as Alt and Ctrl).

Italic

> Indicates new terms, URLs, email addresses, filenames, file extensions, pathnames, directories, and Unix utilities.

 This icon signifies a tip, suggestion, or general note.

 This icon indicates a warning or caution.

Using Code Examples

This book is here to help you get your job done. In general, you may use the code in this book in your programs and documentation. You do not need to contact us for permission unless you're reproducing a significant portion of the code. For example, writing a program that uses several chunks of code from this book does not require permission. Selling or distributing a CD-ROM of examples from O'Reilly books does require permission. Answering a question by citing this book and quoting example code does not require permission. Incorporating a significant amount of example code from this book into your product's documentation does require permission.

We appreciate, but do not require, attribution. An attribution usually includes the title, author, publisher, and ISBN. For example: "*Statistics in a Nutshell* by Sarah Boslaugh (O'Reilly). Copyright 2013 Sarah Boslaugh, 978-1-449-31682-2."

If you feel your use of code examples falls outside fair use or the permission given above, feel free to contact us at *permissions@oreilly.com*.

Safari® Books Online

Safari Books Online (*www.safaribooksonline.com*) is an on-demand digital library that delivers expert content in both book and video form from the world's leading authors in technology and business.

Technology professionals, software developers, web designers, and business and creative professionals use Safari Books Online as their primary resource for research, problem solving, learning, and certification training.

Safari Books Online offers a range of product mixes and pricing programs for organizations, government agencies, and individuals. Subscribers have access to thousands of books, training videos, and prepublication manuscripts in one fully searchable database from publishers like O'Reilly Media, Prentice Hall Professional, Addison-Wesley Professional, Microsoft Press, Sams, Que, Peachpit Press, Focal Press, Cisco Press, John Wiley & Sons, Syngress, Morgan Kaufmann, IBM Redbooks, Packt, Adobe Press, FT Press, Apress, Manning, New Riders, McGraw-Hill, Jones & Bartlett, Course Technology, and dozens more. For more information about Safari Books Online, please visit us online.

How to Contact Us

Please address comments and questions concerning this book to the publisher:

> O'Reilly Media, Inc.
> 1005 Gravenstein Highway North
> Sebastopol, CA 95472
> 800-998-9938 (in the United States or Canada)
> 707-829-0515 (international or local)
> 707-829-0104 (fax)

We have a web page for this book, where we list errata, examples, and any additional information. You can access this page at *http://oreil.ly/stats-nutshell*.

To comment or ask technical questions about this book, send email to *bookquestions@oreilly.com*.

For more information about our books, courses, conferences, and news, see our website at *http://www.oreilly.com*.

Find us on Facebook: *http://facebook.com/oreilly*

Follow us on Twitter: *http://twitter.com/oreillymedia*

Watch us on YouTube: *http://www.youtube.com/oreillymedia*

Acknowledgments

Only one author is listed on the cover of this book, but the contributions of many people played a role in its creation.

I would like to thank my agent, Neil Salkind, for his continued guidance and support; the crew at O'Reilly, including Mary Treseler, Sarah Schneider, and Meghan Blanchette; and all the statisticians who assisted in the technical review process. I would also like to thank my nonstatistician friends who kept pestering me to explain statistical concepts to them and thus encouraged me to write this book, and my colleagues at the Center for Sustainable Journalism at Kennesaw State University for their forbearance and tolerance while I have been working on this revision. On a personal note, I would like to thank my former colleague Rand Ross at Washington University in St. Louis for helping me remain sane throughout the writing process for the first edition and my husband Dan Peck for being the very model of a modern supportive spouse.

Basic Concepts of Measurement

Before you can use statistics to analyze a problem, you must convert information about the problem into data. That is, you must establish or adopt a system of assigning values, most often numbers, to the objects or concepts that are central to the problem in question. This is not an esoteric process but something people do every day. For instance, when you buy something at the store, the price you pay is a measurement: it assigns a number signifying the amount of money that you must pay to buy the item. Similarly, when you step on the bathroom scale in the morning, the number you see is a measurement of your body weight. Depending on where you live, this number may be expressed in either pounds or kilograms, but the principle of assigning a number to a physical quantity (weight) holds true in either case.

Data need not be inherently numeric to be useful in an analysis. For instance, the categories *male* and *female* are commonly used in both science and everyday life to classify people, and there is nothing inherently numeric about these two categories. Similarly, we often speak of the colors of objects in broad classes such as *red* and *blue*, and there is nothing inherently numeric about these categories either. (Although you could make an argument about different wavelengths of light, it's not necessary to have this knowledge to classify objects by color.)

This kind of thinking in categories is a completely ordinary, everyday experience, and we are seldom bothered by the fact that different categories may be applied in different situations. For instance, an artist might differentiate among colors such as *carmine*, *crimson*, and *garnet*, whereas a layperson would be satisfied to refer to all of them as *red*. Similarly, a social scientist might be interested in collecting information about a person's marital status in terms such as *single—never married*, *single—divorced*, and *single—widowed*, whereas to someone else, a person in any of those three categories could simply be considered *single*. The point is that the level of detail used in a system of classification should be appropriate, based on the reasons for making the classification and the uses to which the information will be put.

Measurement

Measurement is the process of systematically assigning numbers to objects and their properties to facilitate the use of mathematics in studying and describing objects and their relationships. Some types of measurement are fairly concrete: for instance, measuring a person's weight in pounds or kilograms or his height in feet and inches or in meters. Note that the particular system of measurement used is not as important as the fact that we apply a consistent set of rules: we can easily convert a weight expressed in kilograms to the equivalent weight in pounds, for instance. Although any system of units may seem arbitrary (try defending feet and inches to someone who grew up with the metric system!), as long as the system has a consistent relationship with the property being measured, we can use the results in calculations.

Measurement is not limited to physical qualities such as height and weight. Tests to measure abstract constructs such as intelligence or scholastic aptitude are commonly used in education and psychology, and the field of psychometrics is largely concerned with the development and refinement of methods to study these types of constructs. Establishing that a particular measurement is accurate and meaningful is more difficult when it can't be observed directly. Although you can test the accuracy of one scale by comparing results with those obtained from another scale known to be accurate, and you can see the obvious use of knowing the weight of an object, the situation is more complex if you are interested in measuring a construct such as intelligence. In this case, not only are there no universally accepted measures of intelligence against which you can compare a new measure, there is not even common agreement about what "intelligence" means. To put it another way, it's difficult to say with confidence what someone's actual intelligence is because there is no certain way to measure it, and in fact, there might not even be common agreement on what it is. These issues are particularly relevant to the social sciences and education, where a great deal of research focuses on just such abstract concepts.

Levels of Measurement

Statisticians commonly distinguish four types or levels of measurement, and the same terms can refer to data measured at each level. The levels of measurement differ both in terms of the meaning of the numbers used in the measurement system and in the types of statistical procedures that can be applied appropriately to data measured at each level.

Nominal Data

With *nominal* data, as the name implies, the numbers function as a *name* or label and do not have numeric meaning. For instance, you might create a variable for gender, which takes the value 1 if the person is male and 0 if the person is female. The 0 and 1 have no numeric meaning but function simply as labels in the same way that you might record the values as M or F. However, researchers often prefer numeric coding systems for several reasons. First, it can simplify analyzing the data because some statistical packages will not accept nonnumeric values for use in certain procedures. (Hence, any data coded nonnumerically would have to be re-coded before analysis.) Second, coding with numbers bypasses some issues in data entry, such as the conflict between upper- and lowercase letters (to a computer, *M* is a different value than *m*, but a person doing data entry might treat the two characters as equivalent).

Nominal data is not limited to two categories. For instance, if you were studying the relationship between years of experience and salary in baseball players, you might classify the players according to their primary position by using the traditional system whereby 1 is assigned to the pitchers, 2 to the catchers, 3 to first basemen, and so on.

If you can't decide whether your data is nominal or some other level of measurement, ask yourself this question: do the numbers assigned to this data represent some quality such that a higher value indicates that the object has more of that quality than a lower value? Consider the example of coding gender so 0 signifies a female and 1 signifies a male. Is there some quality of gender-ness of which men have more than women? Clearly not, and the coding scheme would work as well if women were coded as 1 and men as 0. The same principle applies in the baseball example: there is no quality of baseball-ness of which outfielders have more than pitchers. The numbers are merely a convenient way to label subjects in the study, and the most important point is that every position is assigned a distinct value. Another name for nominal data is *categorical* data, referring to the fact that the measurements place objects into categories (male or female, catcher or first baseman) rather than measuring some intrinsic quality in them. Chapter 5 discusses methods of analysis appropriate for this type of data, and some of the techniques covered in Chapter 13 on nonparametric statistics are also appropriate for categorical data.

When data can take on only two values, as in the male/female example, it can also be called *binary* data. This type of data is so common that special techniques have been developed to study it, including logistic regression (discussed in Chapter 11), which has applications in many fields. Many medical statistics, such as the odds ratio and the risk ratio (discussed in Chapter 15), were developed to describe the relationship between two binary variables because binary variables occur so frequently in medical research.

Ordinal Data

Ordinal data refers to data that has some meaningful *order*, so that higher values represent more of some characteristic than lower values. For instance, in medical practice, burns are commonly described by their degree, which describes the amount of tissue damage caused by the burn. A first-degree burn is characterized by redness of the skin, minor pain, and damage to the epidermis (outer layer of skin) only. A second-degree burn includes blistering and involves the superficial layer of the dermis (the layer of skin between the epidermis and the subcutaneous tissues), and a third-degree burn extends through the dermis and is characterized by charring of the skin and possibly destruction of nerve endings. These categories may be ranked in a logical order: first-degree burns are the least serious in terms of tissue damage, second-degree burns more serious, and third-degree burns the most serious.

However, there is no metric analogous to a ruler or scale to quantify how great the distance between categories is, nor is it possible to determine whether the difference between first- and second-degree burns is the same as the difference between second- and third-degree burns.

Many ordinal scales involve ranks. For instance, candidates applying for a job may be ranked by the personnel department in order of desirability as a new hire. This ranking tells you who is the preferred candidate, the second most preferred, and so on, but does not tell you whether the first and second candidates are in fact very similar to each other or the first-ranked candidate is much more preferable than the second. You could also rank countries of the world in order of their population, creating a meaningful order without saying anything about whether, say, the difference between the 30th and 31st countries was similar to that between the 31st and 32nd countries. The numbers used for measurement with ordinal data carry more meaning than those used in nominal data, and many statistical techniques have been developed to make full use of the information carried in the ordering while not assuming any further properties of the scales. For instance, it is appropriate to calculate the median (central value) of ordinal data but not the mean because it assumes equal intervals and requires division, which requires ratio-level data.

Interval Data

Interval data has a meaningful order and has the quality of *equal intervals* between measurements, representing equal changes in the quantity of whatever is being measured. The most common example of the interval level of measurement is the Fahrenheit temperature scale. If you describe temperature using the Fahrenheit scale, the difference between 10 degrees and 25 degrees (a difference of 15 degrees) represents the same amount of temperature change as the difference between 60 and 75 degrees. Addition and subtraction are appropriate with interval scales because a difference of 10 degrees represents the same amount of change in temperature over the entire scale. However, the Fahrenheit scale has no natural zero point because 0 on the Fahrenheit scale does not represent an absence of temperature but simply a location relative to other temperatures. Multiplication and division are not appropriate with interval data: there is no mathematical sense in the statement that 80 degrees is twice as hot as 40 degrees, for instance (although it is valid to say that 80 degrees is 40 degrees hotter than 40 degrees). Interval scales are a rarity, and it's difficult to think of a common example other than the Fahrenheit scale. For this reason, the term "interval data" is sometimes used to describe both interval and ratio data (discussed in the next section).

Ratio Data

Ratio data has all the qualities of interval data (meaningful order, equal intervals) and a natural zero point. Many physical measurements are ratio data: for instance, height, weight, and age all qualify. So does income: you can certainly earn 0 dollars in a year or have 0 dollars in your bank account, and this signifies an absence of money. With ratio-level data, it is appropriate to multiply and divide as well as add and subtract; it makes sense to say that someone with $100 has twice as much money as someone with $50 or that a person who is 30 years old is 3 times as old as someone who is 10.

It should be noted that although many physical measurements are interval-level, most psychological measurements are ordinal. This is particularly true of measures of value or preference, which are often measured by a Likert scale. For instance, a person might be presented with a statement (e.g., "The federal government should increase aid to education") and asked to choose from an ordered set of responses (e.g., strongly agree, agree, no opinion, disagree, strongly disagree). These choices are sometimes assigned numbers (e.g., 1—strongly agree, 2—agree, etc.), and this sometimes gives people the impression that it is appropriate to apply interval or ratio techniques (e.g., computation of means, which involves division and is therefore a ratio technique) to such data. Is this correct? Not from the point of view of a statistician, but sometimes you do have to go with what the boss wants rather than what you believe to be true in absolute terms.

Continuous and Discrete Data

Another important distinction is that between *continuous* and *discrete* data. Continuous data can take any value or any value within a range. Most data measured by interval and ratio scales, other than that based on counting, is continuous: for instance, weight, height, distance, and income are all continuous.

In the course of data analysis and model building, researchers sometimes recode continuous data in categories or larger units. For instance, weight may be recorded in pounds but analyzed in 10-pound increments, or age recorded in years but analyzed in terms of the categories of *0–17, 18–65*, and *over 65*. From a statistical point of view, there is no absolute point at which data becomes continuous or discrete for the purposes of using particular analytic techniques (and it's worth remembering that if you record age in years, you are still imposing discrete categories on a continuous variable). Various rules of thumb have been proposed. For instance, some researchers say that when a variable has 10 or more categories (or, alternatively, 16 or more categories), it can safely be analyzed as continuous. This is a decision to be made based on the context, informed by the usual standards and practices of your particular discipline and the type of analysis proposed.

Discrete variables can take on only particular values, and there are clear boundaries between those values. As the old joke goes, you can have 2 children or 3 children but not 2.37 children, so "number of children" is a discrete variable. In fact, any variable based on counting is discrete, whether you are counting the number of books purchased in a year or the number of prenatal care visits made during a pregnancy. Data measured on the nominal scale is always discrete, as is binary and rank-ordered data.

Operationalization

People just starting out in a field of study often think that the difficulties of research rest primarily in statistical analysis, so they focus their efforts on learning mathematical formulas and computer programming techniques to carry out statistical calculations. However, one major problem in research has very little to do with either mathematics or statistics and everything to do with knowing your field of study and thinking carefully through practical problems of measurement. This is the problem of *operationalization*, which means the process of specifying how a concept will be defined and measured.

Operationalization is always necessary when a quality of interest cannot be measured directly. An obvious example is intelligence. There is no way to measure intelligence directly, so in the place of such a direct measurement, we accept something that we can measure, such as the score on an IQ test. Similarly, there is no direct way to measure "disaster preparedness" for a city, but we can operationalize the concept by creating a checklist of tasks that should be performed and giving each city a disaster-preparedness score based on the number of tasks completed and the quality or thoroughness of completion. For a third example, suppose you wish to measure the amount of physical activity performed by individual subjects in a study. If you do not have the capacity to monitor their exercise behavior directly, you can operationalize "amount of physical activity" as the amount indicated on a self-reported questionnaire or recorded in a diary.

Because many of the qualities studied in the social sciences are abstract, operationalization is a common topic of discussion in those fields. However, it is applicable to many other fields as well. For instance, the ultimate goals of the medical profession include reducing mortality (death) and reducing the burden of disease and suffering. Mortality is easily verified and quantified but is frequently too blunt an instrument to be useful since it is a thankfully rare outcome for most diseases. "Burden of disease" and "suffering," on the other hand, are concepts that could be used to define appropriate outcomes for many studies but that have no direct means of measurement and must therefore be operationalized. Examples of operationalization of burden of disease include measurement of viral levels in the bloodstream for patients with AIDS and measurement of tumor size for people with cancer. Decreased levels of suffering or improved quality of life may be operationalized as a higher self-reported health state, a higher score on a survey instrument designed to measure quality of life, an improved mood state as measured through a personal interview, or reduction in the amount of morphine requested for pain relief.

Some argue that measurement of even physical quantities such as length require operationalization because there are different ways to measure even concrete properties such as length. (A ruler might be the appropriate instrument in some circumstances, a micrometer in others.) Even if you concede this point, it seems clear that the problem of operationalization is much greater in the human sciences, when the objects or qualities of interest often cannot be measured directly.

Proxy Measurement

The term *proxy measurement* refers to the process of substituting one measurement for another. Although deciding on proxy measurements can be considered as a subclass of operationalization, this book will consider it as a separate topic. The most common use of proxy measurement is that of substituting a measurement that is inexpensive and easily obtainable for a different measurement that would be more difficult or costly, if not impossible, to collect. Another example is collecting information about one person by asking another, for instance, by asking a parent to rate her child's mood state.

For a simple example of proxy measurement, consider some of the methods police officers use to evaluate the sobriety of individuals while in the field. Lacking a portable medical lab, an officer can't measure a driver's blood alcohol content directly to determine whether the driver is legally drunk. Instead, the officer might rely on observable signs associated with drunkenness, simple field tests that are believed to correlate well with blood alcohol content, a breath alcohol test, or all of these. Observational signs of alcohol intoxication include breath smelling of alcohol, slurred speech, and flushed skin. Field tests used to evaluate alcohol intoxication quickly generally require the subjects to perform tasks such as standing on one leg or tracking a moving object with their eyes. A Breathalyzer test measures the amount of alcohol in the breath. None of these evaluation methods provides a direct test of the amount of alcohol in the blood, but they are accepted as reasonable approximations that are quick and easy to administer in the field.

To look at another common use of proxy measurement, consider the various methods used in the United States to evaluate the quality of health care provided by hospitals and physicians. It is difficult to think of a direct way to measure quality of care, short of perhaps directly observing the care provided and evaluating it in relation to accepted standards (although you could also argue that the measurement involved in such an evaluation process would still be an operationalization of the abstract concept of "quality of care"). Implementing such an evaluation method would be prohibitively expensive, would rely on training a large crew of evaluators and relying on their consistency, and would be an invasion of patients' right to privacy. A solution commonly adopted instead is to measure processes that are assumed to reflect higher quality of care: for instance, whether anti-tobacco counseling was appropriately provided in an office visit or whether appropriate medications were administered promptly after a patient was admitted to the hospital.

Proxy measurements are most useful if, in addition to being relatively easy to obtain, they are good indicators of the true focus of interest. For instance, if correct execution of prescribed processes of medical care for a particular treatment is closely related to good patient outcomes for that condition, and if poor or nonexistent execution of those processes is closely related to poor patient outcomes, then execution of these processes may be a useful proxy for quality. If that close relationship does not exist, then the usefulness of the proxy measurements is less certain. No mathematical test will tell you whether one measure is a good proxy for another, although computing statistics such as correlations or chi-squares between the measures might help evaluate this issue. In addition, proxy measurements can pose their own difficulties. To take the example of evaluating medical care in terms of procedures performed, this method assumes that it is possible to determine, without knowledge of individual cases, what constitutes appropriate treatment and that records are available that contain the information needed to determine what procedures were performed. Like many measurement issues, choosing good proxy measurements is a matter of judgment informed by knowledge of the subject area, usual practices in the field in question, and common sense.

Surrogate Endpoints

A surrogate endpoint is a type of proxy measurement sometimes used in clinical trials as a substitute for a true clinical endpoint. For instance, a treatment might be intended to prevent death (a true clinical endpoint), but because death from the condition being treated might be rare, a surrogate endpoint may be used to accrue evidence more quickly about the treatment's effectiveness. A surrogate endpoint is usually a biomarker that is correlated with a true clinical endpoint. For instance, if a drug is intended to prevent death from prostate cancer, a surrogate endpoint might be tumor shrinkage or reduction in levels of prostate-specific antigens.

The problem with using surrogate endpoints is that although a treatment might be effective in producing improvement in these endpoints, it does not necessarily mean that it will be successful in achieving the clinical outcome of interest. For instance, a meta-analysis by Stefan Michiels and colleagues (listed in Appendix C) found that for locally advanced head and neck squamous-cell carcinoma, the correlation between locoregional control (a surrogate endpoint) and overall survival (the true clinical endpoint) ranged from 0.65 to 0.76 (if results had been identical for both endpoints, the correlation would have been 1.00), whereas the correlation between event-free survival (a surrogate endpoint) and overall survival ranged from 0.82 to 0.90.

Surrogate endpoints are sometimes misused by being added after the fact to a clinical trial, being used as substitutes for outcomes defined before the trial begins, or both. Because a surrogate endpoint might be easier to achieve (e.g., improvement in progression-free survival in the trial for an anti-cancer drug rather than improvement in overall survival), this can lead to a new drug being approved on the basis of effectiveness when it might have little effect on the true endpoint or even have a deleterious effect. For further general discussion of issues relating to surrogate endpoints, see the article by Thomas R. Fleming cited in Appendix C.

True and Error Scores

We can safely assume that few, if any, measurements are completely accurate. This is true not only because measurements are made and recorded by human beings but also because the process of measurement often involves assigning discrete numbers to a continuous world. One concern of measurement theory is conceptualizing and quantifying the degree of error present in a particular set of measurements and evaluating the sources and consequences of that error.

Classical measurement theory conceives of any measurement or observed score as consisting of two parts: true score (T) and error (E). This is expressed in the following formula:

$$X = T + E$$

where X is the observed measurement, T is the true score, and E is the error. For instance, a bathroom scale might measure someone's weight as 120 pounds when that person's true weight is 118 pounds, and the error of 2 pounds is due to the inaccuracy of the scale. This would be expressed, using the preceding formula, as:

$$120 = 118 + 2$$

which is simply a mathematical equality expressing the relationship among the three components. However, both T and E are hypothetical constructs. In the real world, we seldom know the precise value of the true score and therefore cannot know the exact value of the error score either. Much of the process of measurement involves estimating both quantities and maximizing the true component while minimizing error. For instance, if you took a number of measurements of one person's body weight in a short period (so that his true weight could be assumed to have remained constant), using a recently calibrated scale, you might accept the average of all those measurements as a good estimate of that individual's true weight. You could then consider the variance between this average and each individual measurement as the error due to the measurement process, such as slight malfunctioning in the scale or the technician's imprecision in reading and recording the results.

Random and Systematic Error

Because we live in the real world rather than a Platonic universe, we assume that all measurements contain some error. However, not all error is created equal, and we can learn to live with *random error* while doing whatever we can to avoid *systematic error*. Random error is error due to chance: it has no particular pattern and is assumed to cancel itself out over repeated measurements. For instance, the error scores over a number of measurements of the same object are assumed to have a mean of zero. Therefore, if someone is weighed 10 times in succession on the same scale, you may observe slight differences in the number returned to you: some will be higher than the true value, and some will be lower. Assuming the true weight is 120 pounds, perhaps the first measurement will return an observed weight of 119 pounds (including an error of –1 pound), the second an observed weight of 122 pounds (for an error of +2 pounds), the third an observed weight of 118.5 pounds (an error of –1.5 pounds), and so on. If the scale is accurate and the only error is random, the average error over many trials will be 0, and the average observed weight will be 120 pounds. You can strive to reduce the amount of random error by using more accurate instruments, training your technicians to use them correctly, and so on, but you cannot expect to eliminate random error entirely.

Two other conditions are assumed to apply to random error: it is unrelated to the true score, and the error component of one measurement is unrelated to the error component of any other measurement. The first condition means that the value of the error component of any measurement is not related to the value of the true score for that measurement. For instance, if you measure the weights of a number of individuals whose true weights differ, you would not expect the error component of each measurement to have any relationship to each individual's true weight. This means that, for example, the error component should not systematically be larger when the true score (the individual's actual weight) is larger. The second condition means that the error component of each score is independent and unrelated to the error component for any other score. For instance, in a series of measurements, a pattern of the size of the error component should not be increasing over time so that later measurements have larger errors, or errors in a consistent direction, relative to earlier measurements. The first requirement is sometimes expressed by saying that the correlation of true and error scores is 0, whereas the second is sometimes expressed by saying that the correlation of the error components is 0 (correlation is discussed in more detail in Chapter 7).

In contrast, systematic error has an observable pattern, is not due to chance, and often has a cause or causes that can be identified and remedied. For instance, a scale might be incorrectly calibrated to show a result that is 5 pounds over the true weight, so the average of multiple measurements of a person whose true weight is 120 pounds would be 125 pounds, not 120. Systematic error can also be due to human factors: perhaps the technician is reading the scale's display at an angle so that she sees the needle as registering higher than it is truly indicating. If a pattern is detected with systematic error, for instance, measurements drifting higher over time (so the error components are random at the beginning of the experiment, but later on are consistently high), this is useful information because we can intervene and recalibrate the scale. A great deal of effort has been expended to identify sources of systematic error and devise methods to identify and eliminate them: this is discussed further in the upcoming section "Measurement Bias" on page 14.

Reliability and Validity

There are many ways to assign numbers or categories to data, and not all are equally useful. Two standards we commonly use to evaluate methods of measurement (for instance, a survey or a test) are *reliability* and *validity*. Ideally, we would like every method we use to be both reliable and valid. In reality, these qualities are not absolutes but are matters of degree and often specific to circumstance. For instance, a survey that is highly reliable when used with demographic groups might be unreliable when used with a different group. For this reason, rather than discussing reliability and validity as absolutes, it is often more useful to evaluate how valid and reliable a method of measurement is for a particular purpose and whether particular levels of reliability and validity are acceptable in a specific context. Reliability and validity are also discussed in Chapter 18 in the context of research design, and in Chapter 16 in the context of educational and psychological testing.

Reliability

Reliability refers to how consistent or repeatable measurements are. For instance, if we give the same person the same test on two occasions, will the scores be similar on both occasions? If we train three people to use a rating scale designed to measure the quality of social interaction among individuals, then show each of them the same film of a group of people interacting and ask them to evaluate the social interaction exhibited, will their ratings be similar? If we have a technician weigh the same part 10 times using the same instrument, will the measurements be similar each time? In each case, if the answer is yes, we can say the test, scale, or rater is reliable.

Much of the theory of reliability was developed in the field of educational psychology, and for this reason, measures of reliability are often described in terms of evaluating the reliability of tests. However, considerations of reliability are not limited to educational testing; the same concepts apply to many other types of measurements, including polling, surveys, and behavioral ratings.

The discussion in this chapter will remain at a basic level. Information about calculating specific measures of reliability is discussed in more detail in Chapter 16 in the context of test theory. Many of the measures of reliability draw on the *correlation coefficient* (also called simply the *correlation*), which is discussed in detail in Chapter 7, so beginning statisticians might want to concentrate on the logic of reliability and validity and leave the details of evaluating them until after they have mastered the concept of the correlation coefficient.

There are three primary approaches to measuring reliability, each useful in particular contexts and each having particular advantages and disadvantages:

- Multiple-occasions reliability
- Multiple-forms reliability
- Internal consistency reliability

Multiple-occasions reliability, sometimes called *test-retest reliability*, refers to how similarly a test or scale performs over repeated administration. For this reason, it is sometimes referred to as an index of *temporal stability*, meaning stability over time. For instance, you might have the same person do two psychological assessments of a patient based on a videotaped interview, with the assessments performed two weeks apart, and compare the results. For this type of reliability to make sense, you must assume that the quantity being measured has not changed, hence the use of the same videotaped interview rather than separate live interviews with a patient whose psychological state might have changed over the two-week period. Multiple-occasions reliability is not a suitable measure for volatile qualities, such as mood state, or if the quality or quantity being measured could have changed in the time between the two measurements (for instance, a student's knowledge of a subject she is actively studying). A common technique for assessing multiple-occasions reliability is to compute the correlation coefficient between the scores from each occasion of testing; this is called the *coefficient of stability*.

Multiple-forms reliability (also called *parallel-forms reliability*) refers to how similarly different versions of a test or questionnaire perform in measuring the same entity. A common type of multiple-forms reliability is *split-half reliability* in which a pool of items believed to be homogeneous is created, then half the items are allocated to form A and half to form B. If the two (or more) forms of the test are administered to the same people on the same occasion, the correlation between the scores received on each form is an estimate of multiple-forms reliability. This correlation is sometimes called the *coefficient of equivalence*. Multiple-forms reliability is particularly important for standardized tests that exist in multiple versions. For instance, different forms of the SAT (Scholastic Aptitude Test, used to measure academic ability among students applying to American colleges and universities) are calibrated so the scores achieved are equivalent no matter which form a particular student takes.

Internal consistency reliability refers to how well the items that make up an instrument (for instance, a test or survey) reflect the same construct. To put it another way, internal consistency reliability measures how much the items on an instrument are measuring the same thing. Unlike multiple-forms and multiple-occasions reliability, internal consistency reliability can be assessed by administering a single instrument on a single occasion. Internal consistency reliability is a more complex quantity to measure than multiple-occasions or parallel-forms reliability, and several methods have been developed to evaluate it; these are further discussed in Chapter 16. However, all these techniques depend primarily on the inter-item correlation, that is, the correlation of each item on a scale or a test with each other item. If such correlations are high, that is interpreted as evidence that the items are measuring the same thing, and the various statistics used to measure internal consistency reliability will all be high. If the inter-item correlations are low or inconsistent, the internal consistency reliability statistics will be lower, and this is interpreted as evidence that the items are not measuring the same thing.

Two simple measures of internal consistency are most useful for tests made up of multiple items covering the same topic, of similar difficulty, and that will be scored as a composite: the *average inter-item correlation* and the *average item-total correlation*. To calculate the average inter-item correlation, you find the correlation between each pair of items and take the average of all these correlations. To calculate the average item-total correlation, you create a total score by adding up scores on each individual item on the scale and then compute the correlation of each item with the total. The average item-total correlation is the average of those individual item-total correlations.

Split-half reliability, described previously, is another method of determining internal consistency. This method has the disadvantage that, if the items are not truly homogeneous, different splits will create forms of disparate difficulty, and the reliability coefficient will be different for each pair of forms. A method that overcomes this difficulty is *Cronbach's alpha* (also called *coefficient alpha*), which is equivalent to the average of all possible split-half estimates. For more about Cronbach's alpha, including a demonstration of how to compute it, see Chapter 16.

Validity

Validity refers to how well a test or rating scale measures what it is supposed to measure. Some researchers describe validation as the process of gathering evidence to support the types of inferences intended to be drawn from the measurements in question. Researchers disagree about how many types of validity there are, and scholarly consensus has varied over the years as different types of validity are subsumed under a single heading one year and then separated and treated as distinct the next. To keep things simple, this book will adhere to a commonly accepted categorization of validity that recognizes four types: content validity, construct validity, concurrent validity, and predictive validity. The face validity, which is closely related to content validity, will also be discussed. These types of validity are discussed further in the context of research design in Chapter 18.

Content validity refers to how well the process of measurement reflects the important content of the domain of interest and is of particular concern when the purpose of the measurement is to draw inferences about a larger domain of interest. For instance, potential employees seeking jobs as computer programmers might be asked to complete an examination that requires them to write or interpret programs in the languages they would use on the job if hired. Due to time restrictions, only limited content and programming competencies may be included on such an examination, relative to what might actually be required for a professional programming job. However, if the subset of content and competencies is well chosen, the score on such an exam can be a good indication of the individual's ability on all the important types of programming required by the job. If this is the case, we may say the examination has content validity.

A closely related concept to content validity is known as *face validity*. A measure with good face validity appears (to a member of the general public or a typical person who may be evaluated by the measure) to be a fair assessment of the qualities under study. For instance, if a high school geometry test is judged by parents of the students taking the test to be a fair test of algebra, the test has good face validity. Face validity is important in establishing credibility; if you claim to be measuring students' geometry achievement but the parents of your students do not agree, they might be inclined to ignore your statements about their children's levels of achievement in this subject. In addition, if students are told they are taking a geometry test that appears to them to be something else entirely, they might not be motivated to cooperate and put forth their best efforts, so their answers might not be a true reflection of their abilities.

Concurrent validity refers to how well inferences drawn from a measurement can be used to predict some other behavior or performance that is measured at approximately the same time. For instance, if an achievement test score is highly related to contemporaneous school performance or to scores on similar tests, it has high concurrent validity. *Predictive validity* is similar but concerns the ability to draw inferences about some event in the future. To continue with the previous example, if the score on an achievement test is highly related to school performance the following year or to success on a job undertaken in the future, it has high predictive validity.

Triangulation

Because every system of measurement has its flaws, researchers often use several approaches to measure the same thing. For instance, American universities often use multiple types of information to evaluate high school seniors' scholastic ability and the likelihood that they will do well in university studies. Measurements used for this purpose can include scores on standardized exams such as the SAT, high school grades, a personal statement or essay, and recommendations from teachers. In a similar vein, hiring decisions in a company are usually made after consideration of several types of information, including an evaluation of each applicant's work experience, his education, the impression he makes during an interview, and possibly a work sample and one or more competency or personality tests.

This process of combining information from multiple sources to arrive at a true or at least more accurate value is called *triangulation*, a loose analogy to the process in geometry of determining the location of a point in terms of its relationship to two other known points. The key idea behind triangulation is that, although a single measurement of a concept might contain too much error (of either known or unknown types) to be either reliable or valid by itself, by combining information from several types of measurements, at least some of whose characteristics are already known, we can arrive at an acceptable measurement of the unknown quantity. We expect that each measurement contains error, but we hope it does not include the *same type* of error, so that through multiple types of measurement, we can get a reasonable estimate of the quantity or quality of interest.

Establishing a method for triangulation is not a simple matter. One historical attempt to do this is the multitrait, multimethod matrix (MTMM) developed by Campbell and Fiske (1959). Their particular concern was to separate the part of a measurement due to the quality of interest from that part due to the method of measurement used. Although their specific methodology is used less today and full discussion of the MTMM technique is beyond the scope of a beginning text, the concept remains useful as an example of one way to think about measurement error and validity.

The MTMM is a matrix of correlations among measures of several concepts (the traits), each measured in several ways (the methods). Ideally, the same several methods will be used for each trait. Within this matrix, we expect different measures of the same trait to be highly related; for instance, scores of intelligence measured by several methods, such as a pencil-and-paper test, practical problem solving, and a structured interview, should all be highly correlated. By the same logic, scores reflecting different constructs that are measured in the same way should not be highly related; for instance, scores on intelligence, deportment, and sociability as measured by pencil-and-paper questionnaires should not be highly correlated.

Measurement Bias

Consideration of *measurement bias* is important in almost every field, but it is a particular concern in the human sciences. Many specific types of bias have been identified and defined. They won't all be named here, but a few common types will be discussed. Most research design textbooks treat measurement bias in great detail and can be consulted for further discussion of this topic. The most important point is that the researcher must always be alert to the possibility of bias because failure to consider and deal with issues related to bias can invalidate the results of an otherwise exemplary study.

Bias can enter studies in two primary ways: during the selection and retention of the subjects of study or in the way information is collected about the subjects. In either case, the defining feature of bias is that it is a source of *systematic* rather than *random* error. The result of bias is that the data analyzed in a study is incorrect in a systematic fashion, which can lead to false conclusions despite the application of correct statistical procedures and techniques. The next two sections discuss some of the more common types of bias, organized into two major categories: bias in sample selection and retention and bias resulting from information collection and recording.

Bias in Sample Selection and Retention

Most studies take place on samples of subjects, whether patients with leukemia or widgets produced by a factory, because it would be prohibitively expensive if not entirely impossible to study the entire population of interest. The sample needs to be a good representation of the study population (the population to which the results are meant to apply) for the researcher to be comfortable using the results from the sample to describe the population. If the sample is biased, meaning it is not representative of the study population, conclusions drawn from the study sample might not apply to the study population.

Selection bias exists if some potential subjects are more likely than others to be selected for the study sample. This term is usually reserved for bias that occurs due to the process of sampling. For instance, telephone surveys conducted using numbers from published directories by design remove from the pool of potential respondents people with unpublished numbers or those who have changed phone numbers since the directory was published. Random-digit-dialing (RDD) techniques overcome these problems but still fail to include people living in households without telephones or who have only a cell (mobile) phone. This is a problem for a research study because if the people excluded differ systematically on a characteristic of interest (and this is a very common occurrence), the results of the survey will be biased. For instance, people living in households with no telephone service tend to be poorer than those who have a telephone, and people who have only a cell phone (i.e., no land line) tend to be younger than those who have residential phone service. If poverty or youth are related to the subject being studied, excluding these individuals from the sample will introduce bias into the study.

Volunteer bias refers to the fact that people who volunteer to be in studies are usually not representative of the population as a whole. For this reason, results from entirely volunteer samples, such as the phone-in polls featured on some television programs, are not useful for scientific purposes (unless, of course, the population of interest is people who volunteer to participate in such polls). Multiple layers of nonrandom selection might be at work in this example. For instance, to respond, the person needs to be watching the television program in question. This means she is probably at home; hence, responses to polls conducted during the normal workday might draw an audience largely of retired people, housewives, and the unemployed. To respond, a person also needs to have ready access to a telephone and to have whatever personality traits would influence him to pick up the telephone and call a number he sees on the television screen. The problems with telephone polls have already been discussed, and the probability that personality traits are related to other qualities being studied is too high to ignore.

Nonresponse bias refers to the other side of volunteer bias. Just as people who volunteer to take part in a study are likely to differ systematically from those who do not, so people who decline to participate in a study when invited to do so very likely differ from those who consent to participate. You probably know people who refuse to participate in any type of telephone survey. (I'm such a person myself.) Do they seem to be a random selection from the general population? Probably not; for instance, the Joint Canada/U.S. Survey of Health found not only different response rates for Canadians versus Americans but found nonresponse bias for nearly all major health status and health care access measures [results are summarized here (*http://bit.ly/TfJ6um*)].

Informative censoring can create bias in any longitudinal study (a study in which subjects are followed over a period of time). Losing subjects during a long-term study is a common occurrence, but the real problem comes when subjects do not drop out at random but for reasons related to the study's purpose. Suppose we are comparing two medical treatments for a chronic disease by conducting a clinical trial in which subjects are randomly assigned to one of several treatment groups and followed for five years to see how their disease progresses. Thanks to our use of a randomized design, we begin with a perfectly balanced pool of subjects. However, over time, subjects for whom the assigned treatment is not proving effective will be more likely to drop out of the study, possibly to seek treatment elsewhere, leading to bias. If the final sample of subjects we analyze consists only of those who remain in the trial until its conclusion, and if those who drop out of the study are not a random selection of those who began it, the sample we analyze will no longer be the nicely randomized sample we began with. Instead, if dropping out was related to treatment ineffectiveness, the final subject pool will be biased in favor of those who responded effectively to their assigned treatment.

Information Bias

Even if the perfect sample is selected and retained, bias can enter a study through the methods used to collect and record data. This type of bias is often called *information bias* because it affects the validity of the information upon which the study is based, which can in turn invalidate the results of the study.

When data is collected using in-person or telephone interviews, a social relationship exists between the interviewer and the subject for the course of the interview. This relationship can adversely affect the quality of the data collected. When bias is introduced into the data collected because of the attitudes or behavior of the interviewer, this is known as *interviewer bias*. This type of bias might be created unintentionally when the interviewer knows the purpose of the study or the status of the individuals being interviewed. For instance, interviewers might ask more probing questions to encourage the subject to recall chemical exposures if they know the subject is suffering from a rare type of cancer related to chemical exposure. Interviewer bias might also be created if the interviewer displays personal attitudes or opinions that signal to the subject that she disapproves of the behaviors being studied, such as promiscuity or drug use, making the subject less likely to report those behaviors.

Recall bias refers to the fact that people with a life experience such as suffering from a serious disease or injury are more likely to remember events that they believe are related to that experience. For instance, women who suffered a miscarriage are likely to have spent a great deal of time probing their memories for exposures or incidents that they believe could have caused the miscarriage. Women who had a normal birth may have had similar exposures but have not given them as much thought and thus will not recall them when asked on a survey.

Detection bias refers to the fact that certain characteristics may be more likely to be detected or reported in some people than in others. For instance, athletes in some sports are subject to regular testing for performance-enhancing drugs, and test results are publicly reported. World-class swimmers are regularly tested for anabolic steroids, for instance, and positive tests are officially recorded and often released to the news media as well. Athletes competing at a lower level or in other sports may be using the same drugs but because they are not tested as regularly, or because the test results are not publicly reported, there is no record of their drug use. It would be incorrect to assume, for instance, that because *reported* anabolic steroid use is higher in swimming than in baseball, the *actual* rate of steroid use is higher in swimming than in baseball. The observed difference in steroid use could be due to more aggressive testing on the part of swimming officials and more public disclosure of the test results.

Social desirability bias is caused by people's desire to present themselves in a favorable light. This often motivates them to give responses that they believe will please the person asking the question. Note that this type of bias can operate even if the questioner is not actually present, for instance when subjects complete a pencil-and-paper survey. Social desirability bias is a particular problem in surveys that ask about behaviors or attitudes that are subject to societal disapproval, such as criminal behavior, or that are considered embarrassing, such as incontinence. Social desirability bias can also influence responses in surveys if questions are asked in a way that signals what the "right," that is, socially desirable, answer is.

Exercises

Here's a review of the topics covered in this chapter.

Problem

What potential types of bias should you be aware of in each of the following scenarios, and what is the likely effect on the results?

1. A university reports the average annual salary of its graduates as $120,000, based on responses to a survey of contributors to the alumni fund.

2. A program intended to improve scholastic achievement in high school students reports success because the 40 students who completed the year-long program (of the 100 who began it) all showed significant improvement in their grades and scores on standardized tests of achievement.

3. A manager is concerned about the health of his employees, so he institutes a series of lunchtime lectures on topics such as healthy eating, the importance of exercise, and the deleterious health effects of smoking and drinking. He conducts an anonymous survey (using a paper-and-pencil questionnaire) of

employees before and after the lecture series and finds that the series has been effective in increasing healthy behaviors and decreasing unhealthy behaviors.

Solution

1. Selection bias and nonresponse bias, both of which affect the quality of the sample analyzed. The reported average annual salary is probably an overestimate of the true value because subscribers to the alumni magazine were probably among the more successful graduates, and people who felt embarrassed about their low salary were less likely to respond. One could also argue a type of social desirability bias that would result in calculating an overly high average annual salary because graduates might be tempted to report higher salaries than they really earn because it is desirable to have a high income.

2. Informative censoring, which affects the quality of the sample analyzed. The estimate of the program's effect on high school students is probably overestimated. The program certainly seems to have been successful for those who completed it, but because more than half the original participants dropped out, we can't say how successful it would be for the average student. It might be that the students who completed the program were more intelligent or motivated than those who dropped out or that those who dropped out were not being helped by the program.

3. Social desirability bias, which affects the quality of information collected. This will probably result in an overestimate of the effectiveness of the lecture program. Because the manager has made it clear that he cares about the health habits of his employees, they are likely to report making more improvements in their health behaviors than they have actually made to please the boss.

The Likert Scale

The Likert scale might be the most common type of rating scale used in human-subject research. This type of scale was first described in 1932 by Rensis Likert (1903–1981), an organizational psychologist who served as director of the University of Michigan Institute for Social Research from 1946 to 1970. Questions using the Likert scale typically present a statement, and subjects are invited to choose their response to it from an ordered, odd-numbered set of choices (most often five but sometimes seven or nine). An example follows.

The United States should adopt a national system of health insurance.

1. Strongly agree
2. Agree
3. Neither agree nor disagree
4. Disagree
5. Strongly disagree

Sometimes an even number of responses is provided, so that there is no neutral middle choice: this is called the forced choice method because the respondent is forced to make the choice to agree or disagree with the statement. Often the order of responses is changed one or more times within a questionnaire so that

---

sometimes 1 = Strongly disagree and sometimes 1 = Strongly agree to detect whether people are automatically selecting the first or last choices without reading the items.

Data gathered by Likert scale is ordinal because although the choices are ordered, there is no reason to believe that there are equal intervals between them. For instance, we have no way of knowing whether the distance between "Strongly agree" and "Agree" is the same as the distance between "Agree" and "Neither agree nor disagree."

Dewey Defeats Truman

Several United States presidential elections have featured inaccurate predictions based on biased samples. It's always humorous to see a respected publication or organization get it completely wrong, but these incidents also serve as a cautionary tale of what can happen when statistics conducted on a biased sample are assumed to apply to the general population.

In 1936, the *Literary Digest* magazine, which had correctly predicted the winner of the U.S. presidential elections of 1916, 1920, 1924, 1928, and 1932, predicted that Republican Alf Landon would defeat Democrat Franklin Roosevelt by a landslide. However, history shows that Roosevelt won the 1936 election in a landslide. The problem with the *Literary Digest* prediction was that although it was based on a large sample (over 2.3 million respondents out of 10 million invited to take part), the sample was biased because it consisted of people who owned automobiles or telephones or who subscribed to the *Literary Digest*. In 1936, such individuals tended to be wealthier than the general population and more likely to be Republican. Because it was necessary to return a postcard to participate in the poll, the *Literary Digest* sample was subject to volunteer bias as well.

In 1948, every major poll predicted that the Republican Thomas Dewey would defeat the Democrat Harry S. Truman for president. The *Chicago Tribune* even printed papers with the front-page headline, "Dewey Defeats Truman." Although polling techniques had improved since 1936, several sources of bias were still present in the polls, which led to this inaccurate prediction. One problem was that telephone surveys were used without statistical correction for the fact that telephone ownership was far more common among the affluent, who were also more likely to support Dewey. Another factor was that there were large numbers of undecided voters in the days leading up to the election, and none of the polls had a good method for predicting for whom these individuals would ultimately vote and how. A third problem was that Dewey's support was stronger in the eastern U.S. than in the western states. Due to the different time zones, the results from eastern states were reported first, and the *Tribune* decided to print papers announcing the result based on those early returns. What the *Tribune* did not anticipate was that Truman would carry many western states, including California, and thus amass sufficient electoral votes to win the election.

2

Probability

Probability theory is fundamental to statistics. Some people find probability to be an intimidating topic, but there's no reason anyone willing to put in the time can't come to understand it at the level necessary to succeed in statistics. As is the case in many fields of study, advanced probability theory can become very complex and difficult to understand, but the basic principles of probability are intuitive and easy to comprehend. What's more, most people are already familiar with probabilistic statements, from the weather report that tells you there is a 30% chance of rain this afternoon to the warning on cigarette packages that smoking increases your risk of developing lung cancer.

If, like most adults, you hold one or more insurance policies, you are already engaged in an enterprise based on probabilistic reasoning. If you drive or own an automobile, for instance, you probably have an automobile insurance policy, which should really be called an automobile expenses insurance policy because it protects the policy-holder against the extreme expenses that can be incurred due to an accident. People don't purchase insurance policies because they are planning to get into a crash; rather, they acknowledge that there is a nonzero probability of such an event occurring in the future.

Governments often require automobile owners to have insurance policies for the same reason; this requirement is not a judgment that you are a bad driver, just an acknowledgment that accidents do happen and few individuals would be able to cover the costs of a major accident out of their own pocket. The insurance industry employs a cadre of statisticians to calculate how much you should be charged for a policy, taking into consideration (among other things) the probability that you will be in an accident or file a claim for any other reason, and the amount that such a claim would cost the company.

You need no more mathematical expertise than that usually covered in high school to understand the basics of probability as presented in this chapter, and under-standing these concepts provides the basis for understanding the statistical techni-ques presented in subsequent chapters. Mastering the content of this chapter will also enable you to understand a large proportion of the statistics you are ever likely to encounter unless you are doing advanced work or have decided to make statistics your field of study. In addition, you will be able to understand probabilistic statements as used in everyday speech and to recognize when they are used incorrectly.

About Formulas

People who haven't done well in math classes in the past often dislike formulas, feeling they are an arcane system of communication invented by mathematicians as a barrier to keep the uninitiated away and reserve all the good jobs for themselves. Although I would never argue that math and statistics are easy subjects, the assumption that formulas are a barrier to understanding is wrong. In fact, formulas are a condensed and unambiguous way of communicating important information and can be considered as a set of instructions written in the language of mathematics. As one of my calculus professors used to say, "Look at the formula, then do what the formula tells you to do."

Mathematical formulas have the advantage of not depending on language, so mathematics can be communicated and understood among people regardless of their native language or national origin. It doesn't matter if you grew up speaking English or Russian or Farsi; as long as you understand the language of mathematics, you can communicate with your colleagues about mathematical topics somewhat independently of the barriers imposed by human languages.

Consider the example of the formula for calculating the arithmetic mean, known in common language as the average of a set of numbers, presented in Figure 2-1.

$$\bar{x} = \frac{1}{n} \sum_{i=1}^{n} x_i$$

Figure 2-1. Formula for calculating the mean

It may look like Greek to you (in fact, some of it is!), but it's really just a set of directions telling you how to do the necessary calculations. Let's break it down into parts:

- x is the number whose mean we are calculating.
- The symbol $\bar{x}$ (read as "x-bar") means the mean of x, which is what we are calculating.
- The symbol x_i (read as "x sub i") means a particular value of x.
- n means the number of values of x being used to compute the mean.
- The summation symbol, Σ, means to add together a number of cases, in this case all values of x. The notations above and below the summation symbol mean to add together all values of x, starting with the first value (x_1) and going to the last value (xn).

The formula tells you to calculate the mean by adding together all the values of x, then dividing by the number of cases that you just added together. Note that multiplying by $1/n$ is the same as dividing by n.

Suppose we want to calculate the mean of three numbers: 1, 3, and 5. In terms of variable notation, we would call them x_1, x_2, and x_3. In this example, $n = 3$ because we have three numbers, so to execute the formula, we add the numbers from x_1 to x_3 and multiply by 1/3, as presented in Figure 2-2.

$$\bar{x} = \frac{1}{3} \sum_{i=1}^{3} x_i = \frac{1}{3}(1 + 3 + 5) = 3$$

Figure 2-2. Calculating the mean of three numbers

You will encounter more complicated formulas as you progress in your statistical studies, but the process for using them is the same:

1. Identify the meaning of each symbol used and the operation required.
2. Identify the values to be substituted for each symbol.
3. Substitute the values into the equation, perform the specified operations, and you have your result.

Basic Definitions

Here are some basic concepts to know for a discussion of probability.

Trials

Probability is concerned with the outcome of *trials*, which are also called *experiments* or *observations*. The crucial fact, whichever term is used, is that they refer to an event whose outcome is unknown. If the outcome of a trial were known, after all, there would be no need to consider its probability. A trial can be as simple as flipping a coin or drawing a card from a deck, or as complex as observing whether a person diagnosed with breast cancer is still alive five years after the diagnosis. We will reserve the term "trial" for a single observation, such as one coin flip, and the term "experiment" for multiple trials, such as the results from flipping one coin five times.

Sample Space

The sample space, signified by S, is the set of all possible elementary outcomes of a trial. If the trial is flipping a coin once, then the sample space is $S = \{heads, tails\}$ (often abbreviated $S = \{h, t\}$) because those two alternatives represent all the possible outcomes for the experiment. The flip may come up either heads (h) or tails (t). If the experiment is rolling a single six-faced die (the plural is *dice*), the sample space is $S = \{1, 2, 3, 4, 5, 6\}$, representing the six faces of the die that may turn up in a single roll. These elementary outcomes are also referred to as sample points. If the experiment consists of multiple trials, all possible combinations of outcomes of the trials must be specified as part of the sample space. For instance, if the trial consists of flipping a coin twice, the sample space is $S = \{(h, h), (h, t), (t, h), (t, t)\}$ because the results could be heads on both flips, heads on the first and tails on the second, tails on the first and heads on the second, or tails on both flips.

Events

An *event*, usually signified by E or any capital letter other than S, is the specification of the outcome of a trial and can consist of a single outcome or a set of outcomes. If the outcome or set of outcomes occurs, we say "the outcome satisfied the event" or "the event occurred." For instance, the event of "heads in flipping one coin" could be specified as $E = \{heads\}$, whereas the event of "odd number in rolling one die" could be specified as $E = \{1, 3, 5\}$. A *simple event* is the outcome of a single experiment or observation, such as a single coin flip. Simple events can be combined into *compound events*, as in the union and intersection examples below. Events can be defined by listing the outcomes or by defining them logically. For instance, if the trial is rolling two dice, and we are interested in how often the sum is less than 6, we could specify this as either $E = \{2, 3, 4, 5\}$ or $E = \{sum\ is\ less\ than\ 6\}$.

A common way to portray the probability of events and combinations of events graphically is through Venn diagrams in which a rectangle represents the sample space and circles represent particular events. Venn diagrams are used in Figures 2-3 through 2-6.

Venn Diagrams

Anyone who was brought up on the new math probably remembers Venn diagrams from elementary school math textbooks. Although the wisdom of introducing set theory to grade schoolers might be debatable, that is surely no fault of the British mathematician John Venn (1834–1923) or his diagrams. Venn diagrams are widely used in mathematics and related fields to display the logical relationship between sets of objects, and they have been adapted by other disciplines, such as literature, as well. Venn spent most of his adult life teaching at Caius College, Cambridge University, where his primary interest was logic, and he published three textbooks, including *Symbolic Logic* (1881), which introduced Venn diagrams. Caius students and faculty today have a daily reminder of Venn's accomplishments: he has been immortalized by stained glass windows in the college dining hall, which portray a Venn diagram with three overlapping sets signified by three circles of different colors.

Union

The *union* of several simple events creates a compound event that occurs if one or more of the events occur. The union of E and F is written $E \cup F$ and means "either E or F or both E and F." Note that the union symbol is similar to a capital letter U. The union of E and F is the shaded area in the Venn diagram in Figure 2-3. Note that this figure portrays two complete circles that partially overlap; the meaning of

---

this diagram is that any point in the shaded area (any point in E, F, or both E and F) satisfies the condition $E \cup F$. To take an example, suppose the event is rolling a six-sided die and that $E = \{1, 3\}$ and $F = \{1, 2\}$. The event $E \cup F$ is satisfied with an outcome of 1, 2, or 3; we can also say that $E \cup F = \{1, 2, 3\}$.

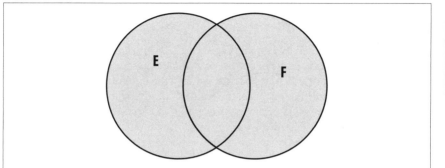

Figure 2-3. The union of E and F (shaded area)

Intersection

The intersection of two or more simple events creates a compound event that occurs only if all the simple events occur. The intersection of E and F is written $E \cap F$ and means "both E and F." The intersection of E and F is the shaded area in the Venn diagram in Figure 2-4; note that only points that belong to both E and F satisfy the condition. To continue with our example, if the event is rolling a six-sided die, and $E = \{1, 3\}$ and $F = \{1, 2\}$, the event $E \cap F$ is satisfied only with the outcome of 1 because 1 is a member of both sets, so $E \cap F = \{1\}$.

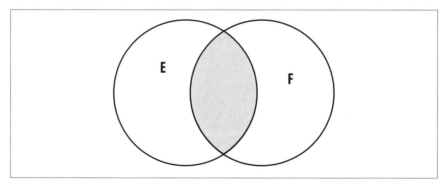

Figure 2-4. The intersection of E and F (shaded area)

Complement

The *complement* of an event means everything in the sample space that is not that event. The complement of event E is written variously as $\sim E$, E^c, or $\bar{E}$, and is read as "not E" or "E complement." For instance, if $E = (numbers > 0)$, $\sim E = (numbers$

≤ 0). Continuing with our example, if the event is rolling a six-sided die and E = {1, 3}, ~E = {2, 4, 5, 6}. The complement of F is the shaded area in the Venn diagram in Figure 2-5.

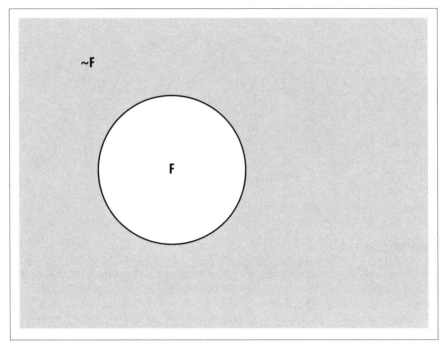

Figure 2-5. The complement of F (shaded area)

Mutual Exclusivity

If events cannot occur together, they are *mutually exclusive*. To put it another way, if two sets have no events in common, they are mutually exclusive. For instance, the event A = (salary is greater than $100K) and event B = (salary is less than or equal to $100K) are mutually exclusive, as are the sets A = (even integers) and B = (odd integers). The mutually exclusive sets E and F are presented in the Venn diagram in Figure 2-6; note that they have no points in common.

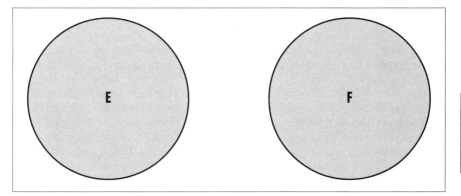

Figure 2-6. E and F are mutually exclusive; they have no points in common

Independence

If two trials are *independent*, the outcome of one trial does not influence the outcome of another. To put it another way, if the trials are independent, knowing the outcome of one trial gives you no information about the outcome of the other. The classic example of independence is flipping an ordinary coin; if you flip the coin twice, the outcome of the first trial has no influence on the outcome of the second trial.

Permutations

In probability theory, *permutations* are all the possible ways elements in a set can be arranged. For instance, if a set consists of the elements (a, b, c), then the permutations of this set are (a, b, c), (a, c, b), (b, a, c), (b, c, a), (c, a, b), and (c, b, a). Note that the order of elements is important in permutations: (a, b, c) is a different permutation than (a, c, b).

You can calculate the number of permutations of any set of distinct elements (meaning that none of the elements repeats within the set) by using *factorials*, which are signified by a number followed by an exclamation point. Many calculators have an $x!$ key to calculate factorials, but factorials can also be calculated by multiplying the number by all lower integers down to 1. Here's an example:

$3! = 3 \times 2 \times 1 = 6$

3! is read as "3 factorial." For a set of three nonrepeated elements, there are 3! or 6 permutations, which agrees with the result we found by listing the preceding different permutations. This makes logical sense because if you have three elements, you have three choices for the first element (a, b, c in our example), two choices for the second element (minus whatever was chosen for the first element), and one choice for the third element (whatever element remains after the first two are chosen). Therefore, you have $3 \times 2 \times 1 = 6$ different ways of arranging the elements. Permutations become large very quickly. For instance, 5! = 120 and 10! = 3,628,800. 20! is so large that it cannot be displayed on most calculators except through scientific notation: 20! = 2.432902008E18.

Scientific Notation

Scientific notation is used to indicate the value of numbers that are very large or very small. Using scientific notation not only saves space (because you don't have to write out lots of zeros) but improves accuracy in communication because it is easy to misread a number including a lot of zeros. The concept behind scientific notation is that any number can be written as a number greater than or equal to 1 and less than 10 (called the *coefficient*) multiplied by a power of 10 (called the *base*). So the number 1234 can be written as 1.234E3 (the E stands for *exponent*), which means 1.234×10^3, that is, 1.234×1000. Similarly, 1.234E – 4 means 1.234 $\times 10^{-4}$ or 1.234×0.0001, which is 0.0001234. Another way to interpret E is as an indication of how many places to the left or right to move the decimal point. Therefore, 1.234E3 tells you to move it three places to the right, producing 1,234, whereas 1.234E – 4 tells you to move it four places to the left for 0.0001234.

Combinations

Combinations are similar to permutations with the difference that the order of elements is not significant in combinations. Therefore, (a, b, c) is the same combination as (b, a, c). For this reason, there is only one combination of the set (a, b, c).

One use of combinations and permutations in statistics is to calculate the number of ways a subset of specified size can be drawn from a set, which allows the calculation of the probability of drawing any particular subset from a set. The general case is that the set in question contains no duplicates, and you will use this assumption in the following discussion. There are several ways to denote permutations and combinations; these are demonstrated in Appendix A along with a few problems. This section will stick to a simple system of notation, using P for permutations and C for combinations. Using this notation, the number of permutations possible when drawing 2 elements from a set of 3 is written 3P2, and the number of combinations of 2 elements from a set of 3 is as 3C2. Continuing with the preceding example, for the set (a, b, c), 3P2 = 6 because there are 6 permutations of 2 elements drawn from a set of 3: (a, b), (a, c), (b, c), (b, a), (c, a), and (c, b). Three combinations of 2 are possible from this set, so 3C2 = 3: (a, b), (a, c), and (b, c).

The number of permutations of subsets of size k drawn from a set of size n is calculated as shown in Figure 2-7.

$$nPk = \frac{n!}{(n-k)!}$$

Figure 2-7. The formula for calculating a permutation

Using this formula, the number of permutations of size 2 that can be drawn from a set of size 8 is shown in Figure 2-8.

$$8P2 = \frac{8!}{(8-2)!} = \frac{8!}{6!} = 56$$

Figure 2-8. Calculating the permutation 8P2

If you have to calculate a permutation by hand, it helps to remember the principle of canceling factors: if you express the numerator and denominator as the product of factors, you can cancel those that appear in both the numerator and denominator. For instance:

12/6 = (2 × 2 × 3)/(2 × 3) = 2

because you can cancel (2 × 3) from both the numerator and denominator.

In the case of the 8P2 permutation, it's not necessary to multiply out each factorial before dividing because you can cancel many of the terms. In this example:

8! = 8 × 7 × 6 × 5 × 4 × 3 × 2 × 1

and

6! = 6 × 5 × 4 × 3 × 2 × 1

so you can cancel most of the numerator, leaving you with:

8P2 = 8 × 7 = 56

Given the same values for n and k, there will always be fewer combinations than permutations because a different order of the same elements counts as a different permutation but not as a different combination. This is clear in the formula for a combination, which is the formula for the permutation divided by the factorial of the number of objects selected, as shown in Figure 2-9.

$$nCk = \frac{n!}{k!(n-k)!} = \frac{nPk}{k!}$$

Figure 2-9. The formula for calculating a combination

Using this formula, you calculate the number of combinations of size 2 that can be drawn from a set of size 8, as shown in Figure 2-10.

$$8C2 = \frac{8!}{2!(8-2)!} = \frac{8P2}{2!} = \frac{56}{2} = 28$$

Figure 2-10. Calculating the combination 8C2

Defining Probability

There are several technical ways to define probability, but a definition useful for statistics is that probability tells us how often something is likely to occur when an experiment is repeated. For instance, the probability that a coin will come up heads can be estimated by executing a number of coin flips and observing how many times it is heads rather than tails. Perhaps the most important single fact about probability is this:

> The probability of an event is always between 0 and 1.

If the probability of an event is 0, that means there is no chance that it will occur, whereas if the probability of an event is 1, that means it is certain to occur. It is conventional in mathematics to specify probability using decimals, so we say that the probability of an event is between 0 and 1, but it is equally acceptable (and more common in everyday speech) to speak in terms of percentages, so it is equally correct to say that the probability of an event is always between 0% and 100%. To move from decimals to percent, multiply by 100 (per cent = per 100), so a probability of 0.4 is also a probability of 40% ($0.4 \times 100 = 40$), and a probability of 0.85 may also be stated as 85% probability.

Negative probability and probabilities greater than 100% are logical impossibilities that exist only as figures of speech. The fact that probability is bounded by 0 and 1 has mathematical implications that are explored further when considering logistic regression in Chapter 11. This fact also provides a useful check on your calculations. If you come up with a probability lower than 0 or greater than 1, you have certainly made a mistake somewhere along the way. Furthermore, if someone tells you there is a 200% chance that you will make a killing in the stock market if you follow his system, you should probably look for a new investment advisor.

Another useful fact about probability is that:

> The probability of the sample space is always 1.

Because the sample space represents all possible outcomes of a trial, the total probability of the sample space must add up to 1. This is a useful fact because although we may know the probability of some events in a sample space, there can be others about which we have no information. However, because we know that the probability of the total sample space equals 1, we can assign a probability to those events about which we have no information based on what probability remains after the known probabilities are considered.

A third useful fact that follows from the first two is that:

> The probability of an event and its complement is always 1.

This fact follows from the definition of a complement: everything in the sample space that is not the event E is the complement of E. Therefore, E and $\sim E$ together must make up the entire sample space, and the probability of E and $\sim E$ together must equal 1. This should be clear from Figure 2-5: the rectangular box represents the sample space, the circle the event E, and the shaded area within the box but outside the circle $\sim E$. Together, E and $\sim E$ comprise the entire sample space, and their union ($E \cup \sim E$) has a probability of 1.

Expressing the Probability of an Event

It is typical to write probability statements as follows:

$P(E) = 0.5$

This is read as "the probability of event E is 0.5" or "there is a 50% probability of event E" (or just "the probability of E is 0.5" or "there is a 50% probability of E"). Using this format, you can write your first fact about probability, that the probability of an event is always between 0 and 1, as:

$0 \leq P(E) \leq 1$

The second fact about probability, which follows from the definition of the sample space S as including all possible outcomes of a trial, may be written as:

$P(S) = 1$

The third fact about probability, that the probability of an event and its complement is always equal to 1, can be written as:

$P(E) + P(\sim E) = 1$

which provides us with the important corollary:

$P(\sim E) = 1 - P(E)$

This will prove very handy in later calculations. If we know the probability of E, we automatically know the probability of $\sim E$, which is $1 - P(E)$. So, if $P(E) = 0.4$, $P(\sim E) = 1 - 0.4 = 0.6$.

Conditional Probabilities

Often we want to know the probability of some event, given that another event has occurred. This is expressed symbolically as $P(E|F)$ and read as "the probability of E given F." The second event is known as the condition, and the process is sometimes referred to as "conditioning on F." Conditional probability is an important concept in statistics because often we are trying to establish that a factor has a relationship with an outcome, for instance that people who smoke cigarettes are more likely to develop lung cancer. Another way to say that a factor has a relationship with an outcome is to say that the probability of the outcomes differs depending on the presence or absence of the factor. To express symbolically that the probability of developing lung cancer (the outcome) is higher for those who smoke (the factor) than for those who do not, we can write:

$$P(\text{lung cancer}|\text{smoker}) > P(\text{lung cancer}|\text{nonsmoker})$$

Conditional probabilities can also be used to define independence. Two variables are said to be independent if the following relationship holds:

$$P(E|F) = P(E)$$

This equation states that the probability of E is the same whether or not variable F is present. To continue with the same example, the equation to state that the probability of having lung cancer is unrelated to smoking would be:

$$P(\text{lung cancer}|\text{smoker}) = P(\text{lung cancer})$$

This equation states that the probability of lung cancer for a person who smokes is the same as the probability for the population in general, smokers and nonsmokers alike. This is just an example, and I'm not implying that it is true; many studies have shown that the probability of lung cancer for a smoker is much higher than the rate in the general population.

Calculating the Probability of Multiple Events

To calculate the probability of any of several events occurring (the union of several events), add the probabilities of the individual events. The specific equation used will depend on whether the events are mutually exclusive (meaning both cannot occur).

Union of mutually exclusive events

If the events are mutually exclusive, as in Figure 2-6, the equation is simply:

$$P(E \cup F) = P(E) + P(F)$$

For a practical example, imagine a college that does not allow double majors. Define the event E = (English major) as having a probability of 0.2 and F = (French major) as having a probability of 0.1. These events are mutually exclusive because students are allowed only one major, so you would calculate the probability of the event (either English or French major) as:

$$P(E \cup F) = 0.2 + 0.1 = 0.3$$

Union of events that are not mutually exclusive

Often, events are not mutually exclusive. For instance, at a college that does allow double majors, the events (English major) and (French major) are not mutually exclusive because conceivably one person could be both an English major and a French major. In this situation, the equation calculating P(English major or French major) must include a term correcting for this overlap. Looking at Figure 2-4, the overlap is the area contained in both circles E and F (their intersection, represented by the shaded area). If you fail to take into consideration that a college that allows students to elect more than one major could have people majoring in both English and French, you risk counting some people twice. (Those with double majors in French and English would be counted both as French majors and as English majors.)

To correct for this potential overlap, use the following equation to calculate the probability of the occurrence of either of two events that are not mutually exclusive:

$$P(E \cup F) = P(E) + P(F) - P(E \cap F)$$

Suppose that P(English major) = 0.2, P(French major) = 0.1, and P(double major in French and English) = 0.05. The probability of a student being either an English or a French major is therefore:

$$P(E \cup F) = 0.2 + 0.1 - 0.05 = 0.25$$

Intersection of independent events

To calculate the probability of all of several events occurring (the intersection of several events), multiply their individual probabilities. The specific formula used depends on whether the events are independent.

If the two events E and F are independent, the probability of both E and F occurring is calculated as simply:

$$P(E \cap F) = P(E) \times P(F).$$

Suppose you are flipping a fair coin (one whose probability of heads is 0.5, whose probability of tails is 0.5, and whose results on each flip is independent). Label the trials so that E = (heads on first flip) and F = (heads on second flip). You have already specified that the probability of heads on either flip is 0.5, and the two trials are independent, so you can compute the probability (heads on both flips) as:

$$P(E \cap F) = 0.5 \times 0.5 = 0.25$$

Intersection of nonindependent events

If two events are not independent, you have to know their conditional probability to calculate the probability of both occurring. The formula to use is:

$$P(E \cap F) = P(E) \times P(F|E)$$

Suppose you are drawing two cards without replacement from a standard deck of 52, meaning that you do not put the cards back in the deck after you draw them. Half of all cards in a standard deck are red and half are black. These events (your first and second draws) are not independent because the probability for the second draw depends on the result of the first draw. If you are interested in the probability of drawing two black cards in these two trials, you can calculate this as follows:

$P(E) = P$ (black card drawn on first trial) = 26/52 = 0.5
$P(F|E) = P$(black card drawn on second trial|black card drawn on first trial) = 25/51 = 0.49

Note that because you are drawing without replacement, there are only 51 cards in the deck for the second draw, and only 25 black cards because you removed 1 black card in the first draw. Using this information, you can calculate the probability that you will draw black cards on both trials as (the intersection of E and F):

$$P(E \cap F) = 0.50 \times 0.49 = 0.245$$

Bayes' Theorem

Bayes' theorem, also known as Bayes' formula, is one of the most common applications of conditional probabilities. A typical use of Bayes' theorem in the medical field is to calculate the probability that a person who tests positive on a screening test for a particular disease actually has the disease. Bayes' theorem also uses several of the basic concepts of probability previously introduced, so careful study of Bayes' formula is a good review for the entire chapter as well. Bayes' theorem for any two events A and B is presented in Figure 2-11.

$$P(A \mid B) = \frac{P(A \cap B)}{P(B)} = \frac{P(B \mid A)P(A)}{P(B \mid A)P(A) + P(B \mid \sim A)P(\sim A)}$$

Figure 2-11. Bayes' theorem

You would use this formula if you know $P(A)$, $P(B)$, and $P(B|A)$ but want to know $P(A|B)$. The numerator of Bayes' theorem uses the fact that the probability of the intersection of two events is the probability of the first event multiplied by the conditional probability of the second event given the first. In this example, the conditional probability of B given A is multiplied by the probability of A, giving the probability of the intersection of A and B, that is, of both A and B occurring.

The denominator uses this same fact plus the fact that any event plus its complement comprises the entire sample space, and together an event and its complement have a probability of 1, so the sum of the conditional probabilities of (B given A) times the probability of A, and (B given ~A) times the probability of ~A, equals the probability of B.

Suppose you have a screening test that is 95% effective in detecting disease in those who have the disease and 99% effective in not falsely diagnosing disease in those who are free of it. Clinicians would say that this test has 95% sensitivity and 99% specificity. Suppose also that the rate of disease in the population is 1%. Using the symbols D for disease, ~D for absence of disease, T for a positive test, and ~T for a negative test, these probabilities can be stated as:

Sensitivity = $P(T|D)$ = 0.95
Specificity = $P(\sim T|\sim D)$ = 0.99
Probability of disease in the population = $P(D)$ = 0.01

These are very high values for sensitivity and specificity. Many commonly used tests and procedures are less accurate. However, these tests are not perfect, and it is possible that a person who tests positive will not in fact have the disease (a false positive) and that a person who tests negative can in fact have the disease (a false negative). Often what you really want to know is, for an individual who has tested positive, what is the probability that he actually has the disease? Using conditional probability notation, you want to know $P(D|T)$. You can calculate this probability by using Bayes' theorem plus the information about sensitivity, specificity, and disease rate in the population previously given, as shown in Figure 2-12.

$$P(D|T) = \frac{P(D \cap T)}{P(T)} = \frac{P(T|D)P(D)}{P(T|D)P(D) + P(T|\sim D)P(\sim D)}$$

Figure 2-12. Bayes' theorem, expressed in terms of disease and test results

Looking at this formula, it is clear that the probability of having the disease, given a positive test, is simply the probability of having both a positive test and the disease divided by the probability of having a positive test (whether or not the person has the disease).

Using the fact that an event plus its complement constitutes the entire sample space, and together they have a probability of 1, you know that the false positive rate is 1 – specificity:

$P(T|\sim D) = 1 - 0.99 = 0.01.$

For the same reason, you know that the probability in the population of not having the disease is 1 – the probability of having the disease:

$P(\sim D) = 1 - P(D) = 1 - 0.01 = 0.99.$

Using these facts plus the information previously supplied, we can calculate $P(D \mid T)$, as shown in Figure 2-13.

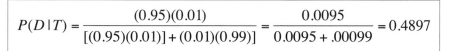

$$P(D|T) = \frac{(0.95)(0.01)}{[(0.95)(0.01)] + (0.01)(0.99)]} = \frac{0.0095}{0.0095 + .00099} = 0.4897$$

Figure 2-13. Using Bayes' theorem to calculate the possibility of having a disease, given a positive test

This example demonstrates an important and underappreciated (at least by the public) fact about screening tests. Even with a highly specific and sensitive screening test, if the disease is rare, the false positive rate will be high relative to the true positive rate. In this example, you expect that about half the people who test positive will be false positives, that is that they won't have the disease. This is not necessarily a reason not to use the test, particularly if the disease has serious consequences and there is an accurate follow-up test to separate the true and false positives. However, any proposal to institute universal screening (whether for a disease or in some other context such as luggage screening at the airport) should always consider the false positive rate and its potential consequences.

It should be noted that the false positive rate depends on the rate of disease in the population as well as the sensitivity and specificity of the screening test. If the disease rate were 0.005 instead of 0.01, fewer of the positives would be true positives and more would be false positives, as shown in the calculations in Figure 2-14.

$$P(D \mid T) = \frac{(0.95)(0.005)}{[(0.95)(0.005)] + (0.01)(0.995)]} = \frac{0.00475}{0.00475 + .00995} = 0.3231$$

Figure 2-14. Another example of using Bayes' theorem to calculate the probability of disease, given a positive test; note the lower rate of true positives, due to a lower rate of disease in the population

In this example, less than one third of the positives are true positives.

The Reverend Thomas Bayes

Bayes' theorem was developed by a British Nonconformist minister, the Reverend Thomas Bayes (1702–1761). Bayes studied logic and theology at the University of Edinburgh and earned his livelihood as a minister in Holborn and Tunbridge Wells, England. However, his fame today rests on his theory of probability, which was developed in his essay, published after his death by the Royal Society of London. There is an entire field of study today known as Bayesian statistics, which is based on the notion of probability as a statement of strength of belief rather than as a frequency of occurrence. However, it is uncertain whether Bayes himself would have embraced this definition because he published relatively little on mathematics during his lifetime.

Enough Exposition, Let's Do Some Statistics!

Statistics is something you do, not something you read about, so the real purpose of the preceding theoretical presentation is to give you the information you need to perform calculations about the probability of events and to use the concepts introduced to be able to reason using your knowledge of statistics. This chapter also introduced concepts, such as independence and mutual exclusivity, which you will need to understand to use more advanced statistical procedures.

The purpose of the problems that follow is to give you some experience in working with the concepts of basic probability. If you are a person who likes to work through many problems to understand a topic, many excellent textbooks focus on probability; several are suggested in Appendix C.

If you are new to solving problems in elementary probability, it may help to follow this procedure:

1. Define the trial, experiment, or both.
2. Define the sample space.
3. Define the event.
4. Specify the relevant probabilities, and do the calculations.

At some point, you might not feel it is necessary to go through all these steps, but they may help you get started working with the exercises. In some cases, an alternative solution, using a different approach to the problem, is provided.

Dice, Coins, and Playing Cards

Because many of the examples in this chapter use dice, coins, and playing cards, this section starts by describing their characteristics.

Dice

The standard die (the singular of dice) used in the Western world is a cube with six sides, each displaying a different number of dots from 1 to 6. A standard assumption in probability calculations is that all sides of the die are equally likely to land facing up when the die is rolled or thrown, so one roll of the die has six equally likely outcomes: 1, 2, 3, 4, 5, and 6. In technical terms, the set of outcomes from rolling one die has a discrete uniform distribution because the possible outcomes can be enumerated, and each outcome is equally likely. The results of two or more dice thrown at once (or multiple throws of the same die) are assumed to be independent of each other, so the probabilities of each combination of numbers are calculated by multiplying the probability of each result.

In the interests of precision, remember that the "equal probability for all sides" holds only for casino dice, in which the pips (circles used to mark the numbers on each side) are painted on. You might be more familiar with dice in which the pips are drilled into the cube face rather than painted on, resulting in unequal weight and, thus, unequal probabilities for the different sides. However, in theoretical discussions of probability, this distinction is usually ignored, and you assume that all sides of the dice are equally probable.

Coins

The standard coin used in probability experiments has two sides, heads and tails. Often, a fair coin is assumed, meaning it is equally likely to come up heads or tails on any toss or flip. For any coin, fair or not, the probability of heads and tails is assumed to be constant on each flip so that the results of previous flips have no influence on later flips, and the results of multiple flips are independent of each other. As with dice, the probability of an actual coin landing heads or tails is seldom exactly 50–50 for a number of physical reasons, including coin design and wear and off-center technique on the part of the person performing the flip, but for the sake of probability exercises, ignore these details unless they are specified in the problem. Sometimes, in the interests of safety, experiments are conducted by spinning coins rather than flipping them (resulting in fewer projectiles flying through the air in a crowded classroom). However, the 50–50 assumption applies even less here, although for the purposes of doing calculations (as opposed to actually spinning coins and recording the results), assume that it applies equally. For more on these issues, see this website (*http://www.sciencenews.org/articles/20040228/fob2.asp*).

Playing cards

The standard pack or deck of playing cards today has 52 cards in four suits: spades, clubs, diamonds, and hearts. Spades and clubs are black cards, and diamonds and hearts are red cards. There are 13 cards in each suit: an ace, numbered cards from 2 through 10, and 3 face cards—the jack, queen, and king. In experiments involving drawing cards from the deck, it is assumed that the cards have been shuffled so any card in the deck is equally likely on a given draw.

Exercises

Problem

If I draw one card from an ordinary deck of 52 playing cards, what is the probability that it will be a red card?

Solution

1. The trial is a single draw of one card from a deck of 52.
2. The sample space is all the possible cards, each of which has an equal probability of being drawn.
3. The event is E = {red card}.
4. Because there are 52 cards in the deck and half (26) are red, the probability of drawing a red card is 26/52 or 0.5. The answer is that you have a 50% probability of drawing a red card on a single draw from a full deck of cards.

Problem

If I roll a die once, what is the probability of getting a number lower than 5?

Solution

1. The trial is a single roll of a six-sided die.
2. The sample space is the numbers (1, 2, 3, 4, 5, 6), all of which are equally likely.
3. The event is E = (any of 1, 2, 3, 4), which can also be considered the union of four simple events, that is, $E = (E = 1) \cup (E = 2 \cup (E = 3) \cup (E = 4)$.
4. Four of the six simple events or possible outcomes that constitute the sample space satisfy the event E, so the probability of E is 4/6 or 0.67 (rounded).

Alternative solution

Another way to look at this is to calculate the probability of each simple event that satisfies the event E and then add them together because the events are mutually exclusive. Using this approach, the probability of each simple event in E is 1/6; that is, there is a 1 in 6 chance that the number will be 1, 1 in 6 that the number will be 2, and so on. Using this approach, the probability of E is 1/6 + 1/6 + 1/6 + 1/6 or 4/6, which is the same answer as the preceding one.

Problem

If I flip a fair coin twice, what is the probability that I will get at least one head?

Solution

1. The experiment is two flips of a fair ($P = 0.5$ for either heads or tails) coin, that is, two independent trials, each with a probability of 0.5.
2. The sample space is $\{(h, h), (h, t), (t, h), (t, t)\}$, all of which are equally likely.
3. The event is E = (at least one head). Three of the events in the sample space satisfy this condition: (h, h), (h, t), and (t, h).
4. Each of the outcomes is equally likely, and three of the four satisfy the event E, so the probability of E is 3/4 or 0.75.

Alternative solution

You can also find this result mathematically by calculating the probability of the complement of this event and then subtracting it from 1 to get the probability of the event. If the event E is (at least one head), its complement is $\sim E$ = (no heads, that is, two tails). You know that the probability of getting a tail on any flip of a fair coin is 0.5, and the flips are independent, so the probability of (t, t) is 0.5×0.5 or 0.25. Using the definition of a complement, $1 - P(\sim E) = P(E)$, so $1 - 0.25 = 0.75$, or $P(E)$. The probability of at least one head from two flips is 0.75, the same answer as the previous solution.

Problem

If I draw one card from a standard 52-card deck, what is the probability that it will be a black (clubs or spades) face card (king, queen, or jack)?

Solution

1. The trial is drawing one card from a 52-card deck.
2. The sample space is all 52 cards, each of which has equal probability of being drawn.
3. The event is E = (black face card); six cards satisfy this condition, the jack, queen, or king of either spades or clubs.
4. The probability is 6/52 or 0.115.

Mathematical solution

P(face card) = 12/52 or 0.231 P(black card) = 26/52 or 0.5 P(black face card) = P(face card) $\times$ P(black card) = $0.231 \times 0.5 = 0.116$

Note that this mathematical solution is possible because the probability of drawing a black card and the probability of drawing a face card are independent.

Problem

If I draw one card from a standard 52-card deck, what is the probability that it will be either black (clubs or spades) or a face card (king, queen, or jack)?

Solution

1. The trial is drawing one card from a 52-card deck.
2. The sample space is all 52 cards, each of which has an equal probability of being drawn.
3. The event is E = (either black card or face card), meaning any of the 26 black cards or any of the 12 face cards will satisfy the event.
4. The two types of cards that will satisfy the condition are not mutually exclusive; some black cards are also face cards and vice versa. There are 26 black cards: ace through king of spades (13) and ace through king of clubs (13). There are 12 face cards: jack, king, and queen for each of hearts, diamonds, clubs, and spades. There are six cards that are members of both categories: the jack, king, and queen of spades, and the jack, king, and queen of clubs, so 26 + 12 – 6 = 32 cards satisfy this event, and the probability is 32/52 or 0.615.

Mathematical solution

P(black card) = 26/52 or 0.500 P(face card) = 12/52 or 0.231 P (black face card) = 6/52 or 0.115 P(black card or face card) = 0.500 + 0.231 – 0.115 = 0.616

The slight difference in solutions (0.615 versus 0.616) is due to rounding error.

Problem:

If I draw a single card from a 52-card deck and it is black, what is the probability that its suit is clubs?

Solution

1. The trial is drawing one card from a 52-card deck.
2. The sample space is all black cards because we are interested in the conditional probability of a card being a club, given that it is a black card. Our sample space is therefore the 26 black cards.
3. The event is E = (club|black card).
4. The probability of the card being a club, given that it is a black card, is 13/26 or 0.5.

 Note that in this example we are calculating a conditional probability (the probability of clubs, conditioned on the fact that the card is black). The unconditional probability of the card being a club, if we had no information about its color, is 13/52 or 0.25.

Mathematical solution

$P(\text{clubs}|\text{black card}) = P(\text{clubs and a black card})/P(\text{black card}) = 0.25/0.5 = 0.5$

Note that clubs are by definition black cards.

Problem

If order is not significant, how many ways are there to select a subset of 5 students from a classroom of 20?

Solution

This is a combinatorial problem that is too lengthy to solve by listing all possible subsets. Instead, use the combinational formula nCk. In this case, $n = 20$ and $k = 5$; apply the formula shown in Figure 2-15.

$$nCk = \frac{20!}{5!(20-5)!} = 15{,}504$$

Figure 2-15. Using the combination formula to determine the number of ways to choose a subset of 5 individuals from a set of 20

Problem

Eighty students are attending a conference: 40 boys and 40 girls. Thirty of the boys are majoring in math, as are 20 of the girls. You know that if you pick a boy at random, there is a 75% chance that he is a math major. You want to know, however, if you pick a math major at random, the probability that the student is male? Hint: use Bayes' theorem.

Solution

$P(male) = 40/80 = 0.5$
$P(\sim male) = 40/80 = 0.5$
$P(math|male) = 30/40 = 0.75$
$P(math|\sim male) = 20/40 = 0.5$

The calculations are shown in Figure 2-16.

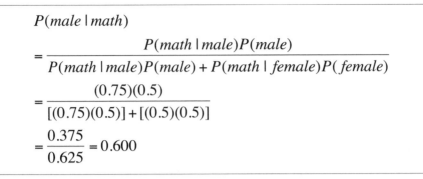

$$P(male \mid math)$$

$$= \frac{P(math \mid male)P(male)}{P(math \mid male)P(male) + P(math \mid female)P(female)}$$

$$= \frac{(0.75)(0.5)}{[(0.75)(0.5)] + [(0.5)(0.5)]}$$

$$= \frac{0.375}{0.625} = 0.600$$

Figure 2-16. Using Bayes' theorem to find the probability that a math major, selected at random, will be male

The probability is 60% that a math major, chosen at random, is male.

Closing Note: The Connection between Statistics and Gambling

Statisticians like to illustrate probability by using dice, coin flips, and playing cards as examples, objects that are also used in gambling (or gaming, in the industry's preferred terminology). One reason is that these objects are familiar to most people. Another is that probabilities of the different outcomes are known and unchanging and, thus, can be used to create simple examples to illustrate the basic concepts of probability, including independence and mutual exclusivity. Their advantage also is that problems can be solved using the concrete objects in question (for instance, by selecting from a standard deck of cards), as well as through mathematical equations.

However, there is also a historical connection because many of the laws of probability were discovered in connection with games of chance and skill involving dice and playing cards. In fact, gambling has been the motivation for many inquiries into the probabilities of different events and combinations of events because the ability of a gambler to win rather than lose money depends in large part on her understanding the probability of different events within the chosen game.

Many historians trace the beginning of modern probability theory to the Chevalier de Mere, a gentleman gambler in seventeenth-century France. He was fond of betting that he would roll at least one six in four rolls of a single die: the wisdom of this bet will be demonstrated in the following paragraphs. However, he also believed that it was a good bet to propose that he would roll one or more double sixes in 24 rolls of a pair of dice: this turned out to be a losing proposition. Fortunately for future statisticians, the Chevalier took this problem to his friend, the philosopher Blaise Pascal, who discussed it with his friend, the mathematician Pierre de Fermat. Consideration of this type of question led to the development of, among other things, Pascal's triangle, the binomial distribution, and the modern concept of probability.

In an even bet among friends, when there is no "house" taking a percentage of the proceeds, a good bet is one you are likely to win more than 50% of the time. To put it another way, a good bet is one in which your likelihood of winning is 0.5 or greater.

The Chevalier's first bet met this standard: the probability of rolling at least one six in four rolls of a die is 0.518. This is easily calculated by considering the probability of rolling no sixes in four trials, which is $(5/6)^4$. Rolling at least one six is the complement of rolling no sixes, so the P(at least one six in four trials) is $1 - (5/6)^4$ or $1 - 0.482$, which is 0.518. This means that about 52% of the time, the Chevalier won this bet.

However, betting that you will roll at least one double six in 24 rolls of a pair of dice is not a wise bet. There are 36 combinations of numbers in each of 2 rolls of a pair of dice, and only one combination is double sixes; therefore, on each roll, the probability is 35/36 that double sixes will not come up. Because each roll of the dice is independent, you can multiply the probabilities for each roll together. Because the probabilities do not change, this means multiplying (35/36) by itself 24 times, which is the same as raising (35/36) to the power of 24. The probability of rolling at least one double six is $1 - P$(no double sixes) or $1 - 0.509$, which is 0.491. Because this probability is less than 0.5, this is a losing bet.

If you are interested in learning more about how probability theory applies to games of chance and skill such as roulette, craps, blackjack, horse racing, and poker, take a look at Edward Packel's *The Mathematics of Games and Gambling*, published by the Mathematical Association of America and listed in Appendix C.

3

Inferential Statistics

Statistical inference is the science of characterizing or making decisions about a population by using information from a sample drawn from that population. Most of the practice of statistics is concerned with inferential statistics, and many sophisticated techniques have been developed to facilitate this type of inference. The concept of inferential statistics can be a bit tricky, so it's worth taking a few minutes to think about what it means to use statistics for inferential reasoning.

The term "inference" is given two definitions by the Merriam-Webster online dictionary (*http://www.m-w.com/dictionary/inference*):

a) The act of passing from one proposition, statement, or judgment considered as true to another whose truth is believed to follow from that of the former

b) The act of passing from statistical sample data to generalizations (as of the value of population parameters) usually with calculated degrees of certainty

The second meaning, which is specific to statistics, is closely related to the first. Inference in general is a method of making judgments about an unknown, drawing on what is already known to be true. Statistical inference is a specific kind of inference in which you make judgments about a population, as stated earlier.

People are sometimes confused about the difference between descriptive statistics (discussed in Chapter 4) and inferential statistics, in part because some statistical procedures are used in both types of statistics, although there can be subtle differences in the formulas as well as in the interpretation of the results. For instance, the same basic procedure is used for calculating the mean of a set of data, whether the data represent a population or a sample: add up all the data values and divide by the number of values. However, there are differences in the way the formula to calculate the mean is written. For a population, you use the Greek letter μ (*mu*) to represent the mean (which is properly called a *parameter* because it is a number that describes a population), whereas for a sample, you use the Latin letter x with a bar over it, $\bar{x}$ (pronounced "x-bar"), to represent the mean (properly called a *statistic* because it is a number that describes a sample).

In other cases, there are more important differences between the formulas used for populations and samples. One well-known example is the formula for the variance. When dealing with a population, you divide by n (the number of cases), but if dealing with a sample, you divide by $n-1$ (one less than the number of cases). These two formulas are explained in detail in Chapter 4 (in the "Measures of Dispersion" on page 90), and if you are new to the study of statistics, read that entire chapter before tackling this one because descriptive statistics are conceptually simpler than inferential statistics.

You might use both kinds of statistics within the same project (for example, descriptive statistics to describe your study sample and then inferential statistics to address the primary questions of your study), but you need to be clear about which type you are using in any particular analysis. It can help to think about the purpose of your analysis: is it merely to describe the data set upon which you are performing the calculations? Or is it to generalize to a larger group that you can't study directly? In the first case, you should be doing descriptive statistics, and in the second, inferential statistics. Here are two rules that state the same information slightly differently:

> If the cases you are studying represent the entire population of interest, and you do not wish to generalize beyond those cases, you should be using descriptive statistics.
> If the cases you are studying do not represent the entire population of interest, and you do wish to generalize beyond those cases, you should be doing inferential statistics.

Probability Distributions

The practice of statistical inference frequently relies on making assumptions about the way data is distributed, so much so that it is common in statistical work to transform data to make it fit some known distribution better. For this reason, this topic of statistical inference begins with a presentation of the concept of a theoretical probability distribution and a review of two commonly used distributions.

A theoretical probability distribution is defined by a formula that specifies what values can be taken by data points within the distribution and how common each value will be (or, in the case of continuous distributions, how common a given range of values will be). Theoretical probability distributions are often presented in graphical form as well; the familiar bell curve of the normal distribution is one example.

Theoretical probability distributions are useful in inferential statistics because their properties and characteristics are known. If the actual distribution of a given data set is reasonably close to that of a theoretical probability distribution, many calculations can be performed on the actual data by using assumptions drawn from the theoretical distribution. In addition, thanks to the central limit theorem (discussed later in this chapter), under certain circumstances you can assume that the distribution of sample means is normal even if the population from which the samples were drawn is not normally distributed.

Probability distributions are commonly classified as *continuous*, meaning the data can take any value within a specified range, or *discrete*, meaning the data can take only certain values. This chapter examines the normal distribution as an example of a continuous distribution and the binomial distribution as an example of a discrete distribution.

The Normal Distribution

The normal distribution is arguably the most commonly used distribution in statistics. This is partly because the normal distribution is a reasonable description of how many continuous variables are distributed in reality, from industrial process variation to intelligence test scores. A second reason for the widespread use of the normal distribution is that under specified conditions, we may assume that sampling distributions of statistics such as the sample mean are normally distributed, even if the samples are drawn from populations that are not normally distributed. This is discussed further in the section on the central limit theorem later in this chapter. The normal distribution is also referred to as the bell curve due to its characteristic shape, and as the Gaussian distribution in honor of the eighteenth-century physicist and mathematician Karl Gauss, who used this distribution to analyze astronomical data.

Inferential Statistics

There is an infinite number of normal distributions, all of which have the same basic shape but differ according to their mean μ (the Greek letter *mu*) and standard deviation σ (the Greek letter *sigma*). Examples of three normal distributions with different means and standard deviations are displayed in Figure 3-1.

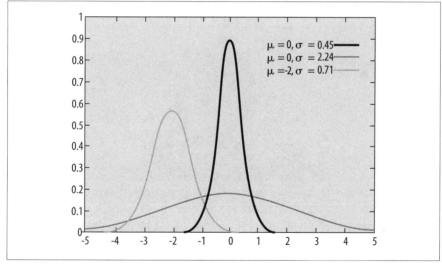

Figure 3-1. Three normal distributions

The normal distribution with a mean of 0 and standard deviation of 1 is known as the *standard normal distribution* or Z *distribution*. Any normal distribution can be transformed to the standard normal distribution by converting the original values to standardized scores (a process discussed later in this chapter and further in

Chapter 16), a procedure that facilitates comparison among populations with different means and standard deviations.

All normal distributions, regardless of their mean and standard deviation, share certain characteristics. These include:

- Symmetry
- Unimodality (a single most common value)
- A continuous range from −∞ to +∞ (from negative infinity to positive infinity)
- A total area under the curve of 1
- A common value for the mean, median, and mode

As noted earlier, there is an infinite number of normal distributions, but they all share certain properties. For the sake of convenience, we often describe normal distributions in terms of units of standard deviation rather than raw numbers because that allows us to apply the same description to any normal distribution.

Because all normal distributions have the same basic shape, we can make some assumptions about how data is distributed within any normal distribution. The empirical rule states that for any normal distribution:

- About 68% of the data will fall within one standard deviation of the mean.
- About 95% of the data will fall within two standard deviations of the mean.
- About 99% of the data will fall within three standard deviations of the mean.

This is illustrated in Figure 3-2, which expresses values in units of standard deviation.

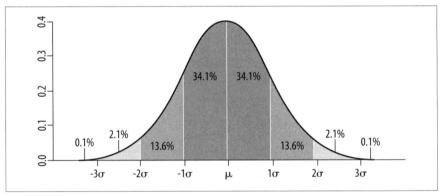

Figure 3-2. Percent of data falling into specified ranges of the normal distribution

Knowledge of these properties of the normal distribution offers a way to judge whether a particular value is typical or atypical compared to other values in the population. The process of making such comparisons is facilitated by converting raw scores (scores in their natural metric, for instance, weight measured in pounds or kilograms) into Z-scores, which express the value of the score in terms of units of the standard deviation. Converting all the values in a data set to Z-scores is analogous to transforming a normally distributed population to the standard normal distribution. For this reason, Z-scores are sometimes referred to as *normalized scores*, the process of converting raw scores to Z-scores as *normalizing* the scores, and the standard normal distribution as the *Z distribution*.

A Z-score is the distance of a data point from the mean, expressed in units of standard deviation. The formula to calculate a Z-score for a value from a population with a known mean and standard deviation is shown in Figure 3-3.

$$Z = \frac{x - \mu}{\sigma}$$

Figure 3-3. The formula for calculating a Z-score

If the variable x is distributed normally with a mean of 100 and standard deviation of 5, that is, $x \sim N\,(100, 5)$, a value of 105 has a Z-score of 1, as shown in Figure 3-4.

$$Z = \frac{105 - 100}{5} = 1.00$$

Figure 3-4. The Z-score for a value of 105 from a population ~N(100, 5)

This tells us that the value of 105 is located one standard deviation above the population mean. Similarly, a value of 110 from this population has a Z-score of 2.00, and a value of 85 has a Z-score of –3. Using the empirical rule previously cited, we classify the value of 105 as above average but not remarkable among the population (about 15.9% of the population would be expected to have higher Z-scores). A score of 110 is more unusual (about 2.5% of the population would be expected to have higher Z-scores), and a score of 85 is below average and quite unusual (less than half of 1% of the population would be expected to have scores this low or lower).

One great advantage of Z-scores is that they facilitate comparison of scores from populations with different means and standard deviations. For instance, looking at one population $x \sim N\,(100, 5)$ and another population $y \sim N\,(50, 10)$, we can't immediately say whether a score of 95 from the first population is more or less unusual than a score of 35 from the second population. However, this comparison is easily made using Z-scores, as shown in Figures 3-5 and 3-6.

$$Z = \frac{95 - 100}{5} = -1.00$$

Figure 3-5. The Z-score for a value of 95 from a population ~N(100, 5)

$$Z = \frac{35 - 50}{10} = -1.50$$

Figure 3-6. The Z-score for a value of 35 from a population ~N(50, 10)

Conversion to Z-scores places both populations on the same metric, and we can see that although both scores are below average for their populations, the second score is more extreme because –1.5 is further from 0, the mean of the standard normal distribution, than –1.0.

The Binomial Distribution

We will use the binomial distribution as an example of a discrete distribution, that is, a distribution for a variable for which only certain values are possible. Consider the case of flipping a coin five times: the number of times the coin comes up heads can take integer values such as 0, 1, 2, 3, 4, or 5 but not values such as 3.2 or 4.6. The variable "number of heads in five coin flips" is therefore a discrete variable. The binomial distribution applies to many types of real-life data with dichotomous outcomes (outcomes that can take only two values), from machine parts that are either defective or acceptable to students who either pass or fail a class.

Events in a binomial distribution are generated by a *Bernoulli process*. A single trial within a Bernoulli process is called a *Bernoulli trial*. The binomial distribution describes the number of successes in *n* trials of a Bernoulli process. "Success" in this case doesn't necessarily mean something good, just that the outcome we are looking for has occurred. For instance, if we were describing how many machine parts out of a sample of 10 were defective, each part would be considered a separate trial, and the trial would be classified as a success if the part were defective. The binomial distribution describes how likely it is that a particular number of parts from the sample of 10 will be defective, given some estimate of the overall rate of defective parts.

Data represented by the binomial distribution must meet four requirements:

1. The outcome of each trial is one of two mutually exclusive outcomes.
2. Each trial is independent, so the result of one trial has no influence on the result of any other trial.
3. The probability of success, denoted as *p*, is constant for every trial.
4. There is a fixed number of trials, denoted as *n*.

Examples of the type of data that could be described by the binomial distribution include the number of heads in 10 flips of a coin, where the probability of heads on any toss is known to be 50%; the number of males in a sample of 5 drawn from a large population known to be 65% male (the population must be large enough for the proportion of males not to change appreciably by the removal of 5 people from the total); and the number of defective items in a sample of 20, drawn from a large population whose defect rate is known to be 1%.

The formula to calculate the probability of a particular number of successes on a particular number of trials is shown in Figure 3-7.

$$\binom{n}{k} p^k (1 - p)^{n-k}$$

Figure 3-7. The formula for the binomial distribution

The formula for a combination is shown in Figure 3-8.

$$\binom{n}{k} = nCk = \frac{n!}{k!(n - k)!}$$

Figure 3-8. The formula for calculating a combination

A combination, as discussed in Chapter 2, expresses the number of ways k items can be chosen from a set of n objects, ignoring order. Note that when the binomial formula is written, the parentheses form specifies the combination to make the entire formula easier to read, but the meaning is the same as the nCk notation we used in Chapter 2.

The symbol ! in this equation means factorial: $n! = (n)(n - 1)(n - 2) \ldots (1)$. For instance, $5! = 5 \times 4 \times 3 \times 2 \times 1 = 120$. n is the number of trials. If we are flipping a coin 10 times, $n = 10$. k is the number of successes. If we want to know the probability of 5 successes in 10 trials, $k = 5$. p, a number between 0 and 1, is the probability of success. If we are flipping a fair coin and the event is heads, $p = 0.5$ (meaning the probability of heads on each flip is 0.50 or 50%).

The binomial formula can be used to calculate the probability of getting a particular number of successes given a fixed probability of success per trial and a fixed number of trials. The abbreviated way to specify a binomial probability is $b(k;n,p)$ or $P(k = k;n,p)$, where k is the number of successes in n trials, each of which has probability p of success. If we wanted to calculate the probability of 2 successes in 20 trials, with $p = 0.4$, we could write $b(2;20,0.4)$ or $P(k = 2;20,0.4)$.

Figure 3-9 shows the graph for three binomial distributions. (Note that each combination of p and n will produce a different distribution.)

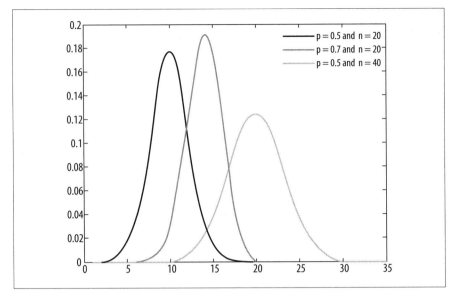

Figure 3-9. Three binomial distributions

As *n* increases, holding *p* constant, the binomial distribution more closely resembles the normal distribution. A common rule of thumb is that if both *np* and $n(1 - p)$ are greater than or equal to 5, the binomial distribution may be approximated by the normal distribution. In Figure 3-9, the distribution ($p = 0.5$, $n = 40$) qualifies for the normal approximation according to this rule because:

$np = 40(0.5) = 20$ $n (1 - p) = 40(1 - 0.5) = 20$

However, a distribution with $p = 0.1$ and $n = 40$ does not qualify for use of the normal approximation to the binomial because:

$np = 40(0.1) = 4$

Complex calculations based on the binomial distribution are usually done using computer software, but a simple example will demonstrate how the formula works. Suppose we are flipping a fair coin five times; what is the probability that we will get exactly one head? We will define "heads" as a success and use the binomial formula to solve this problem. In this example:

$p = 0.5$ (the definition of a fair coin is that heads and tails are equally likely)
$n = 5$ (because we are conducting five trials)
$k = 1$ (because we are calculating the probability of exactly one success)

The probability of exactly one success in five trials, given a probability of success on each trial as 0.5, is calculated in Figure 3-10.

$$P(k = 1;5,0.5) = \binom{5}{1} 0.5^{1}(1 - 0.5)^{5-1} = 0.16$$

Figure 3-10. Calculating b(1;5,0.5)

Breaking down the steps, Figure 3-11 shows how to calculate the combination.

$$\binom{5}{1} = \frac{5!}{1!(5-1)!} = \frac{5 \times 4 \times 3 \times 2 \times 1}{1 \times (4 \times 3 \times 2 \times 1)} = 5$$

Figure 3-11. Calculating 5C1

And Figure 3-12 shows the entire calculation.

$$P(k = 1;5,0.5) = \binom{5}{1} 0.5^{1}(1 - 0.5)^{5-1} = 5 \times (0.5)^{1} \times (0.5)^{4} = 0.16$$

Figure 3-12. Detailed calculation of b(1;5,0.5)

We can also get this result by using the binomial table in Figure D-8, Appendix D.

Independent and Dependent Variables

There are many ways to characterize variables: one of the most common is by the roles they play in a research design or data analysis. Using these criteria, a simple way to describe variables is as either *dependent*, if they represent an outcome of the study, or *independent*, if they are presumed to influence the value of the dependent variable(s). Many study designs include a third category, *control variables*, which might influence the dependent variable but are not the main focus of interest.

Note that the labels "independent," "dependent," and "control" relate to the roles played by the variables in a given design or experiment. This means that a given variable (for instance, weight) could be an independent variable in one study, a dependent variable in another, and a control variable in a third. In addition, other labels are used to describe dependent and independent variables, with some authors preferring to reserve specific labels for particular types of studies. Control variables are particularly problematic because many types of control variables have been defined, depending on their relationship to the independent and dependent variables of interest and the type of study design employed. Control variables are discussed further in Chapter 18, but this discussion will concentrate on independent and dependent variables.

We will use the example of a regression equation to illustrate the concept of independent and dependent variables. This is just a brief introduction; the topic of regression is covered in detail in Chapters 8, 10, and 11.

In a standard linear model such as an OLS regression equation, the outcome or dependent variable is customarily indicated by the letter Y, whereas the independent variables are indicated by X. Subscripts identify each individual X variable: X_1, X_2, and so on. (OLS means Ordinary Least Squares, the most common type of regression; if not otherwise specified, in this book "regression equation" means "OLS regression equation.")

This should be clear from the conventional way of notating a regression equation, as shown in Figure 3-13.

$$Y = \beta_0 + \beta_1 X_1 + \beta_2 X_2 + \beta_3 X_3 + ... + e$$

Figure 3-13. A regression equation

The e in this equation means "error" and reflects the fact that we don't assume any regression equation will perfectly predict Y; instead, we expect that there will always be some error of prediction. Note that each X in the equation is preceded by a β, which is called its *regression coefficient*: β_1 is the regression coefficient for X_1, β_2 is the regression coefficient for X_2, and so on. The values for these regression coefficients are determined through a mathematical process to create the best possible equation for predicting the value of Y from the values of the Xs in a given data set.

Because of this notational convention, the dependent variable is also referred to as the "Y variable" and the independent variables as the "X variables." Other terms used for the dependent variable include the *outcome variable*, the *response variable*, and the *explained variable*. Other names for independent variables include *regressors*, *predictor variables*, and *explanatory variables*.

Some researchers believe that the terms "independent" and "dependent" should be reserved for experimental studies (for instance, a randomized controlled drug trial). In this interpretation, the terms "independent" and "dependent" imply causality, that is, that the value of the dependent variable *depends* at least in part on the values of the independent variables, a statement that is difficult if not impossible to establish in a nonexperimental study. (The distinction between experimental and nonexperimental studies is discussed in detail in Chapter 18.) This book does not embrace this rule because questions of causality are far more complex than the distinction between experimental and nonexperimental studies; thus, we will use "independent variable" to identify the variables that reflect the outcome of a study and "dependent variable" to mean the variables believed to influence the outcome.

Populations and Samples

The concept of populations and samples, which is also discussed in Chapter 4, is crucial to understanding inferential statistics. The process of defining the population and selecting an appropriate sampling method can be quite complex (in fact, many doctoral-level statisticians specialize in this type of work) and requires more study than can be covered here. Instead, the basic issues and concepts will be discussed, and the reader interested in further information on the subject should consult a specialized textbook (several are listed in Appendix C) or take an advanced course in sampling theory.

The population of interest (often called simply "the population") consists of all the people or other entities (for instance, airplane parts or Atlantic salmon) that the researchers would like to study if they had infinite resources. To look at it another way, the population of interest is all the entities about which the researchers would like to generalize their results. Defining the population of interest is the first step in drawing a sample. It might be, for instance, everyone living in the United States in 2007 or men aged 65–75 with a diagnosis of congestive heart failure.

Samples and Censuses

Almost all statistical research is based on a study sample drawn from a population rather than the population itself. The rare exceptions are studies based on data collected from entire populations. When data is systematically collected from an entire population, the result is a *census*. Many national governments conduct a regular census of their population. For instance, the United States conducts a census every 10 years, and the results are used for a variety of purposes, including allocating seats in the House of Representatives. Even though a census intends to collect information from every individual in a population, in practice this is rarely achieved; some people are never counted, and some are counted more than once. For these reasons, some statisticians argue that a well-chosen sample can yield a more accurate estimate of population characteristics than that produced by census data, or that the census data should be supplemented by sample data. For a readable discussion of these issues and a good list of references to more detailed information, see the article by Ivars Peterson listed in Appendix C.

Nonprobability Sampling

There are many ways to draw a sample. Unfortunately, some of the most convenient are based on nonprobability sampling, which leaves them subject to sampling bias. This means there is a high probability that the sample drawn using a nonprobability method will not be representative of the population of interest, and there is no way to correct the sample statistically, so any conclusions about the population based on sample calculations will be questionable. Nonprobability sampling methods are popular because the researcher can bypass the more cumbersome process of drawing a probability sample, but a price is paid for this convenience. Conclusions based on data using nonprobability sampling methods are of limited usefulness in generalizing to a larger population (the usual reason for drawing a sample in the first place) because there is no way to know how the sample relates to the population of interest, and, thus, little faith may be had in conclusions about that population based on results from the sample.

Volunteer samples are a common type of nonprobability sample. Here's an example: a researcher advertises in the newspaper for study subjects and accepts those who answer the ad and volunteer to take part in the study. This is a convenient way to get subjects, but unfortunately people who volunteer for studies can't be assumed to be representative of any general population. Use of volunteer samples is best reserved for circumstances in which it would be difficult to select a sample randomly from a population, for instance in a study about people who use illegal drugs. Even with limited ability to generalize, useful information can be gained from volunteer samples, particularly in the early stages of a project. For instance, you might use volunteer subjects to gather information about drug use within a community, information that you could then use to construct a questionnaire that would be administered to a random sample from the community. Still, results from volunteer samples have limited usefulness if the goal is to generalize beyond the sample.

Convenience samples are another common type of nonprobability sample. Like volunteer samples, convenience samples can be used to collect information in the early stages of a study but have limited usefulness if the goal is to generalize beyond the sample. Here's an example of a convenience sample: you collect information about the shopping habits of people in a particular geographical area by interviewing 50 people shopping at a mall within that area. The problem with this type of sampling is that because those 50 people are not a random selection of area residents, it would not be valid to conclude that their opinions reflect those of the area as a whole. However, you might use the information gained from a survey administered to a convenience sample to construct a questionnaire for a more scientific sample of the area's population.

Quota sampling is a nonprobability sampling method in which the data collector is instructed to get responses from a certain number or proportion of subjects within broad classifications. For instance, in the shopping mall example, the data collector might be instructed to collect data from a sample of 25 men and 25 women or to include at least 20 nonwhite individuals in the sample. Quota sampling is a slight improvement over convenience sampling because it can ensure representation of different demographic groups within the sample. For instance, without the quota requirements, the shopping mall sample might consist of 45 women and 5 men, and no nonwhite individuals at all. However, because quota sampling is a nonprobability sampling method, you still have no way of knowing whether the people in the sample are representative of the population of interest. You might have an even representation of men and women in a quota sample, for instance, but are those in the sample representative of all the men and women who shop at the mall, let alone who live in the area? Quota sampling can also be subject to a particular type of selection bias, which is also a risk in convenience sampling. The data collector might approach people who seem most like himself (for instance in age) or who seem the friendliest or most approachable, rendering the sample even less useful as a means to acquire information about a larger population.

Probability Sampling

In probability sampling, every member of the population has a known probability to be selected for the sample. Although more complex to execute than nonprobability sampling, probability sampling is preferred because the researcher can generalize the results obtained from the sample to the population of interest.

Drawing a probability sample from a population requires devising some type of sampling frame so the researcher can identify and sample members of the population. Sometimes an obvious sampling frame exists. If the population is students enrolled at a particular school, a list of all enrolled students could serve as the sampling frame. Other times, a less optimal sampling frame must be used. For instance, a telephone directory or block of phone numbers in use can be employed for a survey carried out by telephone. A problem with either type of telephone sampling frame is that people without phone service are not included in the population from which the sample is drawn although they might be included in the population of interest. People with unlisted telephone numbers or only callphone service, can also be excluded from a telephone sample drawn using these methods although they might be part of the population of interest. Weighting and other procedures can be used during analysis to make results from the study sample more applicable to the population of interest.

The most basic type of probability sampling is *simple random sampling* (SRS). In SRS, all samples of a given size have an equal probability of being selected. Suppose you wanted to draw a random sample of 50 students attending a particular school. You obtain a list of the students and select 50 at random from the list, using a random number table or random number generator. Because the list represents an enumeration of the entire population and the choice of whom to include in the sample is completely random, every student has an equal probability of being selected for the sample, as does every combination of students. (In this example, all samples of size 50 are equally likely.)

In most cases, SRS has the most desirable statistical properties of any kind of sampling, including the smallest confidence intervals around parameter estimates, and requires the least complex procedures to analyze. However, SRS can be impossible or prohibitively expensive to execute in some contexts, so other methods of probability sampling have been developed when SRS is not possible or practical.

Systematic sampling is similar to SRS. To draw a systematic sample, you need a list or other enumeration of your population. You decide the size of the sample you wish to draw and then compute the number n, which dictates how you will select the sample. You calculate n by dividing the size of the population by the number of subjects you want in your sample. Suppose you have a population of 500 and want to draw a sample of 25; in this case, $n = 20$ because $500/25 = 20$.

You then choose a start number at random between 1 and n and include in your sample the object representing the start number and every nth object following. Suppose you want to draw a random sample of 100 objects from a population of 1,000. The steps to draw a systematic sample are the following:

1. Set $n = 10$ because $1,000/100 = 10$.
2. Choose a number at random between 1 and 10.
3. Select the object with that number and every 10th object thereafter.

If the number chosen at random was 7, your sample would include the 7th object, the 17th, the 27th, and so on up to the 997th object.

Systematic sampling technique is particularly useful when the population accrues over time and there is no predetermined list of population members. For instance, suppose you want to survey people making court appearances in the upcoming year. At the start of your study, you don't know who those people will be, so you make an estimate of the population of interest based on the court caseload in the previous year, decide on your sample size, and calculate n as previously described. You then keep an ordered list of people making court appearances, select your random starting point, and then survey the person corresponding to your random starting point and every nth person afterward who appears in court. If you determine that n is 14 and your random starting point is 10, you would then survey the 10th person, the 24th person, 38th person, and so on until you have your desired sample size.

One caution when using systematic sampling is that you must ensure that the data is not cyclic in a way that corresponds with your random starting point and value of n. For instance, if particular hours or days in court are reserved for particular types of cases, and if your combination of starting point and n means that people whose court dates were scheduled for those times have no possibility of being selected, your sample will not be a random selection of everyone making court appearances.

There are many types of *complex random samples*, an umbrella term for probability sampling methods that impose one or more layers of complexity beyond that of SRS. In a *stratified sample*, the population of interest is divided into nonoverlapping groups or *strata* based on common characteristics. For people, these characteristics might be gender or age; for cities, they might be population size or type of government; and for hospitals, they might be type of governance or number of beds. If comparing different strata or making estimates of the characteristics of subgroups is a primary goal of the study, stratified sampling is a good choice because it can be designed to ensure adequate sampling from each stratum of interest. For instance, a sample drawn using SRS might not include sufficient older adults to estimate their characteristics accurately or compare them with middle-aged people. A stratified sample, in contrast, can be designed to oversample the older adults, and the sample can then be statistically adjusted to correct for the oversampling.

In a *cluster sample*, the population is sampled by using preexisting groups. This technique is often used in national surveys that require in-person interviews or the collection of physical specimens (e.g., blood samples) because sending survey personnel to interview one person in Ruckersville, Virginia, one in Chadron, Nebraska, one in Barrow, Alaska, and so on would be prohibitively expensive. A more economical procedure is to create a sampling plan that incorporates several levels of random selection. On a national level, a cluster sampling plan could be devised that selects geographic regions, then states within regions, cities within states, and so on down to individual households and individuals within households. Precision is decreased with cluster sampling because objects that are clustered within units (for instance, households within cites and cities within states) tend to be more similar than objects selected through SRS. Offsetting this loss of precision is that a larger sample can be collected because the cost savings of cluster sampling are usually substantial.

Cluster sampling can be combined with the technique of *sampling proportional to size*. For instance, you might wish to draw a sample of elementary school students. There is no national list of all elementary school students (at least, not in the United States), but you could compile a list of all elementary schools, and each school would have a list of its students. Therefore, you could select schools at random, possibly in a multistage process, and then draw a random sample from the selected schools. Because schools enroll different numbers of students, you might want to incorporate this information into your sampling plan so that you don't have a disproportionate number of students from small schools (which are more numerous but contain fewer students as compared to large schools). Then you would select a different number of students from each sampled school, based on the number enrolled in the school. This means that you would select twice as many students from a school with an enrollment of 400 as from a school with an enrollment of 200. In this way, your final sample will have a representative proportion of students from both large and small schools.

The Central Limit Theorem

The *central limit theorem* states that the sampling distribution of the sample mean approximates the normal distribution, regardless of the distribution of the population from which the samples are drawn if the sample size is sufficiently large. This fact enables us to make statistical inferences based on the properties of the normal distribution, even if the sample is drawn from a population that is not normally distributed.

The central limit theorem can be stated as follows with regard to the sample mean:

Let $X_1, \ldots Xn$ be a random sample from some population with mean μ and variance σ^2. Then for large n,

$$\overline{X} \sim N(\mu, \frac{\sigma^2}{n})$$

even if the underlying distribution of individual observations in the population is not normal.

The $\sim$ symbol represents "approximately distributed," and the formula can be read as "the mean of X is approximately normally distributed with mean μ and variance σ^2/n".[1]

---

1. Rosner, Bernard. 2000. *Fundamentals of Biostatistics*, 5th ed.; Brooks/Cole, Pacific Grove, CA, 174.

The application of the central limit theorem in practice can be seen through computer simulations that repeatedly draw samples of specified size from a nonnormal population. Figure 3-14 displays a histogram for a population of randomly generated data (100 cases) with a uniform distribution of values ranging from 0 to 100.

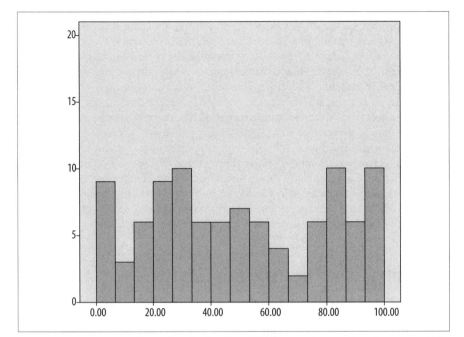

Figure 3-14. Histogram of a uniformly distributed population (N = 100) with range 0–100

The distribution in Figure 3-14 is decidedly not normal. However, the central limit theorem says that when samples of sufficient size are drawn from a nonnormal population, the means of those samples tend to assume a normal distribution. Note that the theorem does not define what constitutes a sufficient size. Analysts have developed rules of thumb regarding this issue, such as the often-repeated rule that the sample size should be 30 or larger, but no absolute rule applies in all cases. For samples drawn from a population that is approximately normal, the sampling distribution of the sample mean might be approximately normal with a sample size as small as 10 or 15, whereas with highly skewed distribution, the sample size required can be 40 or more.

The phrase "sampling distribution of the sample mean" is a mouthful, but its meaning is straightforward. We have already looked at two theoretical distributions (the normal and the binomial), but the fact is that random variables also have distributions. In this case, we are interested in the distribution of means calculated from samples of a given size drawn from a particular population. If we repeatedly draw samples of a given size, calculate the mean of each sample, and plot the distribution of those means, the result is the sampling distribution of the sample mean. We expect that the samples will differ somewhat from each other and thus will have different means, producing a distribution of means. We can predict the general shape that this distribution of sample means will take based on factors such as the population distribution and the sample size.

You can see the influence of sample size on the sampling distribution of the sample mean by comparing Figure 3-15 and Figure 3-16. Figure 3-15 displays the distribution of the means of 100 samples of size $n = 2$ drawn from the population shown in Figure 3-14; Figure 3-16 displays the distribution of the means of 100 samples of size $n = 25$ drawn from the same population. Figure 3-15 still looks much like a uniform distribution, indicating that a sample size of 2 is not sufficient to invoke the central limit theorem for this population.

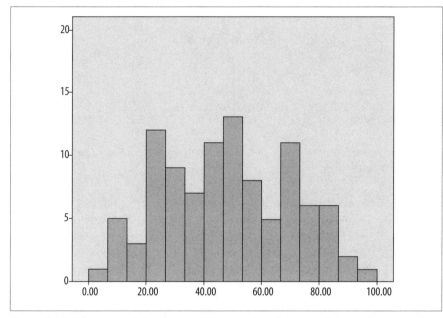

Figure 3-15. Histogram of the means of 100 samples of size n = 2 drawn from a uniform distribution

Figure 3-16 displays the distribution of 100 means calculated from samples of size $n = 25$ drawn from the uniform distribution displayed in Figure 3-14. This distribution is much closer to a normal distribution, so a sample size of 25 appears to be sufficient to invoke the central limit theorem for this population.

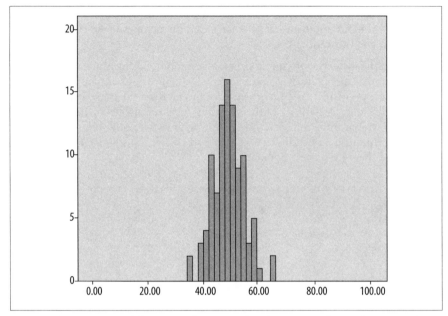

Figure 3-16. Histogram of the means of 100 samples of size n = 25 drawn from a population with uniform distribution

Figures 3-17 to 3-19 demonstrate the same principle by using samples drawn from a skewed (nonsymmetric) population. Figure 3-17 shows the distribution of values for a data set of size 100 with a strongly skewed distribution.

Figures 3-18 and 3-19 demonstrate how the distribution of sample means drawn from this skewed population changes with the size of the samples. Figure 3-18 shows the distribution of means calculated from 100 samples of size $n = 2$, whereas Figure 3-19 shows the distribution of means from 100 samples of size $n = 25$. As with the previous uniform data example, a sample of size $n = 2$ is not sufficient to invoke the central limit theorem for this data, although a sample of 25 seems to be sufficient.

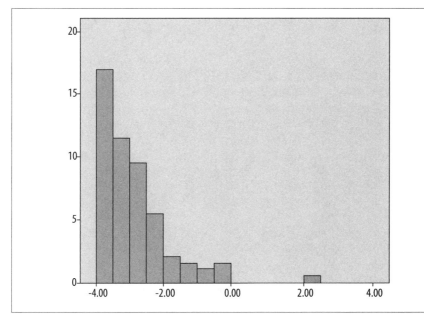

Figure 3-17. Histogram of skewed population (N = 100)

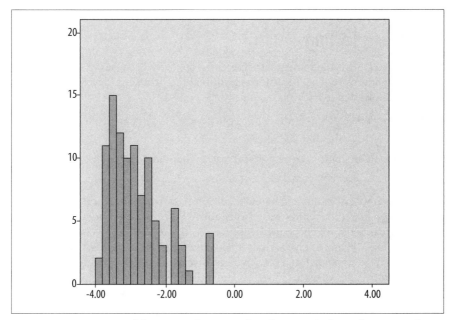

*Figure 3-18. Histogram of the means of 100 samples of size n = 2, drawn from a population
with skewed distribution*

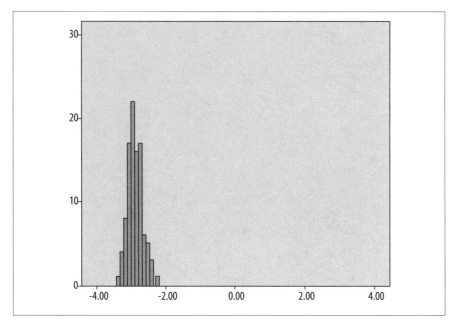

Figure 3-19. Distribution of the means of 100 samples of size n = 25, drawn from a population with skewed distribution

Hypothesis Testing

Hypothesis testing is fundamental to inferential statistics because it allows us to use statistical methods to make decisions about real-life problems. Several conceptual steps are involved in hypothesis testing:

1. Develop a research hypothesis that can be tested mathematically.
2. Formally state the null and alternative hypotheses.
3. Decide on an appropriate statistical test, gather data, and do the calculations.
4. Make your decision based on the results.

Take the example of evaluating a new medication to treat high blood pressure (hypertension). The manufacturer wants to establish that it works better than currently available treatments for the same condition, so the research hypothesis might be something like, "Hypertensive patients treated with the new drug X will show greater lowering of their blood pressure than hypertensive patients treated with the currently available drug Y." If we use μ_1 to signify the mean lowering of blood pressure in the group treated with drug X and μ_2 to signify the mean lowering of blood pressure in the group receiving drug Y, we can state our null and alternative hypotheses as follows:

$H_0: \mu_1 \leq \mu_2$
$H_A: \mu_1 > \mu_2$

H_0 is called the null hypothesis. In this example, the null hypothesis states that drug X is no improvement over drug Y because the lowering of blood pressure achieved by drug X is less than or equal to that achieved by drug Y. H_A, sometimes written as H_1, is called the alternative hypothesis. In this example, the alternative hypothesis states that drug X is more effective than standard treatment because patients treated with drug X show more lowering of their blood pressure than patients treated with drug Y. Note that the null and alternative hypotheses must be both mutually exclusive (no result could satisfy both conditions) and exhaustive (all possible results will satisfy one of the two conditions).

In this example, the alternative hypothesis is *single-tailed*: we state that the group treated with drug X must achieve greater lowering of blood pressure than the group treated with drug Y for the null hypothesis to be rejected. We could also state a *two-tailed* alternative hypothesis if that were more appropriate to our research question. For instance, if we were interested in whether the blood pressure of patients treated with drug X was different, either higher or lower, than that of patients receiving drug Y, we would state this using a two-tailed alternative hypothesis:

$$H_0: \mu_1 = \mu_2$$
$$H_A: \mu_1 \neq \mu_2$$

Two-tailed hypotheses are more common in statistical testing because usually you want to retain the ability to find a difference in either direction.

After the data is collected and the statistics calculated, we can make one of two decisions:

- Reject the null hypothesis.
- Fail to reject the null hypothesis.

Note that if we fail to reject the null hypothesis, this does not mean that we have proven the null hypothesis to be true, only that our study did not find sufficient evidence to reject it.

Rejecting the null hypothesis is sometimes called "finding significance" or "finding significant results" because our statistical analysis must show not only that there are, say, differences in the group means but that those differences are *statistically significant*. The informal meaning of statistically significant is "probably not due to chance," and the process of determining whether results are significant involves not only statistical calculations but the application of customary rules that might vary based on the field of study or other factors.

The process of statistical testing involves choosing a probability level or *p*-value (a topic treated in greater detail later) that defines when sample results will be considered strong enough to support rejection of the null hypothesis. In practice, the *p*-value is most commonly set at 0.05. Why this particular value? It's a somewhat arbitrary cutoff point and dates back to the early twentieth century, when statistics were computed by hand and the results compared to published tables used to determine whether a result was significant. The use of $p < 0.05$ as the standard for significant results has been challenged (see the next sidebar, "Controversies Regarding Hypothesis Testing") but remains a common standard for research in many fields. Alternative lower *p*-values are sometimes used, such as $p < 0.01$ or $p < 0.001$, but no one has been successful in legitimizing the general use of a larger value, such as $p < 0.10$.

Inferential statistics is a powerful tool that allows us to make probabilistic statements about data. However, because those statements are probabilistic rather than absolute, the possibility of error is inherent in the process. Statisticians have defined two types of error possible when making decisions using inferential statistics and have established levels for error rates that are commonly considered acceptable. The two types of error are displayed in Table 3-1.

Controversies Regarding Hypothesis Testing

Despite the ubiquity of hypothesis testing in modern statistical practice and the canonical place that the $\alpha = 0.05$ significance level has achieved, neither practice has gone unchallenged. One of the chief critics is Jacob Cohen, whose arguments are presented in, among other places, his 1994 article, "The Earth Is Round (p < 0.05)".[2] There are valid criticisms of both hypothesis testing in general and the 0.05 value in particular, but neither seems likely to be going away any time soon. On the one hand, we need to establish some standard for statistical significance to minimize the possibility of attributing significance to differences due to sampling error or other chance factors. On the other hand, there's nothing magical about the 0.05 level, even if it is sometimes treated as such. In addition, the significance level of results calculated on a sample is affected by many factors, including the size of the sample involved, and overemphasis on the *p*-value of a result ignores the many reasons a particular study may or may not have found significance. It's a common saying among statisticians that if you have a large enough sample, even a tiny effect will be statistically significant. The take-home message is that statistical methods are powerful tools, but they don't relieve researchers of the need to use their common sense as well.

Table 3-1. Type I and Type II errors

		True state of population	
		H_0 true	H_A true
Decision based on sample statistic	**Fail to reject** H_0	Correct decision: H_0 true and H_0 not rejected	Type II error or β
	Reject H_0	Type I error or α	Correct decision: H_0 false and H_0 rejected

The diagonal boxes represent correct decisions: H_0 is true and is not rejected in the study, or H_0 is false and is rejected in the study. The other two boxes (the off-diagonal boxes) represent Type I and Type II errors. A *Type I error*, also known as *alpha* or α, represents the error made when the null hypothesis is true but is rejected in a study. A *Type II error*, also called *beta* or β, represents the error made when H_0 is false but is not rejected in a study.

---

2. *American Psychologist*, December 1994, 997–1003.

I have set up this matrix to compare the true state of the population (which, of course, is generally unknown to the researcher) with a decision made about the population, based on analysis of a sample. Another way to look at it is to consider a trial in which the null hypothesis is that the defendant is innocent. In a trial situation, there is a true state of affairs (did the defendant commit the crime as charged?), and then there is the jury members' decision based on the information presented to them (did they find the client guilty or not guilty?). The jury doesn't know the true state of affairs any more than a statistician knows the true state of the population, so it might make a correct decision, or it might commit a Type I or Type II error. If the jury finds an innocent client guilty, that is equivalent to a Type I error (it rejects the null hypothesis of innocence when it should not), whereas if it finds a guilty client not guilty, it commits a Type II error (failing to reject the null hypothesis of innocence when it should have rejected it).

The level of acceptability for Type I error is conventionally set at 0.05, as noted previously. Setting alpha at 0.05 means that we accept a 5% probability of Type I error. To put it another way, we understand when setting the alpha level at 0.05 that in our study we have a 5% chance of rejecting the null hypothesis when we should fail to reject it.

Type II error has received less attention in statistical theory because historically it has been considered a less serious error to fail to make an inference that is true (Type II error) than to make an inference that is false (Type I error). Conventional levels of acceptability for Type II error are $\beta = 0.1$ or $\beta = 0.2$. If $\beta = 0.1$, that means the study has a 10% probability of a Type II error; that is, there is a 10% chance that the null hypothesis will be false but will fail to be rejected in the study. To put it another way, it means that in a study that should return significant results based on the true state of the population, there is a 10% chance that the results of the study will not be significant.

The reciprocal of Type II error is *power*, defined as $1 - \beta$. The importance of setting an appropriate power level has become more appreciated in recent years, particularly in the medical field. Researchers and funding agencies have become concerned with power and, thus, with Type II error, in part because they don't want to invest time, effort, and expense in a study unless it has a reasonable probability of finding significant results if it should find them. Power calculations play an important role in planning studies, particularly in determining the sample size required for adequate power; these issues are discussed in more detail in Chapter 15.

Confidence Intervals

When we calculate a single statistic, such as the mean, to describe a sample, that is referred to as calculating a *point estimate* because the number represents a single point on the number line. Although the sample mean is the best unbiased estimate of the population mean, we know that if we drew a different sample, the mean calculated from that sample would probably be different. We certainly don't expect that every sample we draw will have exactly the same mean. It is reasonable to ask how much a point estimate is likely to vary by chance, and for this reason, it has become a common practice in many professional fields to report both point estimates and *interval estimates*. In contrast to a point estimate, which is a single number, an interval estimate is a range of numbers.

One common interval estimate is the *confidence interval*, which is the interval between two values that represent the upper and lower *confidence limits* or *confidence bounds* for a statistic. The formula used to calculate the confidence interval depends on the statistic being used and will be included in the relevant chapters. In this section, our purpose is to convey the concept of the confidence interval. It is calculated using a predetermined significance level, often called α (the Greek letter *alpha*), which is most often set at 0.05, as discussed previously. The *confidence coefficient* is calculated as $(1 - \alpha)$ or, as a percentage, $100(1 - \alpha)\%$. Thus, if $\alpha = 0.05$, the confidence coefficient is 0.95 or 95%. The latter usage is more common; for instance, people often speak of 95% confidence intervals, and professional journals often require you to report the 95% confidence interval along with point estimate statistics.

Confidence intervals are based on the idea that if a study were repeated an infinite number of times, each time drawing a different sample of the same size from the same population, and a confidence interval based on each sample were constructed, $x\%$ of the time the confidence interval would contain the true parameter value that the study seeks to estimate, where x is the size of the confidence interval. For instance, if our test statistic is the mean and we are using a 95% confidence interval, over an infinite number of repetitions of drawing a sample and computing its mean, 95% of the time the confidence interval thus constructed would contain the true mean of the population.

The confidence interval conveys important information about the precision of a point estimate. For instance, suppose we have two samples of students, and in both cases, the mean IQ score for the group is 100 (average intelligence). In one case, however, the 95% confidence interval is (95, 105), whereas in the other case, the 95% confidence interval is (80, 120). Because the first confidence interval is much narrower than the second, the estimate of the mean is more precise for the first sample. In addition, the wider confidence interval for the second sample suggests that those students are drawn from a population with greater variability in IQ than the students in the first sample (although further analysis would be necessary to confirm or reject this hypothesis).

p-values

It is a fact of life when working with inferential statistics that we are generally trying to estimate something that we can't measure directly. For instance, we can't collect data from every hypertensive adult in the world, but we can collect data from a sample of hypertensive adults and make inferences based on that sample. We know that there is always some probability of error in this type of reasoning, including the possibility that significant results can be due to chance factors such as sampling error rather than to the factors of interest in our study.

A *p-value* expresses the probability that results at least as extreme as those obtained in an analysis of sample data are due to chance. The phrase "at least as extreme" is included in the definition because many statistical tests involve comparing the test statistic to some hypothetical distribution, and often (as is the case with the normal distribution), scores closer to the center of the distribution are the most common, whereas scores further from the center of the distribution (the more extreme scores) are less likely. Even if a distribution is not symmetrical (as is the case with the chi-squared distribution, for instance), more extreme results are usually less probable results, so the principle of determining the probability of results at least as extreme as those found in a study remains useful.

This might become clearer by considering a simple illustration. Suppose we are engaged in an experiment involving flipping a coin that we believe to be fair, that is, a coin for which heads (*h*) or tails (*t*) are equally likely outcomes for any single flip. We can express this formally as:

$$P(h) = P(t) = 0.5.$$

We will call each flip a trial. Because the probability of heads on any flip is 0.5, our best guess is that we will get 5 heads on 10 trials, although we also know that on any particular set of 10 trials, we might get a different number of heads. Suppose we flip the coin 10 times, and 8 times it comes up heads. We want to know the *p*-value of this result, that is, how likely is it that a coin with a probability of 0.5 for heads on any single trial would produce 8 heads in 10 trials?

Using a binomial table, computer software, or the binomial formula, we find that the probability of this exact result (8 heads in 10 trials) is 0.0439, meaning that less than 5% of the time would we expect to get exactly 8 heads in 10 flips with a fair coin. The probability for 9 heads in 10 trials is 0.0098, and for 10 heads in 10 trials is 0.0010. This demonstrates that as results move further away from the expected result of 5 heads in 10 trials, they become less likely.

If we are evaluating the probability that the coin is fair, results that are far from our expectation (5 heads in 10 trials) give us strong evidence that it is fair. With this type of question, we usually calculate the probability not just of the result we obtained in our experiment but of results at least as extreme as those we obtained. In this case, the probability of getting 8, 9, or 10 heads in 10 flips of a fair coin is 0.0439 + 0.0098 + 0.0010, or 0.0547. This is the *p*-value for the result of at least 8 heads in 10 trials, using a coin where *P*(heads) = 0.5.

p-values are commonly reported for most research results involving statistical calculations, in part because intuition is a poor guide to how unusual a particular result is. For instance, many people might think it is unusual to get 8 or more heads on 10 trials using a fair coin. There is no statistical definition of what constitutes "unusual" results, so we will use the common standard that the *p*-value for our results must be less than 0.05 for us to reject the null hypothesis (which is, in this case, that the coin is fair). In this example, somewhat surprisingly, this standard is not met. The *p*-value for our result (8 heads in 10 trials) does not allow us to reject the null hypothesis that the coin is fair, that is, that *P*(heads) = 0.5, because 0.0547 is greater than 0.05.

The Z-Statistic

The Z-statistic is analogous to the Z-score discussed earlier, with one important difference: instead of asking what the probability of a particular *score* is, we are now interested in the probability of a particular sample *mean*. The Z-statistic is an important example of the application of the central limit theorem, which allows us to compute the probability of a sample result by using the normal distribution, even if we don't know the distribution of the population from which the sample was drawn.

The formula for calculating the Z-statistic (Figure 3-20) is similar to that for calculating a Z-score (Figure 3-3).

$$Z = \frac{\bar{x} - \mu}{\frac{\sigma}{\sqrt{n}}}$$

Figure 3-20. Formula for the Z-statistic

In this formula, $\bar{x}$ is the mean of our sample,
μ is the population mean,
σ is the population standard deviation, and
n is the sample size.

The big difference between the Z-score and the Z-statistic formulas is in the denominator: for a Z-score we divide by σ, whereas for the Z-statistic we divide by $\sigma/\sqrt{n}$. Note that to calculate the Z-statistic, we must know the population mean and standard deviation; if we know the mean but not the standard deviation, we can calculate the *t*-statistic instead (discussed in Chapter 6). It might help to think of the Z-score as a Z-statistic for a sample of 1, so the denominator is $\sigma/\sqrt{1}$, which is the same as σ and gives us the familiar Z-score formula.

The denominator of the Z-statistic is called the standard error of the mean, sometimes abbreviated SEM or written as $\sigma_{\bar{x}}$. The standard error of the mean is the standard deviation of the sampling distribution of the sample mean. Because the denominator of the standard error of the mean is divided by $\sqrt{n}$, larger samples will tend to produce larger Z-statistics, all else held equal. This will be clear if we calculate the Z-statistic for several samples that differ only in sample size. Suppose we draw three samples from a population with a mean of 50 and a standard deviation of 10:

Sample 1: $\bar{x} = 52$, $n = 30$
Sample 2: $\bar{x} = 52$, $n = 60$
Sample 3: $\bar{x} = 52$, $n = 100$

The calculations for the Z-statistic for each sample are presented in Figures 3-21, 3-22, and 3-23.

$$Z = \frac{52 - 50}{\frac{10}{\sqrt{30}}} = 1.10$$

Figure 3-21. Z-statistic for a sample ($\overline{X}$ = 52, n = 30) from a population ~N(50, 10)

$$Z = \frac{52 - 50}{\frac{10}{\sqrt{60}}} = 1.55$$

Figure 3-22. Z-statistic for a sample ($\overline{X}$ = 52, n = 60) from a population ~N(50, 10)

$$Z = \frac{52 - 50}{\frac{10}{\sqrt{100}}} = 2.00$$

Figure 3-23. Z-statistic for a sample ($\overline{X}$ = 52, n = 100) from a population ~N(50, 10)

It is clear from these examples that sample size has an important influence on our results and that, all else held equal, a larger sample will result in a more extreme Z-score. This topic is taken up in greater detail in the section on sample size and power in Chapter 15, but we will note here that this result makes intuitive sense. The Z-statistic is calculated by dividing a numerator by a denominator, and a larger sample size (larger n) will result in dividing by a smaller denominator, thus a more extreme Z-score (assuming the numerator does not change). We say "more extreme" because if the numerator is negative, the Z-score will be smaller with a larger n (all else held equal) but also further from 0. For instance, in this example, if our sample mean were 48 instead of 52, the Z-values would be –1.10, –1.55, and –2.00.

Suppose we are testing a two-tailed hypothesis with an alpha level of 0.05. In this case, we would also want the *p*-values for each result, which are:

Sample 1: $p = 0.2713$
Sample 2: $p = 0.1211$
Sample 3: $p = 0.0455$

Only the third sample gives us significant results; that is, only the *p*-value from the third sample is less than our alpha level of 0.05 and thus allows us to reject the null hypothesis. This underlines the importance of having adequate sample size when conducting a study.

You can find the *p*-value for a given Z value in several ways: by using statistical software, by using one of the many online calculators (*http://graphpad.com/quick calcs/PValue1.cfm*), or by using probability tables. Probability tables for several of the most common distributions, including the standard normal distribution, are included in Appendix D, along with instructions for using the tables.

Data Transformations

Many of the most common statistical procedures are known as *parametric* statistics, meaning that they make certain assumptions about the distribution of the population from which the sample was drawn. If the sample data indicate that these assumptions have not been met, the researcher has several options for analyzing the data. One is to use alternate, *nonparametric* statistical procedures, which make fewer or no assumptions about the data distribution. Nonparametric statistics are discussed in Chapter 13. Another possibility is to *transform* the data in some way so that the assumptions of the desired parametric statistical procedure are met. There are many ways to transform data, depending on the distribution involved and the assumptions violated. We will examine one case, the transformation of a data set to make it close to a normal distribution, but the principles we discuss apply to other data transformation problems as well. For further information about data transformations, consult a more advanced textbook such as that by Mosteller and Tukey (listed in Appendix C).

The first step in data transformation is to evaluate the data set and decide which, if any, transformations might be appropriate. Two approaches are recommended to evaluate a data set for this purpose. One is to graph the data, for instance, by creating a histogram with a superimposed normal curve. This allows a visual evaluation of the general shape of the data as well as the opportunity to identify outliers (extreme or unusual data values). Observing the general shape of the data can also help suggest which transformations to try. The second approach is to compute one of the statistics designed to test whether the data fits a particular distribution. Two statistics commonly used for this purpose are the Anderson–Darling and the Kolmogorov–Smirnov. Routines to calculate these statistics are included in many statistical packages, and various statistical calculators available on the Internet will also calculate one or both of them. For instance, a statistical calculator to compute the Kolmogorov–Smirnov test is available here (*http://jumk.de/statistic-calculator/*).

Data that is right skewed (assuming a shape in which lower values are more common, and a tail of higher values with lower frequencies extends some distance to the right) may be made more normal by application of the square-root or log transformations. The square-root transformation computes the square root of each value. If the raw data value is 4, the transformed value is 2 because $\sqrt{4} = 2$. The log transformation computes the natural log of each value, so if the raw data value is 4, the transformed value is 1.386 because $ln(4) = 1.386$. Either transformation can be accomplished easily with statistical software, a pocket calculator, or a spreadsheet program.

Figure 3-24 displays a right-skewed data set. Figure 3-25 shows the same data after a square-root transformation (the values graphed are the square roots of the data in Figure 3-24), and Figure 3-26 shows the same data after a log transformation (the values graphed are the natural logs of the data displayed in Figure 3-24).

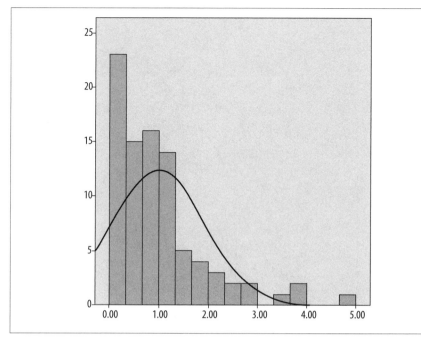

Figure 3-24. Histogram of right-skewed data set (raw values)

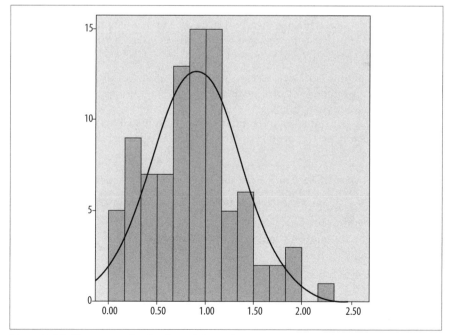

Figure 3-25. Histogram of right-skewed data after square-root transformation

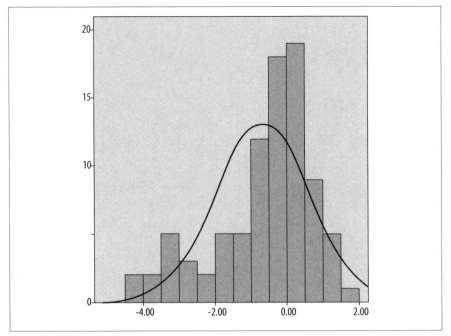

Figure 3-26. Histogram of right-skewed data after natural log transformation

Comparing the three graphs visually, Figure 3-24 is definitely right skewed and does not fit the superimposed normal distribution curve. Figure 3-25 seems to be a much better fit to the normal distribution, and Figure 3-26 seems to have replaced the right skew with a left skew, so it is also nonnormal.

We can also compute statistical tests to see whether either of the transformations has resulted in an acceptably distributed data set. For this purpose, we will calculate the one-sample Kolmogorov–Smirnov (K–S) statistic (using SPSS software, although it is available in other statistical programs as well) to evaluate how closely each data set corresponds to a perfect normal distribution. Results for the three data sets are shown in Table 3-2.

Table 3-2. The Kolmogorov-Smirnov Z statistics and p-values for three data sets

	Raw data	Square-root transformation	Natural log transformation
Kolmogorov–Smirnov Z	1.46	0.66	1.41
p	0.029	0.78	0.04

The null hypothesis for the one-sample K–S test is that the data follow the specified distribution, in this case the normal distribution; the alternative hypothesis is that the data do not follow that distribution. SPSS returns both a K–S statistic (the K–S Z) and a *p*-value for this statistic, and we will apply the rule that we will reject the null hypothesis if $p < 0.05$. For the results in Table 3-2, we reject the null hypothesis for the raw data and the natural log transformed data but fail to reject it for the square-root transformation. Therefore, if we want to use this data in a procedure that calls for normally distributed data, we should use the square-root–transformed data.

If a variable has a left or negative skew, that is, a concentration of high values with a tail of less-frequent lower values to the left, you can reflect the data and then apply the square-root or log transformation. To reflect a variable, add 1 to the largest value in the data set and then subtract each value of the variable from the new number. For instance, if the largest value in the data set is 35, subtract each value from 36 (i.e., 35 + 1) to get the reflected values. This means that the raw value of 1 becomes the reflected value 35 (36 – 1), the raw value of 2 becomes the reflected value of 34 (36 – 2), and so on, up to the raw value of 35, which becomes the reflected value of 1 (36 – 35). Reflection changes a left-skewed distribution to a distribution with a right skew, and then the square root and log transformations can be applied to see whether they improve normality.

Data transformation is not a guaranteed solution to a distribution problem; sometimes it makes the problem worse or introduces a new problem! For this reason, the transformed data should always be evaluated for normality, as we did previously, to see whether the transformation resulted in data with the desired distribution. Note also that a transformation changes the unit of the data. For instance, if you apply the log transformation to a population of blood pressure scores, your unit of measurement becomes the log of blood pressure scores. If you reflect a variable, this reverses the values (what was the highest score is now the lowest), so the interpretation of any statistic based on those values is also reversed. For these reasons, the effects of any data transformations must be kept in mind when reporting and interpreting the statistical results.

Exercises

Problem

In each of the following sets of variables, which are likely candidates to be treated as independent and which as dependent within a research study?

1. Gender, alcohol consumption, and driving record
2. High school GPA (grade point average), university freshman year GPA, choice of university major (selected before enrollment), race/ethnicity, and gender
3. Age, race/ethnicity, smoking habits, and occurrence of breast cancer
4. Accuracy on a coding task, type of instructions given, practice time, and anxiety level

Solution

Note that there is more than one possible answer to these questions. The following answers are simply the most likely research designs.

1. Gender is an independent variable (neither alcohol consumption nor driving record could influence gender). Alcohol consumption would most likely be an independent variable and driving record a dependent variable, so the study

would examine the influence of gender and alcohol consumption on driving record. However, one could conceivably design an experiment in which the roles of alcohol consumption and driving record were reversed, perhaps to test the hypothesis that people are inclined to lower their alcohol consumption after being in a serious accident.

2. University freshman-year GPA is the most likely dependent variable. For temporal reasons, high school GPA would be an independent variable (because high school occurs before university). Race/ethnicity and gender are also independent variables because they are characteristics of a person. For temporal reasons, choice of university major would be an independent variable, if freshman-year GPA is the dependent variable, because the variable description states that the major is chosen before enrolling in university, whereas the freshman-year GPA is calculated after one year of enrollment.

3. Breast cancer is the most likely dependent variable, with age, race/ethnicity, and smoking habits being independent variables.

4. Accuracy is the most likely dependent variable, with type of instructions given, practice time, and anxiety level all independent variables.

Problem

Why is the central limit theorem of primary importance to the practice of inferential statistics?

Solution

The central limit theorem states that the sampling distribution of the sample mean approximates the normal distribution, regardless of the distribution of the population from which the samples are drawn if the sample size is sufficiently large. This is important because if sample size is sufficient, we can use the normal distribution to calculate the probability of results calculated on a sample, even if we don't know the distribution of the population from which the sample was drawn.

Problem

Which type of sampling is described by each of the following scenarios?

1. The goal is to collect information on iron deficiency, obtained through blood tests, on the U.S. population. A sampling plan is devised so that units are selected from successively smaller regions of the country. Regions are selected at random, then states within regions, and so on down to individual households within census block groups.

2. The goal is to find out how elementary school students are reacting to a recently appointed principal. The researcher wants to include equal numbers of male and female students in the sample, so the interviewer is sent to the school with instructions to interview 10 male and 10 female students from among those on the playground after school one day.

3. The goal is to learn more about the domestic life of police officers working in a major city, including how home life is affected when the police officer's spouse is employed outside the home. A complete list of all men and women working

as police officers in this city is available, and a computer draws a random sample of 200 from this list. Members of the sample are then interviewed by telephone.

4. A factory supervisor is concerned that the quality of parts produced might not be equal on all shifts or at all times within a shift. (The factory operates 24 hours per day.) A sampling plan is devised to collect samples of 30 parts at 9 times during the work day, with the times of collection selected randomly within time blocks for each of the three daily shifts. Within each shift, one sample will be drawn within the first 2 hours, one within the middle 6 hours, one within the last 2 hours.

Solution

1. Cluster sampling
2. Quota sampling (and convenience sampling)
3. Simple random sampling
4. Stratified sampling

Problem

You are taking a multiple-choice test with 10 items, in which there is no penalty for wrong answers. There are 5 possible answers for each question, so just by guessing, you have a 20% chance of getting the right answer on each question. Assuming that you simply guess at the right answers, what is the probability that you will get *exactly* 3 answers right?

Solution

This question can be answered by using the binomial distribution with $n = 10$, $k = 3$, and $p = 0.2$, as shown in Figure 3-27.

$$P(k = 3;10,0.20) = \binom{10}{3}0.2^3(1 - 0.2)^7 = 0.20$$

Figure 3-27. Calculating b(3; 10, 0.2)

Therefore, the probability is 0.20 or 20 percent that you will get exactly 3 questions right, under these conditions.

Using Figure D-8 (the binomial probability table in Appendix D), the table probability is 0.20133, which rounds to 0.20.

Problem

Given the same conditions as in the previous question, what is the probability of getting *3 or more* questions right?

Solution

This question can also be answered by using the binomial distribution with $n = 10$, $k = 3$, and $p = 0.20$. It is easier to calculate the probability of getting no more than

2 questions right and then subtracting it from 1, so that is the approach we will use. We can do this because total probability always equals 1, and "at least 3 questions right" and "no more than 2 questions right" together account for all possible outcomes. Applying the binomial formula, we find these probabilities:

$P(k = 0) = 0.11$
$P(k = 1) = 0.27$
$P(k = 2) = 0.30$
$P(k \geq 3) = 1 - P(k \leq 2) = 1 - (0.11 + 0.27 + 0.30) = 0.32$

Therefore, the probability of getting 3 or more answers right, under these conditions, is 0.32 or 32 percent.

Using Figure D-9 (the cumulative binomial probability table in Appendix D), the table probability for $b(2; 10, 0.5)$ is 0.67780; $1 - 0.67780 = 0.3222$, which rounds to 0.32.

Problem

Calculate the Z-scores of the following data values, assuming they came from a normal population with $\mu = 100$ and $\sigma = 2$, and use the standard normal table (Figure D-3 in Appendix D) to find the probability of a score this large or larger for each. Instructions about how to use the probability table are included in Appendix D, along with detailed solutions for each of these problems.

a. 108
b. 95
c. 98

Solution

a. $Z = 4$; $P(Z \geq 4.00) = 1 - (0.50000 + 0.49997) = 0.00003$

$$Z = \frac{108 - 100}{2} = 4.00$$

Figure 3-28. Z-score for value of 108 from population ~N(100, 2)

b. $Z = -2.5$; $P(Z \geq -2.50) = 0.50000 + 0.49379 = 0.99379$

$$Z = \frac{95 - 100}{2} = -2.50$$

Figure 3-29. Z-score for value of 95 from population ~N(100, 2)

c. $Z = -1.0$; $P(Z \geq -1.00) = 0.50000 + 0.34134 = 0.84134$

$$Z = \frac{98 - 100}{2} = -1.00$$

Figure 3-30. Z-score for value of 98 from population ~N(100, 2)

Problem

Which of the following raw scores has a more extreme Z-score, that is, has a Z-score further (in either a positive or negative direction) from 0?

   a. A score of 190, from a population with $\mu = 180$ and $\sigma = 4$

   b. A score of 175, from a population with $\mu = 200$ and $\sigma = 5$

Solution

The second score is more extreme because -5.0 is further from 0 than 2.5 (Figures 3-31 and 3-32).

$$Z = \frac{190 - 180}{4} = 2.50$$

Figure 3-31. Z-score for value of 190 from population ~N(180, 4)

$$Z = \frac{175 - 200}{5} = -5.00$$

Figure 3-32. Z-score for value of 175 from population ~N(200, 5)

Problem

Compute the Z-statistic for each of the following samples, which were drawn from a population with a mean of 40 and a standard deviation of 5. Use the standard normal table (Figure D-3 in Appendix D) to find the probability of a result at least as low as each result.

   a. $\bar{x} = 42$, $n = 35$

   b. $\bar{x} = 42$, $n = 50$

   c. $\bar{x} = 39$, $n = 40$

   d. $\bar{x} = 39$, $n = 80$

Solution

a. $Z = 2.37$; $P(Z \le 2.37) = 0.50000 + 0.49111 = 0.99889$

$$Z = \frac{42 - 40}{\frac{5}{\sqrt{35}}} = 2.37$$

Figure 3-33. Z-statistic for a sample ($\overline{x} = 42$, $n = 35$) from a population ~N(40, 5)

b. $Z = 2.83$; $P(Z \le 2.83) = 0.50000 + 0.49767 = 0.99767$

$$Z = \frac{42 - 40}{\frac{5}{\sqrt{50}}} = 2.83$$

Figure 3-34. Z-statistic for a sample ($\overline{x} = 42$, $n = 50$) from a population ~N(40, 5)

c. $Z = -1.26$; $P(Z \le -1.26) = 1 - P(Z \ge -1.26) = 1 - (0.50000 + 0.39617) = 0.10383$

$$Z = \frac{39 - 40}{\frac{5}{\sqrt{40}}} = -1.26$$

Figure 3-35. Z-statistic for a sample ($\overline{x} = 39$, $n = 40$) from a population ~N(40, 5)

d. $Z = -1.79$; $P(Z \le -1.79) = 1 - P(Z \ge -1.79) = 1 - (0.50000 + 0.46327) = 0.03673$

$$Z = \frac{39 - 40}{\frac{5}{\sqrt{80}}} = -1.79$$

Figure 3-36. Z-statistic for a sample ($\overline{x} = 39$, $n = 80$) from a population ~N(40,5)

Problem

You are a principal in an elementary school. As part of a comprehensive evaluation, one of your students was given an IQ (intelligence) test and received a score of 80. You know that for this student's age group in the population at large, scores on this test are distributed normally ($\mu = 100$, $\sigma = 15$). What statistic will help you interpret this student's score?

Solution

A Z-score will place this student's score of 80 in the context of the distribution of scores for other students of his age. As shown in Figure 3-37, this student scored 1.33 standard deviations below the average for her age group. Although many factors can affect the score on an IQ test (hence, the need for a comprehensive evaluation), a below-average score does suggest that this student might have more difficulty in school than pupils who test higher on IQ tests.

$$Z = \frac{80 - 100}{15} = -1.33$$

Figure 3-37. Z-score for a value of 80 from a population ~N(100, 15)

Using the standard normal distribution table (Figure D-3 in Appendix D), you see that only about 9% of students ($p = 0.09176$) would be expected to have an IQ score this low or lower.

$$P(Z \le -1.33) = 1 - P(Z \ge -1.33) = 1 - (0.50000 + 0.40824) = 0.09176$$

Problem

You are a medical researcher studying the effects of a vegetarian diet on cholesterol levels. Assume the cholesterol level for U.S. men ages 20–65 is distributed normally, with a mean of 210 mg/dL (mg = milligrams, dL = deciliter) and a standard deviation of 45 mg/dL. You are studying a sample of 40 men from this age group who have followed a vegetarian diet for at least one year and find that their mean cholesterol level is 190mg/dL. Which statistic can help you place this result in context?

Solution

You compute the Z-statistic, which places the mean cholesterol level for your vegetarian sample in the context of the total U.S. male population for their age group. As shown in Figure 3-38, the average cholesterol for the vegetarian group is 2.81 standard deviations below the mean for the total population of men in their age group, suggesting that consuming a vegetarian diet is associated with lower cholesterol. As with the IQ example, many factors can affect cholesterol level, and a medical study designed to address this question would normally include more variables; this is a simplified example to illustrate the use of the Z-statistic.

$$Z = \frac{190 - 210}{\dfrac{45}{\sqrt{40}}} = -2.81$$

Figure 3-38. ($\overline{x} = 190$, n = 40) from a population ~N(210, 45)

Using the standard normal distribution table (Figure D-3 in Appendix D), you find that the probability of a result at least this extreme, using a two-tailed test, is 0.00496, so if your alpha value is 0.05, this result is sufficient to reject the null hypothesis (in this case, that a vegetarian diet has no effect on cholesterol).

$(Z \leq -2.81) = 1 - P(Z \geq -2.81) = 1 - (0.50000 + 0.49752) = 0.00248$
$P(Z \geq 2.81) = 0.00248$ (because the Z-distribution is symmetric)
$P[(Z \leq -2.81) \text{ OR } (Z \geq 2.81)] = 2 \times (0.00248) = 0.00496$

4

Descriptive Statistics and Graphic Displays

Most of this book, as is the case with most statistics books, is concerned with *statistical inference*, meaning the practice of drawing conclusions about a population by using statistics calculated on a sample. However, another type of statistics is the concern of this chapter: *descriptive statistics*, meaning the use of statistical and graphic techniques to present information about the data set being studied. Nearly everyone involved in statistical work works with both types of statistics, and often, computing descriptive statistics is a preliminary step in what will ultimately be an inferential statistical analysis. In particular, it is a common practice to begin an analysis by examining graphical displays of a data set and to compute some basic descriptive statistics to get a better sense of the data to be analyzed. You can never be too familiar with your data, and time spent examining it is nearly always time well spent. Descriptive statistics and graphic displays can also be the final product of a statistical analysis. For instance, a business might want to monitor sales volumes for different locations or different sales personnel and wish to present that information using graphics, without any desire to use that information to make inferences (for instance, about other locations or other years) using the data collected.

Populations and Samples

The same data set may be considered as either a population or a sample, depending on the reason for its collection and analysis. For instance, the final exam grades of the students in a class are a population if the purpose of the analysis is to describe the distribution of scores in that class, but they are a sample if the purpose of the analysis is to make some inference from those scores to the scores of other students (perhaps students in different classes or different schools). Analyzing a population means your data set is the complete population of interest, so you are performing your calculations on all members of the group of interest to you and can make direct statements about the characteristics of that group. In contrast, analyzing a sample means you are working with a subset drawn from a larger population, and any statements made about the larger group from which your sample was drawn are probabilistic rather than absolute. (The reasoning behind inferential statistics is discussed further in Chapter 3.) Samples rather than populations are often analyzed for practical reasons because it might be impossible or prohibitively expensive to study all members of a population directly.

The distinction between descriptive and inferential statistics is fundamental, and a set of notational conventions and terminology has been developed to distinguish between the two. Although these conventions differ somewhat from one author to the next, as a general rule, numbers that describe a population are referred to as *parameters* and are signified by Greek letters such as μ (for the population mean) and σ (for the population standard deviation); numbers that describe a sample are referred to as *statistics* and are signified by Latin letters such as $\bar{x}$ (the sample mean) and s (the sample standard deviation).

Measures of Central Tendency

Measures of central tendency, also known as measures of location, are typically among the first statistics computed for the continuous variables in a new data set. The main purpose of computing measures of central tendency is to give you an idea of what a typical or common value for a given variable is. The three most common measures of central tendency are the arithmetic mean, the median, and the mode.

The Mean

The arithmetic *mean*, or simply the mean, is often referred to in ordinary speech as the *average* of a set of values. Calculating the mean as a measure of central tendency is appropriate for interval and ratio data, and the mean of dichotomous variables coded as 0 or 1 provides the proportion of subjects whose value on the variable is 1. For continuous data, for instance measures of height or scores on an IQ test, the mean is simply calculated by adding up all the values and then dividing by the number of values. The mean of a population is denoted by the Greek letter *mu* (μ) whereas the mean of a sample is typically denoted by a bar over the variable symbol: for instance, the mean of x would be written $\bar{x}$ and pronounced "x-bar." Some authors adapt the bar notation for the names of variables also. For instance, some authors denote "the mean of the variable age" by $\overline{age}$, which would be pronounced "age-bar."

Suppose we have a population with only five cases, and these are the values for members of that population for the variable x:

   100, 115, 93, 102, 97

We can calculate the mean of x by adding these values and dividing by 5 (the number of values):

$$\mu = (100 + 115 + 93 + 102 + 97)/5 = 507/5 = 101.4$$

Statisticians often use a convention called *summation notation*, introduced in Chapter 1, which defines a statistic by describing how it is calculated. The computation of the mean is the same whether the numbers are considered to represent a population or a sample; the only difference is the symbol for the mean itself. The mean of a population, as expressed in summation notation, is shown in Figure 4-1.

$$\mu = \frac{1}{n}\sum_{i=1}^{n} x_i$$

Figure 4-1. Formula to calculate the mean

In this formula, μ (the Greek letter *mu*) is the population mean for x, n is the number of cases (the number of values for x), and x_i is the value of x for a particular case. The Greek letter sigma (Σ) means summation (adding together), and the figures above and below the sigma define the range over which the operation should be performed. In this case, the notation says to sum all the values of x from 1 to n. The symbol i designates the position in the data set, so x_1 is the first value in the data set, x_2 the second value, and x_n the last value in the data set. The summation symbol means to add together or sum the values of x from the first (x_1) to the last (x_n). The population mean is therefore calculated by summing all the values for the variable in question and then dividing by the number of values, remembering that dividing by n is the same thing as multiplying by $1/n$.

The mean is an intuitive measure of central tendency that is easy for most people to understand. However, the mean is not an appropriate summary measure for every data set because it is sensitive to extreme values, also known as *outliers* (discussed further later) and can also be misleading for skewed (nonsymmetrical) data.

Consider one simple example. Suppose the last value in our tiny data set was 297 instead of 97. In this case, the mean would be:

$\mu = (100 + 115 + 93 + 102 + 297)/5 = 707/5 = 141.4$

The mean of 141.4 is not a typical value for this data, In fact, 80% of the data (four of the five values) are below the mean, which is distorted by the presence of one extremely high value.

The problem here is not simply theoretical; many large data sets also have a distribution for which the mean is not a good measure of central tendency. This is often true of measures of income, such as household income data in the United States. A few very rich households make the mean household income in the United States a larger value than is truly representative of the average or typical household, and for this reason, the *median* household income is often reported instead (more about medians later).

The mean can also be calculated using data from a *frequency table*, that is, a table displaying data values and how often each occurs. Consider the following simple example in Table 4-1.

Table 4-1. Simple frequency table

Value	Frequency
1	7
2	5
3	12
4	2

To find the mean of these numbers, treat the frequency column as a weighting variable. That is, multiply each value by its frequency. For the denominator, add the frequencies to get the total *n*. The mean is then calculated as shown in Figure 4-2.

$$\mu = \frac{(1 \times 7) + (2 \times 5) + (3 \times 12) + (4 \times 2)}{(7 + 5 + 12 + 2)} = 2.35$$

Figure 4-2. Calculating the mean from a frequency table

This is the same result as you would reach by adding each score (1+1+1+1+ . . .) and dividing by 26.

The mean for *grouped data*, in which data has been tabulated by range and exact values are not known, is calculated in a similar manner. Because we don't know the exact values for each case (we know, for instance, that 5 values fell into the range of 1–20 but not the specific values for those five cases), for the purposes of calculation we use the midpoint of the range as a stand-in for the specific values. Therefore, to calculate the mean, we first calculate this midpoint for each range and then multiply it by the frequency of values in the range. To calculate the midpoint for a range, add the first and last values in the range and divide by 2. For instance, for the 1–20 range, the midpoint is:

(1 + 20)/2 = 10.5

A mean calculated in this way is called a *grouped mean*. A grouped mean is not as precise as the mean calculated from the original data points, but it is often your only option if the original values are not available. Consider the following grouped data set in Table 4-2.

Table 4-2. Grouped data

Range	Frequency	Midpoint
1–20	5	10.5
21–40	25	30.5
41–60	37	50.5
61–80	23	70.5
81–100	8	90.5

The mean is calculated by multiplying the midpoint of each interval by the number of values in the interval (the frequency) and dividing by the total frequency, as shown in Figure 4-3.

$$\mu = \frac{(10.5 \times 5) + (30.5 \times 25) + (50.5 \times 37) + (70.5 \times 23) + (90.5 \times 8)}{(5 + 25 + 37 + 23 + 8)} = 51.32$$

Figure 4-3. Calculating the mean for grouped data

One way to lessen the influence of outliers is by calculating a *trimmed mean*, also known as a *Winsorized mean*. As the name implies, a trimmed mean is calculated by trimming or discarding a certain percentage of the extreme values in a distribution and then calculating the mean of the remaining values. The purpose is to calculate a mean that represents most of the values well and is not unduly influenced by extreme values. Consider the example of the second population with five members previously cited, with values 100, 115, 93, 102, and 297. The mean of this population is distorted by the influence of one very large value, so we calculate a trimmed mean by dropping the highest and lowest values (equivalent to dropping the lowest and highest 20% of values). The trimmed mean is calculated as:

(100 + 115 + 102)/3 = 317/3 = 105.7

The value of 105.7 is much closer to the typical values in the distribution than 141.4, the value of the mean including all the data values. Of course, we seldom would be working with a population with only five members, but the principle applies to large populations as well. Usually, a specific percentage of the data values are trimmed from the extremes of the distribution, and this decision would have to be reported to make it clear what the calculated mean actually represents.

The mean can also be calculated for dichotomous data by using 0–1 coding, in which case the mean is equivalent to the percentage of values with the number 1. Suppose we have a population of 10 subjects, 6 of whom are male and 4 of whom are female, and we have coded males as 1 and females as 0. Computing the mean will give us the percentage of males in the population:

μ= (1+1+1+1+1+1+0+0+0+0)/10 = 6/10 = 0.6 or 60% males

The Median

The *median* of a data set is the middle value when the values are ranked in ascending or descending order. If there are n values, the median is formally defined as the $(n+1)/2$th value, so if $n = 7$, the middle value is the $(7+1)/2$th or fourth value. If there is an even number of values, the median is the average of the two middle values. This is formally defined as the average of the $(n/2)$th and $((n/2)+1)$th value. If there are six values, the median is the average of the $(6/2)$th and $((6/2)+1)$th value, or the third and fourth values. Both techniques are demonstrated here:

Odd number (5) of values: 1, 4, 6, 6, 10; Median = 6 because (5+1)/2 = 3, and 6 is the third value in the ordered list.

Even number (6) of values: 1, 3, 5, 6, 10, 15; Median = (5+6)/2 = 5.5 because 6/2 = 3 and [(6/2) +1] = 4, and 5 and 6 are the third and fourth values in the ordered list.

The median is a better measure of central tendency than the mean for data that is asymmetrical or contains outliers. This is because the median is based on the ranks of data points rather than their actual values, and by definition, half of the data values in a distribution lie below the median and half above the median, without regard to the actual values in question. Therefore, it does not matter whether the data set contains some extremely large or small values because they will not affect the median more than less extreme values. For instance, the median of all three of the following distributions is 4:

Distribution A: 1, 1, 3, 4, 5, 6, 7
Distribution B: 0.01, 3, 3, 4, 5, 5, 5
Distribution C: 1, 1, 2, 4, 5, 100, 2000

Of course, the median is not always an appropriate measure to describe a population or a sample. This is partly a judgment call; in this example, the median seems reasonably representative of the data values in Distributions A and B, but perhaps not for Distribution C, whose values are so disparate that any single summary measure can be misleading.

The Mode

A third common measure of central tendency is the *mode*, which refers to the most frequently occurring value. The mode is most often useful in describing ordinal or categorical data. For instance, imagine that the following numbers reflect the favored news sources of a group of college students, where 1 = newspapers, 2 = television, and 3 = Internet:

1, 1, 2, 2, 2, 2, 3, 3, 3, 3, 3, 3, 3

We can see that the Internet is the most popular source because 3 is the modal (most common) value in this data set.

When modes are cited for continuous data, usually a range of values is referred to as the mode (because with many values, as is typical of continuous data, there might be no single value that occurs substantially more often than any other). If you intend to do this, you should decide on the categories in advance and use standard ranges if they exist. For instance, age for adults is often collected in ranges of 5 or 10 years, so it might be the case that in a given data set, divided into ranges of 10 years, the modal range was ages 40–49 years.

Comparing the Mean, Median, and Mode

In a perfectly symmetrical distribution (such as the normal distribution, discussed in Chapter 3), the mean, median, and mode are identical. In an asymmetrical or skewed distribution, these three measures will differ, as illustrated in the data sets graphed as histograms in Figures 4-4, 4-5, and 4-6. To facilitate calculating the mode, we have also divided each data set into ranges of 5 (35–39.99, 40–44.99, etc.).

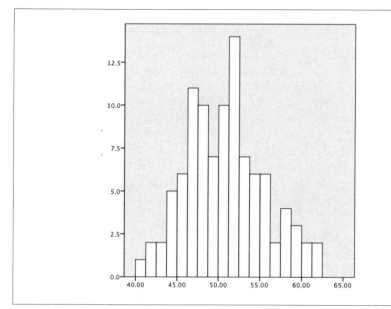

Figure 4-4. Symmetric data

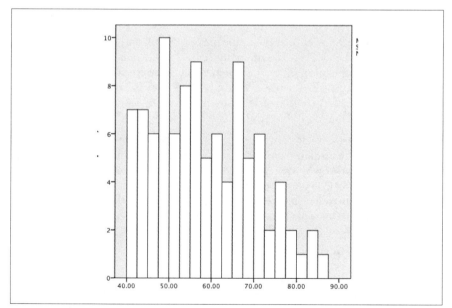

Figure 4-5. Right-skewed data

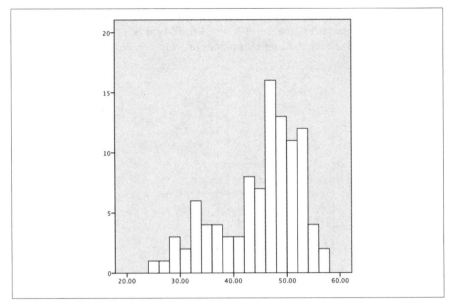

Figure 4-6. Left-skewed data

The data in Figure 4-4 is approximately normal and symmetrical with a mean of 50.88 and a median of 51.02; the most common range is 50.00–54.99 (37 cases), followed by 45.00–49.99 (34 cases). In this distribution, the mean and median are very close to each other, and the two most common ranges also cluster around the mean.

The data in Figure 4-5 is right skewed; the mean is 58.18, and the median is 56.91; a mean higher than a median is common for right-skewed data because the extreme higher values pull the mean up but do not have the same effect on the median. The modal range is 45.00–49.99 with 16 cases; however, several other ranges have 14 cases, making them very close in terms of frequency to the modal range and making the mode less useful in describing this data set.

The data in Figure 4-6 is left skewed; the mean is 44.86, and the median is 47.43. A mean lower than the median is typical of left-skewed data because the extreme lower values pull the mean down, whereas they do not have the same effect on the median. The skew in Figure 4-6 is greater than that in Figure 4-5, and this is reflected in the greater difference between the mean and median in Figure 4-6 as compared to Figure 4-5. The modal range for Figure 4-6 is 45.00–49.99.

Measures of Dispersion

Dispersion refers to how variable or spread out data values are. For this reason, measures of dispersions are sometimes called measures of variability or measures of spread. Knowing the dispersion of data can be as important as knowing its central tendency. For instance, two populations of children may both have mean IQs of

100, but one could have a range of 70 to 130 (from mild retardation to very superior intelligence) whereas the other has a range of 90 to 110 (all within the normal range). The distinction could be important, for instance, to educators, because despite having the same average intelligence, the range of IQ scores for these two groups suggests that they might have different educational and social needs.

The Range and Interquartile Range

The simplest measure of dispersion is the *range*, which is simply the difference between the highest and lowest values. Often the minimum (smallest) and maximum (largest) values are reported as well as the range. For the data set (95, 98, 101, 105), the minimum is 95, the maximum is 105, and the range is 10 (105–95). If there are one or a few outliers in the data set, the range might not be a useful summary measure. For instance, in the data set (95, 98, 101, 105, 210), the range is 115, but most of the numbers lie within a range of 10 (95–105). Inspection of the range for any variable is a good data screening technique; an unusually wide range or extreme minimum or maximum values might warrant further investigation. Extremely high or low values or an unusually wide range of values might be due to reasons such as data entry error or to inclusion of a case that does not belong to the population under study. (Information from an adult might have been included mistakenly in a data set concerned with children.)

The *interquartile range* is an alternative measure of dispersion that is less influenced than the range by extreme values. The interquartile range is the range of the middle 50% of the values in a data set, which is calculated as the difference between the 75th and 25th percentile values. The interquartile range is easily obtained from most statistical computer programs but can also be calculated by hand, using the following rules (n = the number of observations, k the percentile you wish to find):

1. Rank the observations from smallest to largest.
2. If $(nk)/100$ is an integer (a round number with no decimal or fractional part), the kth percentile of the observations is the average of the $((nk)/100)$th and $((nk)/100 + 1)$th largest observations.
3. If $(nk)/100$ is not an integer, the kth percentile of the observation is the $(j + 1)$th largest measurement, where j is the largest integer less than $(nk)/100$.
4. Calculate the interquartile range as the difference between the 75th and 25th percentile measurements.

Consider the following data set with 13 observations (1, 2, 3, 5, 7, 8, 11, 12, 15, 15, 18, 18, 20):

1. First, we want to find the 25th percentile, so $k = 25$.
2. We have 13 observations, so $n = 13$.
3. $(nk)/100 = (25 \times 13)/100 = 3.25$, which is not an integer, so we will use the second method (#3 in the preceding list).
4. $j = 3$ (the largest integer less than $(nk)/100$, that is, less than 3.25).

5. Therefore, the 25th percentile is the $(j+1)$th or 4th observation, which has the value 5.

We can follow the same steps to find the 75th percentile:

- $(nk)/100 = (75*13)/100 = 9.75$, not an integer.
- $j = 9$, the smallest integer less than 9.75.
- Therefore, the 75th percentile is the $9 + 1$ or 10th observation, which has the value 15.
- Therefore, the interquartile range is $(15 - 5)$ or 10.

The resistance of the interquartile range to outliers should be clear. This data set has a range of 19 $(20 - 1)$ and an interquartile range of 10; however, if the last value was 200 instead of 20, the range would be 199 $(200 - 1)$, but the interquartile range would still be 10, and that number would better represent most of the values in the data set.

The Variance and Standard Deviation

The most common measures of dispersion for continuous data are the *variance* and *standard deviation*. Both describe how much the individual values in a data set vary from the mean or average value. The variance and standard deviation are calculated slightly differently depending on whether a population or a sample is being studied, but basically the variance is the average of the squared deviations from the mean, and the standard deviation is the square root of the variance. The variance of a population is signified by σ^2 (pronounced "sigma-squared"; σ is the Greek letter sigma) and the standard deviation as σ, whereas the sample variance and standard deviation are signified by s^2 and s, respectively.

The deviation from the mean for one value in a data set is calculated as $(x_i - \mu)$ where x_i is value i from the data set and μ is the mean of the data set. If working with sample data, the principle is the same, except that you subtract the mean of the sample $(\bar{x})$ from the individual data values rather than the mean of the population. Written in summation notation, the formula to calculate the sum of all deviations from the mean for the variable x for a population with n members is shown in Figure 4-7.

$$\sum_{i=1}^{n}(x_i - \mu)$$

Figure 4-7. Formula for the sum of the deviations from the mean

Unfortunately, this quantity is not useful because it will always equal zero, a result that is not surprising if you consider that the mean is computed as the average of all the values in the data set. This may be demonstrated with the tiny data set (1, 2, 3, 4, 5). First, we calculate the mean:

$$\mu = (1 + 2 + 3 + 4 + 5)/5 = 3$$

Then we calculate the sum of the deviations from the mean, as shown in Figure 4-8.

$$\sum_{i=1}^{n}(x_i - \mu) = (1 - 3) + (2 - 3) + (3 - 3) + (4 - 3) + (5 - 3)$$
$$= (-2) + (-1) + 0 + 1 + 2 = 0$$

Figure 4-8. Calculating the sum of the deviations from the mean

To get around this problem, we work with squared deviations, which by definition are always positive. To get the average deviation or variance for a population, we square each deviation, add them up, and divide by the number of cases, as shown in Figure 4-9.

$$\sigma^2 = \frac{1}{n}\sum_{i=1}^{n}(x_i - \mu)^2$$

Figure 4-9. Calculating the sum of the squared deviations from the mean

The sample formula for the variance requires dividing by $n - 1$ rather than n; the reasons are technical and have to do with degrees of freedom and unbiased estimation. (For a detailed discussion, see the Wilkins article listed in Appendix C.) The formula for the variance of a sample, notated as s^2, is shown in Figure 4-10.

$$s^2 = \frac{1}{n-1}\sum_{i=1}^{n}(x_i - \bar{x})^2$$

Figure 4-10. The formula for a sample variance

Continuing with our tiny data set with values (1, 2, 3, 4, 5), with a mean value of 3, we can calculate the variance for this population as shown in Figure 4-11.

$$\sigma^2 = \frac{1}{n} \sum_{i=1}^{n} (x_i - \mu)^2$$

$$= \frac{1}{5} [(1-3)^2 + (2-3)^2 + (3-3)^2 + (4-3)^2 + (5-3)^2]$$

$$= \frac{1}{5} [(-2)^2 + (-1)^2 + (0)^2 + (1)^2 + (2)^2]$$

$$= \frac{4+1+0+1+4}{5} = \frac{10}{5} = 2.0$$

Figure 4-11. Calculating the variance for a population

If we consider these numbers to be a sample rather than a population, the variance would be computed as shown in Figure 4-12.

$$s^2 = \frac{1}{n-1} \sum_{i=1}^{n} (x_i - \bar{x})^2$$

$$= \frac{1}{5-1} [(1-3)^2 + (2-3)^2 + (3-3)^2 + (4-3)^2 + (5-3)^2]$$

$$= \frac{1}{4} [(-2)^2 + (-1)^2 + (0)^2 + (1)^2 + (2)^2]$$

$$= \frac{4+1+0+1+4}{4} = \frac{10}{4} = 2.5$$

Figure 4-12. Calculating the variance for a sample

Note that because of the different divisor, the sample formula for the variance will always return a larger result than the population formula, although if the sample size is close to the population size, this difference will be slight.

Because squared numbers are always positive (outside the realm of imaginary numbers), the variance will always be equal to or greater than 0. (The variance would be zero only if all values of a variable were the same, in which case the variable would really be a constant.) However, in calculating the variance, we have changed from our original units to squared units, which might not be convenient to interpret. For instance, if we were measuring weight in pounds, we would probably want measures of central tendency and dispersion expressed in the same units rather than having the mean expressed in pounds and variance in squared pounds. To get back to the original units, we take the square root of the variance; this is called the standard deviation and is signified by σ for a population and s for a sample.

For a population, the formula for the standard deviation is shown in Figure 4-13.

$$\sigma = \sqrt{\frac{1}{n}\sum_{i=1}^{n}(x_i - \mu)^2}$$

Figure 4-13. Formula for the standard deviation for a population

Note that this is simply the square root of the formula for variance. In the preceding example, the standard deviation can be found as shown in Figure 4-14.

$$\sigma = \sqrt{\sigma^2} = \sqrt{2} = 1.41$$

Figure 4-14. The relationship between the standard deviation and the variance

The formula for the sample standard deviation is shown in Figure 4-15.

$$s = \sqrt{\frac{1}{n-1}\sum_{i=1}^{n}(x_i - \bar{x})^2}$$

Figure 4-15. Formula for the standard deviation of a sample

As with the population standard deviation, the sample standard deviation is the square root of the sample variance (Figure 4-16).

$$s = \sqrt{s^2} = \sqrt{2.5} = 1.58$$

Figure 4-16. The relationship between the standard deviation and the variance

In general, for two groups of the same size and measured with the same units (e.g., two groups of people, each of size $n = 30$ and both weighed in pounds), we can say that the group with the larger variance and standard deviation has more variability among their scores. However, the unit of measure affects the size of the variance, which can make it tricky to compare the variability of factors measured in different units. To take an obvious example, a set of weights expressed in ounces would have a larger variance and standard deviation than the same weights measured in pounds. When comparing completely different units, such as height in inches and weight in pounds, it is even more difficult to compare variability. The *coefficient of variation* (CV), a measure of relative variability, gets around this difficulty and makes it possible to compare variability across variables measured in different units. The CV is shown here using sample notation but could be calculated for a population by substituting σ for s. The CV is calculated by dividing the standard deviation by the mean and then multiplying by 100, as shown in Figure 4-17.

$$CV = \frac{s}{\bar{x}} \times 100$$

Figure 4-17. The formula for the coefficient of variation (CV)

For the previous example, this would be calculated as shown in Figure 4-18.

$$CV = \frac{1.58}{3} \times 100 = 52.7$$

Figure 4-18. Calculating the coefficient of variation (CV)

The CV cannot be calculated if the mean of the data is 0 (because you cannot divide by 0) and is most useful when the variable in question has only positive values. If a variable has both positive and negative values, the mean can be close to zero although the data actually has quite a broad range, and this can produce a misleading CV value because the denominator will be a small number, potentially producing a large CV value even if the standard deviation is fairly moderate.

The usefulness of the CV should be clear by considering the same data set as expressed in feet and inches; for instance, 60 inches is the same as 5 feet. The data as expressed in feet has a mean of 5.5566 and a standard deviation of 0.2288; the same data as expressed in inches has a mean of 66.6790 and a standard deviation of 2.7453. However, the CV is not affected by the change in units and produces the same result either way, except for rounding error:

5.5566/0.2288 = 24.2858 (data in feet)
66.6790/2.7453 = 24.2884 (data in inches)

Outliers

There is no absolute agreement among statisticians about how to define *outliers*, but nearly everyone agrees that it is important that they be identified and that appropriate analytical techniques be used for data sets that contain outliers. An outlier is a data point or observation whose value is quite different from the others in the data set being analyzed. This is sometimes described as a data point that seems to come from a different population or is outside the typical pattern of the other data points. Suppose you are studying educational achievement in a sample or population, and most of your subjects have completed from 12 to 16 years of schooling (12 years = high school graduation, 16 years = university graduation). However, one of your subjects has a value of 0 for this variable (implying that he has no formal education at all) and another has a value of 26 (implying many years of post-graduate education). You will probably consider these two cases to be outliers because they have values far removed from the other data in your sample of population. Identification and analysis of outliers is an important preliminary step in many types of data analysis because the presence of just one or two outliers can completely distort the value of some common statistics, such as the mean.

It's also important to identify outliers because sometimes they represent data entry errors. In the preceding example, the first thing to do is check whether the data was entered correctly; perhaps the correct values are 10 and 16, respectively. The second thing to do is investigate whether the cases in question actually belong to the same population as the other cases. For instance, does the 0 refer to the years of education of an infant when the data set was supposed to contain only information about adults?

If neither of these simple fixes solves the problem, it is necessary to make a judgment call (possibly in consultation with others involved in the research) about what to do with the outliers. It is possible to delete cases with outliers from the data set before analysis, but the acceptability of this practice varies from field to field. Sometimes a statistical fix already exists, such as the trimmed mean previously described, although the acceptability of such fixes also varies from one field to the next. Other possibilities are to transform the data (discussed in Chapter 3) or use nonparametric statistical techniques (discussed in Chapter 13), which are less influenced by outliers.

Various rules of thumb have been developed to make the identification of outliers more consistent. One common definition of an outlier, which uses the concept of the interquartile range (IQR), is that mild outliers are those lower than the 25th quartile minus 1.5 × IQR or greater than the 75th quartile plus 1.5 × IQR. Cases this extreme are expected in about 1 in 150 observations in normally distributed data. Extreme outliers are similarly defined with the substitution of 3 × IQR for 1.5 × IQR; values this extreme are expected about once per 425,000 observations in normally distributed data.

Graphic Methods

There are innumerable graphic methods to present data, from the basic techniques included with spreadsheet software such as Microsoft Excel to the extremely specific and complex methods available in computer languages such as R. Entire books have been written on the use and misuse of graphics in presenting data, and the leading (if also controversial) expert in this field is Edward Tufte, a Yale professor (with a Master's degree in statistics and a PhD in political science). His most famous work is *The Visual Display of Quantitative Information* (listed in Appendix C), but all of Tufte's books are worthwhile reading for anyone seriously interested in the graphic display of data. It would be impossible to cover even a fraction of the available methods to display data in this section, so instead, a few of the most common methods are presented, including a discussion of issues concerning each.

It's easy to get carried away with fancy graphical presentations, particularly because spreadsheets and statistical programs have built-in routines to create many types of graphs and charts. Tufte's term for graphic material that does not convey information is "chartjunk," which concisely conveys his opinion of such presentations. The standards for what is considered junk vary from one field of endeavor to another, but as a general rule, it is wise to use the simplest type of chart that clearly presents your information while remaining aware of the expectations and standards within your chosen profession or field of study.

Frequency Tables

The first question to ask when considering how best to display data is whether a graphical method is needed at all. It's true that in some circumstances a picture may be worth a thousand words, but at other times, *frequency tables* do a better job than graphs at presenting information. This is particularly true when the actual values of the numbers in different categories, rather than the general pattern among the categories, are of primary interest. Frequency tables are often an efficient way to present large quantities of data and represent a middle ground between text (paragraphs describing the data values) and pure graphics (such as a histogram).

Suppose a university is interested in collecting data on the general health of their entering classes of freshmen. Because obesity is a matter of growing concern in the United States, one of the statistics they collect is the Body Mass Index (BMI), calculated as weight in kilograms divided by squared height in meters. The BMI is not an infallible measure. For instance, athletes often measure as either underweight (distance runners, gymnasts) or overweight or obese (football players, weight throwers), but it's an easily calculated measurement that is a reliable indicator of a healthy or unhealthy body weight for many people.

The BMI is a continuous measure, but it is often interpreted in terms of categories, using commonly accepted ranges. The ranges for the BMI shown in Table 4-3, established by the Centers for Disease Control and Prevention (CDC) and the World Health Organization (WHO), are generally accepted as useful and valid.

Table 4-3. CDC/WHO categories for BMI

BMI range	Category
< 18.5	Underweight
18.5–24.9	Normal weight
25.0–29.9	Overweight
30.0 and above	Obese

Now consider Table 4-4, an entirely fictitious list of BMI classifications for entering freshmen.

Table 4-4. Distribution of BMI in the freshman class of 2005

BMI range	Number
< 18.5	25
18.5–24.9	500
25.0–29.9	175
30.0 and above	50

This simple table tells us at a glance that most of the freshman are of normal body weight or are moderately overweight, with a few who are underweight or obese. Note that this table presents raw numbers or counts for each category, which are sometimes referred to as *absolute frequencies*; these numbers tell you how often each value appears, which can be useful if you are interested in, for instance, how many students might require obesity counseling. However, absolute frequencies don't place the number of cases in each category into any kind of context. We can make this table more useful by adding a column for *relative frequency*, which displays the percent of the total represented by each category. The relative frequency is calculated by dividing the number of cases in each category by the total number of cases (750) and multiplying by 100. Table 4-5 shows the both the absolute and the relative frequencies for this data.

Table 4-5. Absolute and relative frequency of BMI categories for the freshmen class of 2005

BMI range	Number	Relative frequency
< 18.5	25	3.3%
18.5–24.9	500	66.7%
25.0–29.9	175	23.3%
30.0 and above	50	6.7%

Note that relative frequencies should add up to approximately 100%, although the total might be slightly higher or lower due to rounding error.

We can also add a column for *cumulative frequency*, which shows the relative frequency for each category and those below it, as in Table 4-6. The cumulative frequency for the final category should always be 100% except for rounding error.

Table 4-6. Cumulative frequency of BMI in the freshman class of 2005

BMI range	Number	Relative frequency	Cumulative frequency
< 18.5	25	3.3%	3.3%
18.5–24.9	500	66.7%	70.0%
25.0–29.9	175	23.3%	93.3%
30.0 and above	50	6.7%	100%

Cumulative frequency tells us at a glance, for instance, that 70% of the entering class is normal weight or underweight. This is particularly useful in tables with many categories because it allows the reader to ascertain specific points in the distribution quickly, such as the lowest 10%, the median (50% of the cumulative frequency), or the top 5%.

You can also construct frequency tables to make comparisons between groups. You might be interested, for instance, in comparing the distribution of BMI in male and female freshmen or for the class that entered in 2005 versus the entering classes of 2000 and 1995. When making comparisons of this type, raw numbers are less useful (because the size of the classes can differ) and relative and cumulative frequencies more useful. Another possibility is to create graphic presentations such as the charts described in the next section, which can make such comparisons clearer.

Bar Charts

The *bar chart* is particularly appropriate for displaying discrete data with only a few categories, as in our example of BMI among the freshman class. The bars in a bar chart are customarily separated from each other so they do not suggest continuity; although in this case, our categories are based on categorizing a continuous variable, they could equally well be completely nominal categories such as favorite sport or major field of study. Figure 4-19 shows the freshman BMI information presented in a bar chart. (Unless otherwise noted, the charts presented in this chapter were created using Microsoft Excel.)

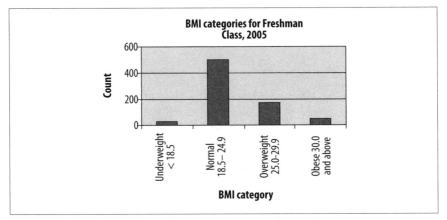

Figure 4-19. Absolute frequency of BMI categories in freshman class

Absolute frequencies are useful when you need to know the number of people in a particular category, whereas relative frequencies are more useful when you need to know the relationship of the numbers in each category. Relative frequencies are particularly useful, as we will see, when comparing multiple groups, for instance whether the proportion of obese students is rising or falling over the years. For a simple bar chart, the absolute versus relative frequencies question is less critical, as can be seen by comparing a bar chart of the student BMI data, presented as relative frequencies in Figure 4-20 with the same data presented as absolute frequencies in Figure 4-19. Note that the two charts are identical except for the y-axis (vertical axis) labels, which are frequencies in Figure 4-19 and percentages in Figure 4-20.

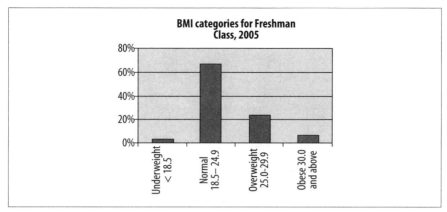

Figure 4-20. Relative frequency of BMI categories in freshman class

The concept of relative frequencies becomes even more useful if we compare the distribution of BMI categories over several years. Consider the fictitious frequency information in Table 4-7.

Table 4-7. Absolute and relative frequencies of BMI for three entering classes

BMI range	1995		2000		2005	
Underweight < 18.5	50	8.9%	45	6.8%	25	3.3%
Normal 18.5–24.9	400	71.4%	450	67.7%	500	66.7%
Overweight	100	17.9%	130	19.5%	175	23.3%
25.0–29.9						
Obese 30.0 and above	10	1.8%	40	6.0%	50	6.7%
Total	560	100.0%	665	100.0%	750	100.0%

Because the class size is different in each year, the relative frequencies (percentages) are most useful in observing trends in weight category distribution. In this case, there has been a clear decrease in the proportion of underweight students and an increase in the number of overweight and obese students. This information can also be displayed using a bar chart, as in Figure 4-21.

This is a *grouped bar chart*, which shows that there is a small but definite trend over 10 years toward fewer underweight and normal weight students and more overweight and obese students (reflecting changes in the American population at large). Bear in mind that creating a chart is not the same thing as conducting a statistical test, so we can't tell from this chart alone whether these differences are statistically significant.

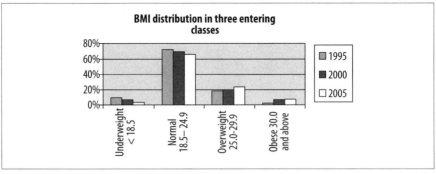

Figure 4-21. Bar chart of BMI distribution in three entering classes

Another type of bar chart, which emphasizes the relative distribution of values within each group (in this case, the relative distribution of BMI categories in three entering classes), is the *stacked bar chart*, illustrated in Figure 4-22.

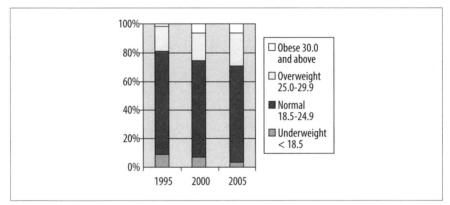

Figure 4-22. Stacked bar chart of BMI distribution in three entering classes

In this type of chart, each bar represents one year of data, and each bar totals to 100%. The relative proportion of students in each category can be seen at a glance by comparing the proportion of area within each bar allocated to each category. This arrangement facilitates comparison in multiple data series (in this case, the three years). It is immediately clear that the proportion of underweight students has declined, and the proportion of overweight and obese students has increased over time.

Pie Charts

The familiar *pie chart* presents data in a manner similar to the stacked bar chart: it shows graphically what proportion each part occupies of the whole. Pie charts, like stacked bar charts, are most useful when there are only a few categories of information and the differences among those categories are fairly large. Many people have particularly strong opinions about pie charts, and although pie charts are still commonly used in some fields, they have also been aggressively denounced in others as uninformative at best and potentially misleading at worst. So you must make your own decision based on context and convention; I will present the same BMI information in pie chart form (Figure 4-23), and you may be the judge of whether this is a useful way to present the data. Note that this is a single pie chart, showing one year of data, but other options are available, including side-by-side charts (to facilitate comparison of the proportions of different groups) and exploded sections (to show a more detailed breakdown of categories within a segment).

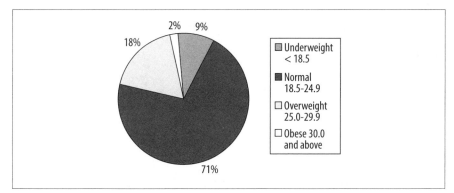

Figure 4-23. Pie chart showing BMI distribution for freshmen entering in 2005

Florence Nightingale and Statistical Graphics

Most people are at least vaguely familiar with Florence Nightingale's role in establishing nursing as a profession and with her heroic efforts to improve hygiene and the quality of nursing provided to British soldiers during the Crimean War. Fewer are aware of her contributions to statistical graphics, including her effective use of graphs and charts to communicate medical information. Nightingale also developed a new type of graph, the polar area diagram (which she called a coxcomb chart and others have termed a Nightingale rose diagram), to display comparative information such as the causes of death (from wounds received in battle, disease, and other causes) each month for British soldiers. Nightingale's charts brought attention to the high proportion of soldiers' deaths caused by disease and enabled her to make her case for the importance of improved sanitation and hygiene to the military authorities. Many of Nightingale's graphics are available for viewing on the Internet along with a discussion of her accomplishments in this field. One example is Julie Rehmeyer's *Science News* article from November 26, 2008, "Florence Nightingale: The Passionate Statistician" (*http://bit.ly/PvLvSS*).

Pareto Charts

The *Pareto chart* or *Pareto diagram* combines the properties of a bar chart and a line chart; the bars display frequency and relative frequency, whereas the line displays cumulative frequency. The great advantage of a Pareto chart is that it is easy to see which factors are most important in a situation and, therefore, to which factors most attention should be directed. For instance, Pareto charts are often used in industrial contexts to identify factors that are responsible for the preponderance of delays or defects in the manufacturing process. In a Pareto chart, the bars are ordered in descending frequency from left to right (so the most common cause is the furthest to the left and the least common the furthest to the right), and a cumulative frequency line is superimposed over the bars (so you see, for instance, how many factors are involved in 80% of production delays). Consider the hypothetical data set shown in Table 4-8, which displays the number of defects traceable to different aspects of the manufacturing process in an automobile factory.

Table 4-8. Manufacturing defects by department

Department	Number of defects
Accessory	350
Body	500
Electrical	120
Engine	150
Transmission	80

Although we can see that the Accessory and Body departments are responsible for the greatest number of defects, it is not immediately obvious what proportion of defects can be traced to them. Figure 4-24, which displays the same information presented in a Pareto chart (produced using SPSS), makes this clearer.

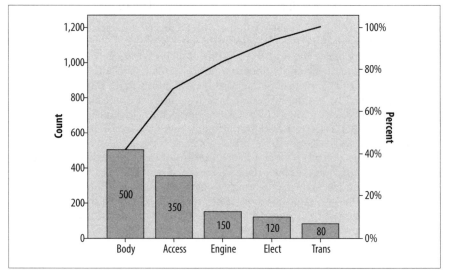

Figure 4-24. Major causes of manufacturing defects

This chart tells us not only that the most common causes of defects are in the Body and Accessory manufacturing processes but also that together they account for about 75% of defects. We can see this by drawing a straight line from the bend in the cumulative frequency line (which represents the cumulative number of defects from the two largest sources, Body and Accessory) to the right-hand *y*-axis. This is a simplified example and violates the 80:20 rule (discussed in the next sidebar about Vilfredo Pareto) because only a few major causes of defects are shown. In a more realistic example, there might be 30 or more competing causes, and the Pareto chart is a simple way to sort them out and decide which processes should be the focus of improvement efforts. This simple example does serve to display the typical characteristics of a Pareto chart. The bars are sorted from highest to lowest, the frequency is displayed on the left-hand *y*-axis and the percent on the right, and the actual number of cases for each cause are displayed within each bar.

Vilfredo Pareto

Vilfredo Pareto (1843–1923) was an Italian economist who discovered what is now called the Pareto principle, also known as the principle of "the vital few and the trivial many" or "the 80:20 rule." The Pareto principle states that in many circumstances, 80% of the activity or outcomes stem from 20% of the causes. For instance, in many countries, approximately 80% of the wealth is owned by approximately 20% of the people; it is often the case in industrial production that 20% of production errors are responsible for 80% of the defects in manufactured products; and in health services usage, 20% of the patients typically use 80% of medical services. The vital few in the Pareto principle are the 20% of people, errors, and so on that account for most of the activity, and the trivial many are the other 80% that collectively account for only 20% of the activity. Pareto is best known today for the Pareto chart, which is commonly used in quality control to help identify which processes are causing most of the difficulties, whether customer complaints or defective products.

The Stem-and-Leaf Plot

The types of charts discussed so far are most appropriate for displaying categorical data. Continuous data has its own set of graphic display methods. One of the simplest ways to display continuous data graphically is the *stem-and-leaf plot*, which can easily be created by hand and presents a quick snapshot of a data distribution. To make a stem-and-leaf plot, divide your data into intervals (using your common sense and the level of detail appropriate to your purpose) and display each data point by using two columns. The stem is the leftmost column and contains one value per row, and the leaf is the rightmost column and contains one digit for each case belonging to that row. This creates a plot that displays the actual values of the data set but also assumes a shape indicating which ranges of values are most common. The numbers can represent multiples of other numbers (for instance, units of 10,000 or of 0.01) if appropriate, given the data values in question.

Here's a simple example. Suppose we have the final exam grades for 26 students and want to present them graphically. These are the grades:

61, 64, 68, 70, 70, 71, 73, 74, 74, 76, 79, 80, 80, 83, 84, 84, 87, 89, 89, 89, 90 92, 95, 95, 98, 100

The logical division is units of 10 points, for example, 60–69, 70–79, and so on, so we construct the stem of the digits 6, 7, 8, 9 (the tens place for those of you who remember your grade school math) and create the leaf for each number with the digit in the ones place, ordered left to right from smallest to largest. Figure 4-25 shows the final plot.

Stem	Leaf
6	148
7	00134469
8	003447999
9	02558
10	0

Figure 4-25. Stem-and-leaf plot of final exam grades

This display not only tells us the actual values of the scores and their range (61–100) but the basic shape of their distribution as well. In this case, most scores are in the 70s and 80s, with a few in the 60s and 90s, and one is 100. The shape of the leaf side is in fact a crude sort of histogram (discussed later) rotated 90 degrees, with the bars being units of 10.

The Boxplot

The *boxplot*, also known as the hinge plot or the box-and-whiskers plot, was devised by the statistician John Tukey as a compact way to summarize and display the distribution of a set of continuous data. Although boxplots can be drawn by hand (as can many other graphics, including bar charts and histograms), in practice they are usually created using software. Interestingly, the exact methods used to construct boxplots vary from one software package to another, but they are always constructed to highlight five important characteristics of a data set: the median, the first and third quartiles (and hence the interquartile range as well), and the minimum and maximum. The central tendency, range, symmetry, and presence of outliers in a data set are visible at a glance from a boxplot, whereas side-by-side boxplots make it easy to make comparisons among different distributions of data. Figure 4-26 is a boxplot of the final exam grades used in the preceding stem-and-leaf plot.

The dark line represents the median value, in this case, 81.5. The shaded box encloses the interquartile range, so the lower boundary is the first quartile (25th percentile) of 72.5, and the upper boundary is the third quartile (75th percentile) of 87.75. Tukey called these quartiles hinges, hence the name hinge plot. The short horizontal lines at 61 and 100 represent the minimum and maximum values, and together with the lines connecting them to the interquartile range box, they are called

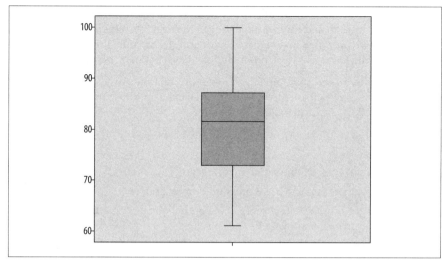

Figure 4-26. Boxplot of exam data (created in SPSS)

whiskers, hence the name box-and-whiskers plot. We can see at a glance that this data set is symmetrical because the median is approximately centered within the interquartile range, and the interquartile range is located approximately centrally within the complete range of the data.

This data set contains no outliers, that is, no numbers that are far outside the range of the other data points. To demonstrate a boxplot that contains outliers, I have changed the score of 100 in this data set to 10. Figure 4-27 shows the boxplots of the two data sets side by side. (The boxplot for the correct data is labeled "final," whereas the boxplot with the changed value is labeled "error.")

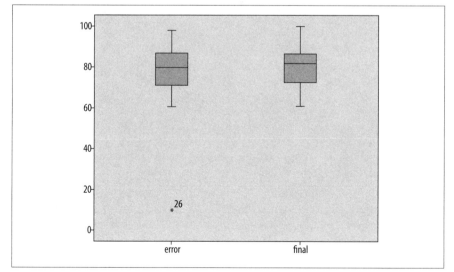

Figure 4-27. Boxplot with outlier (created in SPSS)

Note that except for the single outlier value, the two data sets look very similar; this is because the median and interquartile range are resistant to influence by extreme values. The outlying value is designated with an asterisk and labeled with its case number (26); the latter feature is not included in every statistical package.

Boxplots are often used to compare two or more real data sets side by side. Figure 4-28 shows a comparison of two years of final exam grades from 2007 and 2008, labeled "final2007" and "final2008," respectively.

Without looking at any of the actual grades, I can see several differences between the two years:

- The highest scores are the same in both years.
- The lowest score is much lower in 2008 than in 2007.
- There is a greater range of scores in 2008, both in the interquartile range (middle 50% of the scores) and overall.
- The median is slightly lower in 2008.

That the highest score was the same in both years is not surprising because this exam had a range of 0–100, and at least one student achieved the highest score in both years. This is an example of a *ceiling effect*, which exists when scores or measurements can be no higher than a particular number and people actually achieve that score. The analogous condition, if a score can be no lower than a specified number, is called a *floor effect*. In this case, the exam had a floor of 0 (the lowest possible score), but because no one achieved that score, no floor effect is present in the data.

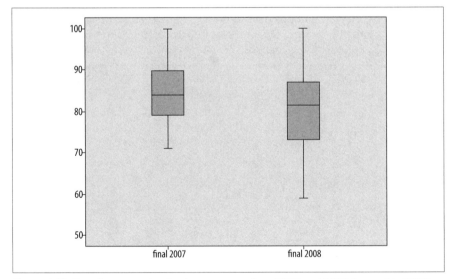

Figure 4-28. Boxplot comparing final exam scores from 2007 and 2008 (created in SPSS)

The Histogram

The *histogram* is another popular choice for displaying continuous data. A histogram looks similar to a bar chart, but in a histogram, the bars (also known as bins because you can think of them as bins into which values from a continuous distribution are sorted) touch each other, unlike the bars in a bar chart. Histograms also tend to have a larger number of bars than do bar charts. Bars in a histogram do not have to be the same width, although frequently they are. The x-axis (vertical axis) in a histogram represents a scale rather than simply a series of labels, and the area of each bar represents the proportion of values that are contained in that range.

Figure 4-29 shows the final exam data presented as a histogram created in SPSS with four bars of width ten and with a normal distribution superimposed. Note that the shape of this histogram looks quite similar to the shape of the stem-and-leaf plot of the same data (Figure 4-25), but rotated 90 degrees.

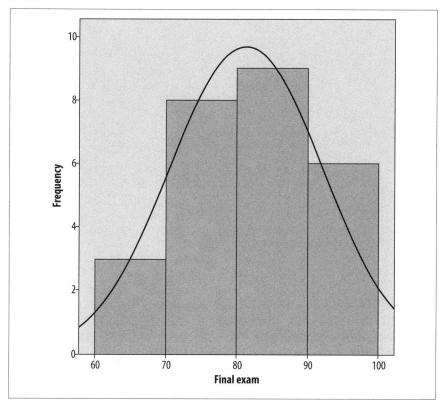

Figure 4-29. Histogram with a bin width of 10

The normal distribution is discussed in detail in Chapter 3; for now, it is a commonly used theoretical distribution that has the familiar bell shape shown here. The normal distribution is often superimposed on histograms as a visual reference so we can judge how similar the values in a data set are to a normal distribution.

For better or for worse, the choice of the number and width of bars can drastically affect the appearance of the histogram. Usually, histograms have more than four bars; Figure 4-30 shows the same data with eight bars, each with a width of five.

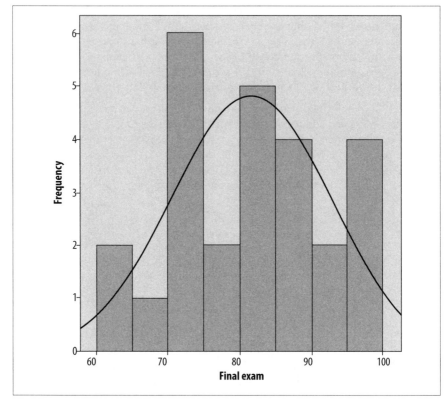

Figure 4-30. Histogram with a bin width of 5

It's the same data, but it doesn't look nearly as normal, does it? Figure 4-31 shows the same data with a bin width of two.

It's clear that the selection of bin width is important to the histogram's appearance, but how do you decide how many bins to use? This question has been explored in mathematical detail without producing any absolute answers. (If you're up for a very technical discussion, see the Wand article listed in Appendix C.). There is no absolute answer to this question, but there are some rules of thumb. First, the bins need to encompass the full range of data values. Beyond that, one common rule of thumb is that the number of bins should equal the square root of the number of points in the data set. Another is that the number of bins should never be fewer than about six. These rules clearly conflict in our data set because $\sqrt{26} = 5.1$, which is less than 6, so common sense also comes into play, as does trying different numbers of bins and bin widths. If the choice drastically changes the appearance of the data, further investigation is in order.

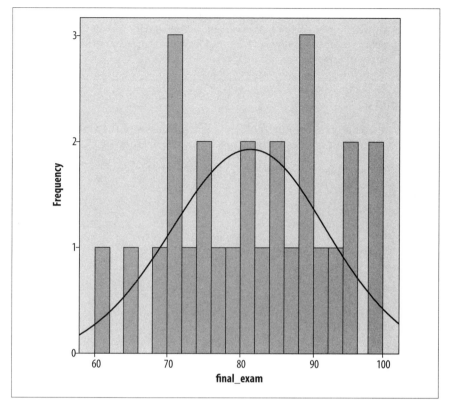

Figure 4-31. Histogram with a bin width of two

Bivariate Charts

Charts that display information about the relationship between two variables are called *bivariate charts*: the most common example is the *scatterplot*. Scatterplots define each point in a data set by two values, commonly referred to as *x* and *y*, and plot each point on a pair of axes; this method should be familiar if you ever worked with Cartesian coordinates in math class. Conventionally the vertical axis is called the *y*-axis and represents the *y*-value for each point. The horizontal axis is called the *x*-axis and represents the *x*-value. Scatterplots are a very important tool for examining bivariate relationships among variables, a topic further discussed in Chapter 7.

Univariate, Bivariate, Multivariate

People sometimes get confused about the meaning of terms such as *univariate* and *bivariate*. However, it's easy to keep them straight if you recall that *uni-* means one and *bi-* means two. Think of a *unicycle*, which has one wheel, and a *bicycle*, which has two. *Multi-* means many and in statistics, it often means more than two. Univariate statistics such as the mean therefore describe characteristics of one variable, and the bar chart and histogram are examples of univariate graphic displays. Bivariate statistics such as Pearson's correlation coefficient describe the relationship between two variables, and bivariate graphs such as the scatterplot display the relationship between two variables. Multivariate statistics such as the multiple correlation and multivariate regression describe the relationship between more than two variables.

Scatterplots

Consider the data set shown in Table 4-9, which consists of the verbal and math SAT (Scholastic Aptitude Test) scores for a hypothetical group of 15 students.

Table 4-9. SAT scores for 15 students

Math	Verbal
750	750
700	710
720	700
790	780
700	680
750	700
620	610
640	630
700	710
710	680
540	550
570	600
580	600
790	750
710	720

Other than the fact that most of these scores are fairly high (the SAT is calibrated so that the median score is 500, and most of these scores are well above that), it's difficult to discern much of a pattern between the math and verbal scores from the raw data. Sometimes the math score is higher, sometimes the verbal score is higher, and often both are similar. However, creating a scatterplot of the two variables, as in Figure 4-32, with math SAT score on the *y*-axis (vertical axis) and verbal SAT score on the *x*-axis (horizontal axis), makes the relationship between scores much clearer.

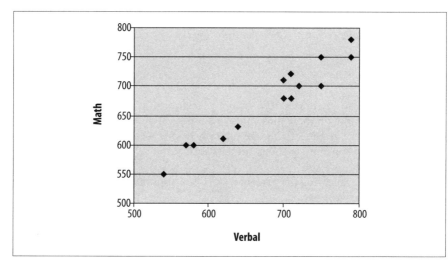

Figure 4-32. Scatterplot of verbal and math SAT scores

Despite some small inconsistencies, verbal and math scores have a strong linear relationship. People with high verbal scores tend to have high math scores and vice versa, and those with lower scores in one area tend to have lower scores in the other.

Not all strong relationships between two variables are linear, however. Figure 4-33 shows a scatterplot of variables that are highly related but for which the relationship is quadratic rather than linear.

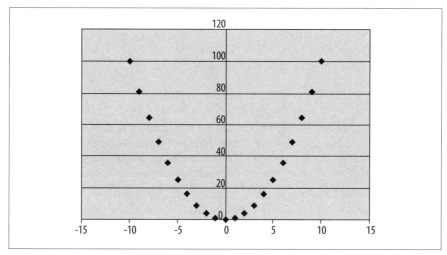

Figure 4-33. Quadratic relationship among variables

In the data presented in this scatterplot, the x-values in each pair are the integers from –10 to 10, and the y-values are the squares of the x-values, producing the familiar quadratic plot. Many statistical techniques assume a linear relationship between variables, and it's hard to see if this is true or not simply by looking at the raw data, so making a scatterplot of all important data pairs is a simple way to check this assumption.

Line Graphs

Line graphs are also often used to display the relationship between two variables, usually between time on the *x*-axis and some other variable on the *y*-axis. One requirement for a line graph is that there can only be one *y*-value for each *x*-value, so it would not be an appropriate choice for data such as the SAT data presented above. Consider the data in Table 4-10 from the U.S. Centers for Disease Control and Prevention (CDC), showing the percentage of obesity among U.S. adults, measured annually over a 13-year period.

Table 4-10. Percentage of obesity among U.S. adults, 1990–2002 (CDC)

Year	Percent obese
1990	11.6%
1991	12.6%
1992	12.6%
1993	13.7%
1994	14.4%
1995	15.8%
1996	16.8%
1997	16.6%
1998	18.3%
1999	19.7%
2000	20.1%
2001	21.0%
2002	22.1%

We can see from this table that obesity has been increasing at a steady pace; occasionally, there is a decrease from one year to the next, but more often there is a small increase in the range of 1% to 2%. This information can also be presented as a line chart, as in Figure 4-34, which makes this pattern of steady increase over the years even clearer.

Although this graph represents a straightforward presentation of the data, the visual impact depends partially on the scale and range used for the *y*-axis (which in this case shows percentage of obesity). Figure 4-34 is a sensible representation of the data, but if we wanted to increase the effect, we could choose a larger scale and smaller range for the *y*-axis (vertical axis), as in Figure 4-35.

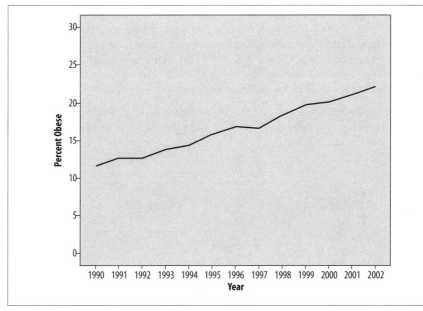

Figure 4-34. Obesity among U.S. adults, 1990–2002 (CDC)

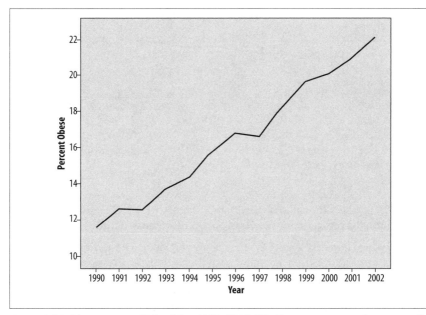

*Figure 4-35. Obesity among U.S. adults, 1990–2002 (CDC), using a restricted range to inflate
the visual impact of the trend*

Figure 4-35 presents exactly the same data as Figure 4-34, but a smaller range was
chosen for the y-axis (10%–22.5% versus 0%–30%), and the narrower range makes

the differences between years look larger. Figure 4-35 is not necessarily an incorrect way to present the data (although many argue that you should also include the 0 point in a graph displaying percent), but it does point out how easy it is to manipulate the appearance of an entirely valid data set. In fact, choosing a misleading range is one of the time-honored ways to "lie with statistics." (See the sidebar "How to Lie with Statistics" on page 117 for more on this topic.)

The same trick works in reverse; if we graph the same data by using a wide range for the vertical axis, the changes over the entire period seem much smaller, as in Figure 4-36.

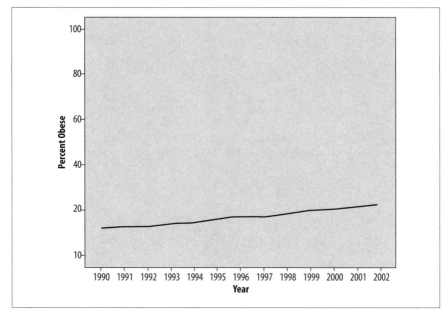

Figure 4-36. Obesity among U.S. adults, 1990–2002 (CDC), using a wide range on the y-axis to decrease the visual impact of the trend

Figure 4-36 presents the same obesity data as Figure 4-34 and Figure 4-35, with a large range on the vertical axis (0%–100%) to decrease the visual impact of the trend.

So which scale should be chosen? There is no perfect answer to this question; all present the same information, and none, strictly speaking, are incorrect. In this case, if I were presenting this chart without reference to any other graphics, the scale would be 7–34 because it shows the true floor for the data (0%, which is the lowest possible value) and includes a reasonable range above the highest data point. Independent of the issues involved with choosing the range for an individual chart, one principle that should be observed if multiple charts are compared to each other (for instance, charts showing the percent obesity in different countries over the same time period or charts of different health risks for the same period), they should all use the same scale to avoid misleading the reader.

Exercises

Like any other aspect of statistics, learning the techniques of descriptive statistics requires practice. The data sets provided are deliberately simple because if you can apply a technique correctly with 10 cases, you can also apply it with 1,000.

My advice is to try solving the problems several ways, for instance, by hand, using a calculator, and using whatever software is available to you. Even spreadsheet programs such as Microsoft Excel offer many simple mathematical and statistical functions. (Although the usefulness of such functions for serious statistical research is questionable, they might be adequate for initial exploratory work; see the references on Excel in Appendix C for more on this.) In addition, by solving a problem several ways, you will have more confidence that you are using the hardware and software correctly.

Most graphic presentations are created using software, and although each package has good and bad points, most can produce most, if not all, of the graphics presented in this chapter and quite a few other types of graphs as well. The best way to become familiar with graphics is to investigate whatever software you have access to and practice graphing data you currently work with. (If you don't currently work with data, plenty that you can experiment with is available for free download from the Internet.) Remember that graphic displays are a form of communication, and keep in mind the point you are trying to make with any graphic.

Problem

When is each of the following an appropriate measure of central tendency? Think of some examples for each from your work or studies.

- Mean
- Median
- Mode

Solution

- The mean is appropriate for interval or ratio data that is continuous, symmetrical, and lacks significant outliers.
- The median is appropriate for continuous data that might be skewed (asymmetrical), based on ranks, or contain extreme values.
- The mode is most appropriate for categorical variables or for continuous data sets where one value dominates the others.

Problem

Find some examples of the misleading use of statistical graphics, and explain what the problem is with each.

Solution

This shouldn't be a difficult task for anyone who follows the news media, but if you get stuck, try searching on the Internet for phrases like "misleading graphics."

Problem

One of the following data sets could be appropriately displayed as a bar chart and one as a histogram; decide which method is appropriate for each and explain why.

a. A data set of the heights (in centimeters) of 10,000 entering freshmen at a university
b. A data set of the majors elected by 10,000 entering freshmen at a university

Solution

a. The height data would be best displayed as a histogram because these measurements are continuous and have a large number of possible values.
b. The majors data would be more appropriately displayed as a bar chart because this type of information is categorical and has a restricted set of possible values (although if there is a large number of majors, the less frequent majors might be combined for the sake of clarity).

Problem

One of the following data sets is appropriate for a pie chart, and one is not. Identify which is which, and explain why.

a. Influenza cases for the past two years, broken down by month
b. The number of days missed due to the five leading causes for absenteeism at a hospital (the fifth category is "all other," including all absences attributed to causes other than the first four)

Solution

a. A pie chart would not be a good choice for the influenza data set because it would have too many categories (24), many of the categories are probably similar in size (because influenza cases are rare in the summer months), and the data doesn't really reflect parts making up a whole. A better choice might be a bar chart or line chart showing the number of cases by month or season.

b. The absenteeism data would be a good candidate for a pie chart because there are only five categories, and the parts do add up to 100% of a whole. One question that can't be answered from this description is whether the different categories (or slices of the pie) are clearly of different size; if so, that would be a further argument in favor of the use of a pie chart.

Problem

What is the median of this data set?

8 3 2 7 6 9 1 2 1

Solution

3. The data set has 9 values, which is an odd number; the median is therefore the middle value when the values are arranged in order. To look at this question more mathematically, because there are $n = 9$ values, the median is the $(n + 1)/2$th value; thus, the median is the $(9 + 1)/2$th or fifth value.

Problem

What is the median of this data set?

7 15 2 6 12 0

Solution

6.5. The data set has 6 values, which is an even number; the median is therefore the average of the middle two values when the values are arranged in order, in this case, 6 and 7. To look at this question more mathematically, the median for an even-numbered set of values is the average of the $(n/2)$th and $(n/2)$th + 1 value; $n = 6$ in this case, so the median is the average of the $(6/2)$th and $(6/2)$th + 1 values, that is, the third and fourth values.

Problem

What are the mean and median of the following (admittedly bizarre) data set?

1, 7, 21, 3, –17

Solution

The mean is $((1 + 7 + 21 + 3 + (-17))/5 = 15/5 = 3$.

The median, because there is an odd number of values, is the $(n + 1)/2$th value, that is, the third value. The data values in order are (–17, 1, 3, 7, 21), so the median is the third value, or 3.

Problem

What are the variance and standard deviation of the following data set? Calculate this by using both the population and sample formulas. Assume $\mu = 3$.

    1 3 5

Solution

The population formula to calculate variance is shown in Figure 4-37.

$$\sigma^2 = \frac{1}{n} \sum_{i=1}^{n} (x_i - \mu)^2$$

Figure 4-37. Formula for population variance

The sample formula is shown in Figure 4-38.

$$s^2 = \frac{1}{n-1} \sum_{i=1}^{n} (x_i - \bar{x})^2$$

Figure 4-38. Formula for sample variance

In this case, $n = 3$, $\bar{x} = 3$, and the sum of the squared deviation scores = $(-2)^2 + 0^2 + 2^2 = 8$. The population variance is 8/3, or 2.67, and the population standard deviation is the square root of the variance, or 1.63. The sample variance is 8/2, or 4, and the sample standard deviation is the square root of the variance, or 2.

5

Categorical Data

A categorical variable is a variable in which the possible responses consist of a set of categories rather than numbers that measure an amount or quantity of something on a continuous scale. For instance, a person might describe his or her gender in terms of male or female, or a machine part might be classified as acceptable or defective. More than two categories are also possible. For instance, a person in the United States might describe his political affiliation as Republican, Democrat, or independent.

Categorical variables may be inherently categorical (such as political party affiliation), with no numeric scale underlying their measurement, or they may be created by categorizing a continuous or discrete variable. Blood pressure is a measure of the pressure exerted on the walls of the blood vessels, measured in millimeters of mercury (Hg). Blood pressure is usually measured continuously and recorded with specific measurements such as 120/80 mmHg, but it is often analyzed using categories such as low, normal, prehypertensive, and hypertensive. Discrete variables (those that can be taken only on specific values within a range) may also be grouped into categorical variables. A researcher might collect exact information on the number of children per household (0 children, 1 child, 2 children, 3 children, etc.) but choose to group this data into categories for the purpose of analysis, such as 0 children, 1–2 children, and 3 or more children. This type of grouping is often used if there are large numbers of categories and some of them contain sparse data. In the case of the number of children in a household, for instance, a data set might include a relatively few households with large numbers of children, and the low frequencies in those categories can adversely affect the power of the study or make it impossible to use certain analytical techniques.

Although the wisdom of classifying continuous or discrete measurements into categories is sometimes debatable (some researchers refer to it as throwing away information because it discards all the information about variability within the categories), it is a common practice in many fields. Categorizing continuous data is done for many reasons, including custom (if certain categorizations may have become accepted in a professional field), and as a means to solve distribution problems within a data set.

Categorical data techniques can also be applied to ordinal variables, meaning those measured on a scale in which the categories might be ranked in order but do not meet the requirement of equal distance between each category. (Ordinal variables are discussed at more length in Chapter 1.) The well-known Likert scale, in which people choose their responses to questions from a set of ordered categories (such as Strongly Agree, Agree, Neutral, Disagree, and Strongly Disagree) is a classic example of an ordinal variable. A special set of analytic techniques, discussed later in this chapter, has been developed for ordinal data that retain the information about the order of the categories. Given a choice, specific ordinal techniques are preferred over categorical techniques for the analysis of ordinal data because they are generally more powerful.

A host of specific techniques has been developed to analyze categorical and ordinal data. This chapter discusses the most common techniques used for categorical and ordinal data, and a few techniques for these types of data are included in other chapters as well. The odds ratio, risk ratio, and the Mantel-Haenszel test are covered in Chapter 15, and some of the nonparametric methods covered in Chapter 13 are applicable to ordinal or categorical data.

The R×C Table

When an analysis concerns the relationship of two categorical variables, their distribution in the data set is often displayed in an R×C *table*, also referred to as a *contingency table*. The R in R×C refers to *row* and the C to column, and a specific table can be described by the number of rows and columns it contains. Rows and columns are always named in this order, a convention also followed in describing matrixes and in subscript notation. Sometimes, a distinction is made between 2×2 tables, which display the joint distribution of two binary variables, and tables of larger dimensions. Although a 2×2 table can be thought of as an R×C table where R and C both equal 2, the separate classification can be useful when discussing techniques developed specifically for 2×2 tables. The phrase "R×C" is read as "R by C," and the same convention applies to specific table sizes, so "3×2" is read as "3 by 2."

Suppose we are interested in studying the relationship between broad categories of age and health, the latter defined by the familiar five-category general health scale. We decide on the categories to be used for age and collect data from a sample of individuals, classifying them according to age (using our predefined categories) and health status (using the five-point scale). We then display this information in a contingency table, arranged like Table 5-1.

This would be described as a 4×5 table because it contains four rows and five columns. Each cell would contain the count of people from the sample with the pair of characteristics described: the number of people under 18 years in excellent health, the number aged 18–39 years in excellent health, and so on.

Table 5-1. Contingency table displaying health status by age category

	Excellent	Very good	Good	Fair	Poor
< 18 Years					
18–35 Years					
40–64 Years					
≥ 65 Years					

Measures of Agreement

The types of reliability described here are useful primarily for continuous measurements. When a measurement problem concerns categorical judgments, for instance classifying machine parts as acceptable or defective, measurements of agreement are more appropriate. For instance, we might want to evaluate the consistency of results from two diagnostic tests for the presence or absence of disease, or we might want to evaluate the consistency of results from three raters who are classifying the classroom behavior of particular students as acceptable or unacceptable. In each case, a rater assigns a single score from a limited set of choices, and we are interested in how well these scores agree across the tests or raters.

Percent agreement is the simplest measure of agreement; it is calculated by dividing the number of cases in which the raters agreed by the total number of ratings. For instance, if 100 ratings are made and the raters agree 80% of the time, the percent agreement is 80/100 or 0.80. A major disadvantage of simple percent agreement is that a high degree of agreement can be obtained simply by chance; thus, it is difficult to compare percent agreement across different situations when agreement due to chance can vary.

This shortcoming can be overcome by using another common measure of agreement called *Cohen's kappa*, the *kappa coefficient*, or simply *kappa*. This measure was originally devised to compare two raters or tests and has since been extended for use with larger numbers of raters. Kappa is preferable to percent agreement because it is corrected for agreement due to chance (although statisticians argue about how successful this correction really is; see the following sidebar for a brief introduction to the issues). Kappa is easily computed by sorting the responses into a symmetrical grid and performing calculations as indicated in Table 5-2. This hypothetical example concerns the agreement of two tests for the presence (D+) or absence (D−) of disease.

Table 5-2. Agreement of two tests on a dichotomous outcome

		Test 2		
		+	−	
Test 1	+	50	10	60
	−	10	30	40
		60	40	100

The four cells containing data are commonly identified as follows:

	+	–
+	a	b
–	c	d

Cells a and d represent agreement (a contains the cases classified as having the disease by both tests, d contains the cases classified as not having the disease by both tests), whereas cells b and c represent disagreement.

The formula for kappa is:

$$K = \frac{P_o - P_e}{1 - P_e}$$

where P_o = observed agreement, and P_e = expected agreement.

$$P_o = (a + d)/(a + b + c + d)$$

that is, the number of cases in agreement divided by the total number of cases. In this case,

$$P_o = 80/100 = 0.80$$
$$P_e = [(a + c)(a + b)]/(a + b + c + d)^2 + [(b + d)(c + d)]/(a + b + c + d)^2$$

and is the number of cases in agreement expected by chance. Expected agreement in this example is:

$$(60*60)/(100*100) + (40*40)/(100*100) = 0.36 + 0.16 = 0.52$$

Kappa, in this case, is therefore calculated as:

$$K = \frac{0.80 - 0.52}{1 - 0.52} = 0.58$$

Kappa has a range of –1 to +1; the value would be 0 if observed agreement were the same as chance agreement and 1 if all cases were in agreement. There are no absolute standards by which to judge a particular kappa value as high or low; however, some researchers use the guidelines published by Landis and Koch (1977):

< 0 Poor
0–0.20 Slight
0.21–0.40 Fair
0.41–0.60 Moderate
0.61–0.81 Substantial
0.81–1.0 Almost perfect

By this standard, our two tests exhibit moderate agreement. Note that the percent agreement in this example is 0.80, but kappa is 0.58. Kappa is always less than or equal to the percent agreement because kappa is corrected for chance agreement.

For an alternative view of kappa (intended for more advanced statisticians), see the following sidebar.

Controversies over Kappa

Cohen's kappa is a commonly taught and widely used statistic, but its application is not without controversy. Kappa is usually defined as representing agreement beyond that expected by chance or, simply, agreement corrected for chance. It has two uses: as a test statistic to determine whether two sets of ratings agree more often than would be expected by chance (which is a dichotomous, yes/no decision) and as a measure of the level of agreement (which is expressed as a number between 0 and 1).

Although most researchers have no problem with the first use of kappa, some object to the second. The problem is that calculating agreement expected by chance between any two entities, such as raters, is based on the assumption that the ratings are independent, a condition not usually met in practice. Because kappa is often used to quantify agreement for multiple individuals rating the same case, whether it is a child's classroom behavior or a chest X-ray from a person who might have tuberculosis, we would tend to expect more than chance agreement. In these cases, kappa overestimates the agreement among tests, raters, and so on by underestimating the amount of observed agreement that is in fact due to chance.

Criticisms of kappa, including a lengthy bibliography of relevant articles, can be found on the website (*http://www.john-uebersax.com/stat/kappa.htm*) of John Uebersax, PhD.

The Chi-Square Distribution

When we do hypothesis testing with categorical variables, we need some way to evaluate whether our results are significant. With $R{\times}C$ tables, the statistic of choice is often one of the *chi-square tests*, which draw on the known properties of the *chi-square distribution*. The chi-square distribution is a continuous theoretical probability distribution that is widely used in significance testing because many test statistics follow this distribution when the null hypothesis is true. The ability to relate a computed statistic to a known distribution makes it easy to determine the probability of a particular test result.

The chi-square distribution is a special case of the gamma distribution and has only one parameter, k, which specifies the degrees of freedom. The chi-square distribution has only positive values because it is based on the sum of squared quantities, as you will see, and is right-skewed. Its shape varies according to the value of k, most radically when k is a low value, as appears in the four chi-square distributions presented in Figure 5-1. As k approaches infinity, the chi-square distribution approaches (becomes very similar to) a normal distribution.

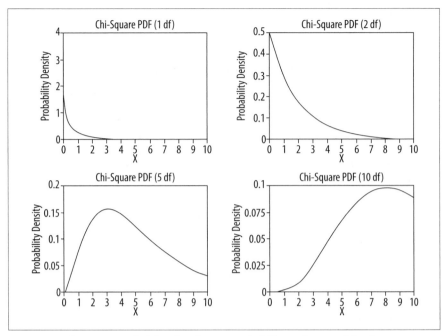

Figure 5-1. Chi-square probability distributions with different degrees of freedom

Figure D-11 contains a list of critical values for the chi-square distribution, which can be used to determine whether the results of a study are significant. For instance, the critical value, assuming $\alpha = 0.05$, for the chi-square distribution with one degree of freedom is 3.84. Any test result above this value will be considered significant for a chi-square test of independence for a 2×2 table (described next).

Note that $3.84 = 1.96^2$ and that 1.96 is the critical value for the Z-distribution (standard normal distribution) for a two-tailed test when $\alpha = 0.05$. This result is not coincidental but is due to a mathematical relationship between the Z and chi-square distributions.

Stated formally: if X_i are independent, standard normally distributed variables with $\mu = 0$ and $\sigma = 1$, and the random variable Q is defined as:

$$Q = \sum_{i=1}^{k} X_i^2$$

Q will follow a chi-square distribution with k degrees of freedom.

Two important points to remember are that you must know the degrees of freedom to evaluate a chi-square value and that the critical values generally increase with the number of degrees of freedom. If $\alpha = 0.05$, the critical value for a one-tailed chi-square test with one degree of freedom is 3.84, whereas for 10 degrees of freedom, it is 18.31.

The Chi-Square Test

The chi-square test is one of the most common ways to examine relationships between two or more categorical variables. Performing the chi-square test involves calculating the chi-square statistic and then comparing the value with that of the chi-square distribution to find the probability of the test results. There are several types of chi-square test; unless otherwise indicated, in this chapter "chi-square test" means the Pearson's chi-square test, which is the most common type.

There are three versions of the chi-square test. The first is called the *chi-square test for independence*. For a study with two variables, the chi-square test for independence tests the null hypothesis that the variables are independent of each other, that is, that there is no relationship between them. The alternative hypothesis is that the variables are related, so they are dependent rather than independent.

For instance, we might collect data on smoking status and diagnosis with lung cancer from a random sample of adults. Each of these variables is dichotomous: a person currently smokes or does not and has a lung cancer diagnosis or does not. We arrange our data in a frequency table as shown in Table 5-3.

Table 5-3. Smoking status and lung cancer diagnosis

	Lung cancer diagnosis	No lung cancer diagnosis
Currently smoke	60	300
Do not currently smoke	10	390

Just looking at this data, it seems plausible that there is a relationship between smoking and lung cancer: 20% of the smokers have been diagnosed with lung cancer versus only about 2.5% of the nonsmokers. Appearances can be deceiving, however, so we will conduct a chi-square test for independence. Our hypotheses are:

H_0: smoking status and lung cancer diagnosis are independent.
H_A: smoking status and lung cancer diagnosis are not independent.

Although chi-square tests are usually performed using a computer, particularly for larger tables, it is worthwhile to go through the steps of calculation for a simple example by hand. The chi-square test relies on the difference between *observed* and *expected* values in each of the cells of the 2×2 table. The observed values are simply what you found (observed) in your sample or data set, whereas the expected values are what you would expect to find if the two variables were independent. To calculate the expected value for a given cell, use the formula shown in Figure 5-2.

$$E_{ij} = \frac{i\text{th row total} \times j\text{th row total}}{\text{grand total}}$$

Figure 5-2. Calculating the expected value for a cell

In this formula, E_{ij} is the expected value for cell ij, and i and j designate the rows and columns of the cell. This subscript notation is used throughout statistics, so it's worth reviewing here. Table 5-4 shows how subscript notation is used to identify the parts of a 2×2 table.

Table 5-4. Subscript notation for a 2×2 table

Cell$_{11}$	Cell$_{12}$	Row 1 ($i = 1$)
Cell$_{21}$	Cell$_{21}$	Row 2 ($i = 2$)
Column 1 ($j = 1$)	Column 2 ($j = 2$)	

Table 5-5 adds row and column totals to our smoking/lung cancer example.

Table 5-5. Smoking and lung cancer data with row and column totals

	Lung cancer diagnosis	No lung cancer diagnosis	Total
Currently smoke	60	300	360
Do not currently smoke	10	390	400
Total	70	690	760

The frequency for cell$_{11}$ is 60, the value for cell$_{12}$ is 300, the total for row 1 is 360, the total for column 1 is 70, and so on. Using dot notation, the total for row 1 is designated as 1., the total for row 2 is 2., the total for column 1 is .1, and .2 is the total for column 2. The logic of this notation is that, for instance, the total for row 1 includes the values for both columns 1 and 2, so the column place is replaced with a dot. Similarly, a column total includes the values for both rows, so the row place is replaced by a dot. In this example, 1. = 360, 2. = 400, .1 = 70, and .2 = 690.

The values for column and row totals are called *marginals* because they are on the margin of the table. They reflect the frequency of one variable in the study without regard to its relationship with the other variable, so the marginal frequency for lung cancer diagnosis in this table is 70, and the marginal frequency for smoking is 360. The numbers within the table (60, 300, 10, and 390 in this example) are called *joint frequencies* because they reflect the number of cases having specified values on both variables. For instance, the joint frequency for smokers with a lung cancer diagnosis is 60 in this table.

If the two variables are not related, we would expect that the frequency of each cell would be the product of its marginals divided by the sample size. To put it another way, we would expect the joint frequencies to be affected only by the distribution of the marginals. This means that if smoking and lung cancer were unrelated, we would expect the number of people who smoke and have lung cancer to be determined only by the number of smokers and the number of people with lung cancer in the sample. By this logic, the probability of lung cancer should be about the same in smokers and nonsmokers if it is true that smoking is not related to the development of lung cancer.

Using the preceding formula, we can calculate the expected values for each of the cells as shown in Figure 5-3.

$$E_{11} = \frac{360 \times 70}{760} = 33.16$$

$$E_{12} = \frac{360 \times 690}{760} = 326.84$$

$$E_{21} = \frac{400 \times 70}{760} = 36.84$$

$$E_{22} = \frac{400 \times 690}{760} = 363.16$$

Figure 5-3. Computing the expected cell frequencies

The observed and expected values for the lung cancer data are presented in Table 5-6; expected values for each cell are in parentheses. We need some way to determine whether the discrepancies can be attributed to chance or represent a significant result. We can make this determination using the chi-square test.

Table 5-6. Observed and expected values for the smoking and lung cancer data

	Lung cancer diagnosis	No lung cancer diagnosis	Total
Currently smoke	60 (33.16)	300 (326.84)	360
Do not currently smoke	10 (36.84)	390 (363.16)	400
Total	70	690	760

The chi-square test is based on the squared difference between observed and expected values in each cell, using the formula shown in Figure 5-4.

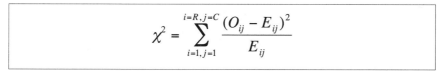

$$\chi^2 = \sum_{i=1, j=1}^{i=R, j=C} \frac{(O_{ij} - E_{ij})^2}{E_{ij}}$$

Figure 5-4. The formula for calculating a chi-square value

The steps for using this formula are:

1. Calculate the observed/expected values for cell$_{11}$.
2. Square the difference, and divide by the expected value.
3. Do the same for the remaining cells.
4. Add the numbers calculated in steps 1–3.

Continuing with our example, for cell_{11}, this quantity is:

$$\frac{(O_{ij} - E_{ij})^2}{E_{ij}} = \frac{(60 - 33.16)^2}{33.16} = 21.72$$

Continuing with the other cells, we find values of 2.2 for cell_{12}, 19.6 for cell_{21}, and 2.0 for cell_{22}. The total is 45.5, which is within rounding error for the value we found using the SPSS statistical analysis program, (45.474).

To interpret a chi-square statistic, you need to know its degrees of freedom. Each chi-square distribution has a different number of degrees of freedom and thus has different critical values. For a simple chi-square test, the degrees of freedom are $(r - 1)(c - 1)$, that is, (the number of rows minus 1) times (the number of columns minus 1). For a 2×2 table, the degrees of freedom are $(2 - 1)(2 - 1)$, or 1; for a 3×5 table, they are $(3 - 1)(5 - 1)$, or 8.

Having calculated the chi-square value and degrees of freedom by hand, we can consult a chi-square table to see whether the chi-squared value calculated from our data exceeds the critical value for the relevant distribution. According to Figure D-11 in Appendix D, the critical value for $\alpha = 0.05$ is 3.841, whereas our value of 45.5 is much larger, so if we are working with $\alpha = 0.05$, we have sufficient evidence to reject the null hypothesis that the variables are independent. If you are not familiar with the process of hypothesis testing, you might want to review that section of Chapter 3 before continuing with this chapter. Computer programs usually return a *p*-value along with the chi-square value and degrees of freedom, and if the *p*-value is less than our alpha level, we can reject the null hypothesis. In this example, assume we are using an alpha value of 0.05. According to SPSS, the *p*-value for our result of 45.474 is less than 0.0001, which is much less than 0.05 and indicates that we should reject the null hypothesis that there is no relationship between smoking and lung cancer.

The *chi-square test for equality of proportions* is computed exactly the same way as the chi-square test for independence, but it tests a different kind of hypothesis. The test for equality of proportions is used for data that has been drawn from multiple independent populations, and the null hypothesis is that the distribution of some variable is the same in all the populations. For instance, we could draw random samples from different ethnic groups and test whether the rates of lung cancer diagnosis are the same or different across the populations; our null hypothesis would be that they are the same. The calculations would proceed as in the preceding example: people would be classified by ethnic group and lung cancer status, expected values would be computed, the value of the chi-square statistic and degrees of freedom computed, and the statistic compared to a table of chi-square values for the appropriate degrees of freedom, or the exact *p*-value obtained from a statistical software package.

The *chi-square test of goodness of fit* is used to test the hypothesis that the distribution of a categorical variable within a population follows a specific pattern of proportions, whereas the alternative hypothesis is that the distribution of the variable follows some other pattern. This test is calculated using expected values based on hypothesized proportions, and the different categories or groups are designated with the subscript i, from 1 to g (as shown in Figure 5-5).

$$\chi^2 = \sum_{i=1}^{g} \frac{(O_i - E_i)^2}{E_i}$$

Figure 5-5. Formula for the chi-square test of goodness of fit

Note that in this formula, there are only single subscripts, for instance E_i rather than E_{ij}. This is because data for a chi-square goodness of fit is usually arranged into a single row, hence the need for only one subscript. The degrees of freedom for a chi-square test of goodness of fit is $(g - 1)$.

Suppose we believe that 10% of a particular population has low blood pressure (hypotension), 40% normal blood pressure, 30% prehypertension, and 20% hypertension. We can test this hypothesis by drawing a sample and comparing the observed proportions to those of our hypothesis (which are the expected values); we will use alpha = 0.05. Table 5-7 shows an example using hypothetical data.

Table 5-7. Expected and observed values for the distribution of blood pressure levels

	Hypotension	Normal	Prehypertension	Hypertension	Total
Expected proportion	0.10	0.40	0.30	0.20	1.00
Expected count	10	40	30	20	100
Observed count	12	25	50	13	100

The computed chi-square value for this data is 21.8 with 3 degrees of freedom and is significant. (The critical value for $\alpha = 0.05$ is 7.815, as can be seen from the chi-square table in Figure D-11 in Appendix D.) Because the value calculated on our data exceeds the critical value, we should reject the null hypothesis that the blood pressure levels in the population follow this hypothesized distribution.

The Pearson's chi-square test is suitable for data in which all observations are independent (the same person is not measured twice, for instance) and the categories are mutually exclusive and exhaustive (so that no case may be classified into more than one cell, and all potential cases can be classified into one of the cells). It is also assumed that no cell has an expected value less than 1, and no more than 20% of the cells have an expected value less than 5. The reason for the last two requirements is that the chi-square is an asymptotic test and might not be valid for sparse data (data in which one or more cells have a low expected frequency).

Yates's correction for continuity is a procedure developed by the British statistician Frank Yates for the chi-square test of independence when applied to 2×2 tables. The chi-square distribution is continuous, whereas the data used in a chi-square test is discrete, and Yates's correction is meant to correct for this discrepancy. Yates's correction is easy to apply. You simply subtract 0.5 from the absolute value of

Categorical Data

(observed – expected) in the formula for the chi-square statistic before squaring; this has the effect of slightly reducing the value of the chi-square statistic. The chi-square formula, with Yates's correction for continuity, is shown in Figure 5-6.

$$\chi^2 = \sum_{i=1, j=1}^{i=R, j=C} \frac{(\,|\,O_{ij} - E_{ij}\,|\,-0.5)^2}{E_{ij}}$$

Figure 5-6. The chi-square formula with Yates's correction for continuity

The idea behind Yates's correction is that the smaller chi-square value reduces the probability of Type I error (wrongly rejecting the null hypothesis). Use of Yates's correction is not universally endorsed, however; some researchers feel that it might be an overcorrection leading to a loss of power and increased probability of a Type II error (wrongly failing to reject the null hypothesis). Some statisticians reject the use of Yates's correction entirely, although some find it useful with sparse data, particularly when at least one cell in the table has an expected cell frequency of less than 5. A less controversial remedy for sparse categorical data is to use *Fisher's exact test*, discussed later, instead of the chi-square test, when the distributional assumptions previously named (no more than 20% of cells with an expected value less than 5 and no cell with an expected value of less than 1) are not met.

The chi-square test is often computed for tables larger than 2×2, although computer software is usually used for those analyses because as the number of cells increases, the calculations required quickly become lengthy. There is no theoretical limit on the number of columns and rows that may be included, but two factors impose practical limits: the possibility of making a coherent interpretation of the results (try this with a 30×30 table!) and the necessity to avoid sparse cells, as noted earlier. Sometimes, data is collected in a large number of categories but collapsed into a smaller number to get around the sparse cell problem. For instance, information about marital status may be collected using many categories (married, single never married, divorced, living with partner, widowed, etc.), but for a particular analysis, the statistician may choose to reduce the categories (e.g., to married and unmarried) because of insufficient data in the smaller categories.

Fisher's Exact Test

Fisher's Exact Test (often called simply Fisher's) is a nonparametric test similar to the chi-square test, but it can be used with small or sparsely distributed data sets that do not meet the distributional requirements of the chi-square test. Fisher's is based on the hypergeometric distribution and calculates the exact probability of observing the distribution seen in the table or a more extreme distribution, hence the word "exact" in the title. It is not an asymptotic test and therefore is not subject to the sparseness rules that apply to the chi-square tests. Computer software is usually used to calculate Fisher's, particularly for tables larger than 2×2, because of the repetitive nature of the calculations. A simple example with a 2×2 table follows.

Suppose we are interested in the relationship between use of a particular street drug and sudden cardiac failure in young adults. Because the drug is both illegal and new to our area, and because sudden cardiac death is rare in young adults, we were not able to collect enough data to conduct a chi-square test. Table 5-8 shows the data for analysis.

Table 5-8. Fisher's Exact Test: calculating the relationship between the use of a novel street drug and sudden cardiac death in young adults

	Cardiac death	No cardiac death	Total
Used drug	7	2	9
Didn't use drug	5	6	11
Total	12	8	20

Our hypotheses are:

H_0: risk of sudden cardiac death is no more common among users of the new drug than in nonusers. H_1: risk of sudden cardiac death is greater in people using the new drug.

Fisher's Exact Test calculates the probability of results at least as extreme as those found in the study. A more extreme result in this study would be one in which the difference in proportion of drug users versus nondrug users suffering sudden cardiac death was even greater than in the actual data (keeping the same sample size). One more extreme result is shown in Table 5-9.

Table 5-9. More extreme data distribution for drug use/cardiac death example

	Cardiac death	No cardiac death	Total
Used drug	8	1	9
Didn't use drug	4	7	11
Total	12	8	20

The formula to calculate the exact probability for a 2×2 table is shown in Figure 5-7.

$$p = \frac{r_1! r_2! c_1! c_2!}{n! a! b! c! d!}$$

Figure 5-7. Formula for Fisher's Exact Test

In this formula, ! means factorial (4! = 4×3×2×1), and cells and marginals are identified using the notation shown in Table 5-10.

Table 5-10. Table notation

a	b	r_1
c	d	r_2
c_1	c_2	n

In this case, $a = 8$, $b = 1$, $c = 4$, $d = 7$, $r_1 = 9$, $r_2 = 11$, $c_1 = 12$, $c_2 = 8$, and $n = 20$. Why is this table more extreme than our observed results? Because if there were no relationship between use of the drug and sudden cardiac death, we would expect to see the distribution in Table 5-11.

Table 5-11. Expected data, assuming independence

	Cardiac death	No cardiac death	Total
Used drug	5.4	3.6	9
Didn't use drug	6.6	4.4	11
Total	12	8	20

In our observed data, there is a stronger relationship between using the drug and cardiac death (more deaths than the expected value for drug users), so any table in which that relationship is even stronger than in our observed data is more extreme and hence less probable if drug use and cardiac death are independent.

To find the *p*-value for Fisher's Exact Test by hand, we would have to find the probability of all the more extreme tables and add them up. Fortunately, algorithms to calculate Fisher's are included in most statistical software packages, and many online calculators also can calculate this statistic for you. Using the calculator available on a page maintained by John C. Pezzullo, a retired professor of pharmacology and biostatistics (*http://statpages.org/ctab2x2.html*), we find that the one-tailed *p*-value for Fisher's Exact Test for the data in Table 5-7 is 0.157. We use a one-tailed test because our hypothesis is one-tailed; our interest is in whether use of the new drug *increases* the risk for cardiac death. Using an alpha level of 0.05, this result is not significant, so we do not reject our null hypothesis that the new drug does not increase the risk of cardiac death.

McNemar's Test for Matched Pairs

McNemar's test is a type of chi-square test used when the data comes from *paired samples*, also known as *matched samples* or *related samples*. For instance, we might use McNemar's to examine the results of an opinion poll on some issue before and after a group of individuals viewed a political advertisement. In this example, each person would contribute two opinions, one before and one after viewing the advertisement. We cannot treat two opinions on the same issue as independent, so we can't use a Pearson's chi-square; instead, we assume that two opinions collected from the same person will be more closely related than two opinions collected from two people. The McNemar's test would also be appropriate if we collected opinions from pairs of siblings or husband–wife pairs on some issue. In siblings and husband–

wife examples, although information is collected from different individuals, the individuals in each pair are so closely related or affiliated that we would expect them to be more similar than two people chosen at random from the population. McNemar's can also be used to analyze data collected from groups of individuals who have been so closely matched on important characteristics that they can no longer be considered independent. For instance, medical studies sometimes look at the occurrence of a particular disease related to a risk factor among groups of individuals matched on multiple characteristics such as age, gender, and race/ethnicity, and use paired data techniques such as McNemar's because the individuals are so closely matched that they are considered related rather than independent samples.

Suppose we want to measure the effectiveness of a political advertisement in changing people's opinions about capital punishment. One way to do this would be to ask people whether they are for or against capital punishment, collecting their opinions both before and after they view a 30-second commercial advocating the abolition of capital punishment. Consider the hypothetical data set in Table 5-12.

Table 5-12. McNemar's test of opinions on capital punishment before and after viewing a television commercial

		After viewing the commercial		
		For capital punishment	Against capital punishment	Total
Before viewing the commercial	For capital punishment	15	25	40
	Against capital punishment	10	20	30
		25	45	70

More people were against capital punishment after viewing the commercial as compared to the same people before viewing the commercial, but is this difference significant? We can test this using McNemar's chi-square test, calculated using the formula in Figure 5-8.

$$\chi^2 = \frac{(b-c)^2}{b+c}$$

Figure 5-8. Formula for McNemar's chi-square test

This formula uses a method of referring to cells by letters, using the plan shown in Table 5-13.

Table 5-13. Method of referring to cells in a 2×2 table by letters

a	b
c	d

Note that this formula is based exclusively on the distribution of discordant pairs (b and c), in this case those in which a person changed his or her opinion after viewing the commercial. McNemar's has a chi-squared distribution with one degree of freedom. The calculations are shown in Figure 5-9.

$$\chi^2 = \frac{(25 - 10)^2}{25 + 10} = \frac{225}{35} = 6.43$$

Figure 5-9. Computing McNemar's chi-square test

As you can see from the chi-square table (Figure D-11 in Appendix D), when alpha = 0.05, the critical value for a chi-square distribution is 3.84, so this result provides evidence that we should reject the null hypothesis that viewing the commercial has no effect on people's opinions about capital punishment. I also determined from a computer analysis that the exact probability of getting a chi-square statistic with one degree of freedom at least as extreme as 6.43 is 0.017 if people's opinions did not change before and after viewing the commercial, reinforcing the fact that the result from this study is significant, and we should reject the null hypothesis.

Proportions: The Large Sample Case

A proportion is a fraction in which all the cases in the numerator are also in the denominator. For instance, we could speak of the proportion of female students in a particular university. The numerator would be the number of female students, and the denominator would be all students (both male and female) at the university. Or we could speak of the number of students majoring in chemistry at a particular university. The numerator would be the number of chemistry majors, and the denominator all students at the university (of whatever major). Proportions are discussed in more detail in Chapter 15. Data that can be described in terms of proportions is a special case of categorical data in which there are only two categories: male and female in the first example, chemistry major and non-chemistry major in the second.

Many of the statistics discussed in this chapter, such as Fisher's Exact Test and the chi-square tests, can be used to test hypotheses about proportions. However, if the data sample is sufficiently large, additional types of tests can be performed using the normal approximation to the binomial distribution; this is possible because, as discussed in Chapter 3, the binomial distribution comes to resemble the normal distribution as n (the sample size) increases. How large a sample is large enough? One rule of thumb is that both np and $n(1 - p)$ must be greater than or equal to 5.

Suppose you are a factory manager, and you claim that 95% of a particular type of screw produced by your plant has a diameter between 0.50 and 0.52 centimeters. A customer complains that a recent shipment of screws contains too many outside the specified dimensions, so you draw a sample of 100 screws and measure them to see how many meet the standard. You will conduct a one-sample Z-test to see whether your hypothesized proportion of 95% of screws meeting the specified standard is correct with the following hypotheses:

H_0: $\pi \geq 0.95$, H_1: $\pi < 0.95$

where π is the proportion of screws in the population meeting the standard (diameter between 0.50 and 0.52 centimeters). Note that this is a one-tailed test; you will be happy if at least 95% of the screws meet the standard and happy if more than 95% meet it. (You would be happiest, of course, if 100% met the standard, but no manufacturing process is perfectly precise.) In your sample of 100 screws, 91 were within the specified dimensions. Is this result sufficient, using the standard of alpha = 0.05, to reject the null hypothesis that at least 95% of the screws of this type manufactured in our plant meet the standard?

The formula to calculate the one-sample Z-test for a proportion is given in Figure 5-10.

$$Z = \frac{p - \pi_0}{\sqrt{\dfrac{\pi_0(1 - \pi_0)}{n}}}$$

Figure 5-10. Formula for the one-sample Z statistic for a proportion

In this formula, π_0 is the hypothesized population proportion,
p is the sample proportion, and
n is the sample size.

Plugging the numbers into this formula gives us a Z-score of –1.835, as shown in Figure 5-11.

$$Z = \frac{0.91 - 0.95}{\sqrt{\dfrac{(0.95)(0.05)}{100}}} = \frac{-0.0400}{0.0218} = -1.835$$

Figure 5-11. Calculating the one-sample Z statistic for a proportion

The critical value for a one-tailed Z-test, given our hypotheses and alpha-level, is –1.645. Our value of –1.835 is more extreme than this value, so we will reject our null hypothesis and conclude that less than 95% of this type of screw, as manufactured in our plant, meet the specified standard.

We can also test for differences in population proportions in the large-sample case. Suppose we are interested in the proportion of high school students who are current tobacco smokers and want to compare this proportion across two countries. Our null hypothesis is that the proportion is the same in both countries, so this will be a two-sided test with the hypotheses:

$H_0: \pi_1 = \pi_2, H_1: \pi_1 \neq \pi_2$

Assuming that our assumptions about sample size are met ($np \geq 5$, $n(1-p) \geq 5$ for both samples), we can use the formula in Figure 5-12 to compute a Z-statistic for the differences in proportions between two populations.

$$Z = \frac{p_1 - p_2}{\sqrt{\dfrac{\hat{p}(1-\hat{p})}{n_1} + \dfrac{\hat{p}(1-\hat{p})}{n_2}}}$$

Figure 5-12. Formula for the Z-statistic for the difference in two proportions

In this formula, p_1 is the proportion in sample 1,
p_2 is the proportion in sample 2,
n_1 is the size of sample 1,
n_2 is the size of sample 2, and
$\bar{x}$ is the pooled proportion, calculated as the sum of successes in both samples (in this case, the number of smokers), divided by the sum of the sample sizes.

Suppose we drew a sample of 500 high school students from each of two countries; in country 1, the sample included 90 current smokers; in country 2, it included 70 current smokers. Given this data, do we have sufficient information to reject our null hypotheses that the same proportion of high school students smoke in each country? We can test this by calculating the two-sample Z test, as shown in Figure 5-13.

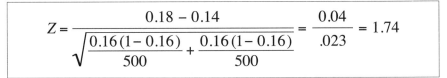

Figure 5-13. Calculating the Z-statistic for the difference in two proportions

Note that our pooled proportion is:

(90 + 70)/(500 + 500) = 160/1000 = 0.16

This Z-value is less extreme than 1.96 (the value needed to reject the null hypothesis at alpha = 0.05; you can confirm this using the normal table [Figure D-3 in Appendix D]), so we fail to reject the null hypothesis that the proportion of smokers among high school students in the two countries is the same.

Correlation Statistics for Categorical Data

The most common measure of association for two variables, Pearson's correlation coefficient (discussed in Chapter 7) requires variables measured on at least the interval level. However, several measures of association have been developed for categorical and ordinal data, and they are interpreted similarly to the Pearson correlation coefficient. These measures are often produced using a statistical software package or an online calculator, although they can also be calculated by hand.

As with Pearson's correlation coefficient, the correlation statistics discussed in this section are measures of association only, and statements about causality cannot be supported by a correlation coefficient alone. There is a plethora of these measures, some of which are known under several names; a few of the most common are discussed here. A good approach if you're using a new statistical software package is to see which measures are supported by that package and then investigate which of those measures are appropriate for your data because there are so many correlation statistics.

Binary Variables

Phi is a measure of the degree of association between two binary variables (two categorical variables, each of which can have only one of two values). Phi is calculated for 2×2 tables; *Cramer's V* is analogous to phi for tables larger than 2×2. Using the method of cell identification described in Table 5-10, the formula to calculate phi is shown in Figure 5-14.

$$\phi = \frac{ad - bc}{\sqrt{(a+b)(c+d)(a+c)(b+d)}}$$

Figure 5-14. Formula for the phi statistic

We can calculate phi for the smoking/lung cancer data in Table 5-3 as shown in Figure 5-15.

$$\phi = \frac{(60)(390) - (300)(10)}{\sqrt{360 + 400 + 70 + 690}} = 0.24$$

Figure 5-15. Calculating the phi statistic

Phi can also be calculated by dividing the chi-square statistic by *n* and then taking the square root of the result as shown in Figure 5-16.

$$\phi = \sqrt{\frac{\chi^2}{n}}$$

Figure 5-16. An alternative formula for the phi statistic

Note that in the first method of calculation, the result can be either positive or negative, whereas in the second, it can only be positive because the chi-square statistic is always positive. The value of phi using the chi-square statistic found using the second formula can be thought of as the absolute value of the value found using the first formula. This is clear from considering the data in Table 5-14.

Table 5-14. Phi example

10	20
20	10

Calculating phi by the first method, we get –0.33, and by the second method, 0.33. You can confirm this using a statistical computer package or an online calculator (*http://statpages.org/ctab2x2.html*), or by performing the calculations by hand. Of course, if we changed the order of the two columns, we would get a positive result using either method. If the columns have no natural order (e.g., if they represent nonordered categories such as color), we might not care about the direction of the association but only its absolute value. In other cases, we might, for instance if the columns represent the presence or absence of disease. In the latter case, we need to be careful about how we arrange the data in the table to avoid producing a misleading result.

Interpreting phi is less straightforward than interpreting the Pearson's correlation coefficient because the range of phi depends on the marginal distribution of the data. If both variables have a 50-50 split (half one value, half the other), the range of phi is (–1, +1), using the first method, or (0, 1), using the second method. If the variables have any other distribution, the potential range of phi is less. This is discussed further in the article by Davenport and El-Sanhurry listed in Appendix C. Keeping this limitation in mind, the interpretation of phi is similar to that of the Pearson correlation coefficient, so a value of –0.33 would indicate a moderate negative relationship (also keeping in mind that there is no absolute definition of "a moderate relationship" and that this result might be considered large in one field of study and rather small in another).

Cramer's V is an extension of phi for tables larger than 2×2. The formula for Cramer's V is similar to the second method for calculating phi, as shown in Figure 5-17.

$$V = \sqrt{\frac{\chi^2}{n(\min \ r-1, \ c-1)}}$$

Figure 5-17. The formula for Cramer's V

where the denominator is n (sample size) times the minimum of $(r - 1)$ and $(c - 1)$, that is, the minimum of two values: the number of rows minus 1, and the number of columns minus 1. For a 4×3 table, this number would be 2, that is, 3 – 1. For a 2×2 table, the formula for Cramer's V is identical to the formula for the second way of calculating phi.

Suppose the chi-square value for a 3×4 table with an n of 200 is 16.70. Cramer's V for this data is shown in Figure 5-18.

$$V = \sqrt{\frac{16.70}{200(2)}} = 0.20$$

Figure 5-18. Calculating Cramer's V

The Point-Biserial Correlation Coefficient

The point-biserial correlation coefficient is a measure of association between a dichotomous variable and a continuous variable. Mathematically, it is equivalent to the Pearson correlation coefficient (discussed in detail in Chapter 7), but because one of the variables is dichotomous, a different formula can be used to calculate it.

Suppose we are interested in the strength of association between gender (dichotomous) and adult height (continuous). The point-biserial correlation is symmetric, like the Pearson correlation coefficient, but for ease of notation we designate height as X and gender as Y and code Y so 0 = males and 1 = females. We draw a sample of men and women and calculate the point-biserial correlation by using the formula shown in Figure 5-19.

$$r_{pb} = \frac{\overline{X}_1 - \overline{X}_0 \sqrt{p(1-p)}}{s_x}$$

Figure 5-19. Formula for the point-biserial correlation coefficient

In this formula, $\overline{X}_1$ = the mean height for females and $\overline{X}_0$ = the mean height for males

p = the proportion of females
s_x = the standard deviation of X

Suppose in our sample, the mean height for males is 69.0 inches, for females 64.0 inches, the standard deviation for height is 3.0 inches, and the sample is 55% female. We calculate the correlation between gender and adult height as shown in Figure 5-20.

$$r_{pb} = \frac{\overline{X}_1 - \overline{X}_0 \sqrt{p(1-p)}}{s_x} = \frac{(64-69)\sqrt{0.55(0.45)}}{3} = -0.829$$

Figure 5-20. The point-biserial correlation of gender and height

A correlation of –0.829 is a strong relationship, indicating that there is a close relationship between gender and adult height in the U.S. population. The correlation is negative because we coded females (who are on average shorter) as 1 and males as 0; had we coded our cases the other way around, our correlation would have been

0.829. Note that the means and standard deviation used in this equation are close to the actual values for the U.S. population, so a strong relationship between gender and height exists in reality as well as in this exercise.

Ordinal Variables

The most common correlation statistic for ordinal data (in which data is ordered but cannot be assumed to have equal distance between values) is *Spearman's rank-order coefficient*, also called *Spearman's rho* or *Spearman's r*, and also designated by r_s. Spearman's rho is based on the ranks of data points (first, second, third, and so on) rather than on their values. Class rank in a school is an example of ratio-level data; the person with the highest GPA (grade point average) is ranked first, the person with the next-highest GPA is ranked second, and so on, but you don't know whether the difference between the 1st and 2nd students is the same as the difference between the 2nd and 3rd. Even if you have data measured on a ratio scale, such as GPA in high school, class ranks are sometimes used in college admissions and scholarship decisions because of the difficulty of comparing grading systems across different classes and different schools.

To calculate Spearman's rho, rank the values of each variable separately, averaging the ranks of any tied values. Then calculate the difference in ranks for each pair of values, and calculate Spearman's rho by using the formula shown in Figure 5-21.

$$r_s = 1 - \frac{6\sum d_i^2}{n(n^2 - 1)}$$

Figure 5-21. Formula for Spearman's rho

Suppose we are interested in the relationship between weekly hours of study and score on a final exam. We collect data for both variables as shown in Table 5-15 (a data set for illustrative purposes to minimize the hand-calculations needed):

Table 5-15. Weekly hours of study and final exam score

Student	Hours of study	Rank	Final exam score	Rank	d_i	d_i squared
1	10	7	93	7	0	0
2	12	9	98	8	1	1
3	8	5	99	9	−4	16
4	15	10	100	10	0	0
5	4	1	92	6	−5	25
6	11	8	90	5	3	9
7	6	3	80	2	1	1
8	7	4	82	3	1	1
9	9	6	84	4	2	4

Student	Hours of study	Rank	Final exam score	Rank	d_i	d_i squared
10	5	2	75	1	1	1

It looks like more studying is associated with a higher grade, although the relationship is not perfect. (Student #3 got a high grade with only an average amount of studying, and student #5 got a good grade with a relatively low amount of studying.) We will calculate Spearman's rho to get a more precise estimate of this relationship. Note that because we square the rank difference, it doesn't matter whether you subtract study rank from exam rank (as we did) or the other way around. The sum of d_i^2 is 58, and Spearman's rho for this data is shown in Figure 5-22.

$$r_s = 1 - \frac{(6)(58)}{10(99)} = 1 - 0.35 = 0.65$$

Figure 5-22. Calculating Spearman's rho

This confirms what we guessed from just looking at the data: there is a strong but imperfect relationship between the amount of time spent studying and the outcome on a test.

Goodman and Kruskal's gamma, often called simply *gamma*, is a measure of association for ordinal variables that is based on the number of concordant and discordant pairs between two variables. It is sometimes called a measure of *monotonicity* because it tells you how often the variables have values in the order expected. If I tell you that two variables in a data set have a positive relationship and that case 2 has a higher value on the first variable than does case 1, you would expect that case 2 also has a higher value on the second variable. This would be a *concordant pair*. If case 2 had a lower value on the second variable, it would be a *discordant pair*. To calculate gamma by hand, we would first create a frequency distribution for the two variables, retaining their natural order.

Consider a hypothetical data set relating BMI (body mass index, a measure of weight relative to height) and blood pressure levels. In general, high BMI is associated with high blood pressure, but this is not the case for every individual. Some overweight people have normal blood pressure, and some normal-weight people have high blood pressure. Is there a strong relationship between weight and blood pressure in the data set shown in Table 5-16?

Table 5-16. Example data to calculate gamma

		Blood pressure		
		Normal	Prehypertensive	Hypertensive
BMI	Normal	25	15	5
	Overweight	10	10	25

The equations to calculate gamma rely on the cell designations shown in Table 5-17.

Table 5-17. Cell designations to compute gamma

a	b	c
d	e	f

First, we have to find the number of concordant pairs (P) and discordant pairs (Q), as follows:

$$P = a\,(e + f) + bf = 25(10 + 25) + 15(25) = 875 + 375 = 1250,$$
$$Q = c\,(d + e) + bd = 5(10 + 10) + 15(10) = 100 + 150 = 250$$

Gamma is then calculated as shown in Figure 5-23.

$$\gamma = \frac{P - Q}{P + Q} = \frac{1250 - 250}{1250 + 250} = 0.67$$

Figure 5-23. Calculating Goodman and Kruskal's gamma

The reasoning behind gamma is clear: if there is a strong relationship between the two variables, there should be a higher proportion of concordant pairs; thus, gamma will have a larger value than if the relationship were weaker. Gamma is a symmetrical measure because it does not matter which variable is considered the predictor and which the outcome; the value of gamma will be the same in either case. Gamma does not correct for tied ranks within the data.

Maurice Kendall developed three slightly different types of ordinal correlation as alternatives to gamma. Statistical computer packages sometimes use more complex formulas to calculate these statistics, so the exact formula any particular program uses should be confirmed with the software manual. All Kendall's tau statistics, like gamma, are symmetrical measures.

Kendall's tau-a is based on the number of concordant versus discordant pairs, divided by a measure based on the total number of pairs (n = the sample size), as shown in Figure 5-24.

$$\tau_a = \frac{P - Q}{\left(\dfrac{n(n - 1)}{2}\right)}$$

Figure 5-24. Formula for Kendall's tau-a

Kendall's tau-b is a similar measure of association based on concordant and discordant pairs, adjusted for the number of ties in ranks. Assuming our two variables are X and Y, tau-b is calculated as ($P - Q$) divided by the geometric mean of the number of pairs not tied on X (X_0) and the number of pairs not tied on Y (Y_0). Tau-b can approach 1.0 or –1.0 only for square tables (tables with the same number of rows and columns). The formula for Kendall's tau-b is shown in Figure 5-25.

$$\tau_b = \frac{P - Q}{\sqrt{(P + Q + X_0)(P + Q + Y_0)}}$$

Figure 5-25. Formula for Kendall's tau-b

In this formula, X_0 = the number of pairs not tied on X, and Y_0 = the number of pairs not tied on Y.

Kendall's *tau-c* is used for nonsquare tables and is calculated as shown in Figure 5-26.

$$\tau_c = (P - Q)\left[\frac{2m}{n^2(m - 1)}\right]$$

Figure 5-26. Formula for Kendall's tau-c

In this formula, m is the number of rows or columns, whichever is smaller, and n is the sample size.

Somers's d is an asymmetrical version of gamma, so calculation of the statistic varies depending on which variable is considered the predictor and which the outcome. Somers's d also differs from gamma because it is corrected for the number of pairs tied on the predictor variable. If the study is set up with the hypothesis involving X predicting Y, Somers's d is corrected for the number of pairs tied on X. If the hypothesis is that Y predicts X, it is corrected for the number of pairs tied on Y. As in tau-b, in Somers's d, tied pairs are removed from the denominator. Using the notation that X_0 = the number of pairs not tied on X and Y_0 = the number of pairs not tied on Y, Somers's d is calculated as shown in Figure 5-27.

$$d(\text{predicting Y from X}) = \frac{P - Q}{P + Q + X_0}$$

$$d(\text{predicting X from Y}) = \frac{P - Q}{P + Q + Y_0}$$

Figure 5-27. Formulas for Somers's d

A symmetric value of Somers's d can be calculated by averaging the two asymmetric values calculated with these formulas.

The Likert and Semantic Differential Scales

Several types of scales have been developed to measure qualities that have no natural metric, such as opinions, attitudes, and perceptions. The best known of these scales is the Likert scale, introduced by Rensis Likert in 1932 and widely used today in fields ranging from education to health care to business management. In a typical Likert scale question, a statement is presented and the respondent is asked to choose from an ordered list of responses. For instance:

> My classes at Lincoln East High School prepared me for university studies.

> 1. Strongly agree
> 2. Agree
> 3. Neutral
> 4. Disagree
> 5. Strongly disagree

This is a classic ordinal scale; we can be reasonably sure that "strongly agree" represents more agreement than "agree," and "agree" represents more agreement than "neutral," but we can't be sure whether the increment of agreement between "agree" and "strongly agree" is the same as the increment between "neutral" and "agree" or if these increments are the same for each respondent.

Categorical and ordinal methods, as described in this chapter, are appropriate for the analysis of Likert scale data, and so are some of the nonparametric methods described in Chapter 13. The fact that Likert scale responses are often identified with numbers has sometimes led researchers to analyze the data as if it were collected on an interval scale. For instance, you can find published articles that report the mean and variance for data collected using a Likert scale. A researcher choosing to follow this path (treating Likert data as interval) should be aware that this is a controversial approach that will be rejected by many editors and that the burden is on the researcher to justify any departure from ordinal or categorical methods of analysis for Likert scale data.

Five levels of response are commonly used with Likert scales because three is thought not to allow sufficient variation of response, whereas seven is believed to offer too many choices. There is also some evidence that people are reluctant to select the extreme values of a scale when a large number of choices is offered. However, some researchers prefer to use an even number of responses, usually four or six, to avoid a middle category that might be chosen by default by some respondents.

The *semantic differential scale* is similar to the Likert scale except that individual data points are not labeled, merely the extreme values. The preceding Likert question could be rewritten as a semantic differential question as follows:

> Please rate your academic preparation at Lincoln East High School in relation to the demands of university study.

> Excellent preparation 1   2   3   4   5   Inadequate preparation

Because individual data points do not have to be labeled, semantic differential items often offer more data points to the respondent. Ten data points is a popular choice because people are familiar with a 10-point judging scale (hence the popular phrase "a perfect 10"). Like Likert scales, semantic differential scales are by nature ordinal, although when a larger number of data points is offered, some researchers argue that they can be analyzed as interval data.

Rensis Likert (1903–1981)

Rensis Likert (pronounced Lick-urt, with the accent on the first syllable) was an American social scientist who specialized in research on organizational behavior and management theory. Likert received his BA in sociology from the University of Michigan in 1926 and his PhD in psychology from Columbia University in 1932; he developed the Likert scale as part of his dissertation research. Likert was a founder of the University of Michigan Institute for Social Research and served as its director from 1946 to 1970; he spent his later years consulting for corporations and writing books on management theory. A central aspect of his work will endear him to self-motivated students and employees around the world: Likert introduced the concepts of participation management and the human-centered organization, based on his findings that there was an inverse relationship between coercive management supervision and employee productivity.

Exercises

Here are some review questions on the topics covered in this chapter.

Problem

What are the dimensions of Tables 5-18 and 5-19? What would be the degrees of freedom for an independent-samples chi-square test calculated from data of these dimensions?

Table 5-18. R×C table a

Table 5-19. R×C table b

Solution

The table dimensions are 3×4 (table a) and 4×3 (table b). Remember, tables are described as R×C, that is, (number of rows) by (number of columns). The degrees of freedom are 6 for the first table [(3 – 1)(4 – 1)] and 6 for the second [(4 – 1)(3 – 1)] because degrees of freedom for chi-square is calculated as [$(r – 1)(c – 1)$].

Problem

Given the distribution of data in the following table, calculate percent agreement and kappa.

Table 5-20. Agreement between two raters

		Rater 2 +	Rater 2 −	
Rater 1	+	70	15	85
	−	30	25	55
		100	40	140

Solution

Percent agreement = 95/140 = 0.68,
Kappa = 0.30,
P_o = (70 + 25)/140 = 0.68,
P_e = (85*100)/(140*140) + (40*55)/(140*140) = 0.54

$$K = \frac{0.68 - 0.54}{1 - 0.54} = 0.30$$

Figure 5-28. Calculating kappa

Problem

What is the null hypothesis for the chi-square test of independence?

Solution

The variables are independent, which also means that the joint probabilities may be predicted using only the marginal probabilities.

Problem

What is the null hypothesis for the chi-square test for equality of proportions?

Solution

The null hypothesis is that two or more samples drawn from different populations have the same distribution on the variable(s) of interest.

Problem

What is an appropriate statistic to measure the relationship between the two independent variables displayed in Table 5-21? What is the value of that statistic, and what conclusion would you draw from it?

Table 5-21. Two independent variables

	D+	D−
E+	25	10
E−	2	5

Solution

Because this is a 2×2 table and two of the cells have expected values of less than five (cells c and d), Fisher's Exact Test should be used. The value is 0.077 (obtained using computer software), which does not provide sufficient evidence to reject the null hypothesis of no relationship between E and D.

Problem

What are the expected values for the cells in Table 5-22? What is the value of the chi-square statistic? What conclusion would you draw about the relationship between exposure and disease, given this data?

Table 5-22. Calculating expected values

	D+	D−
E+	25	30
E−	15	5

Solution

The expected values are given in Table 5-23.

Table 5-23. Expected values: solution

	D+	D−
E+	29.3	25.7
E−	10.7	9.3

Chi-square (1) = 5.144, $p = 0.023$. This is sufficient evidence to reject the null hypothesis that exposure and disease are unrelated. We draw the same conclusion by using the chi-square table (Figure D-11) in Appendix D: 5.144 exceeds the 0.025 critical value (5.024) for a single-tailed chi-square test with one degree of freedom, indicating that we should reject the null hypothesis if $\alpha = 0.05$.

Problem

Table 5-24 represents political affiliations of married couples. Compute the appropriate statistic to see whether the affiliations of husbands and wives are independent of those of their spouses.

Table 5-24. Political affiliations of husbands and wives

		Wife	
		Republican	Democrat
Husband	Republican	20	30
	Democrat	20	20

Solution

McNemar's test is appropriate because the data comes from related pairs. The calculations are shown in Figure 5-29. The value of McNemar's chi-square is 2.00, which is not above the critical value for chi-square with one degree of freedom, at alpha = 0.05, so we do not have evidence sufficient to reject the null hypothesis that the political affiliations of spouses are independent of the affiliation of the other spouse.

$$\chi^2 = \frac{(30 - 20)^2}{30 + 20} = \frac{100}{50} = 2.00$$

Figure 5-29. Calculating McNemar's test

Problem

Which of Kendall's tau statistics would be appropriate for the data in Table 5-25?

Table 5-25. Educational level and job satisfaction

		Satisfaction with job		
		Dissatisfied	Neutral	Satisfied
Educational Level	<HS	45	20	10
	HS grad	15	15	20
	Some college	30	10	25
	College grad or higher	10	15	30

Solution

Kendall's tau-c should be used because the table is not square (it has four rows and three columns).

Problem

What is the argument against analyzing Likert and similar attitude scales as interval data?

Solution

There is no natural metric for constructs such as attitudes and opinions. We can devise scales that are ordinal (the responses can be ranked in order of strength of agreement, for instance) to measure such constructs, but it is impossible to determine whether the intervals among points on such scales are equally spaced. Therefore, data collected using Likert and similar types of scales should be analyzed at the ordinal or categorical level rather than at the interval or ratio level.

Problem

In what circumstance would you compute the Cramer's V statistic?

Solution

Cramer's V is an extension of the phi statistic and should be calculated to determine the strength of association between two categorical variables that have more than two levels. For binary variables, Cramer's V is equivalent to phi.

Problem

You read about a national poll stating that 30% of university students are dissatisfied with their appearance. You wonder whether the proportion at your local university (enrollment 20,000 students) is the same, so you draw a random sample of 150 students and find that 30 report being dissatisfied with their appearance. Conduct the appropriate test to see whether the proportion of students at your university differs significantly from the national result.

Solution

This question calls for a one-sample Z-statistic with a two-tailed test (because you are interested in whether the proportion at your school differs from the national figure in either direction). The test statistic is shown in Figure 5-30.

$$Z = \frac{0.30 - 0.20}{\sqrt{\dfrac{0.30\,(1 - 0.30\,)}{150}}} = \frac{0.10}{0.037} = 2.70$$

Figure 5-30. Calculating the one-sample Z-statistic for a proportion

Using the standard of alpha = 0.05 and a two-tailed test, the critical Z-value is 1.96 (as you can find using Figure D-3 in Appendix D). The Z-value from your sample is more extreme than this, so you reject the null hypothesis that the proportion of students dissatisfied with their appearance at your school is the same as at the national level.

Simpson's Paradox

Simpson's paradox is a circumstance in which the direction of an association reverses when data from several groups is combined. This paradox is well known among baseball fans. For instance, it is possible for player A to have a higher batting average (proportion of hits) than player B in each of two years, yet player A may have a lower batting average than player B when data from the two years are combined. Consider the example in Table 5-26.

Player B had a higher batting average each year yet, over both years combined, a lower average. This phenomenon occurs due to the different number of cases observed for each player in each year.

Table 5-26. Simpson's paradox in baseball

	2000			2001			Combined		
Player	Hits	At-bats	Average	Hits	At-bats	Average	Hits	At-bats	Average
A	10	50	0.200	200	600	0.333	210	650	0.323
B	85	400	0.213	50	145	0.345	135	545	0.248

Simpson's paradox was at the root of a controversy about gender discrimination in university admissions a few years ago. A lawsuit filed against the University of California was denied when it was shown that apparent gender discrimination (a lower percentage of women than men admitted overall to the university) could be explained by the fact that admissions were determined on a department-by-department basis and that most women applied to departments in which the percentage of applicants accepted was low, whereas most men applied to departments in which the percentage of applicants accepted was higher. In fact, in most departments, a slightly lower percentage of men than women were accepted, but this distinction was reversed when admissions data from all departments was combined.

Simpson's paradox is also apparent in the evaluation of medical treatments when treatment A might be superior to treatment B in each of two samples yet inferior when the samples are combined. Some statisticians argue that circumstances such as this should not be called a paradox at all because to do so implies that there is a causal relationship between the two variables.

Table 5-27. Summary of tests covered in this chapter

Name of test	Type of data	What is being tested
Percent agreement	One categorical variable, two raters	How well do the raters agree?
Cohen's kappa	One categorical variable, two raters	How well do the raters agree after correction for chance?
Chi-square test of independence	Two or more categorical variables	Are the variables independent?
Chi-square test of equality of proportions	One categorical variable, samples from two or more populations	Does the variable have the same distribution in the populations from which the samples were drawn?
Chi-square test for goodness of fit	One categorical variable, a hypothesized distribution for it	Does the variable have the hypothesized distribution in the population from which the sample was drawn?
Fisher's Exact Test	Two categorical variables; data may be sparse	Are the variables independent?
McNemar's test	One dichotomous variable, measured on matched pairs	Are there changes in proportions among the matched pairs?
Large-sample Z test for a proportion	Dichotomous variable, one sample, large sample ($np \geq 5$, $n(1 - p) \geq 5$)	Does a population proportion differ from a specified proportion?

Name of test	Type of data	What is being tested
Large-sample Z test for the difference in two proportions	Dichotomous variable, two samples, both large samples ($np \geq 5$, $n(1-p) \geq 5$)	Does the proportion of some variable differ in the populations from which the samples were drawn?
Phi	Two binary variables	How strongly are the variables associated?
Cramer's V	Two categorical variables	How strongly are the variables associated?
Point-biserial correlation	One dichotomous and one continuous variable	How strongly are the variables associated?
Spearman's rho	Two ranked variables	How strongly are the variables associated?
Goodman and Kruskal's gamma	Two ordinal variables	How strongly are the variables associated (based on concordant and discordant pairs)?
Kendall's tau-a	Two ordinal variables	How strongly are the variables associated (based on concordant and discordant pairs)?
Kendall's tau-b	Two ordinal variables	How strongly are the variables associated (based on concordant and discordant pairs; corrected for ties)?
Kendall's tau-c	Two ordinal variables	How strongly are the variables associated (based on concordant and discordant pairs; may be used for nonsquare tables)?

6

The t-Test

The t distribution was introduced by a chemist working in quality control for the Guinness brewery in Ireland, William Sealy Gosset. Gosset described the t distribution in an article under the pseudonym Student; hence, the t distribution is sometimes called the Student's t distribution and the t-test the Student's t-test. There are three major types of t-test, all of which are concerned with testing the difference between means and involve comparing a test statistic to the t distribution to determine the probability of that statistic if the study's null hypothesis is true. The one-way analysis of variance (ANOVA) procedure with two groups is mathematically equivalent to the t-test, but the t-test is used so commonly that it deserves its own chapter. In addition, understanding the logic of the t-test should make it easier to follow the logic of more complex ANOVA designs.

The t Distribution

If you're not familiar with inferential statistics, it might be wise to review Chapter 3 before continuing with this chapter. One basis for inferential statistics is the use of known probability distributions to make inferences about real data sets. In Chapter 3, we discussed the normal and binomial distributions; in this chapter, we discuss the t distribution. Like the normal distribution, the t distribution is continuous and symmetrical. Unlike the normal distribution, the shape of the t distribution depends on the degrees of freedom for a sample, meaning the number of values that are allowed to vary. For the t distribution, the main influence on degrees of freedom is the sample size, and tests on larger sample sizes generally have more degrees of freedom than smaller sample sizes. Calculation of the degree of freedom for the different types of t-test will be discussed in the section covering each test.

As noted, Gosset developed the t distribution for practical reasons. While employed in quality assurance for the Guinness brewery, he was trying to solve the problem of making inferences from samples of limited size. Gosset's key observation was the influence of sample size in determining the probability that the mean of the population lies within a given distance of the mean of the sample. There are two main reasons for using the *t* distribution to test differences in means: when we are working with small samples from a population we believe has an approximately normal distribution and when we do not know the standard deviation of a population and need to use the standard deviation of the sample as a substitute for the population standard deviation. If we are working with a sample size too small to invoke the central limit theorem and we do not believe that the population from which our sample was drawn has an approximately normal distribution, we need to use a nonparametric method (discussed in Chapter 13) instead.

As Figure 6-1 shows, the *t* distribution looks quite similar to the normal, the main difference being the thicker tails that mean extreme values are more probable in the *t* distribution than in the normal. As sample size (hence degrees of freedom) increases, the *t* distribution comes to look more like the normal distribution.

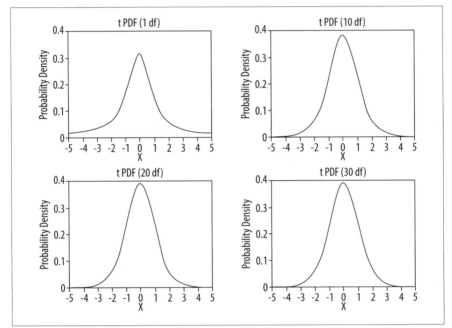

Figure 6-1. Four t distributions

Gosset found that when samples are drawn from a normally distributed population and the sample standard deviation is used to estimate the population variance, the distribution of sample means for some variable *x* drawn from this population can be described by the formula in Figure 6-2.

$$t = \frac{\bar{x} - \mu}{\frac{s}{\sqrt{n}}}$$

Figure 6-2. Formula for the t distribution

In this formula, $\bar{x}$ is the sample mean,

μ is the population mean,

s is the population standard deviation, and

n is the sample size.

This formula is similar to the formula for the Z-statistic presented in Chapter 3; the only difference is that for the t-statistic, we use the sample standard deviation, whereas for the Z-statistic, we use the population standard deviation.

Appendix D includes a table (Figure D-7) with the upper critical values of the t distribution for different degrees of freedom; we say "upper critical values" because the t distribution is symmetric, so there is no need to print the lower critical values. (They would be the negatives of the numbers in this table.) Because only positive values are included in the table, to find the critical value for a two-tailed t-test, we use the column for the α value, which is half of what we want. For a two-tailed test with $\alpha = 0.05$, we use the column for 0.25. Not surprisingly, as sample size increases, the critical values for the t distribution approach those for the standard normal distribution. For instance, we know (from Figure D-7 in Appendix D as well as the discussion in Chapter 3) that the upper critical value in the standard normal distribution for a two-tailed test with alpha = 0.05 is 1.96. For a two-tailed test using the t distribution, with alpha = 0.05, the upper critical value depends on the degrees of freedom (df). For 1 df, the upper critical value is 12.706; for 10 df, the upper critical value is 2.228; for 30 df, 2.042; for 50 df, 2.009; for 100 df, 1.984; and for infinite degrees of freedom, 1.96.

William Sealy Gosset

William Sealy Gosset is often considered the first industrial statistician of modern times. Although his work was motivated by the pragmatic concerns of his employer (Arthur Guinness, Son & Co, the brewers), his applied work gave rise to a set of major inferential statistical tests based on the distribution that he identified. After systematically working through related techniques such as correlation to solve problems at his workplace, he identified the fundamental constraint of small samples and the limitation of techniques that assume large numbers of observations and/or experiments to determine reliability. Later techniques, such as the analysis of variance developed by R. A. Fisher, relied heavily on Gosset's exposition of the t distribution. Gosset's life and work provide excellent examples of the interaction between applied science and theoretical development.

The t-Test

The One-Sample t-Test

One way the t-test is used is to compare the mean of a sample to a population with a known mean. The null hypothesis is usually that there is no significant difference between the mean in the population from which your sample was drawn and the mean of the known population. For instance, you might be interested in the effects of lead exposure on intelligence in children. You know that for 5-year-old children in the United States as a whole, the average score on a particular intelligence test is 100. You have a sample of 15 5-year-old children who have been exposed to lead, and you want to know whether this exposure has affected their intelligence as measured by this particular test. You also know that intelligence scores generally assume a normal distribution in this population. Your null hypothesis is that there is no difference in the intelligence scores of the lead-exposed group and the population as a whole, and you will conduct a two-tailed test with alpha = 0.05.

The formula for the one-sample t-test is shown in Figure 6-3.

$$t = \frac{\bar{x} - \mu_0}{\frac{s}{\sqrt{n}}}$$

Figure 6-3. Formula for the one-sample t-test

In this formula, $\bar{x}$ is the mean of your sample,
μ_0 is the reference mean (in this case, the average intelligence score for all 5-year-olds in the United States),
s is the standard deviation of your sample, and
n is the sample size.

The formulas to calculate the mean and standard deviation of a sample are shown in Figures 6-4 and 6-5.

$$\bar{x} = \frac{\sum_{i=1}^{n} x_i}{n}$$

Figure 6-4. Calculating the sample mean

$$s = \sqrt{\frac{\sum_{i=1}^{n} (x_i - \bar{x})^2}{n-1}}$$

Figure 6-5. Calculating the sample standard deviation

In this formula, x_i is a single x value,
$\bar{x}$ is the sample mean,
s is the sample standard deviation, and
n is the sample size.

There is also a computational formula for the sample standard deviation, which is mathematically identical to the formula in Figure 6-4 but less laborious to calculate if you must do the computations by hand, as shown in Figure 6-6.

$$s = \sqrt{\dfrac{\displaystyle\sum_{i=1}^{n} x_i^2 - \dfrac{\left(\displaystyle\sum_{i=1}^{n} x_i\right)^2}{n}}{n-1}}$$

Figure 6-6. Computational formula for the sample standard deviation

If you want to practice using these formulas, a fully solved example is included at the end of this chapter. For this example, assume the sample mean is 90, the standard deviation is 10, and the sample size is 15, and use this information to calculate the *t*-statistic, as shown in Figure 6-7.

$$t = \dfrac{90 - 100}{\dfrac{10}{\sqrt{15}}} = -3.87$$

Figure 6-7. Calculating the one-sample t-test

The degrees of freedom for the one sample *t*-test is *n*–1; in this example, *df* = 15 – 1 = 14. From the table of upper critical values for the *t* distribution (Figure D-7 in Appendix D), we see that the upper critical value for a two-tailed *t*-test with 14 degrees and alpha = 0.05 is 2.145. Because the absolute value of the *t*-statistic for our data exceeds the upper critical value (|–3.87| > 2.145), we reject the null hypothesis that the lead-exposed children have the same average intelligence test scores as children their age in the entire population. Because the difference in means and the *t*-statistic are negative, we can also say that the mean intelligence score is lower for children exposed to lead as compared to the average for children of the same age in the population as a whole.

Confidence Interval for the One-Sample t-Test

We often want to report a confidence interval as well as a test statistic and significance test. The confidence interval is a range of values around the mean, with the following meaning: if we drew an infinite number of samples of the same size from the same population, *x*% of the time the true population mean would be included in the confidence interval calculated from the samples. If we compute a 95% confidence interval (the most common type), *x* = 95, so we can say that 95% of the confidence intervals calculated from an infinite number of samples of the same size, drawn from the same population, can be expected to contain the true population mean. More generally, a confidence interval gives us information about the precision of a point estimate such as the sample mean. A wide confidence interval tells us that if we had drawn a different sample, we might get a quite different sample mean,

The t-Test

whereas a narrow confidence interval suggests that if we drew a different sample, the sample mean would probably be fairly close to that from the sample we did draw.

The formula to compute a two-tailed confidence interval (CI) for the mean for the one-sample *t*-test is shown in Figure 6-8.

$$CI_{1-\alpha} = \bar{x} \pm \left(t_{\frac{\alpha}{2}, df} \right) \left(\frac{s}{\sqrt{n}} \right)$$

Figure 6-8. Confidence interval formula for a one-sample t-test

In our example, $\alpha = 0.05$,
$\bar{x} = 90$,
$df = n - 1 = 14$,
$s = 10$,
$t_{0.025,14} = 2.145$ (from the table in Figure D-7 in Appendix D), and
$n = 15$.

Putting these values in the formula gives us the result shown in Figure 6-9.

$$CI_{0.95} = 90 \pm (2.145)\left(\frac{10}{\sqrt{15}} \right) = 90 \pm 5.54 = (84.46, 95.54)$$

Figure 6-9. Calculating the confidence interval formula for a one-sample t-test

The 95%, two-tailed confidence interval for our estimate of the population mean is (84.46, 95.54). Note that these numbers are sometimes called the lower boundary and upper boundary of the confidence interval; in this example, the lower boundary is 84.46, and the upper boundary is 95.54.

If you want to calculate a one-sided confidence interval, change the ± to either plus or minus, as appropriate, and use the upper critical value from the *t* table for α rather than $\alpha/2$. To calculate a confidence interval for a different size, use the appropriate upper critical value from the *t* table. For instance, for a one-sided, 90% confidence interval with 20 *df*, the upper critical value for *t* is 1.325.

The Independent Samples t-Test

The *t-test for independent samples*, also known as the *two-sample t-test*, compares the means of two samples. The purpose of this test is to determine whether the means of the populations from which the samples were drawn are the same. The subjects in the two samples are assumed to be unrelated (no one is tested twice, no sibling pairs, etc.) and to have been independently selected from their populations. In addition, we assume that the populations from which the samples were selected have an approximately normal distribution, unless the samples are large enough to invoke the central limit theorem, and that the populations have approximately equal variance. The independent samples t-test is commonly used in many professional fields and is usually calculated using computer software that can include both a test of the assumption of equal variance in the populations (e.g., Levene's test, the Brown-Forsythe test, or Bartlett's test) and a statistical fix if this assumption is not met.

The formula to calculate the independent samples t-test is shown in Figure 6-10.

$$t = \frac{(\bar{x}_1 - \bar{x}_2) - (\mu_1 - \mu_2)}{\sqrt{s_p^2\left(\dfrac{1}{n_1} + \dfrac{1}{n_2}\right)}}$$

where

$$s_p^2 = \frac{(n_1 - 1)s_1^2 + (n_2 - 1)s_2^2}{n_1 + n_2 - 2}$$

Figure 6-10. Formula for the independent samples t-test

In this formula, $\bar{x}_1$ and $\bar{x}_2$ are the means of the two samples,
μ_1 and μ_2 are the means of the two populations,
s_p^2 is the pooled variance,
n_1 and n_2 are the two sample sizes, and
s_1^2 and s_2^2 are the variances of the two samples.

Note that often the null hypothesis for an independent samples t-test is that the difference between the population means is 0, in which case the $(\mu_1 - \mu_2)$ term can be dropped from the equation.

The degrees of freedom for the two-sample t-test is $(n_1 + n_2 - 2)$, that is, 2 fewer than the number of cases when both samples are combined.

This is a complex formula, but it's worth stepping back and looking at the general form of the equation before getting caught up in the details. The formula for the independent samples t-test is similar to the one-sample t-test formula in that the numerator is a difference between means, and the denominator is a measure of variability incorporating both the variability observed within the samples and the size of the samples. The test-statistic for the paired t-test will also follow this basic form, although differing in some details.

Let's look at an example. An age-old physical performance question is whether male football players (soccer players, for American readers) are fitter than male ballet dancers, so a sports physiologist organizes a study in partnership with a local hospital research team to answer the question. The two groups are independent populations because no football player is also a ballet dancer. Two lists of ballet dancers and football players that are maintained by their respective professional associations are also located all over the country, and study members are randomly selected from each group. Because ballet dancers and football players are very busy, only 10 study members from each group can be recruited. All participants are tested on a range of human performance tasks, including walking, running, and stepping, and corresponding physiological measures associated with fitness, including heart-rate variability, pulse-wave velocity, and so on. These measures are then combined to form a single fitness score ranging from 0 to 100. Experience using this method of evaluation has demonstrated that fitness scores calculated using the algorithm used in this study are approximately normally distributed in the population.

The participants are all tested in the same facility at the same time of day, and their responses are assessed and combined using the same clinicians. The results for the two groups are shown in Table 6-1.

Table 6-1. Fitness results for football players and ballet dancers

Ballet dancers	Football players
89.2	79.3
78.2	78.3
89.3	85.3
88.3	79.3
87.3	88.9
90.1	91.2
95.2	87.2
94.3	89.2
78.3	93.3
89.3	79.9

We will use an alpha value of 0.05 for this study. You can calculate the t-statistic entirely by hand, using the formulas for calculating the standard deviation presented earlier in this chapter (and remembering that variance is the square of standard deviation). To speed things along, we calculated the necessary quantities for you, calling the ballet dancers sample 1 and the football players sample 2:

$$\bar{x}_1 = 87.95$$
$$\bar{x}_2 = 85.19$$
$$s_1^2 = 32.38$$
$$s_2^2 = 31.18$$

If we were using a software program, we would check the assumption of equal variance using Levene's test (or one of the alternatives; this is discussed more in the section on the unequal means t-test that follows), which tests the null hypothesis that the two populations do have equal variance. (If we fail to reject the null hypothesis of equal, we can continue with the t-test.)

The pooled sample variance is calculated as shown in Figure 6-11.

$$s_p^2 = \frac{(10-1)32.38 + (10-1)31.18}{10+10-2} = 31.78$$

Figure 6-11. Calculating the pooled variance

The degrees of freedom are $df = n_1 + n_2 - 2 = 18$. Our null hypothesis is that the mean fitness for the two groups is equal, that is, $\mu_1 - \mu_2 = 0$. To test this null hypothesis, we compute the t-statistic, as shown in Figure 6-12.

$$t = \frac{(87.95 - 85.19) - (0)}{\sqrt{37.18\left(\frac{1}{10} + \frac{1}{10}\right)}} = \frac{2.76}{2.73} = 1.01$$

Figure 6-12. Calculating the t-statistic

From Figure D-7 in Appendix D, we see that the upper critical value for a two-tailed t-test with alpha = 0.05 and 18 df is 2.101. The absolute value of our t-value is below that value (i.e., it is closer to 0), so we fail to reject the null hypothesis and conclude that this study did not provide any evidence of a difference in fitness between footballers and ballet dancers.

Confidence Interval for the Independent Samples t-Test

To calculate the two-sided confidence interval for the independent samples t-test, we use the formula shown in Figure 6-13.

$$CI_{1-\alpha} = \left(\overline{x}_1 - \overline{x}_2\right) \pm \left(t_{\frac{\alpha}{2},df}\right)\left(\sqrt{s_p^2\left(\frac{1}{n_1} + \frac{1}{n_2}\right)}\right)$$

where

$$s_p^2 = \frac{\left(n_1 - 1\right)s_1^2 + \left(n_2 - 1\right)s_2^2}{n_1 + n_2 - 2}$$

Figure 6-13. Formula for the confidence interval for the independent samples t-test

There are several points worth noting about this formula:

- It is actually a confidence interval for the difference in the means of the two populations.
- For the value of:

$$t\frac{\alpha}{2}, _{df}$$

we use the upper critical t-value for the df, and half the specified alpha level, from a t-table such as the one in Figure D-7 in Appendix D.

- If this were a one-sided confidence interval, we would use the upper critical t-value for α rather than $\alpha/2$ and would use either plus or minus rather than $\pm$, depending on the direction of the confidence interval.

- The formula includes the denominator of the independent samples t-test as previously calculated.

For our data, we will use alpha = 0.05 and calculate a 95%, two-tailed confidence interval; the result is shown in Figure 6-14.

$$CI_{1-\alpha} = 2.76 \pm (2.10)(2.73) = (-2.97, 8.49)$$

Figure 6-14. Calculating a 95%, two-sided confidence interval for the independent samples t-test

Note that this confidence interval includes 0, which is our null value (the value we posited for the difference in means in our null hypothesis); this result is expected because for this data set, we did not find significant results and thus did not reject the null hypothesis.

Repeated Measures t-Test

With the *repeated measures t-test*, also known as the *related samples t-test*, the *matched samples t-test*, or the *dependent samples t-test*, the units that make up the two samples are not independent but are related in some way. Sometimes, the data in the samples are measurements taken twice from the same people, such as blood pressure before and after taking a prescription drug. Sometimes, the data is collected from people related by affiliation or genetics, such as husbands and wives or siblings. Sometimes, the data is collected from samples of different people who have been closely matched on key characteristics so that they are considered too similar to be treated as independent samples. The measurements are considered as pairs, so the two samples must be of the same size.

The formula to calculate the t-statistic for the repeated measures t-test is based on the difference scores as calculated from each pair of samples. The test statistic is shown in Figure 6-15.

$$t = \frac{\bar{d} - (\mu_1 - \mu_2)}{\frac{s_d}{\sqrt{n}}}$$

Figure 6-15. Formula for the repeated measures t-test

In this formula, $\bar{d}$ = the mean of the difference scores,
μ_1 and μ_2 are the means of the two populations,
s_d is the standard deviation of the difference scores, and
n is the number of pairs.

The null hypothesis for the repeated measures t-test is usually that the mean of the difference scores ($\bar{d}$) is 0, whereas the alternative hypothesis is that this mean is not 0. As with the two-sample t-test, often the quantity ($\mu_1 - \mu_2$) is hypothesized to be 0 and, in that case, may be dropped from the equation.

A difference score is simply the difference between the two values in a pair of measurements, such as the blood pressure after treatment for one person minus the blood pressure before treatment. We calculate a difference score for every pair and then calculate the mean and standard deviation of the difference scores to calculate the t-statistic. Note that n in the context of the repeated measures t-test refers to the number of *pairs*, not the number of measurements. The degrees of freedom is $df = n - 1$.

This might be clearer after working through an example. Suppose we want to test the efficacy of a diet and exercise program in lowering total cholesterol levels in middle-aged men. We decide on a matched pairs t-test because we will test the cholesterol of each subject twice, before they begin the program and again after they complete it. This is sometimes referred to as "using subjects as their own controls" because by measuring the same subjects twice, we hope to remove or minimize the influence of all individual differences other than the one we are interested in, which is how the subject's cholesterol levels respond to the diet and exercise program. We believe that changes in the response to a program such as ours have an approximately normal distribution in the population, and we have only 10 subjects, so the matched pairs t-test is an appropriate measure. Data from this experiment is shown in Table 6-2.

Table 6-2. Cholesterol before and after an exercise and diet program

Before	After	Difference (d) (after-before)
220	200	−20
240	210	−30
225	210	−15
180	170	−10
210	220	10
190	180	−10
195	190	−5
200	190	−10
210	220	10
240	210	−30

Clearly, most subjects had lower cholesterol after completing this program, but was the difference statistically significant? To find out, we compute the repeated measures t-statistic, using the following values calculated from the sample:

$\bar{d} = -11$

$s_d = 13.9$

We will conduct a two-tailed, repeated measures t-test with alpha = 0.05. Our null hypothesis is that the population means are equal, that is, that their difference is 0, so the t-statistic for our data is shown in Figure 6-16.

$$t = \frac{-11-0}{\frac{13.9}{\sqrt{10}}} = -2.50$$

Figure 6-16. Calculating the repeated measures t-test

Because we have 10 pairs, we have 9 degrees of freedom ($df = n - 1$). Using the table of upper critical values for the t distribution (Figure D-7 in Appendix D), we determine that the critical value for a two-tailed t-test with 9 df and alpha = 0.05 is 2.262. The absolute value of our t-statistic exceeds this value, so we reject the null hypothesis and conclude that the exercise and diet program has an effect on total cholesterol. Because the mean difference and the t-statistic are negative, we can also say that the program lowered the total cholesterol of the participants.

You might wonder what the two populations are in this example. The sample measurements drawn before the program began are considered to have been drawn from the general population of middle-aged men, and the sample measurements after completion of the program are considered to have been drawn from the population of middle-aged men who have completed the exercise and diet program. Granted, the second population is theoretical because this is a new program, so what we are really doing is hypothesizing about the changes that would occur in the total cholesterol levels of the first population if it followed the diet and exercise program.

Confidence Interval for the Repeated Measures t-Test

To calculate the confidence interval for the repeated measures t-test, use the formula shown in Figure 6-17.

$$CI_{1-\alpha} = \bar{d} \pm \left(t_{\frac{\alpha}{2}, df} \right) \left(\frac{s_d}{\sqrt{n}} \right)$$

Figure 6-17. Formula for the confidence interval for the repeated measures t-test

For the data in our example, the calculation is shown in Figure 6-18.

$$CI_{0.95} = -11 \pm (2.262)\left(\frac{13.9}{\sqrt{10}}\right) = (-20.94, -1.06)$$

Figure 6-18. Calculating the two-sided, 95% confidence interval for the repeated measures t-test

Note that this confidence interval does not include the null value of 0; this is to be expected because we found significant results with the t-test, that is, we rejected the null hypothesis that the mean difference was 0.

Unequal Variance t-Test

One of the assumptions of the independent samples t-test is that the two populations from which the samples were drawn have approximately equal variance; this is also known as the assumption of homogeneity of variance or, simply, the assumption of homogeneity. If this assumption is not met and the population variances are in fact heterogeneous, the risk of both Type I and Type II errors is increased. This is because the sample variances are pooled in the independent samples t-test, and the results of the test would be seriously distorted if they were not drawn from populations with approximately equal variance. The problem of hypothesis testing between two independent samples in which variances are known to be unequal is the *Behrens-Fisher problem*, and there have been several proposed solutions.

If you are using statistical software to calculate an independent samples t-test, chances are it includes algorithms to calculate one or more tests of the homogeneity of variance. Examples of this type of test include *Levene's test*, the *Brown-Forsythe test*, and the *Bartlett test*. Levene's test is based on the mean, whereas the Brown-Forsythe test is an extension of Levene's test, which uses a trimmed mean or the median. The Bartlett test is the most sensitive to departures from normality (which is not the same thing as unequal variances), so it should be used only if you feel secure that the populations from which your samples were drawn are approximately normally distributed. The important point, however, is to use one of these tests if it is available to you to check whether the assumption of homogeneity is met. The technical details of the different tests, with references to the professional literature regarding them, are available from the *Engineering Statistics Handbook* of the National Institute for Standards and Testing, a public domain document available online (*http://itl.nist.gov/div898/handbook/index.htm*).

If the homogeneity assumption is not met, you can use one of the nonparametric substitutes for the independent samples t-test (discussed in Chapter 13) or use the unequal variance t-test, also known as *Welch's t-test*. Choosing one of these alternatives is particularly wise when you are working with small sample sizes or when you wish to be conservative in drawing inferences. Welch's t-test uses a slightly different formula to calculate the t-statistic and a complex formula to calculate the degrees of freedom.

Welch's t-test uses the formula shown in Figure 6-19 to calculate the t-statistic.

The t-Test

$$t = \frac{\bar{x}_1 - \bar{x}_2}{\sqrt{\dfrac{s_1^2}{n_1} + \dfrac{s_2^2}{n_2}}}$$

Figure 6-19. Formula for Welch's t-test

In this formula, $\bar{x}_1$ and $\bar{x}_2$ are the sample means, $s_1{}^2$ and $s_2{}^2$ are the sample variances, and n_1 and n_2 are the sample sizes.

Note that the formula for Welch's *t*-test does not use pooled variance. The real work comes when calculating the degrees of freedom for Welch's *t*-test, as shown in Figure 6-20.

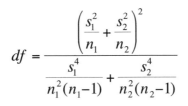

Figure 6-20. Formula for the degrees of freedom for Welch's t-test

Having calculated the *t*-statistic and degrees of freedom, you proceed as you would with any other *t*-statistic, comparing your result with a table of critical values for the *t*-distribution (such as Figure D-7 in Appendix D) and making your decision accordingly.

Exercises

Although you can use a statistical package such as Minitab, SPSS, STATA, or SAS to compute *t*-tests and their significance levels, working through some examples by hand will make the underlying concepts easier to understand. Furthermore, if you consider scenarios from work or school that involve small samples, you might begin to develop a sense of how to approach them inferentially by using *t*-tests. If you understand the details of computing a *t*-test by hand, then using a statistical package will be much easier for you. Also, the output generated by many statistical packages is confusing if you don't understand what you should be looking for, so having worked through some examples by hand can make it easier to spot the information you need in a sea of output.

Problem

A factory manager is disturbed by the number of accidents in the plant she manages, so she institutes a safety program that includes worker education, better lighting in the plant, and incentives for units who improve their safety record. The average number of accidents per week before instituting the safety program was 5, and the distribution was approximately normal. She wants to know whether this has changed since the program began. She draws a sample of 15 post-program weeks and uses administrative records to determine the number of accidents that occurred during each sample week. This data is displayed in Table 6-3. What test should she use to determine whether the average number of accidents per week has changed since the safety program began? What is the test statistic, and what can you conclude from it about the effectiveness of the program?

Table 6-3. Number of accidents per week

Week ID	1	2	3	4	5	6	7	8	9	10	11	12	13	14	15
Number of accidents	5	6	6	4	5	3	2	7	5	4	1	0	3	2	5

Solution

She should compute the one-sample *t*-test, comparing the mean accidents per week as calculated from her sample of 15 post-safety-program weeks, with the population mean before the program. She should use a two-tailed test because it is possible that the accident rate increased after the safety program began, and she would surely want to detect this if it happened. She will therefore conduct a two-tailed, one-sample *t*-test with the null hypothesis that there is no difference in the means of the sample or the population, and she will use the alpha = 0.05 standard.

Here is the information needed to calculate this statistic:

μ_0= 5 (given)
n = 15 (given)
$\bar{x}$= 3.87
s = 2.00

First, we calculate the sample mean and sample standard deviation as shown in Figures 6-21 and 6-22.

$$\bar{x} = \frac{\sum\limits_{i=1}^{n} x_i}{n} = \frac{58}{15} = 3.87$$

Figure 6-21. Calculating the sample mean

$$S = \sqrt{\frac{\sum_{i=1}^{n} x_i^2 - \frac{\left(\sum_{i=1}^{n} x_i\right)^2}{n}}{n-1}} = \sqrt{\frac{280 - \frac{58^2}{15}}{14}} = 2.00$$

Figure 6-22. Calculating the sample standard deviation

Then we plug these numbers into the formula for the one-sample t-statistic, as shown in Figure 6-23.

$$t = \frac{\bar{x} - \mu_0}{\frac{s}{\sqrt{n}}} = \frac{3.87 - 5.00}{\frac{2.00}{\sqrt{15}}} = \frac{-1.13}{0.52} = -2.17$$

Figure 6-23. Calculating the one-sample t-test

We have 14 degrees of freedom ($df = n - 1$). According to Figure D-7 in Appendix D, the upper critical value for a two-tailed test with 14 df if alpha = 0.05 is 2.145. The absolute value of our t-statistic exceeds the critical value, so we reject the null hypothesis that there was no difference in the number of accidents per week after the safety program began. Because the difference between the sample mean and population mean is negative, as is the t-statistic, we can also conclude that the program lowered the accident rate.

Problem

What is the 95%, two-tailed confidence interval for our estimate of the population mean, given these sample results?

Solution

We calculate the 95%, two-tailed confidence as shown in Figure 6-24.

$$CI_{1-\alpha} = \bar{x} \pm \left(t_{\frac{\alpha}{2}, df}\right)\left(\frac{s}{\sqrt{n}}\right) = 3.87 \pm (2.145)\left(\frac{2.00}{\sqrt{15}}\right) = (2.76, 4.98)$$

Figure 6-24. Calculating a 95%, two-sided confidence interval for the one-sample t-test

Note that the upper critical value, 4.97, is very close to the population mean. This is to be expected because our sample t-statistic barely exceeded the critical value for alpha = 0.05; that is, we barely achieved the standard for rejecting the null hypothesis that the difference between the sample and population means is 0.

Problem

What is the 90%, two-tailed confidence interval for our estimate of the population mean, given these sample results?

Solution

To calculate a 90% confidence interval, all we need to change from the formula used for the previous question is the upper critical *t*-value. Using the table in Figure D-7 in Appendix D, we see that the value for alpha = 0.10, two-tailed, with *df* = 14 is 1.761. Plugging this into the formula, we get the result shown in Figure 6-25.

$$CI_{1-\alpha} = \bar{x} \pm \left(t_{\frac{\alpha}{2}, df}\right)\left(\frac{s}{\sqrt{n}}\right) = 3.87 \pm (1.761)\left(\frac{2.00}{\sqrt{15}}\right) = (2.96, 4.78)$$

Figure 6-25. Calculating a 90%, two-sided confidence interval for the one-sample t-test

Note that the 90% confidence interval is narrower than the 95% confidence interval for the same sample data. This is to be expected because of the smaller critical *t*-value used for the 90% confidence interval. To put it another way, the 90% confidence interval includes less of the total probability than the 95% confidence interval, so it's not surprising that it is narrower.

Table 6-4. The different t-tests and their uses

t-test	Data type	Question being answered
One-sample *t*-test	One sample, continuous data, approximate normality	Does the sample come from a population with a specified mean?
Two-sample *t*-test	Two independent samples, continuous data, approximate normality, approximately equal variance	Do the two samples come from populations with equal means?
Repeated measures *t*-test	Two related samples, equal sample sizes, continuous data, approximate normality of difference scores	Do the two samples come from populations with equal means?
Unequal variance *t*-test	Two independent samples, continuous data, approximate normality	Do the two samples come from populations with equal means?

The t-Test

7

The Pearson Correlation Coefficient

The Pearson correlation coefficient is a *measure of linear association* between two interval- or ratio-level variables. Although there are other types of correlation (several are discussed in Chapter 5, including the Spearman rank-order correlation coefficient), the Pearson correlation coefficient is the most common, and often the label "Pearson" is dropped, and we simply speak of "correlation" or "the correlation coefficient." Unless otherwise specified in this book, "correlation" means the Pearson correlation coefficient. Correlations are often computed during the exploratory stage of a research project to see what kinds of relationships the different continuous variables have with each other, and often scatterplots (discussed in Chapter 4) are created to examine these relationships graphically. However, sometimes correlations are statistics of interest in their own right, and they can be tested for significance and reported as inferential statistics as well. Understanding the Pearson correlation coefficient is fundamental to understanding linear regression, so it's worth taking the time to learn this statistic and understand well what it tells you about the relationship between two variables. A key point about correlation is that it is a measure of an observed relationship but cannot by itself prove causation. Many variables in the real world have a strong correlation with each other, yet these relationships can be due to chance, to the influence of other variables, or to other causes not yet identified. Even if there is a causal relationship, the causality might be in the opposite direction of what we assume. For these reasons, even the strongest correlation is not in itself evidence of causality; instead, claims of causation must be established through experimental design (discussed in Chapter 18). In this chapter, we discuss the general meaning of association in the context of statistics and then examine the Pearson correlation coefficient in detail.

Association

Ordinary life is full of variables that appear to be associated with or related to each other, and explicating these relationships is a chief task of the sciences. There's nothing obscure or arcane about thinking about how variables relate to each other, however; people think in terms of associations all the time and often attribute causality to those associations. Parents who order their children to eat more vegetables and less junk food probably do so because they believe there is a relationship between diet and health, and athletes who put in long hours of practice at their sport are most likely doing so because they believe diligent training will lead to success. Sometimes these types of commonsense notions are supported by empirical research, sometimes not, but it seems to be a normal human tendency to take note when things seem to occur together and, often, to believe as well that one is causing the other. As scientists (or just people who understand statistics), we need to be in the habit of questioning whether an apparent association actually exists and, if it does exist, if it is truly causal.

Here are a few examples of conclusions that, although based in some way on observable data, are obviously false:

- There is a strong association between sales of ice cream and the number of deaths by drowning, so the reason must be that people are going in the water too soon after eating ice cream, thus getting cramps and drowning.

- There is a strong association between score on a vocabulary test and shoe size, so the explanation must be that tall people have bigger brains and hence can remember more words.

- There is a strong association between the number of storks in an area and the human birth rate in that area, so obviously storks really do deliver babies.

- A town mayor notes a strong correlation between local sports teams winning championships and ticker-tape parades and decides to hold more parades to improve the performance of the local teams.

Here are the real explanations:

- Both ice cream consumption and swimming are more common in the hotter months of the year, so the apparent relationship is due to the influence of a third variable, that of temperature (or season).

- The data was gathered on schoolchildren and was not controlled for age. We expect that older children will be taller (and have bigger feet) and have acquired larger vocabularies than younger children; hence, the observed association is due to the influence of a third variable, age.

- Storks are more common in rural areas, and birth rates are also higher in rural areas, so the association is due to the influence of a third variable, location.

- This is reversed causality—the parades are held after the championships are won, so the teams' successful seasons are the cause of the parades rather than the parades causing the teams to have good seasons.

It's worth noting that even if two variables have no logical reason at all to be associated, simply by chance they may show some association. This is particularly true in studies with large sample sizes in which a very slight association can be statistically significant, yet have no practical meaning. It's also worth noting that even among variables that are strongly related, such as smoking and lung cancer, there can be significant variation in that relationship among individuals. Some people smoke for years and never get sick, whereas some unfortunate individuals come down with lung cancer despite never having smoked in their lives.

Scatterplots

The *scatterplot* is a useful tool with which to explore the relationships between variables, and usually, creating scatterplots for pairs of continuous variables is part of the exploratory phase of working with a data set. A scatterplot is a graph of two continuous variables. If the research design specifies that one variable is independent and the other is dependent, the explanatory variable is graphed on the *x*-axis (horizontal) and the dependent variable on the *y*-axis (vertical); if no such relationship is specified, it doesn't matter which variable is graphed on which axis. Each member of a sample corresponds to one data point on the graph, described by a set of coordinates (*x*, *y*); if you ever plotted Cartesian coordinates in school, you are already familiar with this process. Scatterplots give you a sense of the overall relationship between the two variables, including direction (positive or negative), strength (strong or weak), and shape (linear, quadratic, etc.). Scatterplots are also a good way to get a general sense of the range of the data and to see whether there are any *outliers*, cases that don't seem to belong with the others.

One important reason for inspecting bivariate relationships (relationships between two variables) is that many common procedures assume that these relationships are linear, an assumption that might not be met with any particular pair of variables in any particular data set. *Linear* in this context means "arranged as a straight line," whereas any other relationship is considered *nonlinear*, although we can also apply a more specific description to nonlinear relationships, such as quadratic or exponential. We don't expect a real data set to perfectly fit any mathematically defined pattern, of course; if the data seems to cluster around a straight line, that's what we mean by a linear relationship.

We can also create a *scatterplot matrix*, which is a display of multiple scatterplots arranged so we can easily see the relationships among pairs of variables. Figure 7-1 displays a scatter plot matrix created by Lloyd Currie of the National Institute of Standards and Technology to inspect the relationships among four pollutants in a data sample: potassium, lead, iron, and sulfur oxide. The scatterplot for each pair of variables is located where the corresponding row and column intersect, so cell (1, 2) (first row, second column) shows the relationship between potassium and lead, cell (1, 3), the relationship between potassium and iron, and so on.

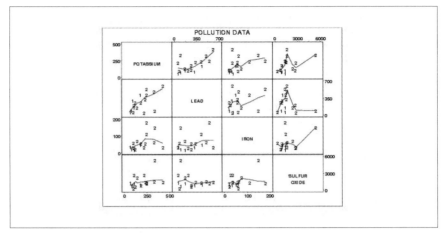

Figure 7-1. Scatterplot matrix of four pollutants

Relationships Between Continuous Variables

In linear algebra, we often describe the relationship between two variables with an equation of the form:

$$y = ax + b$$

In this formula, y is the dependent variable,
x is the independent variable,
a is the *slope*, and
b is the *intercept*.

Note that m is sometimes used in place of a in this equation; this is just a different notational convention and does not change the meaning of the equation. Both a and b can be positive, negative, or 0. To find the value of y for a given value of x, you multiply x by a and then add b. An equation such as this expresses a perfect relationship (given the values of x, a, and b, we can find the exact value of y), whereas equations describing real data generally include an error term, signifying our understanding that the equation gives us a *predicted* value of y that might not be the same as the actual value. It's worth looking at some graphs of data defined by equations, however, to get a sense of what perfect relationships look like when graphed; this should make it easier to spot similar patterns in real data.

Figure 7-2 shows the association between two variables, x and y, that have a perfect positive association: $x = y$. In this equation, $b = 0$, $a = 1$, and for every case, the values of x and y are the same. This equation expresses a positive relationship because as the value of x increases, so does the value of y; in a graph of a positive relationship, the points run from the lower left to the upper right.

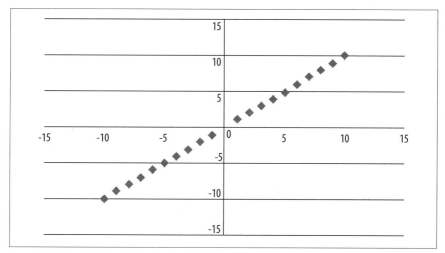

Figure 7-2. Graph of the model y = x

Figure 7-3 shows a negative relationship between x and y: these points are described by the equation $y = -x$. In this equation, $a = -1$, $b = 0$. Note that in a negative relationship such as this one, as the value of x increases, the value of y decreases, and the points in the graph run from the upper left to the lower right.

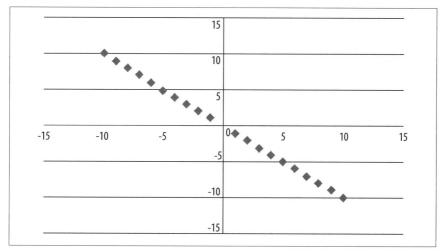

Figure 7-3. Graph of the model y = −x

Figure 7-4 shows a positive relationship between x and y as specified by the model $y = 3x + 2$. Note that this relationship is still perfect (meaning that given the model and a value for x, we can compute the exact value for y) and is represented by a straight line. Unlike the previous two graphs, however, the line no longer runs through the origin (0, 0) because the value for b (the intercept) is 2 rather than 0.

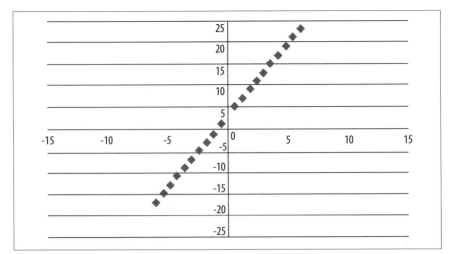

Figure 7-4. Graph of the model y = 3x + 2

In the previous three graphs, the equation of a straight line indicated a strong relationship between the variables. This is not always the case, however; it's possible for the equation of a straight line to indicate no relationship between the variables. When one variable is a constant (meaning it always has the same value) while the value of the other variable varies, this relationship can still be expressed through the equation (and graph) of a straight line, but the variables have no association. Consider the equation $x = -3$, displayed in Figure 7-5; no matter what the value of y, x always has the same value, so there is no association or relationship between the values of x and y. The slope of this equation is undefined because the equation used to calculate the slope has a denominator of 0.

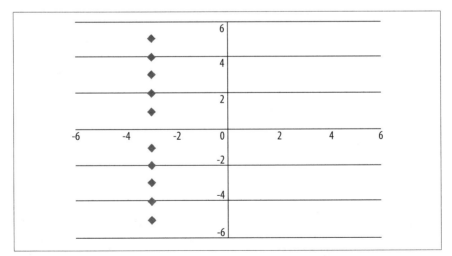

Figure 7-5. Graph of the model x = –3

The equation for calculating the slope of a line is given in Figure 7-6.

$$a = \frac{y_2 - y_1}{x_2 - x_1}$$

Figure 7-6. Equation for calculating the slope of a straight line

where x_1 and x_2 are any two x-values in the data, and y_1 and y_2 are the corresponding y-values. If x_1 and x_2 have the same value, this fraction has a denominator of 0, so the equation and the slope are undefined.

The equation $y = -3$ also expresses no relationship between x and y, in this case because the slope is 0. In this equation, it doesn't matter what the value of x is, the value of y is always -3. The graph of this equation is a horizontal line, as shown in Figure 7-7.

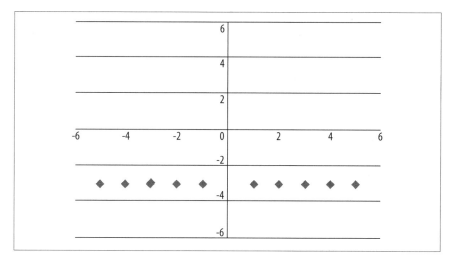

Figure 7-7. Graph of the model $y = -3$

In real data sets, we don't expect that an equation will perfectly describe the relationships among the variables, and we don't expect the graph to be a perfect straight line, even if the linear relationship is quite strong. Consider the graph in Figure 7-8, which displays almost the same data as shown in Figure 7-4; the difference is that we added some random error to the data so the data no longer form a perfect line. The relationship between x and y is still strongly linear and positive, but we can no longer predict the exact value of y, given the value of x, from an equation. To put it slightly differently, knowing the value of x helps us predict the value of y (versus predicting the value of y without any knowledge of the value of x), but we realize that our predicted value for y might be somewhat different than the actual value in the data set.

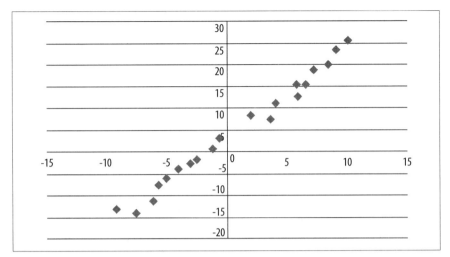

Figure 7-8. Graph of a strong positive relationship

It's unusual to find as close a relationship between x and y in a real data set as is displayed in Figure 7-8. The data in Figure 7-9 is more typical of what we usually find. Note that even though the points are more scattered than in previous examples, the relationship between x and y still seems to be positive and linear.

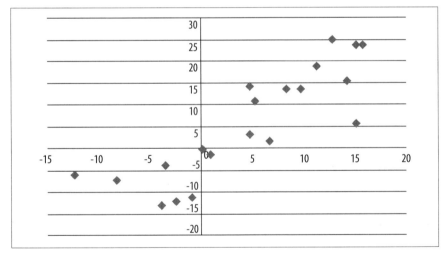

Figure 7-9. Graph of a weaker positive relationship

Two variables may have a strong relationship that is not linear. To take a familiar example, the equation $y = x^2$ expresses a perfect relationship because, given the value of x, we know exactly what the value of y is. However, this relationship is quadratic rather than linear, as can be seen in Figure 7-10. Spotting this type of strong but nonlinear relationship is one of the best reasons for graphing your data.

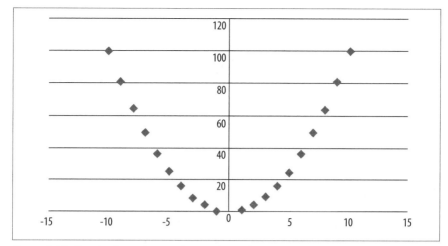

Figure 7-10. Graph of a perfect quadratic relationship

Figure 7-11 shows another common type of nonlinear relationship, a logarithmic relationship defined by the equation $y = LN(x)$, where LN means "the natural logarithm of."

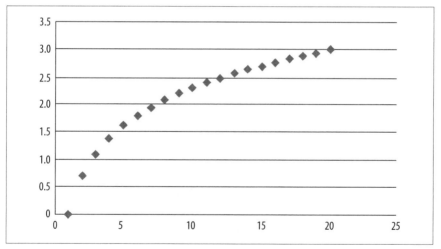

Figure 7-11. Graph of a perfect logarithmic relationship

If your data shows a nonlinear relationship, it might be possible to transform the data to make the relationship more linear; this is discussed further in Chapter 3. Recognizing these nonlinear patterns and knowing different ways to fix them, is an important task for anyone who works with data. For the data in Figure 7-10, if we transformed y by taking its square root and then graphed x and $\sqrt{y}$, we would see that the variables now have a linear relationship. Similarly, for the data graphed in

Figure 7-11, if we transformed y to e^y and then graphed it against x, we would see a linear relationship between the variables.

The Pearson Correlation Coefficient

Scatterplots are an important visual tool for examining the relationships between pairs of variables. However, we might also want a statistical estimate of these relationships and a test of significance for them. For two variables measured on the interval or ratio level, the most common measure of association is the *Pearson correlation coefficient*, also called the *product-moment correlation coefficient*, written as ρ (the Greek letter *rho*) for a population and r for a sample.

Pearson's r has a range of (–1, 1), with 0 indicating no relationship between the variables and the larger absolute values indicating a stronger relationship between the variables (assuming neither variable is a constant, as in the data displayed in Figure 7-5 and Figure 7-7). The value of Pearson's r can be misleading if the data have a nonlinear relationship, which is why you should always graph your data. The labels "strong" and "weak" do not have strict numerical definitions, but a relationship described as strong will have a more linear relationship, with points clustered more closely around a line drawn through the data, than will data with a weak relationship. Some of the definition of strong and weak depends on the field of study or practice, so you will need to learn the conventions for your own field. A few examples of scatterplots of data with different r values are given in Figures 7-12, 7-13, and 7-14 to give you an idea of what different strengths of relationship look like.

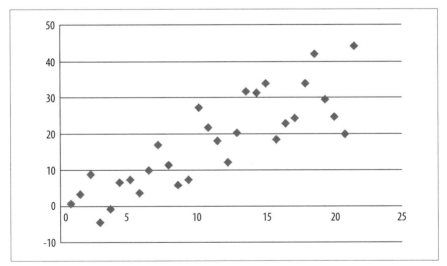

Figure 7-12. Scatterplot (r = 0.84)

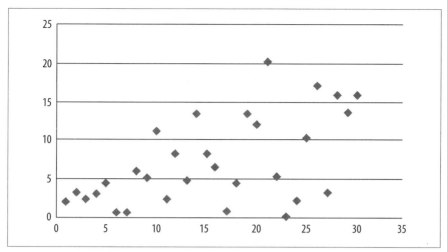

Figure 7-13. Scatterplot (r = 0.55)

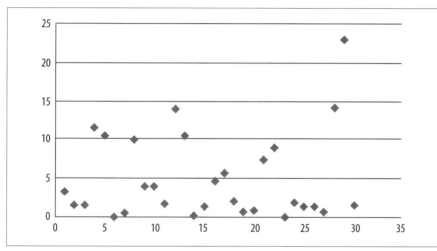

Figure 7-14. Scatterplot (r = 0.09)

Although correlation coefficients are often calculated using computer software, they can also be calculated by hand. The formula for the Pearson correlation coefficient is given in Figure 7-15.

$$r = \frac{SS_{xy}}{\sqrt{SS_x SS_y}}$$

Figure 7-15. Formula for the Pearson correlation coefficient

In this formula, SS_x is the sum of squares of x,
SS_y is the sum of squares of y, and
SS_{xy} is the sum of squares of x and y.

None of the steps in this calculation is difficult, but the process can be laborious, particularly with a large data set. The steps to calculate the sum of squares for x are as follows:

1. For each x score, subtract the mean of x as calculated from the sample. This is called the *deviation score.*

2. Square each deviation score.

3. Add the deviation scores together (hence the name *sum of squares*).

Figure 7-16 shows this written as a formula.

$$SS_x = \sum_{i=1}^{n}(x_i - \bar{x})^2$$

Figure 7-16. Formula for the sum of squares of x

In this formula, x_i is an individual x score,
$\bar{x}$ is the sample mean for x, and
n is the sample size.

This formula makes the meaning of SS_x clear, but it can be time-consuming to calculate. The sum of squares can also be calculated using a computational formula that is mathematically identical but less laborious if the calculations must be carried out by hand, as shown in Figure 7-17.

$$SS_x = \sum_{i=1}^{n} x_i^2 - \frac{\left(\sum_{i=1}^{n} x_i\right)^2}{n}$$

Figure 7-17. Computational formula for the sum of squares of x

The first part of the computational formula tells you to square each x and then add up the squares. The second part tells you to add up all the x scores, square that total, and then divide by the sample size. Then, to get SS_x, subtract the second quantity from the first.

To calculate the sum of squares for y, follow the same process but with the y scores and mean of y.

The process to compute the covariance is similar, but instead of squaring the deviation scores for x or y for each case, you multiply the deviation score for x by the deviation score for y. Written as a formula, it appears as shown in Figure 7-18.

$$SS_{xy} = \sum_{i=1}^{n}(x_i - \bar{x})(y_i - \bar{y})$$

Figure 7-18. Calculating the sum of squares of x and y

There is also a computational formula for the sum of squares of x and y, as shown in Figure 7-19.

$$SS_{xy} = \sum_{i=1}^{n}(x_i y_i) - \frac{\left(\sum_{i=1}^{n}x_i\right)\left(\sum_{i=1}^{n}y_i\right)}{n}$$

Figure 7-19. Computational formula for the sum of squares of x and y

The use of these formulas might become clearer after working through an example. Suppose we drew a sample of 10 American high school seniors and recorded their scores on the verbal and mathematics portions of the Scholastic Aptitude Test (SAT), as shown in Table 7-1. (Each section of the SAT has a range of 200–800.) To make the data easier to read, we have arranged the scores by verbal score in ascending order, but this is not necessary to perform the calculations.

Table 7-1. Verbal and mathematics scores on the SAT

Student	Verbal	Mathematics
1	490	560
2	500	500
3	530	510
4	550	600
5	580	600
6	590	620
7	600	550
8	600	630
9	650	650
10	700	750

Here is the information you need to use the computational formulas (or to check yourself if you calculated these quantities by hand):

$n = 10$

$$\left(\sum_{i=1}^{n}x_i\right) = 5{,}790$$

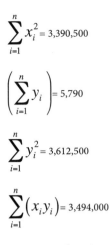

$$\sum_{i=1}^{n} x_i^2 = 3,390,500$$

$$\left(\sum_{i=1}^{n} y_i\right) = 5,790$$

$$\sum_{i=1}^{n} y_i^2 = 3,612,500$$

$$\sum_{i=1}^{n} (x_i y_i) = 3,494,000$$

Next, we plug this information into the computational formulas, as shown in Figure 7-20.

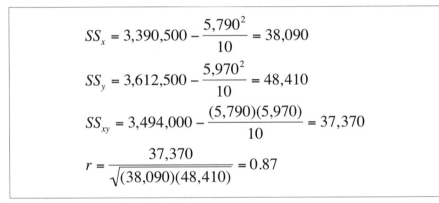

$$SS_x = 3,390,500 - \frac{5,790^2}{10} = 38,090$$

$$SS_y = 3,612,500 - \frac{5,970^2}{10} = 48,410$$

$$SS_{xy} = 3,494,000 - \frac{(5,790)(5,970)}{10} = 37,370$$

$$r = \frac{37,370}{\sqrt{(38,090)(48,410)}} = 0.87$$

Figure 7-20. Calculating r for the SAT verbal and math scores

The correlation between the verbal and math SAT scores is 0.87, a strong positive relationship, indicating that students who score highly on one aspect of the test also tend to score highly on the other. Note that correlation is a symmetrical relationship, so we do not need to posit that one variable causes the other, only that we have observed a relationship between them.

Testing Statistical Significance for the Pearson Correlation

We also want to determine whether this correlation is significant. The null hypothesis for designs involving correlation is usually that the variables are unrelated, that is, $r = 0$, and that is the hypothesis we will test for this example; the alternative hypothesis is that $r \neq 0$. We will use an alpha level of 0.05 and compute the statistic in Figure 7-21 to test whether our results are significantly different from 0. This statistic has a t distribution with $(n - 2)$ degrees of freedom; degrees of freedom is a statistical concept referring to how many things can vary in a given design. It is also a number we need to know to use the correct t-distribution to evaluate our results.

$$t = \frac{r\sqrt{n-2}}{\sqrt{1-r^2}}$$

Figure 7-21. Formula for the significance test for the Pearson correlation coefficient

In Figure 7-21, r is the Pearson correlation for the sample, and n is the sample size. For our data, the calculations are shown in Figure 7-22.

$$t = \frac{0.87\sqrt{10-2}}{\sqrt{1-0.87^2}} = \frac{2.46}{0.49} = 5.02$$

Figure 7-22. Calculating the significance test for the correlation between SAT math and verbal scores

According to the t-table (Figure D-7 in Appendix D), the critical value for a two-tailed t-test with 8 degrees of freedom at $\alpha = 0.05$ is 2.306. Because our computed value of 5.02 exceeds this critical value, we will reject the null hypothesis that the SAT math and verbal scores are unrelated. We also calculated the exact p-value for this data by using an online calculator (*http://faculty.vassar.edu/lowry/corr_stats.html*) and found the two-tailed p-value to be 0.0011, also indicating that our result is highly improbable if the verbal and math scores are truly unrelated in the population from which our sample was drawn.

The Coefficient of Determination

The correlation coefficient indicates the strength and direction of the linear relationship between two variables. You might also want to know how much of the variation in one variable can be accounted for by the other variable. To find this, you can calculate the *coefficient of determination*, which is simply r^2. In our SAT example, $r^2 = 0.87^2 = 0.76$. This means that 76% of the variation in SAT verbal scores can be accounted for by SAT math scores and vice versa. We will expand further on the concept of the coefficient of determination in the chapters on regression because very often one of the purposes in building a regression model is to find a set of predictor variables that can account for a high proportion of the variation in our outcome variable.

Exercises

Problem

Which of the following scatterplots (Figures 7-23, 7-24, and 7-25) suggest that the two graphed variables have a linear relationship? For those that do, identify the direction of the relationship and guess its strength, that is, the Pearson's correlation coefficient for the data. Note that no one expects you to be able to guess an exact correlation coefficient by eye, but it is useful to be able to make a plausible estimate.

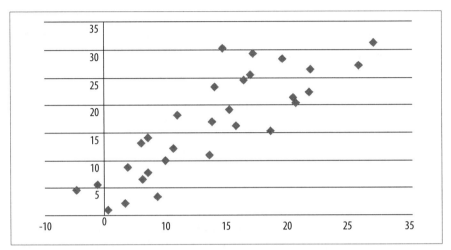

Figure 7-23. Scatterplot a

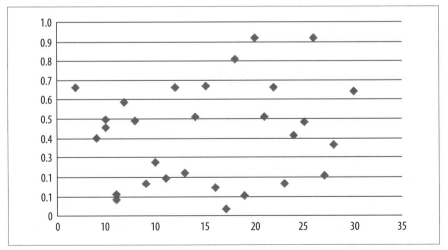

Figure 7-24. Scatterplot b

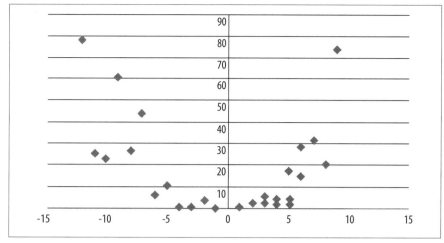

Figure 7-25. Scatterplot c

Solution

a. Strong positive linear relationship ($r = 0.84$).

b. Weak relationship ($r = 0.11$).

c. Nonlinear, quadratic relationship. Note that $r = -0.28$ for this data, a respectable correlation coefficient, so that without the scatterplot, we could easily have missed the nonlinear nature of the relationship between these two variables.

Problem

Find the coefficient of determination for each of the data sets from the previous problem, if appropriate, and interpret them.

Solution

a. $r^2 = 0.84^2 = 0.71$; 71% of the variability in one variable can be explained by the other variable.

b. $r^2 = 0.11^2 = 0.01$; 1% of the variability in one variable can be explained by the other variable. This result points out how weak a correlation of 0.11 really is.

c. r and r^2 are not appropriate measures for variables whose relationship is not linear.

Problem

Several studies have found a weak positive correlation between height and intelligence (the latter as measured by the score on an IQ test), meaning that people who are taller are also on average slightly more intelligent. Using the formulas presented in this chapter, compute the Pearson correlation coefficient for the data presented in Table 7-2, which represent height (in inches) and scores on an IQ test for 10 adult women. Then test the correlation for significance (do a two-tailed test with alpha = 0.05), compute the coefficient of determination, and interpret the results. For the sake of convenience, we will designate height as the x variable and IQ score as the y variable.

Table 7-2. Height and IQ

Student	Height (inches)	IQ score
1	60	103
2	62	100
3	63	98
4	65	95
5	65	110
6	67	108
7	68	104
8	70	110
9	70	97
10	71	100

Solution

The calculations are shown in Figures 7-26 and 7-27.

$n = 10$

$$\sum_{i=1}^{n} x_i = 661$$

$$\sum_{i=1}^{n} x_i^2 = 43,817$$

$$\sum_{i=1}^{n} y_i = 1,025$$

$$\sum_{i=1}^{n} y_i^2 = 105,327$$

$$\sum_{i=1}^{n} \left(x_i y_i \right) = 67,777$$

$$SS_x = 43{,}817 - \frac{661^2}{10} = 124.9$$

$$SS_y = 105{,}327 - \frac{1{,}025^2}{10} = 264.5$$

$$SS_{xy} = 67{,}777 - \frac{(661)(1{,}025)}{10} = 24.5$$

$$r = \frac{24.5}{\sqrt{(124.9)(264.5)}} = 0.135$$

Figure 7-26. Calculating the correlation between height and IQ

Coefficient of determination $= r^2 = 0.018$

$$t = \frac{0.135\sqrt{10-2}}{\sqrt{1-0.135^2}} = \frac{0.382}{0.991} = 0.385$$

Figure 7-27. Calculating the t-statistic for the correlation between height and IQ

In this data, we observe a weak ($r = 0.135$, $r^2 = 0.018$) positive relationship between height and IQ score; however, this relationship is not significant ($t = 0.385$, $p > 0.05$), so we do not reject our null hypothesis of no relationship between the variables.

If you are interested in this issue, see the paper by Case and Pearson in Appendix C; although primarily concerned with the relationship between height and income, it also summarizes research about height and intelligence.

8

Introduction to Regression and ANOVA

Regression and analysis of variance (ANOVA) are two techniques within the general linear model (GLM). If you're not comfortable with the concept of a linear function, you should review the discussion of the Pearson correlation coefficient in Chapter 7. In Chapters 8 through 11, we cover a number of statistical techniques, some of them fairly complex but all built on this basic principle of the linear relationship among two or more variables. This chapter presents the most basic linear models, simple regression and one-way ANOVA; Chapters 9 through 11 present more complex techniques within the GLM family. The types of analysis presented in these chapters are nearly always performed using computer software; fortunately, most of them are common enough to be included in any statistical computing package. Also fortunately, it's usually not difficult to figure out how to use a given package if you understand the theory underlying the model. For this reason, we concentrate on explaining how these models work and keep our advice sufficiently broad that it should apply to most systems.

The General Linear Model

Underlying all techniques within the GLM family is the assumption that a dependent variable is the function of one or more independent variables. We often speak in terms of *predicting* or *explaining* a dependent variable, using a set of independent variables, but step back for a minute to consider what it means for one variable to be a function of another variable (or set of variables—to keep it simple for now we'll stick to the simplest case of one dependent and one independent variable). You probably remember functions of the type $y = f(x)$ from studying algebra; this equation says that if you know the value of x, you can compute the value of y by following the procedure specified by the $f(x)$ function. Here are a few examples of functions:

- $y = x$ means that the value of y is the same as the value of x, so $(x, y) = (1, 1)$, $(2, 2)$, $(3, 3)$. The notation $(x, y) = (1, 1)$, $(2, 2)$, and so on, is just a compact way of saying, "if $x = 1$, $y = 1$; if $x = 2$, $y = 2$;" and so on.

- $y = ax$ means that the value of y is the product of the value of x and the constant a. If $a = 3$, $(x, y) = (1, 3)$, $(2, 6)$, $(3, 9)$, and so on; the value of y is always 3 times the value of x. If $a = 0.5$, $(x, y) = (1, 0.5)$, $(2, 1)$, $(3, 1.5)$, and so on. In this type of model, a is often called the *slope* of the equation.

- $y = ax + b$ means that the value of y is the product of the value of x and the constant a plus the value of the constant b. Note that x is multiplied by a, and then the value of b is added to this product. If $a = 1$ and $b = 5$, $(x, y) = (1, 6)$, $(2, 7)$, $(3, 8)$, and so on. In this type of model, b is often called the *constant* of the equation because its value does not change; whatever the value of x, the value of b is always the same, so the value of b is constant.

- $y = x^2$ means that the value of y is the square of the value of x, that is, the value of x multiplied by itself. Therefore, $(x, y) = (1, 1)$, $(2, 4)$, $(3, 9)$, and so on.

In this chapter, we discuss the bivariate case, equations with only two variables; this type of equation can always be described by $y = ax + b$ (remembering that b is a constant, not a variable).

Writing Linear Equations

There are several ways to write a linear equation, but the important parts of the equation remain the same. For describing a simple linear equation with one predictor and a constant, the $y = ax + b$ method is sufficient. In this equation, y is the *dependent variable* or *outcome*, a is the *slope* or *coefficient*, and b is the *constant* or *intercept*. The term *intercept* refers to the value at which the line described by the equation crosses the y-axis; it's the value of y when $x = 0$. *Slope* refers to the relationship between x and y: how much change in y is predicted for 1 unit change in x? You might remember your algebra textbook referring to slope as rise over run; rise in this case refers to change in the y variable, run to change in the x variable. If you feel you need a review of the algebra of linear equations, you should review "Relationships Between Continuous Variables" on page 176 in Chapter 7 and try a few of the relevant practice problems in Appendix A.

Another type of notation is used more commonly in statistics when writing linear equations, particularly for equations with multiple predictors. In this notation, a simple linear equation is written in the format of $y = \beta_0 + \beta_1 x_1 + e$, where β_0 is the intercept, β_1 the slope or *coefficient*, and e the *residual* or *error* term, which is included because when working with real data (as opposed to manipulating algebraic equations), we don't expect to be able to predict the value of y perfectly from an equation. The residual or error term represents the difference between the value of y as observed and the value of y as predicted from the equation.

In statistics, the term "coefficient" is more often used than slope when referring to the β_1 term because we often work with equations with many predictor variables (multiple linear regression), in which case, no one predictor variable entirely determines the line's slope. The meaning of the coefficient in a multiple linear equation is the predicted change in y for a one-unit change in x, *holding the value of all other x variables constant.* So, in the equation $y = \beta_0 + B_1x_1 + B_2x_2 + B_3x_3 + e$, there are three predictor variables (x_1, x_2, and x_3), and the coefficient B_1 expresses the predicted change in y for a one-unit change in x_1, holding x_2 and x_3 constant.

Linear Regression

Suppose the model $y = ax + b$ describes the relationship between two variables, x and y. In algebra, we can have perfect relationships so that the value of y is always perfectly predicted by the value of x. The examples we gave previously are this type of model. If we say, for instance, that $y = 2x + 7$, we know that if the value of x is 0, the value of y will be 7. In this type of case, the correlation coefficient between x and y will always be 1.00, indicating a perfect relationship—we can always predict the value of y from the value of x without error.

In statistics, however, we are often trying to fit an equation to a real set of data. In this case, we don't expect a perfect relationship between x and y. That is, we don't assume that we will always be able to predict the value of y, given the value of x, without error. Real life is much more variable than the closed system of mathematics, and even the strongest relationships observed in the real world are seldom perfect in the mathematical sense.

Consider the relationship between height and weight in adults. It makes intuitive sense that these two variables should have a strong positive relationship; in general, tall people weigh more than short people. However, this relationship is not perfect; we can all think of short people who are quite heavy and tall people who are quite thin. Similarly, we expect to see a positive relationship between years of education and income among people of working age; in general, people with more years of education earn more money. However, this relationship is not perfect either; one of the richest men in the world, Bill Gates, did not graduate from college, and many university towns are full of people with advanced degrees working at low-paying jobs. When working with real data, we don't expect to find perfect relationships, but we do try to find useful ones. For instance, we don't expect to be able to develop an equation to predict someone's weight perfectly from his height (or even from a much more complex equation including many other predictor variables). Instead, we want to build an equation that is useful for our purposes and that improves our predictive ability, meaning that if we know a person's height, we can use the equation to make a better prediction of his weight than we could if we didn't know his height.

We could explore the relationship between height and weight by using scatterplots and the correlation coefficient, but linear regression takes us a step further. When doing a regression analysis, we imagine drawing a straight line (the regression line) representing the relationship between two variables; such a line is often superimposed over a scatterplot to clarify the relationship between the variables further. Consider the scatterplot in Figure 8-1.

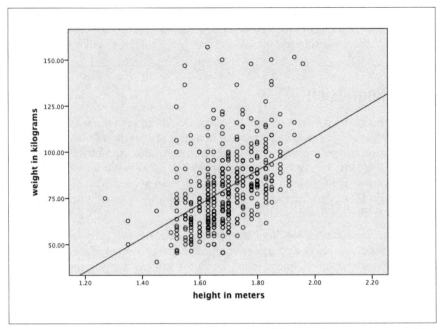

Figure 8-1. Scatterplot of height in meters and weight in kilograms for 436 American adults

This is a scatterplot of the height (in meters) and weight (in kilograms) for 436 American adults; the data comes from a random subsample of data collected for the 2010 BRFSS (Behavioral Risk Factor Surveillance System), a health survey conducted annually in the United States. [You can read more about the BRFSS and download data for your own analyses from this website (*http://www.cdc.gov/brfss/technical_in fodata/surveydata/2010.htm*).] As expected, the relationship is positive and somewhat linear (the data somewhat cluster around a line) but is far from perfect: most data points do not lie on the regression line (the line superimposed over the scatterplot), and some are quite far from it. This is typical of the kind of results you get with real-world data; relationships are not perfect, but if your model is good, they might be strong enough to be useful.

In this case, the correlation (r) between height and weight is 0.47, and the coefficient of determination (r^2) is 0.22. This means that about 22% of the variation in weight can be accounted for by height, not exactly a perfect prediction or explanation but much better than 0. The regression equation for this data is:

$$y = 91x - 74$$

The slope is 91, the constant –74. To find the predicted weight for a person, you replace x with her height in meters and do the math. This equation predicts that a person who is 1.8 meters tall would weigh 89.8 kilograms because:

$$y = 91(1.8) - 74 = 89.8.$$

Of course, if we were really interested in predicting weight, we would develop a more complex model including factors such as gender and age, but this example serves well enough to illustrate the basic concepts of simple regression. You might have noticed that although correlation does not require you to specify which variable is the predictor and which is the outcome, you do have to make that choice when working with regression. I designated weight as the outcome and height as the predictor, which makes logical sense because height is fixed in adulthood and could logically be considered a causal factor for weight. (All else being equal, including build, tall people do tend to weight more than short people.) I don't think I could make an argument that weight has a causal relationship with height.

It is possible to calculate a regression line by hand (I did it in graduate school, and before computers came into wide use, everyone did it that way), but it is much more common to use a statistical computing package to do this calculation. Regression is an extremely common procedure, and almost any statistical package you might be using will probably include the routines to do regression calculations. For those who wish to go through the process of calculating regression parameters by hand, a solved example is included at the end of this chapter.

Even if you are never planning to calculate a regression equation by hand, it is worth considering the logic behind the process. When a statistical package produces a regression line for a set of data, it calculates the equation that will produce the line that is as close as possible to all the data points considered together. This is sometimes described as minimizing the squared deviations, where the squared deviations are the sum of the squared deviations between each data point and the regression line. This is easy to visualize with simple regression because only two dimensions are involved (the predictor and outcome variables); the same principle applies for more complex models (with more variables), but it is more difficult to illustrate because of the greater number of dimensions involved.

Consider Figure 8-2. It's a scatterplot for a small data set, with a superimposed regression line. Note that although the regression line is fairly close to all the points, none of them actually lie on the line; this is not unusual, particularly with small data sets, because the goal is to produce the line that is the closest to *all* the points, even if it contains none of them. In Figure 8-2, you could draw vertical lines from each point to the regression line; the length of each vertical line represents the *error of prediction*, or *deviation*, for each individual point. If you squared the length of each line and added them up, that would be the sum of the squared deviations for this data set. The regression line is drawn so as to minimize all those squared deviations so it is as close as a straight line can be to all the points in this data set. The difference between each point and the regression line is also called the residual because it represents the variability in the actual data points not accounted for by the equation generating the line. "Minimizing the squared deviations" can also be expressed as "minimizing the errors of prediction" or "minimizing the residuals."

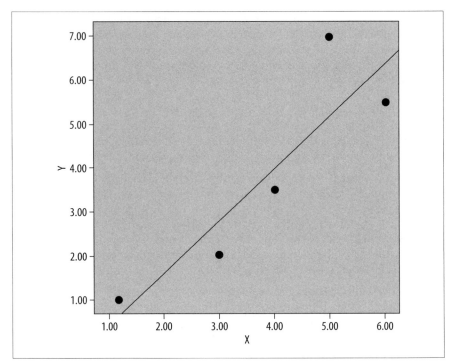

Figure 8-2. Errors of prediction in a small data set

Assumptions

As with most statistical procedures, linear regression makes certain assumptions about the data used in an analysis; if these assumptions are violated, the results of the analysis might not be valid. Key assumptions for simple linear regression include:

Data appropriateness
> The outcome variable should be continuous, measured at the interval or ratio level, and be unbounded (or at least cover a wide range); the predictor variables should be continuous or dichotomous. Categorical predictors with more than two categories can be recoded into a series of dichotomous dummy variables; this is covered in Chapter 10.

Independence
> Each value of the outcome variable is independent of each other value. This would be violated if there were some pattern of time dependency, for instance, or if some of the dependent variables were measured from subjects clustered into larger units (such as members of the same family or children studying in the same classroom) in some way that affected their value on the dependent variable. This assumption is checked by your knowledge of the data and how it was collected.

Linearity

The relationship between the predictor and outcome variable resembles a straight line. This assumption is checked by graphing the data; if it resembles a shape other than a straight line, you might need to transform one or both variables or choose another procedure.

Distribution

The continuous variables are approximately normally distributed and do not have extreme outliers. The distribution of continuous variables may be checked by creating a histogram (eyeballing the data) and by a statistical test for normality such as the Kolmogorov-Smirnov. An outlier is defined as a data value that is far from the other values for the same variable in a data set; sometimes it is described as a data value that doesn't seem to belong with the others. Outlier detection is partly a matter of judgment, is further discussed in Chapter 17, and can be a multistep process. (An unusual data value can be the result of an error in data entry, for instance, or it might be an apparently valid value.)

Homoscedasticity

The errors of prediction are constant over the entire data range. This means that the errors are not, for instance, smaller when the y value is small and larger when the y value is large. This assumption is checked by graphing the standardized residuals against the standardized predicted values; the data should resemble a cloud without any indication that the errors of prediction are not constant over the whole range of the data. Figure 8-3 shows homoscedastic data and Figure 8-4 heteroscedastic data.

Independence and normality of the errors

The error of prediction for each data point should be independent of the error of prediction for each other data point, and the errors should be normally distributed. The independence assumption is checked by the Durbin-Watson test (discussed later), and the normality assumption is checked by graphing the residuals (error terms).

Suppose we are interested in adolescent fertility (the rate of childbirth to women age 15–19 years) and in what factors at the country level are associated with adolescent fertility. Our first idea is that gender inequality might be related to adolescent fertility, and we hypothesize that adolescent fertility is lower in countries where women are treated more equally. We will do a regression analysis to test this hypothesis, using data downloaded from the United Nations Human Development Project (*http://hdr.undp.org/en/statistics/data/*). We will use the Gender Inequality Index as our predictor variable; this index is composed of five variables measuring aspects of women's reproductive health, empowerment, and labor force participation, and has a range of approximately 0–100 (in our data set, from 6.5 to 79.1), with lower numbers signifying greater equality.

Note that this is what is known as ecological or aggregated data; the value for each variable relates to a measurement on a country rather than on an individual. There's nothing wrong with using ecological data, but you must be careful to draw conclusions only for data at the same level of aggregation as the data you analyzed; in this case, our results will apply at the country level, not at the level of the individual.

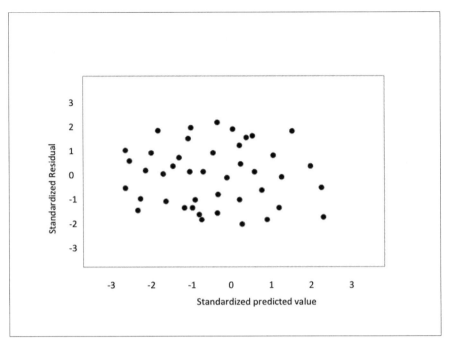

Figure 8-3. Homoscedasticity

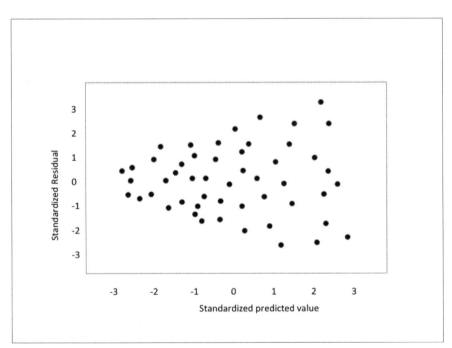

Figure 8-4. Heteroscedasticity

We start by running through our assumptions. A frequency table confirms that both variables are continuous and have substantial ranges, and we have 135 cases with values on both variables, which is more than enough for a simple regression analysis. Our data is also independent because data for each country was collected separately. Our third assumption, linearity, can be examined with a scatterplot. Here, as shown in Figure 8-5, we encounter a problem: the relationship is curvilinear rather than linear.

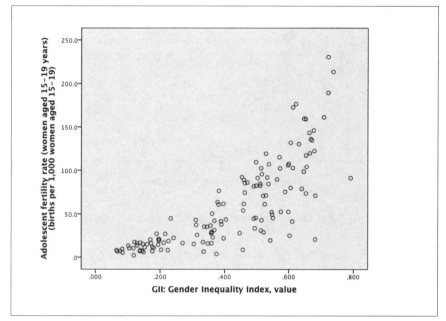

Figure 8-5. Scatterplot of adolescent fertility and the Gender Inequality Index

We perform a natural log transformation (discussed in Chapter 3) of the adolescent fertility rate and find that the relationship between these two variables is much more linear. The scatterplot for the transformed variable and the Gender Inequality Index is presented in Figure 8-6.

We will check the normality of our variables with histograms and the Kolmogorov-Smirnov (K-S) statistic. The K-S statistic compares the distribution of a variable with a reference distribution. (In this case, the reference distribution is the normal distribution.) The null hypothesis for the K-S statistic is that the variable was drawn from the reference distribution, so in this example, if we fail to reject this null hypothesis, we can proceed on the assumption that the variables were drawn from normally distributed populations. Both histograms (not shown) look acceptably normal, and the Kolmogorov-Smirnov statistics are not significant (K-S = 1.139, p = 0.149 for the natural log of adolescent fertility rate; K-S = 1.223, p = 0.101 for the Gender Inequality Index).

We will examine assumptions 5 and 6 after running our regression analysis. We believe that gender inequality influences adolescent fertility, and we have

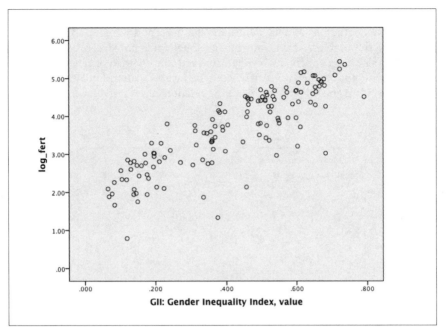

Figure 8-6. Scatterplot of the natural log-transform of adolescent fertility and the Gender Inequality Index

transformed the adolescent fertility rate by taking its natural log (LN), so our regression model is:

$$LN(\text{adolescent fertility}) = \beta_0 + \beta_1(\text{GII}) + e$$

This is a different style of notation than we used earlier in this chapter, but this style is more commonly used when discussing regression, so now is a good time to make a switch. Our Y variable or outcome in this case is LN(adolescent fertility); our intercept or constant, formerly written as b, is now written as β_0; and our slope, formerly written as a, is now written as β_1. This notation will be particularly handy when we discuss regression with multiple predictors because they can be designated as β_1, β_2, and so on; these terms are called *coefficients*.

Different statistical packages produce different output, but there is enough commonality among them that you should be able to produce the basic regression output produced by any of the major systems if you understand how to read the output from any one of them. We will present the most important information from our analyses in simple tables to avoid favoring one system over another.

The first thing we want to do is evaluate the overall fit of our model. Overall model fit is usually expressed in terms of an F statistic and probability value, and evaluates whether the entire model is better than no model. Another way to look at this is that the F statistic and probability evaluates our model against a model in which all the predictor variables have a weight of 0 (the null model). We are also interested in how much variance in the outcome variable is explained by our model; particularly with a large data set, it is possible to have a model that has significant predictors yet explains very little of the variance in the outcome.

This model has an F statistic of 190.964 with 1 and 185 degrees of freedom and a p-value of <0.001; we therefore conclude that it is better than the null model. The R-value or correlation is 0.714 and a coefficient of determination or R^2 value of 0.509; this means that the Gender Inequality Index can explain more than 50% of the variation in adolescent fertility rates among the countries in our data set. Note that although in this case we are working with only two variables, correlations in regression are conventionally denoted with a capital R, and we have followed that convention here. The Durbin-Watson statistic for this data is 2.076, signifying that the error terms in our data are independent (good). The Durbin-Watson statistic ranges from 0 to 4, and a value of 2 indicates complete independence; our value is very close to 2, so we can consider the assumption of the independence of error terms met.

The regression coefficients for this analysis are displayed in Table 8-1.

Table 8-1. Coefficients table for a regression analysis of the Gender Inequality Index and the natural log of the adolescent fertility rate

	Unstandardized coefficients		Standardized coefficients		
	B	Std. error	Beta	t	Sig.
Constant	1.798	0.112		16.118	< 0.001
GII	4.446	0.244	0.845	18.221	< 0.001

The column labeled B under Unstandardized Coefficients gives us the coefficients to write our regression equation. In this case, that equation would be:

$$LN(\text{adolescent fertility}) = 1.798 + 4.446(\text{GII}) + e$$

This tells us that the natural log of the adolescent fertility rate increases about 4.4 units for every 1 unit increase in the Gender Inequality Index; the relationship is positive, confirming our hunch that greater gender inequality was associated with higher adolescent fertility. The Std. Error column presents the standard errors for the coefficient estimates. The Beta column under Standardized Coefficients presents, as the name suggests, the standardized regression coefficient; this can be useful in regression analyses with multiple predictors measured on different scales. The t column shows the t-statistic for each coefficient and is calculated by dividing B by its standard error. For instance, for GII:

$$t = 4.446/0.244 = 18.221$$

The final column is the significance of the t-statistic. We usually aren't concerned with the significance of the constant (all that tells us is whether it is significantly different from 0, which is not ordinarily a question of interest), but we are interested in the significance of the coefficients for our predictors. In this case, GII is a highly significant predictor ($p < 0.001$) of adolescent fertility.

It might be helpful to think about what we are testing with this analysis. Our main interest is whether gender inequality predicts the adolescent fertility rate; if the t-statistic for gender inequality in the coefficients table is not significant, this means that we could drop the term for gender inequality from our equation. To put it another way, a nonsignificant result for gender inequality would mean that the coefficient for that term was not significantly different from 0, so it could be dropped from our equation without harming the equation's ability to predict or explain the outcome variable.

Our final steps are to finish checking our assumptions to be sure that our results are valid. We can check homoscedasticity (assumption 5) by graphing the standardized residuals against the standardized predicted values; results are shown in Figure 8-7.

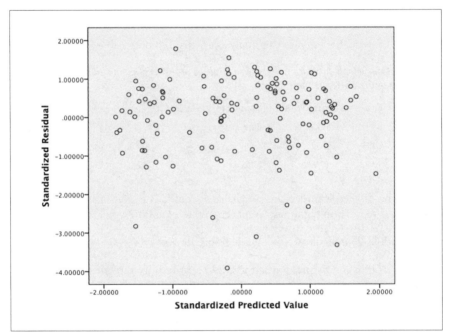

Figure 8-7. Scatterplot of the standardized residuals and standardized predicted values

This is a classic data cloud, providing no evidence that the error of prediction is not constant, so the assumption of homoscedasticity is met. Finally, we will check the normality of our residuals by creating a histogram (not pictured) of them and calculating the Kolmogorov-Smirnov statistic; the value of the Kolmogorov-Smirnov statistic is 1.355 ($p = 0.51$), so our analysis barely passes the test of normality.

Not all statistical analyses produce significant results. In Table 8-2, we present the results of a regression analysis using the size of the female population in a country (measured in 1,000s) to predict the country's Gender Inequality Index score.

Table 8-2. Coefficients table for a regression analysis predicting a country's Gender Inequality Index by the size of its female population (measured in 1,000s)

	Unstandardized coefficients		Standardized coefficients		
	B	Std. error	Beta	t	Sig.
Constant	0.282	0.074		3.806	0.002
Female population (1,000s)	0.000	0.000	0.306	1.285	0.217

We can see from the *t*-value (1.285) and significance (0.217) that the size of the female population is not a significant predictor of gender equality; another clue of this result is the unstandardized coefficient of 0.000 for this predictor, meaning that the value of this coefficient, carried to three decimal places, is essentially zero. A scatterplot of the two variables (Figure 8-8) indicates a basically random relationship between them, and logically speaking, there's no reason countries with large female populations (which means a country with large populations, period) should have a consistently higher or lower level of gender equality than countries with small female populations, so we will not pursue this analysis further.

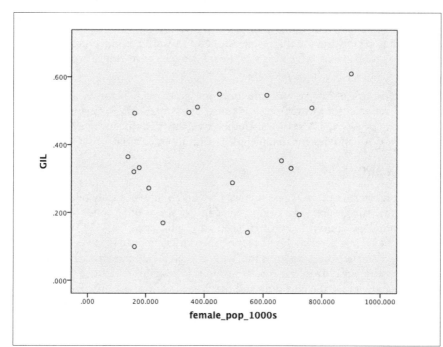

Figure 8-8. Scatterplot of female population (measured in 1000s) and Gender Inequality Index

Analysis of Variance (ANOVA)

Analysis of variance (ANOVA) is a statistical procedure used to compare the mean values on some variable between two or more independent groups. It's called analysis of variance because the procedure involves partitioning variance, attributing the variance observed in a data set to different causes or factors, including group membership. However, because it is usually used to compare the means between groups, many a student has privately thought that the real name should be A-MEAN-A. Nevertheless, ANOVA is a useful technique, particularly when analyzing data from designed experiments (such as the differences between the control and experimental groups in a clinical trial).

The major test statistic for an ANOVA is the *F ratio*, which can be used to determine whether statistically significant differences exist between the groups. For example, we might be interested in testing the efficacy of three drugs intended to lower blood pressure; we could form four groups of hypertensive patients and give each group one of the drugs (plus one group to act as a control, meaning they receive either no medication or standard care). After a period, we would measure the blood pressure on the patients in this study to see whether any of the drugs in the experiment had produced significant reductions in their blood pressure and whether there were differences among the drugs in this result. An ANOVA would produce an F ratio comparing the group means, which we would test for significance using a predetermined standard such as $p < 0.01$ or $p < 0.05$.

The simplest type of ANOVA includes only one group or predictor variable and one outcome variable; for this reason, it is called one-way ANOVA. Chapter 9 covers more complex types of ANOVA, including two-way and three-way ANOVA (factorial ANOVA), and designs that include a continuous covariate (ANCOVA).

One-Way ANOVA

The simplest form of ANOVA is one-way ANOVA, in which only one variable is used to form the groups to be compared. This variable is often called a *factor*, and that terminology is even more often used with more complex ANOVA designs. Suppose we are interested in the efficacy of a new drug intended to lower blood sugar in Type II diabetics; we could test this with an ANOVA, comparing the new drug to another drug already in use. The factor in this design is the drug administered, and it has two levels: the new drug and the drug already in use. The factor used in a one-way ANOVA can have more than two levels: in the previous example of comparing three hypertension drugs and a control group, we had one factor with four levels.

A one-way ANOVA with two levels is equivalent to performing a *t*-test. Our null hypothesis in this type of design is usually that the two groups have the same mean, whereas the alternative is either that they have different means (a two-sided test) or differ in one direction only (a one-sided test). Even if there is a significant difference in group means, we don't expect that there will be no overlap among members of the groups; in fact, it would be unusual if there were no overlap. We also expect that there will be variability within each group, and the calculations for one-way ANOVA are concerned with comparing the variability *within* groups (for instance, the variability observed in blood sugar among the patients receiving the new drug) and the variability *between* groups (the difference between the patients taking the experimental drug and the patients taking the standard drug).

ANOVA also has assumptions that must be met for the technique to be used appropriately. Because linear regression and ANOVA are really just two ways of looking at data using the general linear model, it is not surprising that some of the assumptions of ANOVA are also assumptions of regression.

Data appropriateness
> The outcome variable should be continuous, measured at the interval or ratio level, and be unbounded (or at least cover a wide range); the factors (group variables) should be dichotomous or categorical.

Independence
> Each value of the outcome variable is independent of each other value. This would be violated if there was some pattern of time dependency, for instance, or if some of the dependent variables were measured from subjects clustered into larger units (such as members of the same family or children studying in the same classroom) in some way that affected their value on the dependent variable. This assumption is checked by your knowledge of the data and how it was collected.

Distribution
> The continuous variable is approximately normally distributed *within each group*. The distribution of the continuous variable can be checked by creating a histogram (eyeballing the data) and by a statistical test for normality such as the Kolmogorov-Smirnov.

Homogeneity of variance
> The variance of each of the groups should be approximately equal. This is checked by a procedure such as the Levene statistic; the null hypothesis is that the variance is homogeneous, so if the results of the Levene statistic are *not* statistically significant (normally, the criterion of $\alpha < 0.05$ is used), that means the variances are sufficiently homogeneous to proceed.

ANOVA is considered a robust procedure, meaning that it can produce good results even when some assumptions are violated; for instance, when group sizes are equal, the F-statistic produced by ANOVA is reliable even if the distribution of the continuous variable is nonnormal. Similarly, when group sizes are equal, the F-statistic is robust to violations of the assumption of homogeneity. If you want to read more about the debates surrounding these issues, a relevant article by Glass is listed in Appendix C. However, violations of the assumption of independence can seriously distort your results, so you need to be sure that this assumption is met before using ANOVA with your data.

Suppose we are comparing two methods of weight training, and our measurement is how much improvement we have observed in the total lifted in a full squat after three months of training with one method or the other. Our null hypothesis is that the means are the same in both groups after the training; in other words, that on average neither method of training produces better results than the other. We randomly assigned our subjects at the start of the experiment and measured their maximum squat; it was approximately equal in both groups. The box plots in Figure 8-9 show the improvement in weight lifted after three months; it is clear that people training with the first method made, on average, more improvement because group 1 has a higher median, represented by the black line in the center of the box, and the range is higher. However, it is also clear that there was variability between the two groups and that there were also considerable overlaps between the two groups. It is not the case that everyone in group 1 improved more than everyone in group 2, simply that on average, group 1 showed more improvement.

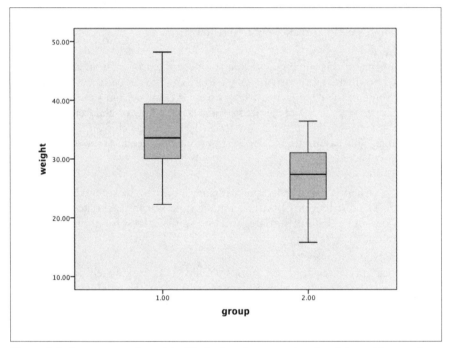

Figure 8-9. Improvement in weight lifted after three months of training with one of two methods

In fact, group 1 improved by an average of 34.21 pounds and group 2 by an average of 26.42 pounds. Is this difference statistically significant? To answer this, we will conduct a one-way ANOVA. First, we compute some basic statistics on this data, as presented in Table 8-3.

Table 8-3. Descriptive statistics for weightlifting data (two methods of training)

Group	N	Mean	Std. dev.	95% CI lower bound	95% CI upper bound
1	15	34.21	7.38	30.13	38.31
2	15	26.42	6.16	23.01	29.83
Total	30	30.32	7.76	27.41	33.22

We note that we have equal numbers of subjects in each group, approximately equal variance, and that the 95% confidence intervals for the mean between the two groups do not overlap (although they come very close). We also computed the Levene statistic to see whether our two groups had homogeneous variance; our results (0.626, $p = 0.435$) tell us not to reject the assumption of homogeneity, so we can proceed to interpret our ANOVA results.

The statistical results for an ANOVA are customarily presented in a table similar to Table 8-4.

Table 8-4. One-way ANOVA results for the weightlifting data (two methods of training)

	Sum of squares	df	Mean square	F	Sig.
Between groups	455.86	1	455.86	9.86	0.004
Within groups	1294.52	28	46.23		
Total	1750.38	29			

By the standard of $\alpha < 0.05$, these are significant results, so we can reject the null hypothesis that the means of the two groups are equal; in fact, method 1 produced significantly better results than method 2. This is a simple ANOVA table because we have only one factor and two levels, but it's worth taking some time to look at the different parts of it because that will help you understand more complex ANOVA tables.

The table has three rows: one presents the data for between groups variance, one for within groups variance, and one for total variance. Adding up the sums of squares and the degrees of freedom (df) for between and within groups gives us the value for the total data set. Between groups variance refers to that attributed to group membership, that is, the variation in individual scores attributed to the method of training used. Within groups variance refers to the variance within each training group; as we saw in the box plot in Figure 8-9, there was substantial variation within each group as well as between the two groups. The degrees of freedom refers to how many things can vary when computing each part of the statistics; total degrees of freedom is $n - 1$ (1 less than the total number of subjects), between groups degrees of freedom is $k - 1$ (one less than the number of groups), and within groups degrees of freedom is $n - k$. The sum of squares (SS) is the sum of the squared deviation scores for between groups, within groups, and the total, whereas the mean square (MS) is the sum of squares divided by the degrees of freedom, so in this example:

SS(between) = 455.86/1 = 455.86
SS(within) = 1294.52/28 = 46.23

The F-statistic is the ratio of the sum of squares between and within groups, so in this example:

$F = 455.86/46.23 = 9.86$

Our statistical package calculated the significance of the F-statistic automatically, but we could have also compared it to the values on an F table (similar to the normal distribution table and other tables included in Appendix D). Because F tables have two degrees of freedom (for the numerator and denominator), they are quite bulky, so we have not included one in this book; however, you can find a public domain F table here (*http://www.itl.nist.gov/div898/handbook/eda/section3/eda3673.htm*).

Post Hoc Tests

If you have only two groups, a significant F-test means that the two groups differ from each other. However, with more than two groups, your ANOVA analysis might have returned a significant F-test (called an *omnibus test*), meaning that the group means are not identical, yet you might still be left wondering which groups differ from which others. To answer this question, it is possible to conduct a *post hoc test*; as the name implies, post hoc tests are conducted after the fact, after you have found a significant omnibus F-test. There are a number of post hoc tests, and some are more typically used in some fields, others in other fields. One good choice is the Scheffe test, which tests all comparisons among the groups for significance and is adjusted statistically to control for conducting multiple tests on the same data. (Using the Scheffe test controls the experiment-wise error rate and does not increase the probability of a Type I error.)

Suppose we were comparing three methods of weight training instead of two. The descriptive statistics for this data are presented in Table 8-5.

Table 8-5. Descriptive statistics for weightlifting data (three methods of training)

Group	N	Mean	Std. dev.	95% CI lower bound	95% CI upper bound
1	15	34.21	7.38	30.13	38.31
2	15	26.42	6.16	23.01	29.83
3	15	30.04	9.22	24.94	35.15
Total	45	30.32	7.76	27.41	33.22

We have the same sample size for all three groups, which is the optimal setup for an ANOVA. Looking at the group means, method 3 seems to produce results lower than group 1 but higher than group 2. The 95% confidence interval for group 3 does overlap with both groups, so it will be interesting to see what our post hoc results tell us about these three methods of training.

The Levene test has a value of 1.447 ($p = 0.247$), so the homogeneity of variance assumption is met. The results of the ANOVA are presented in Table 8-6.

Table 8-6. ANOVA results for the weightlifting data (three methods of training)

	Sum of squares	df	Mean square	F	Sig.
Between groups	456.04	2	228.30	3.86	0.029
Within groups	2483.76	42	59.14		
Total	2940.36	44			

The F-statistic is significant, meaning that the means of the three groups differ. However, we want to know more—are the results from the group using method 1 significantly better than both the groups using methods 2 and 3, for instance? Are the results produced by method 3 significantly better than those produced by method 2? To answer these questions, we conduct a Scheffe post hoc test, producing the results presented in Tables 8-7 and 8-8.

Table 8-7. Results from the Scheffe post hoc test for the weightlifting data (three methods of training)

I Group	J Group	Mean difference (I–J)	Std. error	Sig.	95% CI lower bound	95% CI upper bound
1	2	7.80	2.81	0.029	0.67	14.92
1	3	4.17	2.81	0.341	−2.95	11.30
2	1	−7.80	2.81	0.029	−14.92	−0.67
2	3	−3.62	2.81	0.442	−10.75	3.50
3	1	−4.17	2.81	0.341	−11.30	2.95
3	2	3.62	2.81	0.442	−3.50	10.75

Table 8-8. Homogeneous subsets from the Scheffe post hoc test (three methods of training)

Group	N	Subset for $\alpha = 0.05$ 1	2
2	15	26.42	
3	15	30.04	30.04
1	15		34.22
Sig.		0.442	0.341

Tables 8-7 and 8-8 present the same conclusion, but the information is arranged differently. Looking at either table, we can see that the mean of group 1 differs from the mean of group 2 but that the mean of group 1 does not differ from the mean of group 3, nor does the mean of group 2 differ from the mean of group 3.

Table 8-7 presents all possible pairwise comparisons between the groups; half the table is redundant because both the comparison of group 1 with group 2 and group 2 with group 1 are presented. For instance, the first row presents the comparison of group 1 with group 2 (the notations "I group" and "J group" are conventional). The mean difference in the means of these two groups is 7.80, and the difference is significant ($p = 0.029$). The 95% confidence interval for this difference in means is (0.67, 14.92); note that it does not cross zero. The second row of Table 8-7 presents the comparison between group 1 and group 3; the mean difference is 4.17, and it is not significant ($p = 0.341$). Note for comparison that the confidence interval does cross zero (–2.95, 11.30). The third row compares group 2 with group 1; the results are the same as in row 1, except that the signs are reversed (because in row 3, the mean of group 1 is subtracted from the mean of group 2, whereas in row 1, the mean of group 2 was subtracted from the mean of group 1). Row 4 compares the means for groups 2 and 3; the mean difference is –3.62, and this is not significant ($p =$ 0.442). Rows 5 and 6 are redundant with rows 2 and 4.

Table 8-8 presents a column for each set of groups that form a homogeneous subset; in a homogeneous subset, the means of the groups included do not differ significantly from each other. In this case, groups 2 and 3 form a homogeneous subset (column 1); groups 1 and 3 also form a homogeneous subset (column 2).

Calculating Simple Regression by Hand

Regression coefficients can be calculated by hand, using the sums of squares, variances of X and Y, and a few other quantities that can all be calculated without the use of a computer. The problem with hand calculations is not that any particular step of the process is difficult but that with a data set of any size, the work involved quickly becomes tedious and prone to error. However, going through a modified version of this process can be useful in understanding the meaning of the regression coefficients, and it is in that spirit that the following section is provided.

We noted earlier that, when dealing with real data, we don't expect the predictions made by a regression equation to be perfect. In fact, we assume that there will be some differences between the observed values in a data set and the predicted values as computed using the regression equation. We also discussed the squared deviations, which are the square of the difference between each observed data point and its predicted value according to the regression equation. The sum of these squared deviations is the sum of squares of errors, or SSE, and is computed as shown in Figure 8-10.

$$SSE = \sum_{i=1}^{n}(y_i - \hat{y}_i)^2$$

Figure 8-10. The sum of squares of errors

In this formula, y_i is an observed data value, and $\hat{y}_i$ is the predicted value (according to the regression equation) for that value. Because the value of $\hat{y}_i$ is determined by the regression equation ($ax_i + b$), the sum of squares of errors can also be written as shown in Figure 8-11.

$$SSE = \sum_{i=1}^{n}(y_i - (ax_i + b))^2$$

Figure 8-11. Another way to write the sum of squares of errors

The purpose of the regression equations is to minimize the SSE, that is, to make the predicted values as close as possible to the observed values. The computational formulas for the elements necessary to compute a simple regression equation are given in Figures 8-12 to 8-15. Note that S_{xx} is the variance of x, and S_{xy} is the covariance of x and y.

$$S_{xx} = \sum x^2 - \frac{\left(\sum x\right)^2}{n}$$

Figure 8-12. Computing the variance of x

$$S_{xy} = \sum xy - \frac{\left(\sum x\right)\left(\sum y\right)}{n}$$

Figure 8-13. Computing the covariance of x and y

$$a = \frac{S_{xy}}{S_{xx}}$$

Figure 8-14. Computing the slope of a simple regression equation

$$b = \frac{\sum y}{n} - a\frac{\sum x}{n}$$

Figure 8-15. Computing the intercept of a simple regression equation

Suppose you have been given the values in Figure 8-16 computed from a data set relating IQ (y) to height in meters (x); you can use this information to calculate the regression line for that data set. You could also compute these quantities by hand, but that process is extremely laborious with even a moderately sized data set—so laborious, in fact, that you are apt to forget why you were doing the calculations in the first place.

$$\sum x = 33.25$$

$$\sum y = 2{,}486$$

$$\sum x^2 = 53.01$$

$$\sum y^2 = 299{,}676$$

$$\sum xy = 3{,}973.04$$

$$n = 21$$

Figure 8-16. Data needed to compute a simple regression equation

Using the equations plus the values presented in Figure 8-16, we calculate the regression equation as follows:

$\Sigma x/n = 33.25/21 = 1.58$
$\Sigma y/n = 2{,}486/21 = 118.38$
$S_{xx} = 53.01 - (33.25)^2 / 21 = 0.36$
$S_{xy} = 3{,}973.04 - (33.25)(2{,}486)/21 = 36.87$
$a = 36.87/0.36 = 102.42$
$b = 118.38 - [(102.42)(1.58)] = -43.44$

The regression equation is:

$$y = 102.42x - 43.44 + e$$

or

$$IQ = 102.42(\text{height}) - 43.44 + e$$

For a person of 2 meters in height, the equation predicts an IQ of 161.40 (genius level!) because:

$$102.42(2) - 43.44 = 161.40$$

Needless to say, this is a fictitious example that demonstrates the technique of regression; no slur is intended toward the intelligence of anyone, regardless of stature.

Exercises

Regression

The first set of questions uses data from the United Nations Development Project (*http://hdr.undp.org/en/statistics/data/*) to examine variables related to adolescent fertility (the birth rate for women aged 15–19 in a given country, expressed as the number of births per 1,000 women in this age group). You decided to look at the level of education in the country, using the "mean years of adult schooling" variable, thinking that countries in which the average years of school completed is higher might have a lower adolescent fertility rate.

Problem

Figure 8-17 presents the scatterplot of the two variables (using the natural log of adolescent fertility as discussed in this chapter). What does it suggest about their relationship, and does it seem to support a simple regression analysis with these two variables?

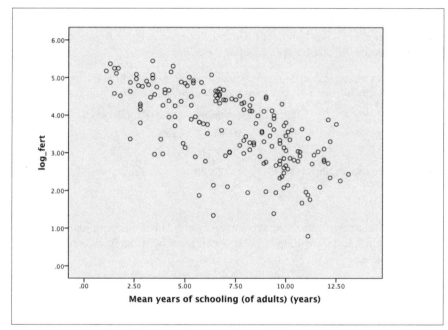

Figure 8-17. Scatterplot of the natural log-transform of adolescent fertility and the mean years of adult schooling

Solution

The scatterplot indicates a moderately strong negative relationship. (Higher levels of education are associated with lower levels of adolescent fertility.) Both variables appear continuous and have a reasonable range to support a regression analysis.

Problem

The regression analysis produced the information in Table 8-9; fill in the missing value for R square and interpret the information provided in Table 8-9.

Table 8-9. Model information

R	R square	Durbin-Watson
0.663		2.199

Solution

The R^2 value is 0.440 (found by squaring 0.663). This is the coefficient of determination for the model and means that 44.0% of the variation in this data set in the log-transformed adolescent fertility rate can be explained by variation in the mean years of adult schooling. The Durbin-Watson statistic tests the assumption that the error terms are independent; a value of 2 signifies absolute independence, and because our value (2.199) is close to 2, we can conclude that this assumption has been met.

Problem

Table 8-10 presents the coefficients table from this same regression analysis. Fill in the values for the missing *t*-statistics, write the regression equation for this analysis, and interpret the information provided in the table.

Table 8-10. Coefficients table for a regression analysis predicting the natural log of the adolescent fertility rate from the mean years of adult schooling in a country

	Unstandardized coefficients		Standardized coefficients		
	B	Std. error	Beta	t	Sig.
Constant	5.248	0.146			< 0.001
Mean years of schooling (adults)	−0.217	0.019	−0.663		< 0.001

Solution

The *t*-statistics are 35.945 for the constant and –11.421 for the mean years of schooling; they are found by dividing the *B* for each term by its standard error. For the constant, this is:

$$5.248/0.146 = 35.945$$

The regression equation for this analysis is:

Log_adoles_fertility = 5.248 – 0.217(mean years adult schooling)

This equation says that the predicted log of adolescent fertility decreases by 0.217 units for every year's increase in mean years of adult schooling in a country. The *t*-statistics and their significance tests tell us that both coefficients are significantly different from zero. The Beta coefficient (–0.663) for mean years of schooling is the standardized value of the regression coefficient for this term (–0.217); it is not particularly useful for a simple regression equation, but for an equation with multiple predictors measured on different scales, it can be used to compare the importance of the different predictors.

This analysis supports the assertion that there is a significant negative relationship between the level of education in a country and the adolescent fertility rate: on average, the adolescent fertility rate is lower in countries where adults have completed more years of schooling.

ANOVA

These questions use data from the 2010 BRFSS (Behavioral Risk Factors Surveillance System, an annual survey of health-related information for the United States). Although you can download data from the BRFSS (*http://www.cdc.gov/brfss/technical _infodata/surveydata.htm*), the analysis in this section is based on a random sample from the 2010 data, so you shouldn't expect to find exactly the same results if you do the analysis yourself.

You are interested in whether there is a relationship between asthma and body weight. You will use an ANOVA to examine whether there is a significant difference in body weight between people who have ever received a diagnosis of asthma (lifetime asthma diagnosis) and those who have not. Your group variable, asthma diagnosis, is dichotomous, and your outcome variable, body weight, is continuous. Because your audience is U.S. officials, you will use weight in pounds rather than kilograms. (Both measurements are provided in the data set.)

Problem

Figure 8-18 presents the box plot for lifetime asthma diagnosis and body weight in pounds. What information can you glean about the data from this box plot? If you are unfamiliar with box plots, you can review the relevant section in Chapter 3.

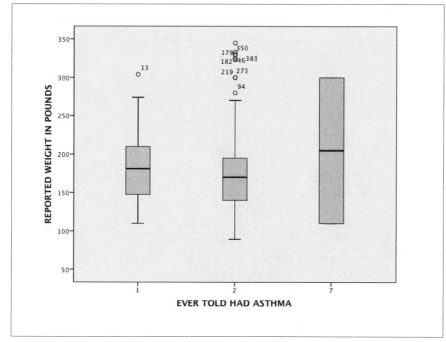

Figure 8-18. Box plots for current asthma diagnosis and body weight in pounds

Solution

All sorts of alarms should be going off in your head from looking at this box plot, which presents a fine example of why data screening is important. First of all, there are three groups for lifetime asthma diagnosis rather than 2; a quick perusal of the code book [also available here (*http://www.cdc.gov/brfss*)] tells us that 7 is a missing value, so we should exclude cases with that value from this analysis. Both valid groups (with and without an asthma diagnosis) have outliers; those are the data points identified by circles, and the number next to each is the case number that has that particular value. This makes us question whether body weight is normally distributed, so we will examine that before continuing. Finally, the median weights for those with and without an asthma diagnosis are almost the same, suggesting that this variable might not be the most promising if our interest is identifying factors strongly related to asthma. Nonetheless, we will continue with our analysis because a finding of nonsignificance can also provide us with useful information.

Problem

We created a histogram for weight and computed the Kolmogorov-Smirnov statistic for this variable; the histogram is presented in Figure 8-19, and the Kolmogorov-Smirnov statistic was 1.898 (p – .001). Together, what do they tell us about the distribution of weight in this data set?

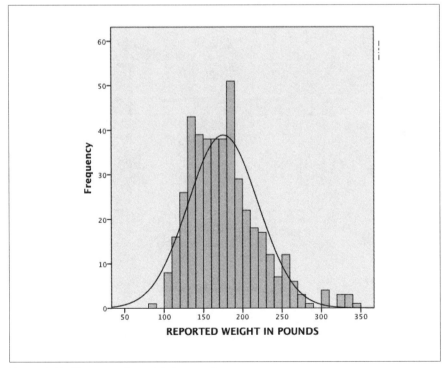

Figure 8-19. Histogram for weight in pounds

Solution

The histogram has a positive skew (high values are more common than would be expected by the normal distribution); this is clear when comparing the distribution of the actual data (the histogram bars) to the superimposed normal curve (representing a perfect normal distribution). The Kolmogorov-Smirnov statistic is highly significant, indicating that the null hypothesis that the variable has a normal distribution should be rejected.

Problem

We did a natural log transformation of weight and checked normality again; in this case, the histogram (not shown) looked approximately normal, and the Kolmogorov-Smirnov statistic was 0.961 ($p = 0.314$), indicating acceptable normality. We also computed the Kolmogorov-Smirnov statistic for each group separately; neither was significant, so we are confident that the distribution within each group is normal. The box plots for the transformed variable are shown in Figure 8-20; what do they suggest about this data?

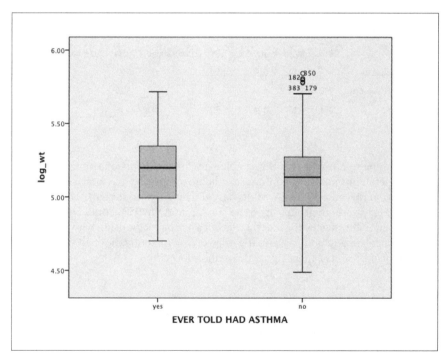

Figure 8-20. Box plots with transformed weight variables

Solution

The group without an asthma diagnosis still has a cluster of outliers, but because the data is acceptably normal, we will continue with our analysis. The group with an asthma diagnosis has a slightly higher median than the group without the diagnosis, but there is considerable overlap between the two groups as well. We will proceed with our ANOVA for this data.

Problem

The Levene statistic for this ANOVA is < 0.001 ($p = 0.983$). Why did we compute the Levene statistic, and what do these results tell us?

Solution

The Levene statistic tests whether the assumption of homogeneity of variance has been met. The null hypothesis is that the variances of the different groups are homogeneous; in this case, we do not have a significant value for this statistic, so we can consider the homogeneity of variance assumption upheld.

Problem

Table 8-11 presents some descriptive data about the transformed weight variable in this data set. What do you notice in this table, and what are the implications for the analysis?

Table 8-11. Descriptive statistics for the transformed weight variable for people with and without a lifetime asthma diagnosis

Group	N	Mean	Std. dev.	95% CI lower bound	95% CI upper bound
Asthma diagnosis	44	5.19	0.24	5.12	5.27
No asthma diagnosis	390	5.13	0.24	5.10	5.15
Total	434	5.16	0.24	5.11	5.16

Solution

The first thing to notice is that the sample sizes are highly unequal, suggesting that this data set might not be the optimal candidate for an ANOVA analysis to answer this question (because ANOVA works best with a balanced design). The second is that the means of both groups are quite similar, and the 95% confidence intervals overlap quite a bit, suggesting that there is not a strong relationship between lifetime asthma diagnosis and body weight. It's still worth completing the analysis, however; nonsignificant results can also provide useful information.

Problem

The ANOVA results are presented in Table 8-12; what do they tell you about the relationship between lifetime asthma diagnosis and body weight? Use the standard of alpha = 0.05 for significance testing.

Table 8-12. One-way ANOVA results for lifetime asthma diagnosis and body weight

	Sum of squares	df	Mean square	F	Sig.
Between groups	456.04	2	228.30	3.86	0.029
Within groups	2483.76	42	59.14		
Total	2940.36	44			

Solution

This analysis found a significant relationship ($F = 3.86$, $p = 0.029$) between lifetime asthma diagnosis and body weight. Referring to Table 8-11, people who have ever been diagnosed as having asthma have on average a higher body weight than those who have never been diagnosed with asthma.

 Because we transformed weight to its natural log, our means table (Table 8-11) reports the natural logs of weight. To make these figures more meaningful to our audience, we need to transform these results back into their original units (pounds). We can do this by taking the antilog of the means in Table 8-11:

$$e^{5.19} = 179.5$$
$$e^{5.13} = 169.0$$

We can then add this information to the second sentence of our answer so it reads, "People who have ever been diagnosed as having asthma have on average a higher body weight (mean = 179.5 lbs) than those who have never been diagnosed with asthma (mean = 169.0 lbs). Converting back to the original units also points up a danger when working with transformed units: a difference that might look small in the transformed units (5.19 versus 5.13) can be much more impressive in the original units (179.5 versus 169.0).

 Because the BRFSS data is a survey collected at a single point in time, it cannot answer questions of causality regarding body weight and obesity. It might be that asthma leads to increased body weight (perhaps by making it more difficult to exercise), or it might be that increased body weight leads to asthma (perhaps by stressing the lungs more). It's also possible that one or more additional factors can explain this observed relationship, for instance, both increased body weight and asthma might be associated with poverty.

Intro to Regression and ANOVA

9

Factorial ANOVA and ANCOVA

Chapter 8 introduced simple regression and ANOVA. In this chapter, we present more complex types of ANOVA: factorial ANOVA (ANOVA with more than one grouping variable, or factor) and ANCOVA, which is an ANOVA design that includes a continuous covariate. Chapter 10 presents similar extensions of the simple regression model introduced in Chapter 8.

In research, most ANOVA designs include at least two grouping variables or factors; these models rely on the same basic principles as one-way ANOVA, but the additional complexity introduces additional concerns, including the evaluation of interactions between the factors. These types of analyses are nearly always done with a computer statistical package, but fortunately, there is enough commonality among those packages that generally if you can read the output from one, you can easily learn to read the output from another. We present information from the analyses as generically as possible to make it understandable no matter what computer program you are using.

Factorial ANOVA

It's relatively rare in real-life studies that we are interested in the influence of a single factor. Instead, we are often interested in the influence of several factors and, possibly, how they interact as well. Factorial designs (ANOVAs including several factors) help us understand the combined effect of multiple factors on a dependent variable. We might be interested in both main effects—the effect of each factor considered alone—and interaction effects—the effects of the different factors in combination. As with one-way ANOVA, factorial ANOVA is most suited for designed experiments and equal cell sizes, i.e., approximately equal numbers of subjects in each subgroup or cell created by any combination of the factors. The major assumptions for factorial ANOVA are the same as for one-way ANOVA, as presented in Chapter 8. Independence of observations and the homogeneity of variance are particularly important; fortunately, statistical packages generally provide statistical tests of homogeneity of variance, such as Levene's test, whereas independence of observations is generally dealt with at the experimental design stage.

The most common factorial designs are $a \times b$ (two-way, two factors) and $a \times b \times c$ (three-way, three factors). It's possible to have more complex designs, but the results become increasingly difficult to interpret, and higher levels of complexity may be more easily accommodated in a linear regression design. As with one-way ANOVA, each factor is a categorical variable with at least two levels, whereas the outcome or dependent variable is a continuous variable measured at the interval or ratio level.

Interaction

With more than one factor, you need to be concerned with interaction among the factors. The definition of interaction is that the effect of one variable depends on the level of another variable; in other words, the effect of one variable is different, depending on the value of the other variable. This might be easier to understand by looking at some graphs showing extreme examples of interaction and lack of interaction; you will seldom find cases this obvious using real data, but the graphs are useful to illustrate the concept.

Let's consider some hypothetical data on the relationship among grip strength (the outcome, measured in PSI, pounds per square inch) and two factors, gender and alcohol consumption. If we graph the data and there is no interaction between the factors, the graph might look something like Figure 9-1.

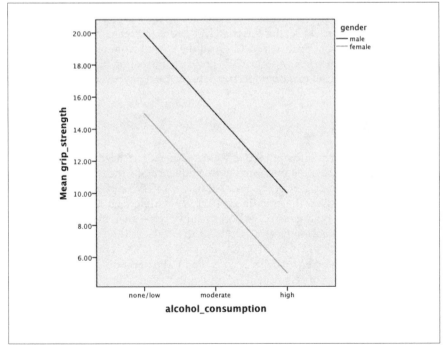

Figure 9-1. Data without interaction

This graph shows no interaction between alcohol consumption and gender; grip strength (the y axis) decreases for both men and women as alcohol consumption increases (the x axis). The rate of decrease is the same for both genders, so the lines are parallel, and men have a stronger grip strength at every level of alcohol consumption.

Figure 9-2 displays data that does contain an interaction; alcohol consumption influences grip strength differently for men than for women. In fact, the effect is opposite: alcohol consumption increases grip strength for women while diminishing it for men.

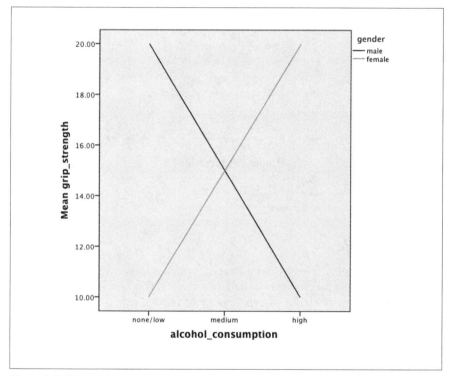

Figure 9-2. Data with an interaction

Lines do not have to cross for there to be an interaction; Figure 9-3 shows an interaction characterized by lines that are not parallel but diverge, showing that the effect of alcohol on grip strength is greater for women than for men.

In both Figure 9-2 and Figure 9-3, we see that the effect of alcohol on grip strength depends on the level or value of a third variable, gender; the relationship between alcohol and grip strength is different for men than for women. Of course, we can't tell by looking at a graph whether an interaction is significant; for that, we need statistical testing.

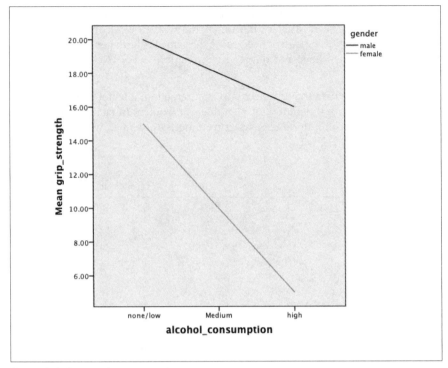

Figure 9-3. Data with an interaction

Two-Way ANOVA

Physical performance measures often vary in populations, and declines in grip strength, for instance, might be correlated with a number of clinical conditions. Your research team is interested in studying how two factors, gender and alcohol consumption, are related to grip strength and in how these factors interact. You have three primary research questions:

1. Does gender influence grip strength?
2. Does alcohol consumption influence grip strength?
3. Do gender and alcohol consumption interact to influence grip strength?

We will treat alcohol consumption as a dichotomous variable, contrasting those that consume alcohol at least weekly with those who do not.

Our hypotheses can be stated verbally as:

Main effect for gender

H_0: there is no difference in grip strength between men and women.
H_1: there is a difference in grip strength between men and women.

Main effect for alcohol

H_0: there is no difference in grip strength between men and women.
H_1: there is a difference in grip strength between men and women.

Interaction of gender and alcohol

H_0: the influence of alcohol consumption on grip strength is the same for men and women.
H_1: the influence of alcohol consumption on grip strength is not the same for men and women.

Table 9-1 shows sample data for the first 12 cases collected in the grip strength lab (total $n = 50$). Six women and six men had their grip strength measured, and each gender group had three drinkers and three nondrinkers (defined as drinking at least weekly or never drinking).

Table 9-1. Relationship between grip strength (DV) and gender and alcohol consumption (IVs)

Gender	Alcohol	Grip strength (psi)
Female	Yes	19
Female	Yes	20
Female	Yes	21
Female	No	30
Female	No	25
Female	No	28
Male	Yes	31
Male	Yes	30
Male	Yes	35
Male	No	32
Male	No	35
Male	No	32

The two main effects are testing mean population differences based on the null hypotheses:

$$\mu_{males} - \mu_{females} = 0$$
$$\mu_{alcohol} - \mu_{noalcohol} = 0$$

Note that the null hypotheses for the main effects are stated in terms of *difference scores*; stating that two quantities are the same is equivalent to stating that their difference is 0. Interaction hypotheses are usually expressed in terms of differences of differences. In this example, saying that there is no difference in the influence of alcohol on grip strength for men and women can be expressed as:

$$\mu_{men/alcohol} - \mu_{men/noalcohol} = \mu_{women/alcohol} - \mu_{women/noalcohol}$$

This study was very close to balanced, with 24 females and 26 males and 24 alcohol drinkers versus 26 abstainers. The R^2 for the model (coefficient of determination) was 0.566, meaning that the two factors plus their interaction accounted for 56.6% of the variation in grip strength observed in this data set. Levene's test ($F = 0.410$, $p = 0.746$) indicated that the assumption of homogeneity was met. The sample means are:

- Gender main effect: female (25.25), male (31.65)
- Alcohol main effect: alcohol (26.71), no alcohol (30.31)

The means by gender and alcohol consumption are presented in Figure 9-4.

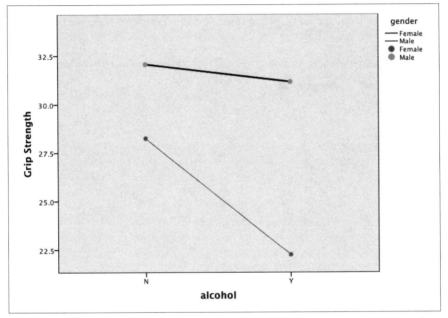

Figure 9-4. Means plot for the effects of gender and alcohol consumption on grip strength

There appear to be both main effects and an interaction effect: in our sample, men have greater grip strength than women, those who do not consume alcohol have greater grip strength than those who do, and the effect of alcohol consumption on grip strength is greater for men than for women. To see whether these differences are statistically significant, we need to perform a two-way ANOVA.

Some statistical packages produce many tables, but a few are particularly useful. In our case, we are interested in testing the significance of the main effects and inter-action effect in this model. The key data from the ANOVA is presented in Table 9-2.

Table 9-2. ANOVA testing differences in grip strength (DV) for gender and alcohol consumption (IVs)

Source	Sum of squares	df	Mean square	F	Sig.
Corrected model	733.085	3	244.362	20.033	< 0.001
Intercept	40426.436	1	40426.436	3299.504	< 0.001
Gender	504.806	1	504.806	41.385	< 0.001
Alcohol	148.325	1	148.325	12.160	0.001
Gender* alcohol	80.769	1	80.769	6.622	0.013
Error	561.095	46	12.198		
Total	42135.000	50			
Corrected total	1294.180	49			

Using the standard of alpha = 0.05, the rows for gender (main effect), alcohol (main effect), and gender*alcohol (interaction effect) tell us that, as we guessed from the means plot, all three effects are significant. A summary of the results of this analysis follow.

Both main effects and the interaction tested in the design are significant:

Gender main effect: F(1, 46) = 41.385, p < 0.001
> The direction of the effect shows that women generally have lower grip strength than men.

Alcohol main effect: F(1, 46) = 12.160, p = 0.001
> The direction of the effect shows that those who consume alcohol generally have a lower grip strength than those who do not consume alcohol.

Gender × alcohol interaction: F(1, 46) = 6.622, p = 0.013
> The interaction shows that gender and alcohol interact, with alcohol consumption associated with a greater loss of grip strength in women as compared with men.

Note that we must be wary of presenting causal statements (alcohol consumption harms grip strength) because this is an observational study—we asked people whether they drank and measured their grip strength, but we did not administer alcohol to them and record the changes in their grip strength. The association between alcohol consumption and grip strength could be due to any number of factors. For instance, athletes might forgo alcohol consumption as part of their training rules and might also have increased grip strength because of their training.

Three-Way ANOVA

The two-way factorial model can easily be extended to three factors. After demonstrating significant main effects for gender and alcohol consumption on grip strength, your research team investigates other possible factors that might influence grip strength. In the literature, there appears to be a lot of discussion about the influence of age on grip strength, with a marked decline appearing after the age of

40. You decide to add an age category (below 40 or above 40) to determine whether age has any influence or as much influence on grip strength as the other factors.

Table 9-3 shows the first 12 cases for this study.

Table 9-3. Relationship among grip strength (DV) and gender, alcohol consumption, and age (IVs)

Gender	Alcohol	Grip strength (psi)	Age
Female	Yes	19	Below 40
Female	Yes	20	Above 40
Female	Yes	21	Below 40
Female	No	30	Above 40
Female	No	25	Below 40
Female	No	28	Above 40
Male	Yes	31	Below 40
Male	Yes	30	Above 40
Male	Yes	35	Below 40
Male	No	32	Above 40
Male	No	35	Below 40
Male	No	32	Above 40

Hypothesis testing becomes more complicated with three factors because we have potentially seven hypotheses: main effects for gender, alcohol, and age; two-way interactions of gender*alcohol, gender*age, and alcohol*age; and a three-way interaction, gender*alcohol*age. We've already demonstrated how to verbalize two-way interactions. The null hypothesis we are testing with our three-way interaction can be stated as "the difference in the influence of alcohol consumption on grip strength between men and women does not vary with age category."

To produce a means plot with three factors, we actually have to produce two plots: one for subjects below 40, the other for subjects above 40. The means plots are displayed in Figure 9-5.

The means plot suggests that age will be an important factor in clarifying the relationships of interest because it appears to interact with both gender and alcohol use. The key results of this analysis are presented in Table 9-4. We will use the standard of alpha = 0.05 to evaluate the significance of the effects in this model.

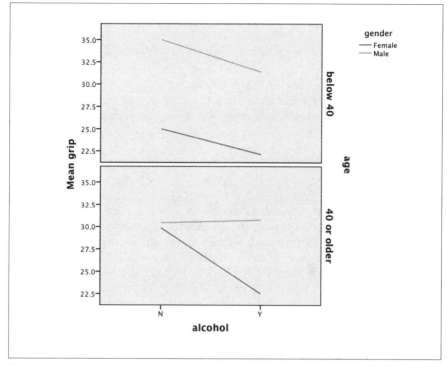

Figure 9-5. Means plot for a three-way ANOVA

Table 9-4. Three-way ANOVA testing differences in grip strength by gender, alcohol consumption, and age

Source	Sum of squares	df	Mean square	F	Sig.
Corrected model	864.583	7	123.512	12.075	<0.001
Intercept	35902.885	1	35902.885	3510.081	<0.001
Gender	548.630	1	548.630	53.637	<0.001
Alcohol	128.214	1	128.214	12.535	0.001
Age	0.003	1	0.003	0.000	0.986
Gender*alcohol	33.446	1	33.446	2.370	0.078
Gender*age	75.758	1	75.758	7.407	0.009
Alcohol*age	0.226	1	0.226	0.022	0.883
Gender*alcohol*age	49.491	1	49.491	4.839	0.033
Error	429.597	42	10.229		
Total	42135.000	50			
Corrected total	1294.180	49			

Two of the three main effects in this model are significant:

Gender main effect: F(1, 42) = 53.637, p < 0.001
> The direction of the effect shows that women generally have lower grip strength than men.

Alcohol main effect: F(1, 42) = 12.535, p = 0.001
> The direction of the effect shows that those who consume alcohol generally have a lower grip strength than those who do not consume alcohol.

Age main effect: F(1, 42) = 0.000, p = 0.986 (not significant)

One of the two-way interactions is significant:

Gender × alcohol interaction: F(1, 42) = 2.370, p = 0.078 (not significant)
Gender × age interaction: F(1, 42) = 7.407, p = 0.009
> The difference in grip strength for those who do and do not use alcohol is much different for men in the two age groups, whereas for women, the pattern does not change much. Grip strength for men age 40 and older is little affected by whether they use alcohol; for men younger than age 40, alcohol use is associated with a decline in grip strength. The decline in grip strength with alcohol use is somewhat greater for women age 40 and older as compared with younger women, but this difference by age category is not as extreme as the difference for men.

Alcohol × age interaction: F(1, 42) = 0.022, p = 0.883 (not significant)

The three-way interaction is significant:

Gender × alcohol × age interaction: F(1, 42) = 4.839, p = 0.033

These results are interesting because although the main effect of age is not significant, one two-way interaction including age is significant (gender*age), as is the three-way interaction gender*alcohol*age. It's also interesting that the gender*alcohol interaction was not significant in the three-way model but was significant in the two-way model. This demonstrates a point, which applies to regression as well: when you add or remove terms from a model, often the significance of other variables in the model will change as well. When reporting results from a complex model, it is always necessary to specify exactly what model was tested because, very often, predictors interact with each other; in a different analysis, perhaps age would be a significant predictor of grip strength.

Even though age does not have a significant main effect in this model, we should keep it in the analysis because it is usual to include any variable that is significant as an interaction as a main effect also. The results of this analysis are both interesting enough and intriguing enough to suggest that we should investigate further. One option that might be helpful would be to switch to a regression equation and include age as a continuous predictor (use age in years as a predictor rather than dichotomizing it to under/over 40). Another possibility is that two categories for age are not sufficient, and perhaps 40 is not the ideal dividing line; we could investigate this with further analyses as well.

ANCOVA

Analysis of covariance (ANCOVA) is a variation of factorial ANOVA that allows a continuous covariate to be included in the model. The most common reason for using this model is to control for a potentially confounding effect of a covariate. For instance, you might be interested in the earnings of college graduates according to their field of study (science, humanities, business, etc.). This could be addressed with a one-way ANOVA with salary as the dependent variable and field of study as the categorical factor. However, if your data set includes not just recent graduates but people who have been in the working world for substantially different lengths of time, you realize this might affect their salaries because, in general, people's salaries increase with age and/or with the number of years they have been working in a field. You could control for years on the job, or age, by adding one of those variables as a continuous covariate to your ANOVA design, giving you an ANCOVA. You can use more than one covariate in an ANCOVA. Although adding covariates to control for confounders is not a perfect solution, it's better than ignoring the potential confounders altogether. One way to think of this use of ANCOVA is that by controlling for the effect of the continuous covariate(s), you are examining what the relationship between the factors and the continuous outcome would be if all cases had the same value for the covariate(s). For instance, in the field of study and salary example, by using age as a continuous covariate, you are examining what the relationship between those two factors would be if all the subjects in your study were the same age.

Another typical use of ANCOVA is to reduce the residual or error variance in a design. We know that one goal of statistical modeling is to explain variance in a data set and that we generally prefer models that can explain more variance, and have lower residual variance, than models that explain less. If we can reduce the residual variance by including one or more continuous covariates in our design, it might be easier to see the relationships between the factors of interest and the dependent variable.

The assumptions of ANOVA apply to ANCOVA, and there are two additional assumptions (numbers 5 and 6) as well for ANCOVA:

Data appropriateness
> The outcome variable should be continuous, measured at the interval or ratio level, and be unbounded (or at least cover a wide range); the factors (group variables) should be dichotomous or categorical; the covariate(s) should be continuous, measured at the interval or ratio level, and be unbounded or cover a wide range. This assumption is checked by inspecting the data through frequency tables, histograms, and so on.

Independence
> Each value of the outcome variable should be independent of each other value. This would be violated if there were some pattern of time dependency, for instance, or if some of the dependent variables were measured from subjects clustered into larger units (e.g., members of the same family or children studying in the same classroom) in some way that affected their value on the dependent

variable. This assumption is checked by your knowledge of the data and how it was collected.

Distribution

The outcome variable should be approximately normally distributed *within each group*. The distribution of outcome variable may be checked by creating a histogram (eyeballing the data) and by a statistical test for normality such as the Kolmogorov-Smirnov.

Homogeneity of variance

The variance of each of the groups should be approximately equal. This is checked by a procedure such as the Levene statistic; the null hypothesis is that the variance is homogeneous, so if the results of the Levene statistic are *not* statistically significant (normally, the criterion of $a < 0.05$), that means the variances are sufficiently homogeneous to proceed.

Independence of the covariates and the effect of the factors

The variance explained by the covariate should be unique and not overlap with the variance explained by the factors. This is most often a problem in observational studies in which random assignment is not used; if the two groups vary on the covariate and it explains some of the variance of the outcome variable, there is no way to separate the variance explained by the factors from that explained by the covariates. If random assignment is not possible, the next best approach is to determine whether the levels of the covariate differ significantly among your groups; if they do, don't use the covariate. Common sense also plays a role here: can you make a reasonable case that the variance explained by your covariate will explain unique variance in the outcome variable? If not, don't use the covariate.

Homogeneity of regression slopes

The relationship between the covariate and the dependent variable should be the same for all groups. This can be checked by creating and plotting regression lines for the covariate and dependent variable, separately for each group, and by creating interaction terms and testing them for significance. The regression lines should be approximately parallel; their slopes should be approximately equal. The interaction terms should not be significant.

Continuing with the grip strength example, the research team becomes concerned that they've left an important variable out of the model: whether a person exercises. It makes intuitive sense that working out could improve grip strength, so they decide to add one more variable to the model: the minutes each week the individual spends in physical activity. This is a continuous variable with a broad range, so it can be added as a continuous covariate to the gender and alcohol consumption (IVs) and grip strength (DV).

The first assumption we have to check is whether our covariate explains unique variance (assumption 5). We can make a logical case that time spent exercising could explain unique variance in grip strength, and we can compute the means of this covariate in the groups; if these means are not significantly different, we will proceed with the analysis. For the purposes of demonstration, we're dropping back to our two-way model with the gender and alcohol consumption factors and adding the covariate exercise, operationalized as the minutes exercised per week.

We did two one-way ANOVAs (analogous to *t*-tests) for the mean of minutes exercised by gender and alcohol consumption; the key results are presented in Table 9-5.

Table 9-5. Results of one-way ANOVAs weekly minutes of exercise by gender and alcohol consumption

Variable	Subgroup	Mean	F	Sig.
Gender	Males	100.74	1.069	0.306
	Females	87.64		
Alcohol consumption	Yes	106.01	3.209	0.080
	No	83.78		

As you can see, although there are differences in the average minutes exercised per week between men and women and between those who do and do not consume alcohol, these differences are not significant at the $\alpha = 0.05$ standard.

We also need to check the homogeneity of the regression slopes (assumption 6). As with evaluating normality, we will do both a graphical and a statistical test of this assumption. For the graphical test, we will create scatterplots with regression lines for the relationship between grip strength (the outcome) and exercise (the covariate) for males versus females and for alcohol versus no alcohol. For each of the pairs, the slopes should be approximately equal. The scatterplots and regression lines for gender are presented in Figure 9-6 and the scatterplots and regression lines for alcohol in Figure 9-7.

There's nothing too alarming in either figure; the slopes appear to be approximately equal, and that's good news for the assumption of homogeneity of slopes. We will also conduct a statistical test of this assumption by creating a model including an interaction term of the covariate and the factor. (We will do separate models for each factor.) If this interaction term is not significant, we will consider the assumption of homogeneity of slopes to be upheld for this data. The data from these analyses is presented in Tables 9-6 and 9-7.

Table 9-6. Testing the homogeneity of slopes assumption for gender and exercise

Source	Sum of squares	df	Mean square	F	Sig.
Corrected model	560.053	3	186.684	11.698	<0.001
Intercept	7807.479	1	7807.479	489.212	<0.001
Gender	69.358	1	69.358	4.346	0.043
Exercise	40.686	1	40.686	2.549	0.117
Gender*exercise	2.363	1	2.363	0.148	0.702
Error	734.127	46	15.959		
Total	42135.000	50			
Corrected total	1294.180	49			

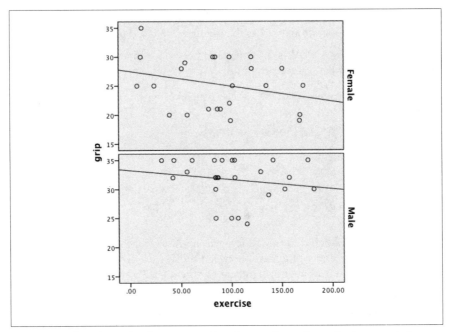

Figure 9-6. The relationship between weekly minutes of exercise and grip strength for men and women

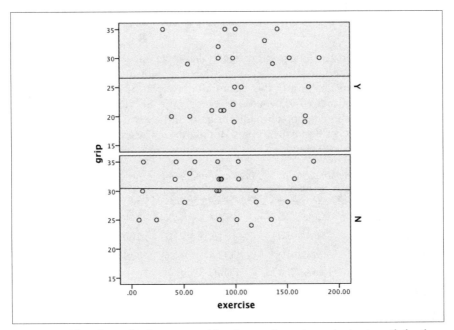

Figure 9-7. The relationship between weekly minutes of exercise and grip strength for those who do and do not consume alcohol

Table 9-7. Testing the homogeneity of slopes assumption for alcohol and exercise

Source	Sum of squares	df	Mean square	F	Sig.
Corrected model	161.863	3	53.954	2.192	0.012
Intercept	6619.891	1	6619.891	268.931	<0.001
Alcohol	29.800	1	29.800	1.211	0.277
Exercise	0.019	1	0.019	0.001	0.978
Alcohol*exercise	0.146	1	0.146	0.006	0.939
Error	1132.317	46	24.616		
Total	42135.000	50			
Corrected total	1294.180	49			

Note that the only reason we are running these models is to check on the significance of the interaction terms; we're not testing a theory, so we don't care about model fit, significance of other terms, and so on. As you can see in Tables 9-6 and 9-7, neither of the interaction terms is significant; for gender*exercise, the *p*-value is 0.702, and for alcohol*exercise, the *p*-value is 0.939. These results tell us that, using the standard of alpha = 0.05, the homogeneity of slopes assumption is upheld for this analysis (and this data set), so we can continue with our ANCOVA.

The Levene's test for our ANCOVA for grip strength including the factors alcohol consumption and gender and the covariate exercise has a value of 0.292 ($p = 0.381$); this is not significant, so the assumption of equal variance is upheld. The R^2 for this model is 0.576, so these factors explain about 57.6% of the variance in grip strength in this data. This is a small improvement on the R^2 of 0.566 for the two-way ANOVA (factors = gender, alcohol) discussed earlier in this chapter. The ANCOVA results are presented in Table 9-8.

Table 9-8. ANCOVA for grip strength, with factors gender and alcohol consumption, and covariate weekly minutes of exercise

Source	Sum of squares	df	Mean square	F	Sig.
Corrected model	745.596	4	186.399	15.290	<0.001
Intercept	7289.554	1	7289.554	597.957	<0.001
Exercise	12.511	1	12.511	1.026	0.316
Gender	517.299	1	517.299	42.434	<0.001
Alcohol	117.498	1	117.498	9.638	0.003
Gender*alcohol	78.573	1	78.573	6.445	0.015
Error	548.584	45	12.191		
Total	42135.000	50			
Corrected total	1294.180	49			

Both our factors and their interaction are significant, but the covariate is not.

- For gender, $F(1, 45) = 42.434$, $p = <0.001$
- For alcohol, $F(1, 45) = 9.638$, $p = 0.003$
- For gender*alcohol, $F(1, 45) = 6.445$, $p = 0.015$
- For exercise $F(1, 45) = 1.026$, $p = 0.316$ (not significant)

Because we didn't improve model fit much by adding this covariate, we might consider whether there is a better way to measure exercise. The type of exercise could be important: those who engage in weightlifting probably improve their grip strength more than those who do distance running, for instance. Perhaps exercise would be more useful as a dichotomous or categorical variable; maybe the difference between doing any exercise and none is more important than the time spent in exercise (in which case, exercise would become a factor rather than a covariate). This illustrates why any research project is usually an ongoing concern: you begin with an idea, test it, and then refine your idea and test it again. Lather, rinse, and repeat, as they say in the advertising world—don't expect to create the perfect model the first time around.

Exercises

Problem

You are planning to conduct a two-way ANOVA; as part of the process, you conduct Levene's test, which has a p-value of 0.045. What does this mean for your analysis?

Solution

Levene's test is a test of the homogeneity assumption for ANOVA, that each group has approximately the same variance. The null hypothesis is that the variances are equal, so if Levene's test is not significant, the assumption of equal variance is upheld, and you can proceed with the ANOVA. In this case, using the conventional standard of $\alpha = 0.05$, Levene's test is significant, meaning that you should reject the assumption of homogeneity and should not proceed with the ANOVA without transforming your data or otherwise remedying the problem.

Problem

You are working on a two-way ANOVA; one of your factors has two levels, the other, three levels. As part of the process of working through this analysis, you create the means plot displayed in Figure 9-8. Interpret this graphic and its significance for your analysis.

Solution

There may be an interaction between your factors. In general, levels 1 and 3 of factor 1 are associated with lower levels of the outcome and level 2 of factor 1 with a higher level of the outcome. However, this effect is greater for cases with level 1 of factor 2, so it seems that the effect of factor 1 might partly depend on the level of factor 2.

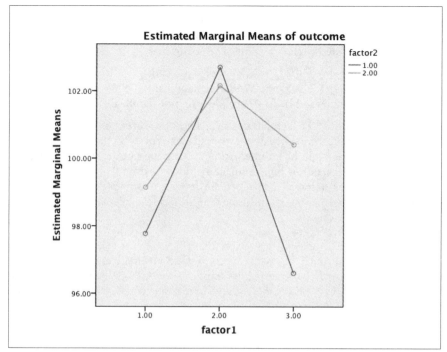

Figure 9-8. Means plots for an ANOVA

Problem

Table 9-9 presents the results from the two-way ANOVA whose means plot was presented in the previous problem. What can you conclude from the table and the means plot together about the relationship between the factors and the outcome in this analysis? Use the standard of alpha = 0.05 for significance testing.

Table 9-9. ANOVA with two factors

Source	Sum of squares	df	Mean square	F	Sig.
Corrected model	145.392	5	29.078	0.172	0.971
Intercept	198801.665	1	298801.665	1766.133	0.000
Factor1	103.782	2	51.891	0.307	0.739
Factor2	17.849	1	17.849	0.105	0.748
Factor1*factor2	23.762	2	11.881	0.070	0.932
Error	4060.418	24	169.184		
Total	303007.475	30			
Corrected total	4205.810	29			

Solution

Neither of the factors is significantly related to the outcome, and their interaction is not significantly related to the outcome. The results are:

Factor1: $F(2, 24) = 0.307$, $p = 0.739$ (not significant)
Factor2: $F(1, 24) = 0.105$, $p = 0.748$ (not significant)
Factor1*Factor2: $F(2, 24) = 0.070$, $p = 0.932$ (not significant)

This is an illustration of the fact that not all analyses produce significant results and that you shouldn't get too excited about reading means plots. In this case, the means plot suggested there might be an interaction in the data, but the ANOVA demonstrates that this interaction is not significantly different from 0 and neither are the main effects of either factor, so it's back to the drawing board for this research team. The R^2 for this model was 0.035, meaning that the model explained less than 4% of the variability in the outcome.

Problem

You are planning an ANCOVA with one continuous covariate and one factor with three levels. As part of checking the assumptions for ANCOVA, you created the graphs in Figure 9-9. What do these graphs represent, what assumption are they checking, and what can you conclude from them?

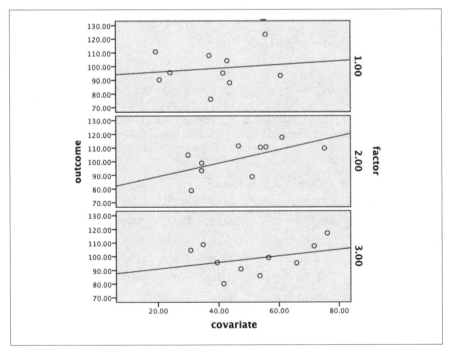

Figure 9-9. Graphs checking an ANCOVA assumption

Solution

These graphs are scatterplots with superimposed regression lines for the outcome variable (y-axis) and the covariate (x-axis); each level of the factor is represented by a different graph. This type of graph is created to test the assumption of homogeneity of slopes, which states that the relationship of the covariate and outcome should be the same for all levels of the factor. If this is true, the slope of the regression line for the covariate and outcome should be approximately the same for all levels of the factor. In this case, the slope for level 2 of the factor is steeper than for level 1 and level 3, but it's hard to tell whether the difference is significant without performing a statistical test.

Problem

Continuing with testing the assumptions for the ANCOVA described in the previous problem, we conducted an analysis that produced the data presented in Table 9-10. Use the standard of alpha = 0.05.

Table 9-10. Data from an analysis testing an assumption of ANCOVA

Source	Sum of squares	df	Mean square	F	Sig.
Corrected model	742.689	5	148.538	1.029	0.453
Intercept	19233.663	1	19233.663	133.292	0.000
Factor	93.367	2	46.683	0.324	0.727
Covariate	487.758	1	487.758	3.380	0.078
Factor*covariate	129.749	2	64.875	0.450	0.643
Error	3463.121	24	144.297		
Total	303007.475	30			
Corrected total	4205.810	29			

Solution

This is a statistical test of the assumption of homogeneity of slopes; if the slopes are homogeneous, the interaction term factor*covariate should not be significant. In these results, the interaction term is not significant ($F = 0.450$, $p = 0.643$), so the difference in slopes is not significant, and we can continue with the ANCOVA.

Problem

Continuing with the problem of trying to predict grip strength discussed throughout this chapter, the research team decides that strength training, rather than exercise in general, might be a better predictor of grip strength than exercise in general. They add a continuous covariate, minutes per week spent in strength training, to the two-way model with the dichotomous factors gender (male/female) and alcohol consumption (yes/no). After checking the ANCOVA assumptions for this model, they proceeded with testing it, producing the results presented in Table 9-11. The R^2 for this ANCOVA was 0.628. Interpret this R^2 and the information in Table 9-11 and these results with those of the ANCOVA with exercise as a covariate, presented in

Table 9-8. Use the standard of alpha = 0.05 to evaluate the significance of the effect in this model.

Table 9-11. ANCOVA for grip strength with factors gender and alcohol consumption and covariate weekly minutes of strength training

Source	Sum of squares	df	Mean square	F	Sig.
Corrected model	813.327	4	203.332	19.029	<0.001
Intercept	6622.003	1	6622.003	619.711	<0.001
Strength train	80.242	1	80.242	7.509	0.009
Gender	388.763	1	388.763	36.382	<0.001
Alcohol	63.086	1	63.086	5.904	0.019
Gender*alcohol	34.597	1	34.597	3.238	0.079
Error	480.853	45	10.686		
Total	42135.000	50			
Corrected total	1294.180	49			

Solution

This model explains somewhat more variance (62.8%) than the model including exercise as a covariate (57.6%). In this model, both factors and the covariate are significantly related to the outcome, grip strength; the interaction of the factors is not significant. The key results are:

- For gender, $F(1, 45) = 36.382$, $p = <0.001$
- For alcohol, $F(1, 45) = 5.094$, $p = 0.019$
- For gender*alcohol, $F(1, 45) = 3.238$, $p = 0.079$ (not significant)
- For strength training $F(1, 45) = 7.509$, $p = 0.009$

10

Multiple Linear Regression

In Chapter 8, we introduced simple linear regression, in which one independent variable is used to predict or explain the value of one dependent variable. This model is useful for introducing the principles of linear regression, but in real-world situations, simple regression is rarely used. Multiple linear regression, in which two or more independent variables are related to a single dependent variable, is much more common. Multiple regression is a common research technique used in many fields, including the sciences, medicine, the social sciences, and education. One attraction of multiple regression is flexibility; predictor variables can be continuous, categorical, or dichotomous, and any combination of these variable types can be used in a single equation. If a categorical variable is used, it must be recoded into dichotomous dummy variables. We cover this technique also in this chapter. With the additional complication of multiple predictor variables, additional assumptions must be met, and these are discussed in this chapter as well. Finally, the ability to use multiple predictors means that model-building strategies are useful to build the best model for a particular purpose; these strategies are also discussed in this chapter.

Multiple Regression Models

The study of simple linear regression models and the bivariate correlation coefficient and its square (the coefficient of determination) are useful as an introduction to the concepts of regression analysis; in reality, very few fields of study spend much time working with regression equations involving only two variables. Consider models used to study climate change, such as General Circulation Models (GCMs) and even more sophisticated Atmosphere-Ocean General Circulation Models (AOGCMs). These models have been developed over the past 30 years to facilitate the increasingly accurate forecast of weather patterns. The models involve understanding and quantifying relationships between potentially hundreds and thousands of variables in many qualitative categories. For example, in the mid-1970s, models focused on variables derived from atmospheric conditions, whereas in the near future, models will be available that are based on atmospheric data combined with land surface, ocean and sea ice, sulphate and nonsulphate aerosol, carbon cycle, dynamic vegetation, and atmospheric chemistry data. By combining these additional sources of variation into a large-scale statistical model, predictions of weather activity of qualitatively different types have been made possible at different spatial and temporal scales.

In this chapter, we will be working with multiple regression on a much smaller scale. This is not unrealistic from a real-world point of view; in fact, useful regression models may be built using a relatively small number of predictor variables (say, from 2 to 10), although the people building the model might consider far more predictors for inclusion before selecting those to keep in the final model. There are many ways to build a regression model and many purposes; there is not one best way to build a model, but there might be a best way to build a particular model for a particular purpose. The advice offered in this chapter will be general, so it's up to you to learn about the conventions and expectations of the particular professional field in which you work. To take a simple example, a regression model can be built on the guiding principle of parsimony (including a relatively small number of variables, each of which explains a large proportion of variance) or on the principle of explaining the maximum amount of variance (in which case more variables will probably be included in the model, each explaining some additional but perhaps small proportion of variance). Neither approach is best in all circumstances, so it's best to know what the expectations are in your field of study or work.

Another difference among fields and places of employment is the extent to which theory is expected to guide statistical work. In academia, theory is highly valued, and building a model based only on the relationships found in a particular data set is greatly frowned on. In the business world, however, building models using automated methods (e.g., forward and backward entry, discussed later in this chapter) can be completely acceptable. I tend toward the theory-driven side of the issue because I've spent most of my career in academia, but there are specific situations in which a less theoretical approach may be called for. The point, once again, is to know the customs and expectations of your field and to be clear about what you are doing and why.

Two general principles apply to regression modeling. First, each variable included in the model should carry its own weight, meaning it should explain unique variance in the outcome variable. Often, the rule applied is that each variable must explain a statistically significant amount of variation. It's a fact that you can't make a regression model worse, in the sense of making it explain less variance, by adding a new variable, but even models built on the principle of maximizing the variance explained generally have some rules to determine whether a particular variable improves the model sufficiently to keep it in the model. Second, when you deal with multiple predictors, you have to expect that some of them will be correlated with one another as well as with the dependent variable; this means that adding or subtracting one will probably change the coefficients of all the other variables in the model. This is very important when interpreting your results because it's not enough to say that variable A is or is not a significant predictor of outcome E; you have to say that variable A is or is not a significant predictor in a model that also includes variables B, C, and D.

Formally, multiple linear regression models take the form:

$$Y = \beta_0 + \beta_1 X_1 + \beta_2 X_2 + \ldots \beta_n X_n + e$$

where Y is the dependent variable, β_0 is the intercept, $X_1, X_2, \ldots X_n$ are the independent variables, $\beta_1, \beta_2, \ldots \beta_n$ are the coefficients, and e is the residual or error term. We introduced this model in Chapter 8, but it's worth reviewing its main features here. The dependent variable (Y) and independent variables ($X_1, X_2, \ldots X_n$) are observed data, whereas the intercept (β_0) and coefficients ($\beta_1, \beta_2, \ldots \beta_n$) are values computed by the multiple linear regression algorithm to minimize the residual or error (e) in the model. For a given case (i), the predicted Y value Y_i is calculated by multiplying the observed values for that case (X_1, X_2, etc.) with their corresponding coefficients (β_1, β_2, etc.) and adding the intercept β_0. The difference between the observed value of Y_i and the predicted value $\hat{Y}_i$ is the error of prediction or residual e_i for that case. The coefficients are determined to minimize the total squared residual. (The residuals have to be squared because some are positive, some are negative, and they sum to 0 if not squared.)

The assumptions of simple regression (discussed in Chapter 8) also hold for multiple regression. In addition, as soon as we use more than one predictor, we have to worry about *multicollinearity*. This means that none of our predictor variables should correlate highly with any other predictor. In particular, no variable can be a linear combination of other variables; this means that you can't include as predictors the variables A, B, and $A + B$. You may laugh, but it's easy to create a new variable and forget to remove its component parts from the predictors list. Predictor variables that are highly correlated tend to explain much the same as variance in an outcome variable, obscuring the relationship of each individual predictor with the outcome. In addition, models containing predictors that are highly correlated tend to be unstable, so adding or removing one variable from the model can radically change the coefficients and significance of the other predictors. (We expect a little change when adding or subtracting a variable, but not major change.) Fortunately, most statistical computing packages have a built-in function to check multicollinearity in regression models, so this can be assessed after the model has been run

We will build a regression model to predict adolescent fertility (the birth rate for girls ages 15–19, expressed as births per 1,000) from a number of other demographic variables. We'll use data from the United Nations Development Project; you can download the same data here (*http://hdr.undp.org/en/statistics/data/*) and try the analysis yourself on whatever statistical system you use, or see whether you can build a better model. We will be working with only a few variables here to keep our demonstration simple, but there's no reason to limit your own analyses to just those variables. One other important note: this is *ecological* data, measured at the country level; therefore, any relationships we see should only be generalized to the country level (not, for instance, to individual people).

Our first step is to look at our candidate variables. As discussed in Chapter 8, the adolescent fertility rate is not normally distributed, but the natural log transformation is, so we will use the transformed variable as our outcome. Figure 10-1 shows a histogram of the natural log transformation of adolescent fertility; it looks fairly normal, and the Kolmogorov-Smirnov statistic (evaluating the probability of the variable coming from a population with a normal distribution, as discussed in Chapter 8) for this variable is 1.139 ($p = 0.149$), so it is acceptably normal.

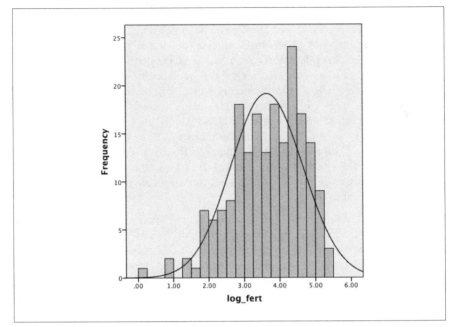

Figure 10-1. Histogram of the natural log transform of the adolescent fertility rate

One variable we think might make a good predictor is life expectancy at birth, which can be interpreted as an indicator of the general level of health in a country. However, the histogram of life expectancy, displayed in Figure 10-2, is definitely not normal. In fact, there seem to be two groups of countries, one with fairly low life expectancy and an almost uniform distribution across a range from the mid-40s to the mid-60s, and another group with high life expectancy with the values approximately normally distributed with a central value in the mid-70s. We believe that the important distinction is between countries with low versus high life expectancy (versus between high and very high life expectancy), so we will dichotomize the cases to reflect this belief. About one-third of cases have values of 66 years or fewer, and this is in the range of where the break seems to occur between a smaller group of low-life-expectancy countries and a larger group of high-life-expectancy countries, so we will use the value of 66.0 years to dichotomize life expectancy into low and high categories.

Another variable that might help our model is GNI (gross national income) per capita, expressed in international dollars in PPP (purchasing power parity) terms; this figure allows us to compare the relative wealth or poverty of different countries. In general, higher-income countries have lower adolescent fertility, so this should be a good predictor for our model. The advantage of using GNI in PPP terms is that it is expressed in terms that express the ability to purchase equivalent goods in the

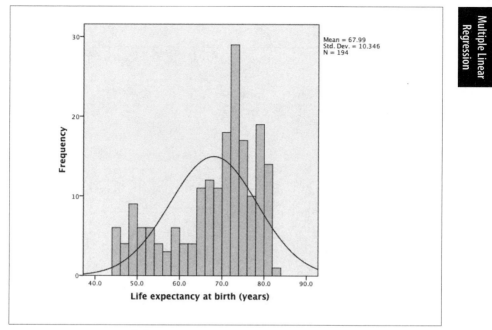

Mean = 67.99
Std. Dev. = 10.346
N = 194

Figure 10-2. Histogram of life expectancy at birth

different countries and, thus, includes information about the different price levels in each country while avoiding issues of fluctuating international exchange rates. The latter would be a problem if a particular currency, such as the U.S. dollar, were used to express income in other countries. A histogram of GNI per capita is presented in Figure 10-3; it has a strong right skew. We compute a natural log transformation of GNI, as presented in Figure 10-4; this looks much closer to a normal distribution, and the Kolmogorov-Smirnov statistic also indicates acceptable normality ($K\text{-}S = 0.737$, $p = 0.649$), so we will use the log-transformed GNI in our model.

Another variable that might be useful to us is expected years of schooling; it's logical to hypothesize that countries willing and able to invest in the education of their children might also have lower rates of adolescent fertility. This variable expresses how many years of school children currently entering primary school are expected to complete based on current age-specific enrollment figures. Figure 10-5 shows the distribution of expected years of school; the gap at the upper right is because this statistic is capped at 18 years. Nevertheless, the Kolmogorov-Smirnov statistic indicates acceptable normality ($K\text{-}S = 0.975$, $p = 0.298$), so we can include it in our model without transformation.

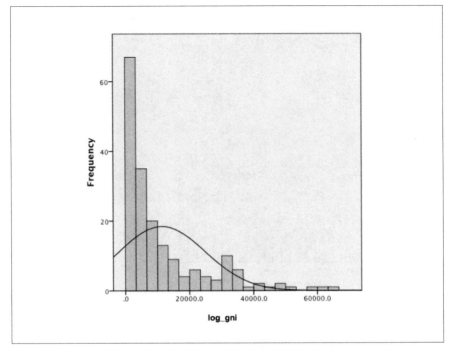

Figure 10-3. Histogram of GNI (PPP 2005 international dollars) per capita

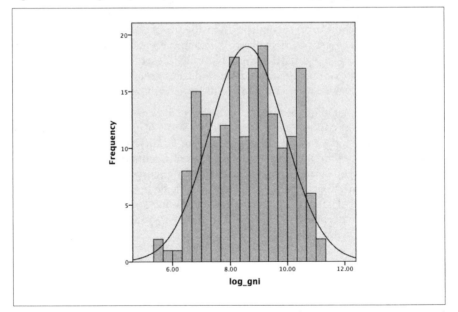

Figure 10-4. Histogram of the natural log transformation of GNI (PPP 2005 international dollars) per capita

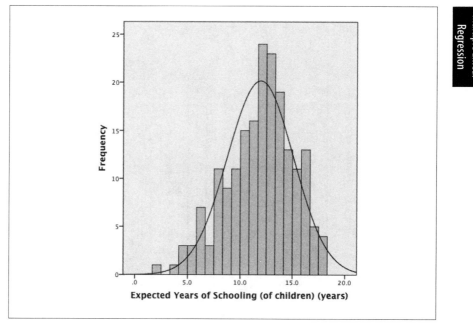

Figure 10-5. Histogram of expected years of schooling

Finally, we consider including the percent urbanized, meaning the percent of the population of a given country that lives in an urban area. This variable is acceptably normal, as shown by its histogram (Figure 10-6) and the Kolmogorov-Smirnov test ($K\text{-}S = 0.893$, $p = 0.403$).

The next thing we need to check is linearity; the relationship between our continuous predictor variables and the outcome should resemble a straight line. The scatterplots (not shown) all indicate linear relationships, so we consider this assumption upheld.

Although our regression analysis will produce multicollinearity statistics, we will also look at the relationships among our predictor variables by creating a correlation matrix; among other things, this will show us whether two of our predictors are closely related. The correlation matrix (upper triangle) for our three continuous predictors is displayed in Table 10-1.

Table 10-1. Correlation matrix for the natural log of GNI, percent urban, and expected years of schooling

	Log_gni	Pct. urban	Exp. yrs schooling
Log_gni	1.000	0.723	0.805
Pct. urban		1.000	0.644
Exp. yrs schooling			1.000

Not surprisingly, all three are highly correlated. We will keep this in mind while building our model. We can also look at the relationship between our dichotomous

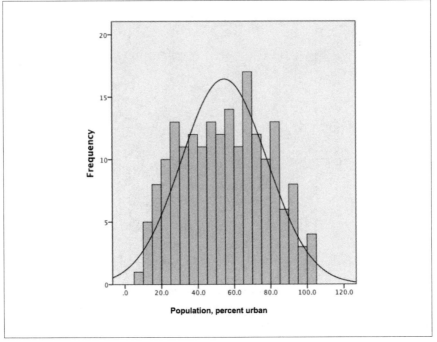

Figure 10-6. Histogram of the percent of the population living in urban areas

variable and the other three by doing one-way ANOVAS for the mean differences in the continuous variables for the two groups. Not surprisingly, all three tests are highly significant, as shown in Table 10-2; countries with higher life expectancy are more urbanized, have higher per capita incomes, and have more expected years of schooling for children.

Table 10-2. Means and one-way ANOVA results in countries with high and low life expectancies on the natural log of GNI, percent urban, and expected years of schooling

Variable	Life exp.	Mean	Std. dev.	F	Sig.
Pct urban	< 66 yrs	35.5	15.8	89.158	< 0.001
	≥ 66 yrs	63.8	21.0		
Log_gni	< 66 yrs	7.3	0.9	188.163	< 0.001
	≥ 66 yrs	9.2	1.0		
Exp. yrs schooling	< 66 yrs	8.6	2.5	206.874	< 0.001
	≥ 66 yrs	13.3	2.0		

Theory suggests that all the variables we are considering are closely related to adolescent fertility, so we begin with a model that includes all of them as predictors. This model is a significant improvement on the null model (F (4, 182) = 53.500, p < 0.001) and has an R of 0.735 and an R^2 of 0.540, meaning that it explains 54% of the variation in the adolescent fertility rate. The key statistics from the regression analysis are presented in Table 10-3.

Table 10-3. Coefficients table for model 1

	Unstandardized coefficients		Standardized coefficients			
	B	Std. error	Beta		*t*	Sig.
Constant	7.706	0.377			20.949	< 0.001
Log_gni	−0.360	0.072	−0.487		−4.993	< 0.001
Pct. urban	0.002	0.003	0.059		0.794	0.428
Exp. yrs. schooling	−0.073	0.029	−0.233		−2.513	0.013
Life_exp_dichot	−0.234	0.159	−0.114		−1.474	0.142

As with simple regression, each line of this table presents the information about one of the predictors in the model; the difference from simple regression is that the influence of each predictor is evaluated in the context of the entire model. Because we know our predictors are highly related to each other, we assume that they overlap in terms of the variance they explain in the outcome variable. In a regression model, when all predictors are entered at the same time (as we did here), each predictor only gets credit for the unique variance it explains; this could explain why variables that seem as though they should be good predictors of adolescent fertility (percent urbanized and life expectancy) are not significant in this model.

The key results for the individual predictors are:

Log_gni: $\beta = -0.360$, $t = -4.993$, $p < 0.001$
Per capita income is a significant predictor of the adolescent fertility rate in a model also including the percent urban, expected years of schooling, and dichotomized life expectancy. The coefficient is negative, indicating that countries with higher per capita incomes have on average lower rates of adolescent fertility.

Pct. urban: $\beta = -0.002$, $t = 0.794$, $p = 0.428$
Percent of population living in an urban area is not a significant predictor of the adolescent fertility rate in a model also including the log of GNI, expected years of schooling, and dichotomized life expectancy.

Expected years of schooling: $\beta = -0.073$, $t = -2.153$, $p = 0.013$
Expected years of schooling is a significant predictor of the adolescent fertility rate in a model also including the percent urban, log of GNI, and dichotomized life expectancy. The coefficient is negative, indicating that countries with more years of expected schooling have on average lower rates of adolescent fertility.

Dichotomized life expectancy: $\beta = -0.234$, $t = -1.474$, $p = 0.142$
Life expectancy at birth (dichotomized into < 66 yrs and ≥ 66 yrs) is not a significant predictor of the adolescent fertility rate in a model including the percent urban, expected years of schooling, and log of GNI.

Because we have multiple predictors in this model, it's worth taking a look at the standardized coefficients (Beta) in this table. The absolute values of these coefficients tell us which predictors are explaining the most variance in the model (something we can't determine directly by the coefficients because they are measured on different scales). By this standard, log_gni is explaining the most variance (Beta = -0.487), followed by expected years of schooling (Beta = -0.233), dichotomized life expectancy (-0.114), and percent urban (Beta = 0.059); not surprisingly, the two significant predictors have the highest absolute Beta coefficients.

We will rerun this model with only the significant predictors, but one other thing is worth noting. In factorial ANOVA (Chapter 9), interactions among variables are automatically tested. This is not the case with regression; if you want an interaction tested, you have to specify it in the model. We deal with this issue after we have decided which predictors we want to include in our model.

We ran a second model, including only log_gni and expected years of schooling. This model is significantly better than the null model ($F (2, 184) = 105.21, p < 0.001$) and has an R of 0.685 and an R^2 of 0.470, so omitting two variables from the model resulted in a decrease of only 7% in the explained variance for the model. This suggests, as we expected, that because our predictors are closely related, they were explaining much of the same variance in the adolescent fertility rate. The key statistics from the regression analysis are presented in Table 10-4.

Table 10-4. Coefficients table for model 2

	Unstandardized coefficients		Standardized coefficients		
	B	Std. error	Beta	t	Sig.
Constant	7.837	0.345		22.730	< 0.001
Log_gni	−0.366	0.063	−0.495	−5.827	< 0.001
Exp. yrs. schooling	−0.085	0.027	−0.271	−3.190	0.002

Both predictors are significant, and the absolute values of their coefficients and t-statistics have increased (particularly for expected years of schooling), further suggesting that they overlapped with the two variables we have omitted from this model. The key results for the individual predictors are:

Log_gni: $\beta = -0.366, t = -5.827, p < 0.001$
  Per capita income is a significant predictor of the adolescent fertility rate in a model including expected years of schooling. The coefficient is negative, indicating that countries with higher per capita incomes have on average lower rates of adolescent fertility.

Expected years of schooling: $\beta = -0.085, t = -3.190, p = 0.002$
  Expected years of schooling is a significant predictor of the adolescent fertility rate in a model including per capita income. The coefficient is negative, indicating that countries with more years of expected schooling have on average lower rates of adolescent fertility.

The next thing we want to do is test for an interaction between per capita income and expected years of schooling. We do this by adding an interaction term, log_gni*exp_schooling, to the model and seeing whether it is significant. This model explains more variance ($R^2 = 0.546$) and produces an interesting result in terms of our predictor variables, as shown in Table 10-5.

Table 10-5. Coefficients table for model 3

	Unstandardized coefficients		Standardized coefficients		
	B	Std. error	Beta	t	Sig.
Constant	5.039	1.280		3.936	< 0.000
Log_gni	−0.019	0.165	0.026	−0.118	0.906
Exp. yrs. schooling	0.159	0.111	0.507	1.436	0.153
Log_gni*expected_schooling	−0.029	0.013	−1.193	−2.267	0.025

Adding the interaction term changes everything. Per capita income and expected years of schooling are no longer significant predictors in a model including the interaction term, and the direction of influence of expected years of schooling has reversed. The interaction term is the only significant predictor in the model, but we will keep all three terms because the interaction is only meaningful in the context of the main effects. A significant interaction term means that the effect of one variable is modified by the level of the other variable; in this case, the effect of per capita national income on adolescent fertility is modified by expected years of schooling, and the influence of expected years of schooling is modified by per capita national income. Explaining interactions when using continuous variables is particularly tricky, but the picture might become clearer by looking at a graphic of this relationship.

Figure 10-7 presents the means plot for the log of adolescent fertility rate (the y-axis) at low, medium, and high levels of expected years of schooling (the separate lines) and log of per capita national income (the x-axis). The low level is defined as the bottom third of values for the given variable, the middle level as the central third of values, and the high level as the upper third of values.

What Figure 10-7 makes clear is that although increased per capita national income and increased expected years of schooling are both associated with lowered levels of adolescent fertility, the amount of decrease does depend on the interaction of the two variables. Note also that for the highest level of schooling, there were no countries in the lowest third of national per capita income, which is why that line only has two data points.

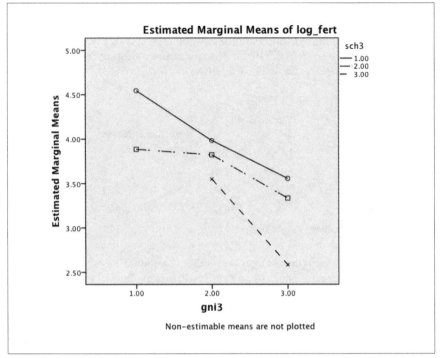

Figure 10-7. Means for the natural log of the adolescent fertility rate for low, medium, and high levels of the log of per capita national income and expected years of schooling

For countries with a low level of expected years of schooling, there is an almost linear decrease in the adolescent fertility rate among the three levels of per capita income. For countries with a medium level of schooling, the decrease from low to medium per capita income is relatively small, whereas the decrease from medium to high per capita income is much greater. For countries with a high level of schooling, the decrease in adolescent fertility from a medium to a high level of per capita income is much greater than the decrease for countries with either a low or a medium level of schooling.

Figure 10-8 shows another way to look at this interaction. In this figure, we have created scatterplots of the expected years of schooling and the natural log of adolescent fertility at low, medium, and high levels of per capita national income. The slope of the regression line (indicating the relationship between the natural log of adolescent fertility and years of expected schooling) is noticeably steeper for the highest level of the natural log of per capita national income, again indicating an interaction between the two predictor variables. Also interesting is the fact that although over the whole range of data the relationship between the natural log of adolescent fertility and expected years of schooling is fairly strong ($R^2 = 0.44$), within any one of the three categories of national income, this relationship is much weaker (0.118 for the lowest income countries, 0.052 for the middle income countries, and

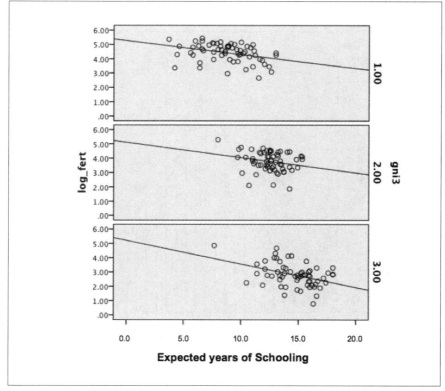

Figure 10-8. The relationship between the natural log of adolescent fertility and expected years of schooling for low, middle, and high income countries

0.168 for the high income countries), indicating the strong relationship between our predictor variables.

Clearly, we haven't exhausted the possibilities for exploring the relationship among per capita national income, expected years of schooling, and adolescent fertility rates. Equally clearly, we won't explain adolescent fertility rates with just two variables, but for the purposes of this demonstration, we have a model we can work with. The Durbin-Watson statistic for this model is 0.195, very close to the null value of 2, so we can assume the assumption of independence of errors is upheld. The Kolmogorov-Smirnov statistic for the standardized residuals for this model is 0.663 ($p = 0.772$), and the histogram presented in Figure 10-9 looks close to normal, so we can consider the assumption of normality of residuals upheld.

We will evaluate the assumption of homoscedasticity by plotting the standardized residuals against the standardized predicted values, as shown in Figure 10-10.

This graph is pretty much a cloud of points with no indications of heteroscedasticity, so we will assume the assumption of homoscedasticity upheld.

We also need to look at multicollinearity among our predictor variables. We do this by calculating the tolerance and VIF (variance inflation factor) for the predictors in

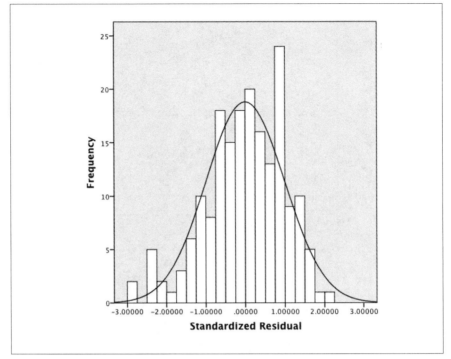

Figure 10-9. Histogram of standardized residuals for model 3

our model; this is an option provided with many regression algorithms. Note that VIF is just the reciprocal of tolerance (VIF = 1/tolerance), so interpreting either statistic will provide the same result. There are various rules of thumb for interpreting tolerance and VIF; one popular rule is that tolerance should not be greater than 10 or VIF lower than 0.10. By that standard, we have a problem with this data, as shown in Table 10-6.

Table 10-6. Multicollinearity diagnostics for model 3

Predictor	Tolerance	VIF
Log_gni	0.50	20.04
Expected years of schooling	0.20	50.35
Log_gni*expected years of schooling	0.01	11.73

However, other scholars believe that the conventional values of VIF and tolerance do not indicate an invalid regression model; see for instance the O'Brien article listed in Appendix C. We know that our predictor variables are highly correlated, so if we were to continue with this analysis, we would consider more variables for inclusion in the model, might drop one or both of these, and might combine these two variables (possibly along with some others) into an index term for inclusion in the model. For the purposes of this example, we will continue to interpret this model.

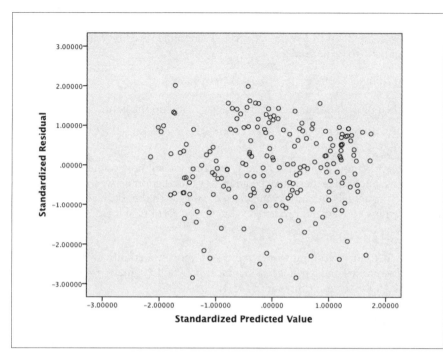

Figure 10-10. Histogram of standardized residuals for model 3

The regression equation for our data is:

> $Log\_fert = 5.039 - 0.019(log\_gni) + 0.159(exp\_schooling) - 0.029(log\_gni*exp\_schooling) + e$

Although the coefficients of log_gni and exp_schooling are not significantly different from 0 in this analysis, we keep them in the equation because the interaction term only has meaning in the context of the equation including the variables that make up the interaction. Note also that it would be a mistake to interpret the coefficients of log_gni or exp_schooling without reference to their interaction; instead, each coefficient has to be interpreted in the context of the entire equation.

We can use this equation to predict values for the adolescent fertility rate for a country, given its values of per capita national income and years of expected schooling. Note that both our fertility and national income variables are natural log transformations, so if we are given these variables in their raw form, we must transform them before putting them into the equation. Our results from this equation will be in terms of the log of the rate of adolescent fertility; because this might not be a meaningful value for most people, we can convert it to the rate of adolescent fertility, which is more easily understood. Note also that we want to keep the input for our predictions within the range of values included in this data set; to do otherwise would be reasoning beyond the range of the values, and we don't want to do that because we can't assume that our regression equation is valid outside the range of values used to create it.

Suppose we want to calculate the predicted adolescent fertility rate for a country with a gross national per capita income of 12,000 (in PPP international dollars, as previously defined) and 12 years of expected schooling. We first need to covert the income statistic to its natural log:

$LN(12,000) = 9.393$

We can then plug these values into the equation and do the math:

$Predicted(Log\_fert) = 5.039 - 0.019(9.393) + 0.159(12) - 0.029(9.393^*12)$
$= 3.500$

Note that we removed the error term (e) because we are now calculating *predicted log_fert*; we know that there may be an error of prediction between this predicted value and the actual, measured value for a country with these values on the X variables. We now convert our predicted value for *log_fert* by taking the antilog:

$e^{3.500} = 33.12$

This tells us that, according to our regression model, a country with 12 years of expected schooling and a per capita income of 12,000 PPP international dollars has a predicted adolescent fertility rate of 33.12 per 1,000.

Dummy Variables

Multiple linear regression can accommodate predictor variables that are either continuous or dichotomous. However, sometimes we need to work with a variable with more than two categories. In this case, we need to recode the categorical variable into a number of dichotomous or dummy variables. Suppose a college wants to do some research into the initial annual salaries for its graduates from the class of 2010; the data is also coded to indicate the student's GPA and general field of study (humanities, sciences, social sciences, or education). GPA is recorded to two decimal places with a defined ceiling of 4.0 (a perfect or straight-A average) and a floor of 0.0 (failure to pass any courses), although the actual range of the data is from 2.5 to 4.0; these are graduates, so we would expect higher than average grades compared to all college students. Salary is expressed in thousands of dollars and has a range of 19.6 to 58.6.

We would like to include field of study in a model predicting initial salary, but first we must recode it into dummy (dichotomous) variables. We can't simply include it in the model because the statistical package will interpret the numbers used to code this variable as having numerical importance (e.g., 2 is greater than 1) when in fact they are simply labels that indicate categories. There are several ways to code dummy variables; we present one of the most common methods here.

We have a categorical variable with four categories; therefore, we need to create three dummy variables to code the information contained in this variable. Speaking more generally, if a variable has k categories, you need $k - 1$ dummy variables to replace it. We need to choose one category to serve as our reference category; the other categories will be compared to this one.

For this analysis, we choose humanities as our reference category because it has the lowest average salary among the four groups, as shown in Table 10-7; choosing the group with the lowest salary will give us positive coefficients for the other categories, which might be easier to explain to general audiences (for instance, the parents of prospective students).

Table 10-7. Average annual starting salaries for college graduates from four general fields of study

Field	Average salary (thousands of dollars)	Standard deviation of salary
Humanities	22.7	11.4
Sciences	56.3	9.3
Social sciences	28.9	10.1
Education	28.0	8.1

Our dummy coding scheme is presented in Table 10-8.

Table 10-8. Dummy coding for field of study

Field	X_1	X_2	X_3
Humanities	0	0	0
Sciences	1	0	0
Social sciences	0	1	0
Education	0	0	1

We create three new dummy variables, X_1, X_2, and X_3, and give them a value of 0 or 1, depending on the value of the field of study. For our reference category, humanities, all three dummy variables have a value of 0. For each of the other three fields, one of the dummy variables has a value of 1, and the others have a value of 0. This combination of three dummy variables uniquely identifies each field of study; if a case has values $X_1 = 0$, $X_2 = 1$, and $X_3 = 0$, we know the field of study is the social sciences.

The regression equation predicting salary from field of study is:

$$Y = \beta_0 + \beta_1 X_1 + \beta_2 X_2 + \beta_3 X_3 + e$$

In this equation, β_0 will be the mean salary for humanities majors, and the other coefficients will be the difference in mean salary between that field of study and the mean salaries for humanities majors. For instance, β_1 will represent the difference between science majors and humanities majors. The regression coefficients table for this data is presented in Table 10-9.

Table 10-9. Regression results for regression equation prediction including dummy variables

	Unstandardized coefficients		Standardized coefficients		
	B	Std. error	Beta	t	Sig.
Constant	22.682	3.102		7.313	<0.000
X_1	33.611	4.386	0.905	7.662	<0.000
X_2	6.247	4.386	0.168	1.434	0.163
X_3	5.288	4.386	0.142	1.206	0.236

The equation for this data is:

$$Average\_salary = 22.682 + 33.611(X_1) + 6.247(X_2) + 5.288(X_3) + e$$

To calculate the average salary for any of our four fields of study, we simply plug the values for the X variables into this equation and solve it. For instance, for someone from the field of education, the relevant values are $X_1 = 0$, $X_2 = 0$, and $X_3 = 1$. Putting these values in the equation gives us:

$$Predicted\_Average\_salary\ (education) = 22.682 + 33.611(0) + 6.247(0) + 5.288(1) = 27.97$$

This is the average salary for graduates from the education field and matches the figure in Table 10-7 (within rounding error). If you go through the same exercise for the other three fields, you will find that the values calculated using the regression equation for these fields also match the values presented in Table 10-7. The *t*-tests for each coefficient are testing whether they are different from 0. Because they are dummy variables, and because we coded them using the humanities field as our reference group, the *t*-test tells us whether starting salaries for students from a particular field are significantly different from the starting salaries in the humanities field. We can see from Table 10-9 that there is a significant difference between initial salaries in the sciences and in the humanities because X_1 is significantly different from 0 ($t = 7.662$, $p < 0.001$), whereas the other two comparisons are not significant. This brings up an important point about dummy coding—if you have particular comparisons in mind, be sure to code your dummy variables to facilitate those comparisons.

Methods for Building Regression Models

We've been looking at fairly simple regression models, but often the model-building process begins with 10, 20, or even more predictor variables under consideration for inclusion, and even with a smaller number of predictors, you might want to use a formal model-building process. Many statistical packages include several choices of algorithms for model-building, and in some systems, you can combine different methods or algorithms within the same model.

There are two categories of model building: *stepwise* methods for considering predictors for inclusion and exclusion and *blocking* methods to designate which predictors should be considered for inclusion in a given step. The term "block" refers to a group of predictor variables that are entered as a group into a model or that are under consideration for inclusion as a group; in the models in this chapter, we've entered all our predictors in a single block, but other choices are available, as we will see. The term "stepwise" refers to how predictor variables are selected for inclusion within a block; stepwise methods are generally automated and select which variables within a block to add to or retain within a model, using criteria you specify.

Automated methods of model building are not accepted in all fields of study, mainly because these methods build a model based on the data in your sample rather than according to theory. Because we often build models with the intent of generalizing beyond our sample, this raises obvious concerns. Another criticism of automated methods is that they amount to conducting many significance tests on the same data without any correction for the inflation of the experiment-wise error rate, increasing the probability of committing a Type I error. However, automated models are considered to be acceptable in some fields of study and work, and if they are acceptable within your specific area of application, there's no reason you shouldn't use them. One thing you should keep in mind, however, is that the three stepwise methods can result in three regression models, so you should have a reason for choosing whichever method you decide to use.

Automated methods of model building rely in part on a measurement called *partial correlation*; this means the correlation between two variables with the effects of one or more other variables removed from the correlation. In automated regression algorithms, partial correlation is used to identify the unique variance explained by a predictor variable to choose the predictor that has the strongest relationship with the outcome when evaluated in the presence of other predictors. Even in a model in which you decide which predictors to include and their order of entry, examining the partial correlations (which can be automatically generated by many statistical packages) can be useful in evaluating the importance of particular predictors in the presence of other predictors.

There are three basic stepwise methods of model building:

Backward removal
> All predictors in a block are added to the model at once and then removed one by one until the removal of a variable significantly damages the fit of the model. This algorithm considers variables for removal according to the amount of unique variance they explain in the full model. The variable that explains the least unique variance (that has the smallest partial correlation) is the first to be considered for removal, and then the variable explaining the least variance in the model that remains after the first variable is removed, and so on. The user specifies the criteria for removal of a variable and for evaluating model fit.

Forward entry
> Predictor variables are added one at a time to the model, beginning with the predictor that has the largest absolute correlation with the dependent variable; for the second and subsequent predictors, the variable is chosen that has the largest partial correlation with the predictor, that is, the variable that explains the most unique variance in the dependent variable. Each variable must meet

user-specified criteria for entering the model, generally based on improving model fit or the individual significance of the predictor.

Stepwise

The stepwise method is a combination of forward entry and backward removal. Predictors are entered in the regression model one at a time based on how much they will improve model fit. Each time a new predictor is added, the predictors already in the model are evaluated and may be removed if they no longer significantly improve the fit of the model.

Blocking methods are not automated but are a way you can enter or test variables in groups. In this chapter, we have been entering all variables in one block, but there are times when you might want to enter variables in separate blocks. One example is when you want to see how much a set of variables can improve model fit after another set of variables is already in the model. For instance, you might have developed an intervention to encourage people to improve their health through exercise. You know that many demographic factors (gender, race or ethnicity, income, etc.) are also related to exercise and health, and you want to separate the variance in people's exercise habits and health that can be attributed to your intervention from the variance that can be attributed to demographic factors. To accomplish this, you would enter the demographic variables into the equation as a block and then enter the variables related to your intervention in a second block; that way, any variance in outcomes explained by your study will be variance above and beyond that explained by the demographic variables. This type of model is particularly useful in observational studies in which you can't use random assignment to attempt to control the influence of variables (such as demographics) that might be related to your outcome.

Blocking can also be combined with automated model building because it is possible to use one automated method in one block and another (or no automated method) in another block. Continuing with the preceding example, you might have measurements of a number of demographic characteristics and not be sure which are most useful in explaining variance in your model. If it is acceptable in your field to use automated processes of model building, you could enter all the demographics in a single block and let the algorithm decide which are most useful in explaining the variance in your outcome variable. You could then enter the variables for your own study in a second block to see how much variance they explain after that explained by the demographic variables; in the second block, you do not need to use any automated model-building method but can simply enter all your variables at once in this block.

Let's look at a simple example to examine the effect of using different stepwise techniques. Imagine you are an educator interested in the relationship between IQ and traditional measures of general ability such as performance on numerical, reading, verbal, and reasoning skills as well as nontraditional measures such as musical and physical performance. A subset of the sample data is shown in Table 10-10.

Table 10-10. Data showing the relationship between traditional measures and nontraditional measures of general ability and IQ

IQ	Numerical	Reading	Verbal	Physical	Musical	Reasoning
85.0	3.0	5.0	7.0	10.0	6.0	10.0
90.0	3.0	6.0	7.0	10.0	6.0	10.0
95.0	4.0	6.0	7.0	9.0	7.0	8.0
100.0	4.0	7.0	8.0	9.0	7.0	5.0
100.0	5.0	7.0	8.0	8.0	8.0	6.0
100.0	5.0	8.0	8.0	7.0	9.0	5.0
105.0	6.0	8.0	8.0	6.0	8.0	4.0
105.0	6.0	8.0	8.0	5.0	7.0	5.0
110.0	7.0	9.0	8.0	4.0	6.0	6.0
110.0	7.0	9.0	8.0	3.0	6.0	9.0
115.0	8.0	10.0	9.0	3.0	5.0	10.0
120.0	9.0	10.0	9.0	1.0	4.0	9.0

You decide to explore the relationships between the variables, calculating all pairwise correlations and their statistical significance, as shown in Table 10-11 (upper triangle only). Unsurprisingly, the most traditional measures (numerical, reading, and verbal) are highly positively correlated with IQ (** = $p < 0.01$). Also not surprisingly, many of these measures of ability are highly correlated with each other, meaning that any regression model including several of them will probably have a high degree of collinearity. However, reasoning does not show a strong relationship with most of the other variables (except musical performance), and physical performance has a strong negative relationship to IQ and several of the other measurements of ability. The lack of a significant bivariate relationship between IQ and musical performance is also surprising.

Table 10-11. Pairwise relationships between traditional measures and nontraditional measures of general ability and IQ

	IQ	Numerical	Reading	Verbal	Physical	Musical	Reasoning
IQ	1.000	0.978**	0.976**	0.914**	−0.955**	−0.427	−0.073
Numerical		1.000	0.963**	.887**	−0.986**	−0.481	0.026
Reading			1.000	.912**	−0.954**	−0.381	−0.055
Verbal				1.000	−0.836**	−0.337	−0.103
Physical					1.000	0.503	−0.062
Musical						1.000	−0.738**
Reasoning							1.000

If you are more interested in exploring relationships among the variables in this data set than in testing a particular model based on theory, you might decide to use an automated method to build your model. You decide to build two models, using two methods (forward entry and backward removal), and then compare these models. For forward entry, you set the criteria for entry at $p \leq 0.05$ (the coefficient for any predictor must meet that standard to be included in the model); for backward entry, you set the criteria for removal at $F \geq 0.100$ (variables will be removed if the level of change in the probability of the F-statistic is not below 0.100).

Forward entry

In the forward entry method, the predictor with the strongest pairwise correlation with IQ ($r = 0.978$), numerical skills, is entered first into the model. For this model, $R^2 = 0.956$, and the overall model is significant, with $F(1, 10) = 217.36$, $p = 0.000$. None of the other predictors makes a significant improvement in model fit, so this is also our final model, with coefficients as shown in Table 10-12. This result is both surprising (because other researchers have found variables such as verbal skills to be closely related to IQ) and not surprising (because most of our predictors are so highly correlated that we would expect a large amount of overlap in any variance they can explain in IQ).

Table 10-12. Final regression model built using the automated forward entry method

| | Unstandardized coefficients | | Standardized coefficients | | |
	B	Std. error	Beta	t	Sig.
Constant	74.318	2.043		36.374	0.000
Numerical	5.122	0.347	0.978	14.743	0.000

Table 10-13 displays information about the variables excluded from the final model. You can see from looking at the t-statistics and significance columns that some of them came very close to inclusion, particularly Reading ($t = 2.239$, p = 0.052), so it's easy to imagine that if you had drawn a different sample of subjects, Reading might have been included in the model and Numerical excluded. The regression model arrived at through forward entry regression is:

$$IQ = 74.318 + 5.122(Numerical) + e$$

Table 10-13. Variables excluded from the final regression model built using the automated forward entry method

Model	Beta in	t	Sig.	Partial correlation	Tolerance
Reading	.467	2.239	0.052	0.598	0.072
Verbal	.219	1.648	0.134	0.482	0.213
Physical	.288	.716	0.492	0.232	0.029
Musical	.057	.737	0.480	0.239	0.768
Reasoning	−.098	−1.594	0.146	−0.469	0.999

One great advantage of using the forward method of model building is that you quickly arrive at the smallest model that explains the greatest amount of variance given your data set. This is particularly useful if you have a large number of predictors, no particular theory about how they relate to each other or to the outcome, and just want the best model for this data. This approach is similar to data mining in that you simply want to know what your data can tell you without any intent to treat it as a sample from a larger population or to generalize from your results to other data. The problem with this approach is that a model built using automated methods can be highly dependent on the specific data set used to build the model; this is a problem if you intend to generalize from your data set to a larger population. When many of your predictors have high correlations with each other and with the outcome variable, as in our example, small differences in those correlations can result in highly unstable models; if you drew a different sample, the model generated using the same automated methods could look quite different from the model generated from your first sample.

Backward removal

Backward removal models begin by putting all the designated predictor variables in the model and then removing them one by one, beginning with the variable that makes the least contribution to prediction; the model is rerun each time a variable is removed, so the contributions made by each variable are calculated for each new model.

Table 10-14 shows the five models produced en route to the final model; after each iteration, one IV is removed, starting with Verbal and proceeding through Physical, Musical, and Reasoning. Table 10-15 shows the coefficients for each model iteration as well as the corresponding *t* values and their significance.

Recall that the forward method resulted in only one IV—Numerical skills—being included in the model; it's interesting that by using the backward method, we arrive at a final model including two predictors, Numerical skills and Reading skills. It's also instructive to observe how the coefficients change as variables are removed from the model; this emphasizes that usually adding or subtracting a variable from a model will change the coefficients for most or all the other variables.

Table 10-14. Backward stepwise model for linear regression

Model	Variables entered	Variables removed	Method
1	Reasoning, numerical, musical, verbal, reading, physical	.	Enter
2	.	Verbal	Backward (criterion: probability of F-to-remove ≥ 0.100).
3	.	Physical	Backward (criterion: probability of F-to-remove ≥ 0.100).
4	.	Musical	Backward (criterion: probability of F-to-remove ≥ 0.100).
5	.	Reasoning	Backward (criterion: probability of F-to-remove ≥ 0.100).

Table 10-15. Standardized coefficients for each model iteration

Model		Unstandardized coefficients		Standardized coefficients		
		B	Std. Error	Beta	t	Sig.
1	Constant	64.480	20.702		3.115	0.026
	Numerical	3.827	2.369	0.731	1.616	0.167
	Reading	3.070	1.749	0.487	1.755	0.140
	Verbal	.048	2.628	0.003	0.018	0.986
	Physical	1.011	1.423	0.305	0.710	0.509
	Musical	−1.222	0.864	−0.167	−1.414	0.216
	Reasoning	−.742	0.445	−0.169	−1.668	0.156
2	Constant	64.514	18.819		3.428	0.014
	Numerical	3.851	1.822	0.735	2.114	0.079
	Reading	3.088	1.301	0.490	2.373	0.055
	Physical	1.026	1.040	0.310	0.986	0.362
	Musical	−1.224	0.777	−0.167	−1.575	0.166
	Reasoning	−.743	0.402	−0.169	−1.848	0.114
3	Constant	80.511	9.530		8.448	0.000
	Numerical	2.449	1.137	0.467	2.153	0.068
	Reading	2.863	1.279	0.454	2.239	0.060
	Musical	−1.179	0.775	−0.161	−1.522	0.172
	Reasoning	−0.785	0.399	−0.179	−1.968	0.090
4	Constant	68.274	5.524		12.360	0.000
	Numerical	3.149	1.122	0.601	2.806	0.023
	Reading	2.476	1.352	0.393	1.831	0.105
	Reasoning	−0.294	0.253	−0.067	−1.161	0.279
5	Constant	64.655	4.649		13.908	0.000
	Numerical	2.765	1.093	0.528	2.529	0.030
	Reading	2.945	1.316	0.467	2.239	0.050

The final regression model produced by the backward elimination method (model #5 in Table 10-14) is:

$$IQ = 64.655 + 2.765(Numerical) + 2.945(Reading) + e$$

This model explains 97.2% of the variance in IQ slightly more than the model created by the forward method (95.6%). Although both models explain almost the same amount of variance, it's interesting to note how the coefficients differ. The model produced using the forward method has a larger intercept and a larger coefficient for Numerical. These differences are probably explained by the fact that some of the variance explained by Numerical in the first model is explained by Reading in the second and that the inclusion of a second predictor naturally lowers the intercept because each IQ score is now explained by two aptitude scores rather than one.

Exercises

Multiple linear regression can be used to investigate a number of types of research questions, as shown in the following examples.

Example 1

As a human resources specialist, you are interested in the motivational factors that are associated with productivity (the outcome) in IT teams, based on the KLOC metric (thousands of lines of code written per week). There are four motivational factors believed to influence productivity; these may be based on either intrinsic or extrinsic motivation and may be either self-reported or observed. Four scales are developed to measure these types of factors and are used as predictor variables in the model:

- Intrinsic self-report (IS)
- Intrinsic observed (IO)
- Extrinsic self-report (ES)
- Extrinsic observed (EO)

KLOC is expressed as thousands of lines of code; the four predictors are measured on a scale from 0 to 100. Descriptive statistics for these variables are presented in Table 10-16.

Table 10-16. Descriptive statistics for four types of motivational factors and KLOC

Variable	n	Mean	Std. dev.
Productivity (KLOC)	50	3.5	2.3
Intrinsic self-report (IS)	50	41.3	14.8
Intrinsic observed (IO)	50	54.7	19.4
Extrinsic self-report (ES)	50	27.1	16.5
Extrinsic observed (EO)	50	40.7	25.5

The upper triangle of the correlation matrix for these variables is shown in Table 10-17; correlations with a *p*-value of 0.05 or less are marked with an asterisk (*).

Table 10-17. Correlation matrix for four types of motivational factor and KLOC

	KLOC	IS	IO	ES	EO
KLOC	1.00	0.25	0.12	0.43*	0.67*
IS		1.00	−3.70*	−1.70	0.35*
IO			1.00	0.18	−0.18
ES				1.00	0.61*
EO					1.00

Problem

What do you notice in the correlation matrix that can help guide your creation of a regression model using this data?

Solution

First, two of the four predictors have a significant bivariate correlation with the outcome: extrinsic self-report ($r = 0.43$, $p = 0.002$) and extrinsic observed ($r = 0.67$, $p < 0.001$); the specific p-values were not included in Table 10-17, but are from the computer printout. Second, some of our predictors have significant correlations with each other, something we should keep in mind while building our model; these pairs of closely related predictors are intrinsic self-report and intrinsic observed ($r = -0.37$, $p = 0.008$), intrinsic self-report and extrinsic observed ($r = 0.35$, $p = 0.013$), and extrinsic self-report and extrinsic observed ($r = 0.612$, $p < 0.001$).

You decide to include all four predictors in your regression model; this model explains 51.5% of the variance in KLOC and produces the coefficients and significance tests presented in Table 10-18. The overall test of fit for this model produces the result $F(4, 45) = 11.927$, $p < 0.001$.

Table 10-18. Coefficients table for a regression analysis predicting KLOC from four types of psychological factor

	Unstandardized coefficients		Standardized coefficients		
	B	Std. error	Beta	t	Sig.
Constant	−0.989	1.253		−0.790	0.434
IS	0.022	0.023	0.129	0.970	0.337
IO	0.023	0.009	0.280	2.370	0.017
ES	0.003	0.023	0.019	0.124	0.902
EO	0.062	0.015	0.660	4.044	< 0.001

Problem

Interpret the information in Table 10-18, write the regression equation, and suggest what the next step might be in your effort to understand the relationship among these variables.

Solution

The regression equation for this data is:

$$KLOC = -0.989 + 0.022(IS) + 0.023(IO) + 0.003(ES) + 0.062(EO) + e$$

This model is significantly better than the null model, but only two of the four predictor variables are significantly different from 0: IO ($t = 2.370$, $p = .017$) and EO ($t = 4.044$, $p < 0.001$). Depending on the purpose for your analysis, you could stop here or continue exploring the data. You know from the correlations table that IS is significantly correlated to both IO and EO in this data set; that might explain its lack of significance in this model, so you might try running a model with just IS as a predictor to see how much variance it explains by itself. You also might add more variables to the model, such as gender; it's possible that men and women have different structures of motivation.

Example 2

You are a management consultant working in the retail sector, conducting a time-in-motion study to determine which of two predictors (barcode scanner size and operator accuracy) has the greatest effect on the outcome of throughput at the checkout counter, measured in items per second. The question is difficult to answer because the units of measurement in each case are different; the scanner size is measured in square centimeters, whereas operator accuracy is measured as the proportion of times the operator scans an item successfully on the first try. Your client wants to increase throughput because customers have complained that queues in the store are long. However, larger scanners are more expensive than smaller ones, and training courses for staff require expenditure while not necessarily increasing accuracy. The manager wants to know whether to spend money on more training (or hiring better staff) or purchasing larger scanners, so you decide to conduct a study see which variable contributes more to throughput: scanner size or operator accuracy.

Throughput and accuracy are continuous variables; although size is theoretically a continuous variable, in this data set it has only three values (2 sq cm, 4 sq cm, and 6 sq cm), so you decide to treat it as a categorical value. There are three scanners of each size in the study; descriptive information about the continuous variables is presented in Table 10-19.

Table 10-19. Descriptive information for throughput and operator accuracy

Variable	N	Mean	Std. dev.	Minimum	Maximum
Throughput	30	0.76	0.36	0.20	1.50
Accuracy	30	81.31	4.38	73.62	91.13

Problem

Your first task is to create a dummy coding scheme for scanner size. Treat the smallest scanner size as the reference category, and assign values to as many X variables as are needed to code the values for scanner size uniquely.

Solution

The most obvious coding scheme is presented in Table 10-20; note that variables maintain the order of values for the variable. The 2 sq cm value must be coded with the value of 0 for both X_1 and X_2, whereas the codes for the 4 sq cm and 6 sq cm sizes could be exchanged, and the coding would still meet the specification of 2 sq cm as the reference category.

Table 10-20. A dummy coding scheme

Size	X_1	X_2
2 sq cm	0	0
4 sq cm	1	0
6 sq cm	0	1

Assume you have checked all the necessary assumptions and run the regression analysis, using the dummy coding scheme previously presented. This model is significantly better than the null model ($F(3, 26) = 21.805$, $p < 0.001$) and explains 68.3% of the variance in throughput. The coefficients table for this analysis is presented in Table 10-19.

Problem

Write the regression equation based on the information in Table 10-21, and make a recommendation to the manager, backed up with information from this analysis.

Table 10-21. Descriptive information for throughput and operator accuracy

	Unstandardized coefficients		Standardized coefficients		
	B	Std. error	Beta	t	Sig.
Constant	0.737	0.917		0.803	0.429
Accuracy	−0.003	0.011	−0.034	−0.246	0.808
X_1	0.071	0.094	0.094	0.756	0.456
X_2	0.685	0.015	0.909	6.491	<0.001

Solution

The regression equation is:

$$Throughput = 0.737 - 0.003(Accuracy) + 0.071(X_1) + 0.685(X_2) + e$$

A regression analysis ($n = 30$) examined the effects of operator accuracy (the proportion of times the operator scans an item successfully on the first try) and scanner size (measured in square centimeters) on throughput (measured in items scanned) per second. The practical context of this study is a retail environment in which the goal is to increase throughput. Three sizes of scanner were included in the study (2 sq cm, 4 sq cm, and 6 sq cm). Throughput and accuracy were both approximately normally distributed; throughput had a range of 0.20 to 1.50, a mean of 0.76, and a standard deviation of 0.36; accuracy had a range of 73.62 to 91.13 with a mean of 81.31 and a standard deviation of 4.38.

The regression model explained 68.3% of the variance in throughput. Operator accuracy was not related to throughput ($t = -0.246$, $p = 0.808$), but scanner size was; the largest scanner (6 sq cm) was a significant improvement over the smallest (2 sq cm) ($t = 6.491$, $p = 0.000$). The medium-sized scanner (4 cm) showed no improvement in accuracy over the smallest ($t = 0.756$, p = 0.456). My recommendation is to purchase a 6 sq cm scanner because that size of scanner is the variable most strongly related to increased throughput.

11

Logistic, Multinomial, and Polynomial Regression

Multiple linear regression is a powerful and flexible technique that can handle many types of data. However, there are many other of types of regression that are more appropriate for particular types of data or to express particular relationships among the data. We discuss a few of these regression techniques in this chapter. Logistic regression is appropriate when the dependent variable is dichotomous rather than continuous, multinomial regression when the outcome variable is categorical (with more than two categories), and polynomial regression is appropriate when the relationship between the predictors and the outcome variable is best expressed through an equation including polynomial terms (such as x^2 or x^3). If you are unfamiliar with odds ratios, it would be good to read the section of Chapter 15 covering them before reading this chapter because the odds ratio plays a key role in interpreting the output of logistic regression.

Logistic Regression

Multiple linear regression may be used to find the relationship between a single, continuous outcome variable and a set of predictor variables that might be continuous, dichotomous, or categorical; if categorical, the predictors must be recoded into a set of dichotomous dummy variables.

Logistic regression is in many ways similar to multiple linear regression, but it's used when the outcome variable is dichotomous (when it can take only two values). The outcome might be dichotomous by nature (a person is either a high school graduate or she is not) or represent a dichotomization of a continuous or categorical variable. (Blood pressure is measured on a continuous scale, but for the purposes of analysis, people might simply be classified as having high blood pressure or not.) Outcome variables in logistic regression are conventionally coded as 0–1, with 0 representing the absence of a characteristic and 1 its presence. The outcome variable in linear regression is a *logit*, which is a transformation of the probability of a case having the characteristic in question; you can easily convert logits to probabilities and back, as will be demonstrated.

You might be wondering why you can't use multiple linear regression with a categorical outcome. There are two reasons:

1. The assumption of homoscedasticity (common variance) is not met with categorical variables.
2. Multiple linear regression can return values outside the permissible range of 0–1 (presence or absence).

The *logit* is also called the *log odds* for reasons that are clear from its definition. If *p* is the probability of a case having some characteristic, then the logit for this case is defined as in Figure 11-1.

$$\text{logit}(p) \;=\; \log \frac{p}{1-p} = \log(p) - \log(1-p)$$

Figure 11-1. Definition of a logit

The natural log (base *e*) is used to convert probabilities to logits.

Apart from having an outcome expressed as a logit, the form of a logistic regression equation with *n* predictor variables looks very similar to that of a linear regression equation, as can be seen in Figure 11-2.

$$\text{logit}(p) \;=\; \beta_0 + \beta_1 X_1 + \beta_2 X_2 ... + \beta_n X_n + e$$

Figure 11-2. The logistic regression equation

As with linear regression, we have measures of model fit for the entire equation (evaluating it against the null model with no predictor variables) and tests for each coefficient (evaluating each against the null hypothesis that the coefficient is not significantly different from 0). The interpretation of the coefficients is different, however; instead of interpreting them in terms of linear changes in the outcome, we interpret them in terms of *odds ratios* (discussed in this chapter and in Chapter 15; note that odds ratios are used frequently in medical and epidemiological statistics).

As with linear regression, logistic regression makes several assumptions about the data:

Independence of cases
    As with multiple linear regression, each case should be independent of other cases, so you should not have multiple measurements on the same person, members of the same family, and so on (if family membership is likely to make cases more related than two cases chosen at random).

Linearity

There is a linear relationship between the logit of the outcome variable and any continuous predictor. This is tested by creating a model with the logit as outcome and as predictors, each continuous predictor, its own natural logs, and an interaction term of each predictor and its natural log. If the interaction terms are not significant, we can assume the linearity criterion has been met.

No multicollinearity

As with multiple linear regression, no predictor should be a linear function of other predictors, and predictors should not be too closely related to each other. The first part of this definition is absolute (usually violated only due to the researcher's absentmindedness, as when including the predictors *a*, *b*, and *a* + *b* in an equation); the latter is open to interpretation and is assessed through multicollinearity statistics produced during the regression analysis. As discussed in Chapter 10, statisticians disagree on how much of a threat less than absolute multicollinearity presents to a regression model.

No complete separation

The value of one variable cannot be perfectly predicted by the value(s) of another variable or set of variables. This is a problem that most frequently arises when you have several dichotomous or categorical variables in your model; you can test for it by doing cross-tabulation tables on these variables and checking that there are no empty cells.

Suppose you are interested in factors to health insurance coverage in the United States. You decide to use a random sample of 500 cases from the 2010 BRFSS data set, an annual survey of U.S. adults. (For more on the BRFSS, see Chapter 8.) Insurance coverage is dichotomous; after examining several potential predictors, you decide to use gender (dichotomous) and age (continuous) as predictors. In this data set, 87.4% of the respondents have health insurance, their mean age is 56.4 years (standard deviation 17.1 years), and the respondents are 61.7% female.

Looking at the assumptions for logistic regression, the first is met because we know the BRFSS data is collected by trained researchers following a national sampling plan. To evaluate linearity between the logit and age, we construct a regression model including age, the natural log of age, and the interaction of those two terms. The results are shown in Table 11-1.

Table 11-1. Testing age for linearity with the logit

	B	Std. error	Wald	df	Sig.	Exp(B)
Age	1.305	1.136	1.321	1	0.250	3.690
Ln(Age)	−9.353	7.884	1.407	1	0.235	0.000
Age*ln(age)	−0.218	0.198	1.209	1	0.271	0.804
Constant	15.862	13.055	1.476	1	0.224	7.74E6

For this analysis, the only thing we are interested in is whether the interaction terms are significant. This is tested by the *Wald statistic*, a type of chi-square. As we can see from the significance column, the interaction term in this model is not significant, so we can consider the assumption of linearity in the logit met.

We will evaluate multicollinearity by running a linear regression model with multi-collinearity diagnostics. We found tolerance values of 0.999 and VIF (variance inflation factor) of 1.001 for both variables, indicating that multicollinarity is not a problem. As discussed in Chapter 10, a standard rule of thumb is that tolerance should not be greater than 10 or VIF less than 0.10. The lack of multicollinearity is not surprising because these data come from a randomized national sample, and over a broad range of ages (18 and older in this case), there is no expected relationship between gender and age.

To check for complete separation, we create a cross-tabulation table between our dichotomous predictor (gender) and outcome (health insurance coverage). The cross-tabulated frequencies are presented in Table 11-2.

Table 11-2. Testing for complete separation

		Gender	
		Female	Male
Insurance	No	32	20
	Yes	234	167

We have no empty cells; in fact, we have no nearly empty cells. The latter can be a problem because although it does not constitute complete separation, it can produce estimates with very high standard errors, hence wide confidence intervals.

To see what complete separation looks like, consider the hypothetical data presented in Table 11-3.

Table 11-3. Hypothetical data showing complete separation

		Gender	
		Female	Male
Insurance	No	62	0
	Yes	234	167

In this example, all persons who do not have insurance are female. Therefore, if we know that a case does not have insurance, we also know that the case's gender is female; that is what is meant by complete separation. In practice, complete separation is more likely to occur when you have many categorical predictors (imagine that we included employment status, marital status, and educational level in this model as well) and some of them are sparsely distributed in some categories. A logistic regression model will not run if the data have complete separation, so the best solution is to see whether you can recode the variable. If marital status has six categories (married, widowed, divorced, single never married, living with same-sex partner, living with opposite-sex partner), perhaps you can combine them to produce only two or three categories, each of which includes sufficient cases to avoid the problem

of separation. Of course, you need to be able to defend your choice of which categories to combine. For instance, if you could make a case that the most important
information is simply whether a person is married versus not married, you should
feel free to recode the variable to reflect that information. Even if you don't have
complete separation, it's wise to avoid a variable with only a few cases in some of
its categories because this can produce estimates with extremely wide confidence
intervals, as noted earlier.

Having met the assumptions, we continue with the analysis. In logistic regression,
overall model fit is evaluated in several ways. First, there is an omnibus test of the
model coefficients, testing whether our entire model is better than the null model
with no coefficients; the model passes this test with a chi-square statistic (2 df) of
16.686 ($p < 0.001$). We also get three measures of model fit: the –2 log likelihood,
the Cox & Snell R^2, and the Nagelkerke R^2. The –2 log likelihood is somewhat
analogous to the residual sum of squares from a linear regression. It's hard to interpret the value of the –2 log likelihood by itself, but it is useful when comparing two
or more nested models (models in which the larger model includes all the predictors
from the smaller model) because a smaller –2 log likelihood indicates better model
fit. We can't compute a Pearson's R or R^2 for a regression model, but two pseudo-
R^2 statistics are possible: the Cox & Snell and Nagelkerke R^2. Both are based on the
log likelihood of the model versus the null model; because the range of the Cox &
Snell R^2 never reaches the theoretical maximum of 1.0, Nagelkerke's R^2 includes a
correction, which results in it having a higher value. Both are interpreted the same
as the coefficient of determination in linear regression, that is, as the amount of
variance in the outcome explained by the model. Because of the correction, Nagelkerke's R^2 generally has a higher value than Cox & Snell's R^2 for a given model. For
this model, the –2 log likelihood is 301.230, the Cox & Snell R^2 is 0.038, and the
Nagelkerke R^2 is 0.073.

The coefficients table for this model is presented in Table 11-4.

Table 11-4. Coefficients table for a logistic regression model predicting insurance status from
age and gender

	B	Std. error	Wald	df	Sig.	Exp(B)	95% CI for Exp(B)	
							Lower	Upper
Male	0.030	0.310	0.010	1	0.922	1.031	0.561	1.893
Age	0.035	0.009	16.006	1	< 0.001	1.036	1.018	1.054
Constant	0.118	0.475	0.062	1	0.804	1.125		

We recoded gender into a new variable, Male, with values 0 for female and 1 for
male; this is easier to interpret because we don't have to remember how the category
was coded. As with linear regression, the value and significance tests for the constant
in logistic regression are usually not our focus of interest. Predictors are evaluated
in logistic regression by the Wald chi-square; the significant values are interpreted
just like the p-values for any other statistic. In this case, we see that age is a significant
predictor of insurance status (Wald chi-square (1 df) = 16.006, p <0.001), whereas
male is not significant (Wald chi-square (1 df) = 0.010, p = 0.922). Remembering

that we coded insurance status so that 0 = no insurance and 1 = insurance, we see that because the coefficient for age is positive (0.035), increasing age is associated with increasing probability of having insurance.

The Exp(B) column gives the odds ratio for each predictor and the outcome variable, adjusted for all the other variables in the model; the last two columns give the 95% confidence interval for the adjusted odds ratio. If you are unfamiliar with odds ratios, you should read the section covering them in Chapter 15 before proceeding because only a brief explanation will be offered here. An odds ratio is, as the name implies, the ratio of the odds for two conditions. In the case of the first row in this table, the odds ratio for *Male* is the ratio of the odds of having insurance, if you are male, to the odds of having insurance if you are female. The neutral value for the odds ratio is 1; values higher than 1 indicate increased odds, values lower than 1, decreased odds. Because the odds ratio for male is higher than 1 (1.031), this indicates that in this data set, men have better odds of having insurance than do females. However, this result is not significant, as can be seen from the *p*-value for the Wald statistic (0.922) and the fact that the 95% confidence interval for the odds ratio (0.561, 1.893) crosses the neutral value of 1. We can therefore say that in a model predicting insurance status from gender and age, gender was not a significant predictor.

Looking at the second line of the table, we see that age is a significant predictor of insurance status in a model also including gender. The adjusted odds ratio is 1.036, and the 95% confidence interval is (1.018, 1.014); note that the confidence interval does not cross 1. The odds ratio for age and male look small (barely higher than 1, in fact), but remember that this is the difference in odds for an increase in age of 1 year; for instance, it is the odds of insurance for someone who is 35 compared to someone who is 34 (adjusted for gender). To find the expected change for a larger number of years, you exponentiate the odds ratio by the number of years. For instance, the predicted change in the odds of having insurance for an age difference of 10 years (adjusted for gender) is:

$$1.036^{10} = 1.424$$

It's often useful to report a few hypothetical examples like this along with your results to help your audience understand the importance of variables measured on a continuous scale. The logistic regression equation for this model is:

$$Logit(p) = 0.118 + 0.030(male) + 0.035(age) + e$$

As noted earlier, though our model is significantly better than the null model at predicting insurance status, it doesn't explain much of the variance in our data (determined by examining the pseudo-R^2 statistics). This is not surprising because there are probably many other variables besides age and gender related to whether a person has insurance; if we were to continue this analysis, we would certainly test the effects of employment and income, for instance. We might also try dichotomizing age to under/over 65 years because we know that virtually everyone over the age of 65 is entitled to insurance coverage under the federal Medicare insurance program. We might also consider running an equation for just people younger than 65 because we don't expect to see much variation in insurance status for people age 65 and older.

Converting Logits to Probabilities

People outside the field of statistics are unlikely to be familiar with the logit, and it's often better to present results to them in units they understand. For logistic regression, the obvious choice is probabilities. Fortunately, a logistic equation for any set of predictors can be converted to a probability by using the following formula:

$$Predicted\ probability = e^{\text{logistic regression equation}} / (1 + e^{\text{logistic regression equation}})$$

Continuing with our previous BRFSS example, we can find the probability of an individual having insurance by plugging his or her X values into our equation and then using that equation in the formula presented earlier. For instance, for a male ($X_1 = 1$) of age 40 ($X_2 = 40$), the predicted logit is:

$$Predicted\ logit(p) = 0.118 + 0.030(1) + 0.035(40) = 1.548$$

We then put this value into the equation to predict probability:

$$Predicted\ probability = e^{1.548} / (1 + e^{1.548}) = 0.825\ or\ 82.5\%$$

Multinomial Logistic Regression

If you have a data set that would be suitable for logistic regression, except that the outcome variable is categorical (with more than two categories), it might be a good candidate for multinomial logistic regression. Returning to the BRFSS data, we are interested in what variable predicts health status. Fortunately for us, the BRFSS includes a variable measuring health status on a scale that is commonly accepted and used in the fields of medicine and public health. Often described as self-reported general health, this variable asks people to indicate which of five categories best describes their general health:

1. Excellent
2. Very good
3. Good
4. Fair
5. Poor

The responses to this question in our sample are displayed in Table 11-5.

Table 11-5. Self-reported general health

	Frequency	Percent	Cumulative percent
Excellent	64	14.7	14.7
Very good	149	34.3	49.0
Good	136	31.3	95.2
Fair	65	14.9	80.2
Poor	21	4.8	100.0

We will use age (continuous) and gender (dichotomous) in a multinomial regression equation to predict self-reported health status. Because we have relatively sparse data in one of our outcome categories, we will do a cross-tabs table with gender to see whether we have empty or nearly empty cells; if so, this would be a problem for the same reason (complete or near-complete separation) discussed in the logistic regression example. The results are shown in Table 11-6.

Table 11-6. Cross-tabulation of general health status and gender

Health status	Female	Male
Excellent	36	28
Very good	92	57
Good	80	56
Fair	45	20
Poor	15	6

This is mixed news: although we don't have any empty cells, a cell of size 6 (males in poor health) might give us quite wide confidence intervals. We decide to combine the two lowest categories and proceed with our analysis. We need to choose one of the categories to serve as a reference category for the analysis; the computer algorithm will compare each of the other categories to this one to see whether there is any significant difference among them. We choose the Excellent category.

The model fit information for multinomial logistic regression is similar to that for binomial logistic regression. The –2 log likelihood for this model is 660.234 (this might come in handy if we want to compare the fit of this model to more complex models), and our model predicts significantly better than the null model without predictor variables (χ^2 (6 df) = 19.194, p = 0.004). The pseudo-R^2 statistics tell us that we aren't explaining much of the variance (Cox & Snell's R^2 = 0.043, Nagelkerke's R^2 = 0.046), but we're not surprised: we expect that many more things than gender and age would influence a person's general health status. We also get likelihood ratio tests, which tell us how model fit changes if one of the predictors is removed. If model fit is significantly lowered (as tested with a chi-square statistic), that means the variable is making a significant contribution to predicting the outcome variable. Data from the likelihood ratio test is presented in Table 11-7.

Table 11-7. Likelihood ratio tests for a multinomial regression model predicting general health status from age and gender

	−2 log likelihood of reduced model	Chi-square	df	Sig.
Intercept	660.234	0.000	0	.
Age	675.719	15.485	3	0.001
Male	660.609	3.375	3	0.337

The "reduced model" in each case is the model lacking the variable tested. From this table, we can see that age is a significant predictor of general health status, but gender is not. The intercept is not tested because removing it does not change the degrees of freedom of the model. As we said earlier, a lower –2 log likelihood indicates better fit, so we are not surprised to see a substantial increase in the –2 log likelihood when *Age* is removed from the model (675.719 vs. 660.234) but very little change (660.608 vs. 660.234) when *Male* is removed.

The parameter estimates for our full model are presented in Table 11-8. Note that this is really three models at once because different coefficients are estimated for each of our comparisons (*very good* versus *excellent*, *good* versus *excellent*, and *fair/poor* versus *excellent*).

Table 11-8. Parameter estimates for a multinomial regression model predicting general health status from age and gender

Gen hlth category		B	Std. error	Wald	df	Sig.	Exp(B)	95% CI for Exp(B) Lower bound	Upper bound
Very good	**Intercept**	0.681	0.519	1.723	1	0.189			
	Age	0.001	0.009	0.004	1	0.949	1.001	0.984	1.018
	Male = 1	0.227	0.003	0.562	1	0.454	1.255	0.693	2.274
Good	**Intercept**	−0.142	0.542	0.068	1	0.794			
	Age	0.015	0.009	2.836	1	0.092	1.015	0.998	1.033
	Male = 1	0.095	0.307	0.096	1	0.057	1.100	0.602	2.009
Fair/ poor	**Intercept**	−1.766	0.638	7.740	1	0.005			
	Age	0.030	0.010	8.701	1	0.003	1.030	1.010	1.051
	Male = 1	0.559	0.348	2.581	1	0.108	1.748	0.884	3.457

Our misgivings at seeing such low pseudo-R^2 statistics are upheld in this model because only one predictor in one of our comparisons is significant: *Age* for the comparison of *Fair/poor* versus *Excellent* health. Because the coefficient is positive (0.030), and the Exp(B) or odds ratio is greater than one, we can see that increased age is associated with increased probability of having *Fair/poor* versus *Excellent* health. Note also that the 95% confidence interval for *Age* in this comparison (1.010, 1.051) does not cross the null value of 1.0, a result expected because the Wald chi-square test is significant for this predictor in this comparison.

Polynomial Regression

So far, you have largely learned about model fitting when the relationship between a DV and one or more IVs is linear, that is, the value of a DV can be predicted by a weighted linear sum of the IVs plus an intercept value. In the two-dimensional plane, such relationships can be viewed as straight lines that have nonzero slope. However, many phenomena have nonlinear relationships, and you need to be able to model these relationships as well. Any relationship that is not entirely linear is, by definition, nonlinear, so any discussion of nonlinear modeling must be very broad indeed. In this section, you learn about two of the most commonly used regression models, which are based on *quadratic* and *cubic* polynomials.

A quadratic model has both a linear and squared term for the IV, whereas the cubic model has a linear, squared, and cubic term for the IV; the principle is that you include all lower order terms as well as the highest order term. Each curve has a number of extreme points equal to the highest order term in the polynomial, so a quadratic model will have a single maximum, whereas a cubic model has both a relative *maximum* and a *minimum*. Figure 11-3 shows a quadratic model ($Y = X^2$) and Figure 11-4, a cubic model ($Y = X^3$).

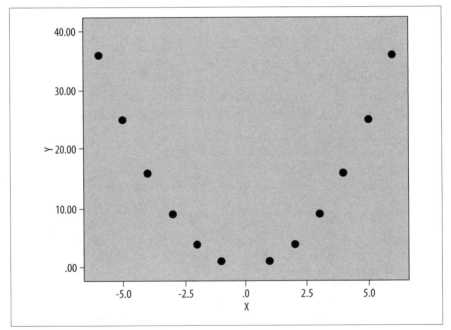

Figure 11-3. Quadratic model ($Y = X^2$).

Let's look at an example from sports psychology. The Yerkes-Dodson Law, first formulated in 1908, predicts a quadratic relationship between arousal (the predictor variable) and performance (the outcome variable). For many athletes, achieving the optimal level of physiological arousal—corresponding to the single maxima of the

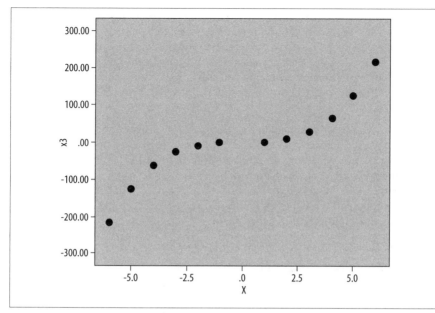

Figure 11-4. Cubic model (Y = X³)

DV—corresponds to their goal of producing their best possible athletic performance. If athletes are not aroused enough, their performance will be poor; conversely, if athletes are over-aroused, their performance will also be poor.

However, if the relationship between arousal and performance is actually cubic, increasing arousal even further might result in improvements in performance, which would be a contrary prediction to the quadratic model. Polynomial regression can be used to determine the goodness of fit for both the quadratic and cubic models, and the one with the best goodness of fit can be taken as the most accurate description of the relationship between arousal and athletic performance.

Watters, Martin, and Schreter (1997)[1] designed an experiment to determine whether there was a quadratic relationship between caffeine (a drug that produces arousal) and cognitive performance on a battery of tests. The experimental setup required a dose of caffeine to be administered at regular intervals in a single session (6 × 100 mg); this would introduce practice effects and would lead to an increase in performance session to session, independent of arousal. Any residual variation accounted for by a quadratic term would then indicate the underlying relationship between arousal and performance.

You might be wondering why participants in the study were not simply invited back several times to complete the test, with the caffeine dosage randomly assigned on each occasion. The reasoning was ethical; the researchers wanted to observe any adverse reactions at low dosages, which would be impossible on the first trial in a

1. Watters, P.A., Martin, F., & Schreter, Z. (1997). "Caffeine and cortical arousal: The nonlinear Yerkes-Dodson Law." *Human psychopharmacology: clinical and experimental*, 12, 249–258.

truly randomized design because some of the subjects would be receiving the highest dosage on their first trial, and the researchers wanted to minimize the number of return visits. To obtain a higher degree of experimental control, a repeated measures design was used, in which each participant attended a placebo and treatment session (single blind). If the experimenter noted an adverse reaction, the experiment would be halted. The order of attendance for either the placebo or treatment condition was randomized.

As designed, the experiment had both a within-subjects and between-subjects comparison, with the former showing the dose-response relationship and the latter confirming that the dose-response relationship observed was not the product of chance (or practice). Only the within-subjects analysis is shown here. The analysis proceeds by adding terms progressively into the model, starting with caffeine, followed by the square and cube of caffeine. Table 11-9 shows some sample data that may be obtained in this type of experiment.

Table 11-9. Relationship between caffeine and cognitive performance

0mg	100mg	200mg	300mg	400mg	500mg	600mg
10.0	15.0	17.0	18.0	15.0	13.0	11.0
8.0	10.0	14.0	16.0	12.0	10.0	9.0
15.0	16.0	18.0	24.0	20.0	17.0	15.0
14.0	17.0	21.0	22.0	21.0	17.0	13.0
15.0	16.0	18.0	20.0	18.0	16.0	12.0
10.0	15.0	17.0	18.0	15.0	13.0	11.0
8.0	10.0	14.0	16.0	12.0	10.0	9.0
15.0	16.0	18.0	24.0	20.0	17.0	15.0
14.0	17.0	21.0	22.0	21.0	17.0	13.0
15.0	16.0	18.0	20.0	18.0	16.0	12.0

For the linear model $Y = \beta_0 + \beta_1 X_1 + e$, where Y is performance and x is caffeine, there was virtually no relationship between the two variables: $R^2 = 0.001$, and the F-statistic also showed that the coefficient for caffeine was not significantly different from 0 ($F(1, 68) = 0.097$, $p = 0.757$).

For the quadratic plus linear model, $Y = \beta_0 + \beta_1 X_1 + \beta_2 X_1^2 + e$ for the same variables found a significant relationship between caffeine and performance. For this model, $R^2 = 0.462$, $F(2, 67) = 28.81$, and $p < 0.001$. The coefficients table for this model, Table 11-10, shows that both the linear and quadratic terms made a significant contribution to the model fit, with a strongly linear effect accompanied by a negative quadratic term. The relative contribution that both terms make to the model, as demonstrated by the absolute value of their beta coefficients, is comparable (β_{linear} = 2.314 versus $\beta_{quadratic}$ = −2.448).

Table 11-10. Quadratic model predicting performance from caffeine consumption

	Unstandardized coefficients		Standardized coefficients			
	B	Std. error	Beta		t	Sig.
Caffeine	0.044	0.006		2.314	7.166	<0.001
Caffeine ** 2	−7.429E−5	0.000		−2.448	−7.580	<0.001
(Constant)	12.014	0.784			15.324	<0.001

The cubic plus quadratic plus linear model $Y = \beta_0 + \beta_1 X_1 + \beta_2 X_1^2 + \beta_3 X_1^3 + e$ did not explain a significant additional amount of variation in performance, and the coefficient for the cubic term was not significant, confirming that the linear and quadratic relationship model best explains the relationship between caffeine consumption and athletic performance.

Overfitting

One of the more amazing features of modern statistical computing packages is that you can automatically specify and perform any number of tedious statistical tests at the click of a button. This ability to run many models quickly can be useful if you are simply exploring the data, or if your *a priori* hypotheses have failed to meet expectations and you are trying to figure out what is actually going on in the data. However, many statisticians frown on building models based purely on a given data set, deeming it "going on a fishing expedition" and, when nonlinear regression is involved, *arbitrary curve-fitting*. We discussed the dangers of mechanical model-building in Chapter 10, but the cautions apply even more here because you are not simply adding and subtracting predictor variables but also changing their form. However, this type of model building is acceptable in some fields, so if that is the case in your workplace or school, there's no reason you shouldn't take advantage of all the possibilities offered by modern computer packages. Some statistical packages allow you to request that all possible linear and nonlinear relationships between two variables be calculated, and then you can simply select the one that does the best job of explaining the data.

If you're going to engage in this type of model fitting, you should be aware of the dangers inherent in the process. We illustrate these with a simple example. Imagine that you are a nutritionist interested in the relationship between smoking and blood pressure, with the results obtained from a small study shown in Table 11-11. You know that there is a relationship between the two, but as an expert witness in a court case, you are under pressure to prove the strongest possible link between the two variables. Some of the data, showing the diastolic blood pressure and daily cigarette consumption for the first few cases, is shown in Table 11-11.

Table 11-11. Relationship between diastolic blood pressure and daily cigarette smoking

DiastolBP	Daily cigs
80.0	0.0
75.0	0.0
90.0	1.0
80.0	0.0
75.0	0.0
95.0	10.0
90.0	20.0
100.0	25.0
110.0	30.0
140.0	35.0

Summary results from a few models (with diastolic blood pressure as the outcome, and daily cigarettes smoked as the predictor) are displayed in Table 11-12. As you can see from Table 11-12, there are many types of relationships two variables can have besides linear. Even more surprising, a model including linear and quadratic terms explains 97% of variability in diastolic blood pressure. No one has ever reported a cubic relationship between the two variables before, so you think you have discovered a very convincing argument.

Table 11-12. Relationship between diastolic blood pressure and daily cigarette smoking

Equation	Model summary					Parameter estimates			
	R square	F	df1	df2	Sig.	Constant	b_1	b_2	b_3
Linear	0.781	28.518	1	8	0.001	78.423	1.246		
Quadratic	0.869	23.118	2	7	0.001	80.984	−0.386	0.053	
Cubic	0.970	64.155	3	6	0.000	79.069	3.975	−0.299	0.007
Compound	0.813	34.853	1	8	0.000	79.007	1.013		
Growth	0.813	34.853	1	8	0.000	4.370	0.012		
Exponential	0.813	34.853	1	8	0.000	79.007	.0120		

Do the R^2 values computed by such an approach have any real meaning? Yes and no; one real risk with fishing expeditions is *overfitting*. This means your data fits your data set too well and explains the random variation in it as well as the significant relationships. Because the purpose of inferential statistical analysis is to find results that will generalize to other samples drawn from the same population, overfitting defeats the whole purpose of conducting the analysis in the first place. You might have a model that fits your particular data set remarkably well, but it won't necessarily fit any other data set, so it hasn't produced any useful knowledge for your field.

The best protection against overfitting is to build your models based on theory. If you decide to use mechanical procedures to build your model, you should test it across multiple samples to be sure you are modeling major relationships within the data instead of random noise. If only limited samples are available, such as in destructive testing environments, resampling techniques such as *bootstrapping* and the *jackknife* may be employed; these are discussed in the Efron book listed in Appendix C.

Exercises

Problem

You are comparing two nested logistic regression models (models in which the larger models include all the predictor variables included in the smaller models). Model A has a –2 log likelihood of 200.465; Model B has a –2 log likelihood of 210.395. Which model fits the data better?

Solution

Model A has the better fit; when comparing two nested models, the model with the smaller –2 log likelihood fits the data better.

Problem

You are planning a logistic regression analysis, using one dichotomous and one categorical predictor. The following table presents cross-tabulation results for the Y variable and the two predictors (X_1 and X_2). Do any alarm bells go off when you read this table? If so, how would you fix the problem?

		$X_1 = 1$	$X_1 = 2$	$X_1 = 3$
$Y = 0$				
	$X_2 = 1$	25	32	20
	$X_2 = 2$	27	17	32
$Y = 1$				
	$X_2 = 1$	34	6	23
	$X_2 = 2$	41	36	5

Solution

Although there are no empty cells, two are sparsely populated (with 6 and 5 cases, respectively), which might result in wide confidence intervals. If possible (and theoretically defensible, based on the meanings of the categories for variable X_1), the best solution may be to combine the second and third categories for this variable.

Problem

You have conducted a logistic regression analysis to predict the probability of high school students becoming dropouts, using their GPA and gender as predictors. This is your regression equation:

Logit(p) = 4.983 + 1.876(*male*) –2.014(*GPA*) + *e*
Dropout (the *Y* variable) is coded so 1 = dropped out, 0 = not dropped out.
GPA is a continuous variable ranging from 0.00 to 4.00.
Male (the gender variable) is coded so 0 = female and 1 = male.

What is the predicted probability of dropout for a female with a 3.0 GPA?

Solution

To calculate this probability, plug the values for female and GPA into the logistic regression equation and then use the following formula to calculate the predicted probability of dropout:

$$Predicted\ probability = e^{\text{logistic regression equation}}/(1 + e^{\text{logistic regression equation}})$$

The predicted logit is:

$$Logit(p) = 4.983 + 1.876(0) - 2.014(3.0) = -1.059$$

The predicted probability of dropping out is:

$$Predicted\ probability = e^{-1.059}/(1 + e^{-1.059}) = 0.258 = 25.8\%.$$

Problem

Continuing with the problem of predicting who will drop out of high school, you decide to add another variable to the equation: whether the student's mother graduated from high school (coded so 0 = no, 1 = yes). After doing the appropriate data checks, you run the equation, which produces the coefficients and significance tests shown in Table 11-13. This model is significantly better than the null model in predicting high school dropouts (chi-square (3) = 28.694, $p < 0.001$); the value for the Cox & Snell R^2 is 0.385 and, for the Nagelkerke R^2, 0.533.

Table 11-13. Coefficients for a logistic regression equation predicting the high school dropout from gender, GPA, and mother's education

	B	Std. error	Wald	df	Sig.	Exp(B)	95% CI for Exp(B) Lower	Upper
Male	2.107	0.770	7.495	1	0.006	8.224	1.819	37.170
GPA	−1.599	0.756	4.466	1	0.035	0.202	0.046	0.890
Mother HS graduate	−2.430	1.104	4.847	1	0.028	0.088	0.010	0.766
Constant	5.021	2.420	4.305	1	0.038	151.526		

Interpret the information in this table, including specifying which predictors are significant, in which direction, and what the Exp(B) column and its 95% confidence interval means.

Solution

All the predictor variables in this model are significantly related to the probability of a student dropping out of high school. Males are more likely than females to drop out (B = 2.107; Wald chi-square (1) = 7.495, p = 0.006). Higher GPA predicts a lower probability of dropping out (B = -1.599; Wald chi-square (1) = 4.466, p = 0.035), as does having a mother who graduated from high school (B = -2.430; Wald chi-square (1) = 4.867, p = 0.028).

The Exp(B) column presents the adjusted odds ratios for each of the predictor variables. As expected, male gender has an odds ratio greater than 1 (8.224), indicating that males are more than 8 times as likely as females to drop out, after adjusting for GPA and mother's education; the 95% confidence interval for male gender is (1.819, 37.170). The odds ratios for GPA and having a mother who graduated from high school are less than 1, indicating that a higher GPA or a mother who is a high school grad are associated with a lower probability of dropping out. The odds ratios and the 95% confidence intervals are 0.202 (0.046, 0.890) for GPA and 0.088 (0.010, 0.766) for having a mother who graduated from high school. Note that none of the confidence intervals cross the neutral value of 1; this is expected because all the predictors are significant.

12

Factor Analysis, Cluster Analysis, and Discriminant Function Analysis

There are more statistical techniques in use today than could possibly be covered in a single book. In fact, there are more types of statistics out there than anyone could hope to master in a lifetime. However, it's often useful to be familiar with a technique, even if you don't know how to perform it yourself. You might need to read articles using techniques you have not mastered yourself, for instance, or you might decide you need to learn a technique or hire a consultant familiar with the technique after reading about how it was used in someone else's research. This chapter introduces several advanced statistical techniques by providing some specific examples of how they have been used; the techniques themselves will not be taught because the intent is to help the reader identify when one of these techniques is appropriate for a given research question. Methodologies covered in this chapter include factor analysis, cluster analysis, and discriminant function analysis.

Factor Analysis

Factor analysis (FA) uses standardized variables to reduce data sets by using *principal components analysis* (PCA), the most widely used data reduction technique. It is based on an *orthogonal decomposition* of an input matrix to yield an output matrix that consists of a set of orthogonal components (or factors) that maximize the amount of variation in the variables from the input matrix. This process usually produces a smaller, more compact number of output components. In linear algebra terms, PCA works from the covariance matrix to produce a set of *eigenvectors* and *eigenvalues*. The components in the output matrix are linear combinations of the input variables; the components are created so the first component maximizes the variance captured, and each subsequent component captures as much of the residual variance as possible while taking on an uncorrelated direction in space. A more general version of PCA is *Hotelling's canonical correlation analysis* (CCA), which—assuming multivariate normality—can be used to test whether two sets of variables are independent.

PCA is primarily used for three major purposes:

- In hypothesis testing, to use techniques based on the general linear model to produce variables that are orthogonal
- To compress a large number of variables into a smaller, more manageable data set
- To identify latent variables in large data sets that are represented by highly correlated input variables

Although the first two purposes are usually achieved by PCA, the third is typically approached using factor analysis (FA), which is also based on orthogonal decomposition but can involve more complex techniques such as variance maximizing rotation (*varimax*). You will learn about some of these techniques in this chapter. Note that in FA, the retained principal components are known as *common factors*, and correlations with the input variables are called *factor loadings*.

Let's look at an example from the field of psychometrics. Historically, FA has been used to test various theories of mental performance and intelligence, including the hypothesis that a single general factor underlies intelligence and the rival hypothesis that multiple orthogonal factors comprise intelligence. In turn, the general findings derived from large-scale studies of intelligence and cognitive function in the population have allowed reliable understanding of individual differences to be determined from a number of test instruments. The process of understanding individual differences, and compensating for them, was heavily influenced by the thinking of Carl Friedrich Gauss, the inventor of the Gaussian or normal distribution, followed by the later work of Bessel, who developed a personal equation to make corrections in observations made by different astronomers.

Early attempts to understand intelligence and measurable variables began with scientists such as James Cattell, who tried to quantify intelligence in terms of a set of mental tests such as reaction times, rate of movement, and grip strength. Later work showed that results from these tests were uncorrelated with actual academic performance. However, work by Charles Spearman on the general intelligence factor, g, extracted from the results of a battery of psychological tests, led to the widespread adoption of FA and PCA-like methods in psychometrics. Later work by Louis Leon Thurstone and others suggested that there must be at least two independent cognitive factors underlying intelligence: a linguistic factor, L, and a quantitative factor, Q. Even today, this characterization of intelligence is seen in standardized tests such as the Scholastic Aptitude Test (SAT), taken by many American students planning to attend university, and the Graduate Record Examination (GRE), taken by many before entering graduate school. Both have three major components—verbal, writing, and math—corresponding roughly to the linguistic (verbal and writing) and quantitative (math) breakdown suggested by Thurstone.

Let's look at a typical psychometric example in which a set of cognitive and mental performance scores is taken as the input matrix, and an output matrix is then produced that is of lower dimension. The term matrix here just refers to numerical information arranged in a pattern corresponding to the meaning of each piece of information. This process might be driven by a hypothesis. For instance, a specific psychological theory might predict two factors (e.g., L and Q), so only the two factors accounting for the highest proportion of variance would be selected. On the other hand, if the work is more exploratory, the data might be allowed to determine partially how many factors are retained, following some standard criterion or rule. The most commonly used criterion for factor retention is the *Guttman-Kaiser criterion*, which only retains eigenvalues greater than 1 (in the case of FA); following this rule, factors are retained if the variation they account for is greater than the average for the variable if variation were equally distributed across the input data set. Other retention criteria include the *Velicer partial correlation procedure*, *Bartlett's test*, and the *broken stick model*, while a more graphical approach is to use the *scree plot* of eigenvalues to determine which factors to retain. With a scree plot, you graph the eigenvalues and retain those before the slope levels out like the scree (loose rock) that accumulates at the foot of a mountain.

Suppose you have a data set containing the results from the administration of a standard battery of tests; data for the first five participants is displayed in Table 12-1. A psychologist is interested in determining whether a general intelligence factor underlies performance across all these components of intelligence or whether there are distinct factors on which individual variables are highly loaded. For example, is an L factor strongly associated with reading and verbal ability, and a separate Q factor associated with arithmetical and geometrical ability?

Table 12-1. Psychometric test results

Reading	Music	Arithmetic	Verbal	Sports	Spelling	Geometry
8	9	6	8	5	9	10
5	6	5	5	6	5	5
2	3	2	6	8	6	4
8	9	10	9	8	10	6
10	7	1	10	5	10	2

The first way to begin exploring the data is to create a matrix for the bivariate correlations among all the variables; Table 12-2 shows the upper triangle of this matrix. This is a good way to see which variables have significant relationships with each other and which do not. The first line for each pair displays the Pearson's r, the second line its significance level.

Table 12-2. Correlations among psychometric test variables

		Reading	Music	Arithmetic	Verbal	Sports	Spelling	Geometry
Reading	r	1.000	0.535	−0.253	0.860	−0.469	0.762	−0.386
	p		0.111	0.481	0.001	0.172	0.010	0.270
Music	r		1.000	0.249	0.262	−0.263	0.380	0.069
	p			0.488	0.464	0.463	0.278	0.850

		Reading	Music	Arithmetic	Verbal	Sports	Spelling	Geometry
Arithmetic	r			1.000	−0.501	0.206	−0.307	0.758
	p				0.140	0.568	0.389	0.011
Verbal	r				1.000	−0.236	0.895	−0.569
	p					0.511	0.001	0.086
Sports	r					1.000	0.054	0.266
	p						0.881	0.458
Spelling	r						1.000	−0.291
	p							0.415
Geometry	r							1.000
	p							

These correlations appear to support the idea of separate Q and L factors.

For L:

- Verbal performance and reading scores appear to be highly correlated ($r = 0.860$, $p = 0.001$).
- Reading and spelling scores are highly correlated ($r = 0.762$, $p = 0.010$).
- Verbal performance and spelling scores are also correlated ($r = 0.895$, $p < 0.001$).

For Q:

- Geometry and arithmetic scores are highly correlated ($r = 0.758$, $p = 0.011$).

None of the other variables (e.g., sporting or musical performance) are significantly correlated with any other variables, so you could expect that two interpretable factors will result from the FA.

The first step after computing PCA is to examine what proportion of variance is accounted for by the factor structure. This is done by examining the *communalities*, as shown in Table 12-3 in the column labeled Extraction. Here, you can see that some variables, such as music, have relatively low communality (0.779), whereas others, such as spelling, have very high communality (0.967). Variables with high communality have a large proportion of their variance explained by the extracted factors, whereas those with a low communality have a lot of unexplained variance remaining.

Table 12-3. Communalities

	Initial	Extraction
Reading	1.000	0.929
Music	1.000	0.779
Arithmetic	1.000	0.868
Verbal	1.000	0.955
Sports	1.000	0.943

	Initial	Extraction
Spelling	1.000	0.967
Geometry	1.000	0.814

Tables 12-4 through 12-6 show the initial eigenvalues, the extraction sums of squared loadings, and the rotation sums of squared loadings resulting from this FA. These tables are the most important as the most significant part of the results for interpretation. In Table 12-4, you can see that the first three factors extracted account for 89.378% of the variance; thus, you can immediately see the power of PCA because it has reduced seven variables to three factors while still accounting for almost all the variation within the data! Table 12-5 shows the three extracted factors before rotation, and Table 12-6 shows the extracted factors after rotation was performed using varimax with the *Kaiser normalization*. The varimax rotation rotates the axes of the factors so that orthogonality is preserved while maximizing the sum of variances of the loadings. Note that this does not affect the total amount of variance accounted for by the three factors, but the relative proportion of variance between factors does change.

Table 12-4. Initial eigenvalues

Component	Initial eigenvalues		
	Total	% of variance	Cumulative %
1	3.488	49.829	49.829
2	1.651	23.591	73.420
3	1.117	15.958	89.378
4	0.425	6.069	95.446
5	0.234	3.343	98.789
6	0.067	0.952	99.742
7	0.018	0.258	100.000

Table 12-5. Extraction sums of squared loadings

Extraction sums of squared loadings		
Total	% of variance	Cumulative %
3.488	49.829	49.829
1.651	23.591	73.420
1.117	15.958	89.378

Table 12-6. Rotation sums of squared loadings

Rotation sums of squared loadings		
Total	% of variance	Cumulative %
2.846	40.653	40.653
2.066	29.517	70.170
1.345	19.208	89.378

People new to FA sometimes feel that there must be some trickery with rotation, especially because it is used as an aid to interpreting factor loadings and to discovering the existence of latent structure. However, it's really a legitimate technique that serves a very useful purpose in helping the researcher tease out which variables are most closely associated with each factor.

The benefits of rotation can be seen by comparing Tables 12-7 and 12-8, which show the component matrixes for this analysis before and after rotation. For component 1, which corresponds to the latent L factor, you can see that rotation has the effect of increasing the relative loadings of the most relevant variables so that spelling, reading, and verbal skills have the highest loadings on this factor. After rotation, component 2, which corresponds to the Q factor, has higher loadings for arithmetic and geometry, whereas the loadings for unrelated variables such as music are now relatively decreased. Component 3 has a high loading only for sports, so although it does represent a distinct factor, it does not reflect any latent structure and will be disregarded in this analysis. Thus, rotation has helped us clarify which test scores (reading, music, etc.) are most closely related to our two components.

Table 12-7. Unrotated component matrix

	Component		
	1	2	3
Reading	0.902	0.328	−0.085
Music	0.386	0.775	−0.174
Arithmetic	−0.582	0.727	0.028
Verbal	0.955	0.009	0.209
Sports	−0.403	−0.059	0.882
Spelling	0.819	0.235	0.491
Geometry	−0.664	0.597	0.130

Table 12-8. Rotated component matrix

	Component		
	1	2	3
Reading	0.859	−0.144	−0.412
Music	0.593	0.490	−0.433
Arithmetic	−0.158	0.917	0.050

	Component		
	1	**2**	**3**
Verbal	0.869	−0.438	−0.088
Sports	−0.046	0.176	0.954
Spelling	0.955	−0.164	0.169
Geometry	−0.246	0.846	0.195

Graphical examination of the data can also help clarify relationships among the variables. Returning to the question of the selection of eigenvalues, Figure 12-1 shows the scree plot resulting from this analysis; each circle responds to one of the eigenvalues in Table 12-4. Higher values explain more variance, and it's clear that after the third eigenvalue, there's not much gain in explanation with each additional eigenvalue. If you picture the eigenvalues as rocks tumbling down a mountain, it's clear that there's a bend toward the horizontal at the third or fourth eigenvalue (there's a subjective element to interpreting scree plots), whereas eigenvalues 4 through 7 are just heaped up at the foot of the mountain. Therefore, we have two or three components worth retaining in this analysis, and this corresponds to the results using the Guttman-Kaiser criterion; component 3 has a value barely above the cutpoint of 1.0.

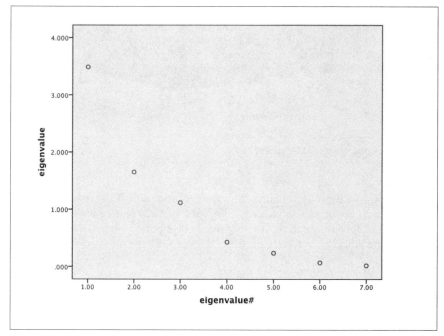

Figure 12-1. Scree plot

Figure 12-2 shows the effect of rotating the data in three-dimensional space; you can see that the variables associated with the *L* factor (spelling, verbal, and reading) are closely clustered in 3D space, as are the variables associated with the Q factor (arithmetic and geometry). Note that the other two variables (sports and music) are then roughly equidistant from the centroids of the two component-oriented clusters. The impact of the rotation is often easier to observe in three-dimensional space than by looking at the loading tables.

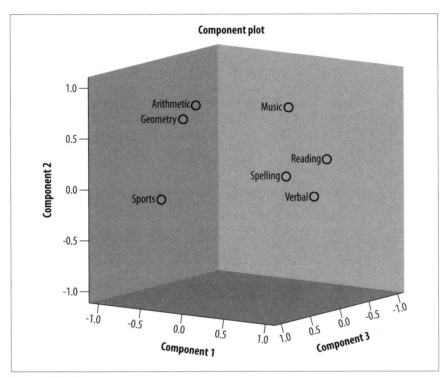

Figure 12-2. Component plot in rotated space

Table 12-9 shows the output matrix from the FA procedure. This shows the scores for the three components computed for the first five study participants; if this were the GRE or SAT, these are the scores that would be reported back to the test takers. Note that the precision of the results depends on your computer package.

Table 12-9. Component scores for each participant

Subject	Component 1 (L)	Component 2 (Q)	Component 3 (Sports)
1	0.518	1.132	−0.095
2	−1.170	−0.128	0.084
3	−1.396	−1.207	1.619
4	1.094	1.198	1.128
5	0.706	−1.049	0.014

As with all the other techniques you have learned about in this book, PCA and FA have some basic prerequisites that must be met for the results to be valid and/or reliable. Large data sets are most commonly used for PCA and FA because in general as the data set grows larger, the results become more reliable. In the case of psychometrics, reliability is usually established when a test has been administered to hundreds of thousands of individuals across different national and linguistic groups. The other main requirement is that the number of cases must always be larger than the number of variables in the input matrix. Normally, tests for statistical significance are not performed with PCA, so outliers and other potential sources of bias are much less likely to cause problems than, say, with ANOVA. For PCA, the assumptions of linear correlation also hold; that is, that variables must be linearly related, and none should have either zero or perfect correlation.

Cluster Analysis

Cluster analysis is a set of techniques that allows cases to be grouped based on their values for one or more variables. Some cluster analysis techniques allocate cases to groups by partition, whereas other techniques provide for hierarchical trees that show the taxonomic relationship between groups and their ancestors. A related technique, *discriminant function analysis* (DFA), can be used to develop rules to assign cases to groups, based on an understanding of the parametric structure of the groups; DFA is better at predicting group membership than cluster analysis alone. Often, the two techniques are used in conjunction with each other. Cluster analysis may be used when the number of groups is initially unknown; once this number has been established, DFA may be used for the prediction of individual group membership for each case.

Cluster analysis is very useful for two scenarios. First, you might already know how many groups you expect to find in the data, so you pass this number of groups to the algorithm and let it take care of the allocation (*k-means*). Alternatively, you might not know how many groups exist, in which case you can ask the algorithm to estimate how many groups there actually are.

Cluster analysis is a highly empirical tool; its success depends largely on the quality of data supplied. Cluster analysis works by taking an input vector of Y, with n cases and p variables, and allocating each of n cases to one of k groups. Each of the p variables measures some aspect of an object under study. Continuing with the psychometric example, each variable may represent a score on a particular type of ability test (reading, spelling, etc.). The algorithm works by randomly creating k clusters, identifying the *centroids* or cluster centers, and assigning each case to the closest centroid. Cases are moved between clusters to minimize within-cluster variability and maximize between-cluster variability. The process continues until it converges according to some predefined criterion. Note that because some randomness is introduced by the initial assignment of centroids, you don't always get the same answer.

The computational goal of cluster analysis is to ensure that all members of groups $1 \ldots k$ are similar to other members of their group and dissimilar to members of other groups. Similarity—or dissimilarity—is determined by the use of a specific distance measure. A number of measures have been developed, including the following:

Euclidean distance
> This is the geometric distance between two points in a multidimensional space.

Manhattan distance
> This is a city-block distance that reduces the influence of outliers.

Mahalanobis distance
> Within-cluster distances tend to be increased, whereas between-cluster distances are decreased.

Let's revisit the psychometric example. Having shown that three factors underlie student abilities, the psychologist is now interested in determining whether there might be some basis for classifying students in different educational groups based on this latent structure because the identified factors for L, Q, and *Sports* were orthogonal. The issue is specialization; if students are identified as being good at sports, linguistic, or quantitative work, they could be streamed appropriately into classes specializing in those areas. (The question of at what age such specialization should occur is a question for another day.) The main problem with this approach is that some students might be good at more than one of these skills, and the idealized view provided by the rotated loading matrix shown in Figure 12-2 might not apply to all cases.

The psychologist decides to use cluster analysis to determine whether three distinct groups are in this data set, corresponding to distinct members of the proposed *Linguistic*, *Quantitative*, and *Sports* tracks. Because we believe there are three groups, we pass $k = 3$ to the algorithm and ask it to identify three groups and then assign each student to a class.

The initial cluster centers are shown in Table 12-10, and after several iterations, the algorithm converges to a solution, with the final cluster membership for the first five cases, the final cluster centers, and pairwise distances between the final clusters shown in Tables 12-11 through 12-13 (upper triangle only for Table 12-13). The initial cluster centers are related to the correlations and the corresponding principal components extracted in the previous analysis. Cluster 1 is strongly associated with reading, verbal, and spelling; cluster 2 with arithmetic and geometry; and cluster 3 with sports. Although there are some changes during the iterative process, these groupings tend not to change. The resulting group allocations are simply a function of the distance from each centroid. The pairwise distances between each centroid are also reasonably consistent with each other; that is, the between-group distances appear to have been successfully maximized, and there does not appear to have been difficulty in separating them. Adding more cases into the analysis would almost certainly improve the reliability of the result.

Table 12-10. Initial cluster centers

	Cluster		
	1	2	3
Reading	10.00	3.00	2.00
Music	9.00	9.00	3.00
Arithmetic	3.00	10.00	2.00
Verbal	10.00	2.00	6.00
Sports	6.00	6.00	8.00
Spelling	10.00	4.00	6.00
Geometry	3.00	9.00	4.00

Table 12-11. Cluster solution: cluster membership

Case number	Cluster	Distance
1	1	6.565
2	3	2.915
3	3	2.915
4	1	7.078
5	1	4.468

Table 12-12. Cluster solution: final cluster centers

	Cluster		
	1	2	3
Reading	8.57	3.00	3.50
Music	8.86	9.00	4.50
Arithmetic	4.00	10.00	3.50
Verbal	9.00	2.00	5.50
Sports	5.14	6.00	7.00
Spelling	9.00	4.00	5.50
Geometry	3.86	9.00	4.50

Table 12-13. Cluster solution: pairwise distances between final cluster centers

Cluster	1	2	3
1		12.971	8.562
2			9.925

Table 12-14 shows the ANOVA results for the significance of each variable in terms of *discriminability*. The results are not intended to be a strict test of statistical significance in the sense of hypothesis testing but are useful in examining which variables helped differentiate among the clusters. Spelling, verbal, and reading scores were all significant (unsurprisingly), but the scores making up the second and third clusters (arithmetic and geometry and sports) were not significant. The first result makes sense because scoring highly on spelling, verbal, and reading does discriminate between the first and second groups, but the lack of discriminability for the third cluster is a surprise (although recall from the preceding PCA that the third factor, *Sports*, had an eigenvalue barely above 1 and accounted for only 15% of the variance).

Table 12-14. ANOVA results for discriminability

	Cluster		Error		F	Sig.
	Mean square	df	Mean square	df		
Reading	28.893	2	1.745	7	16.558	0.002
Music	15.321	2	1.622	7	9.443	0.010
Arithmetic	17.000	2	9.214	7	1.845	0.227
Verbal	26.950	2	0.643	7	41.922	0.000
Sports	2.771	2	4.122	7	0.672	0.541
Spelling	17.550	2	1.786	7	9.828	0.009
Geometry	11.571	2	8.194	7	1.412	0.305

Discriminant Function Analysis

Discriminant function analysis (DFA) is used to construct rules that allow classification of cases into two or more groups, based on a linear combination of variables; the groups are known before the analysis begins, and the goal of the analysis is to find variables that do the best job of predicting the group membership of new cases. I was once involved in a study to predict the racial and ethnic group for university students who had failed to fill out that part of a questionnaire (information needed for reporting to the federal government). In this case, we knew what categories the federal government used for race and ethnicity and had to use other information in the students' records to try to assign them to the appropriate group.

The goal in DFA is to determine a function or functions that maximize the separation between groups, hence achieving the greatest possible accuracy in assigning cases to groups. These functions are typically linear combinations of the input variables and are called *linear discriminant functions* (LDFs). Cluster analysis and classification analysis are in some ways trying to solve the same problem by different means; both want to find the maxima of different functions (e.g., maximizing distance or classification accuracy).

We return to the psychometric example. Given the group allocations provided by cluster analysis, DFA can be used to determine a set of discriminant functions that provides maximum separation between the groups. It is then possible to test the null hypothesis of the equality of group means for each variable. In the two-group case, this can be evaluated using a *t*-test; with more than two groups, an *F*-test can be performed. The results shown in Table 12-15 indicate that there are significant

---

differences for reading, $F(2, 7) = 16.558$, $p = 0.002$; music, $F(2, 7) = 9.443$, $p = 0.010$; verbal, $F(2, 7) = 41.922$, $p = 0.001$; and spelling, $F(2, 7) = 9.828$, $p = 0.009$. Thus, in terms of discriminability, you could retain only the tests for reading, music, verbal, and spelling and still maximize the distance between groups.

Table 12-15. Tests of equality of group means

	Wilks's lambda	F	df_1	df_2	Sig.
Reading	0.174	16.558	2	7	0.002
Music	0.270	9.443	2	7	0.010
Arithmetic	0.655	1.845	2	7	0.227
Verbal	0.077	41.922	2	7	< 0.001
Sports	0.839	0.672	2	7	0.541
Spelling	0.263	9.828	2	7	0.009
Geometry	0.713	1.412	2	7	0.305

Table 12-16 shows the two *canonical discriminant functions* required to classify the cases into groups. Interestingly, the first function captures 96% of the variance, whereas the second function captures only 4%.

Table 12-16. Canonical discriminant functions

Function	Eigenvalue	% of variance	Cumulative %	Canonical correlation
1	79.224	96.0	96.0	0.994
2	3.287	4.0	100.0	0.876

Table 12-17 shows the computed values for *Wilks's lambda*, which can be used to evaluate the significance of the discriminant functions in a multivariate sense. The row labeled 1 through 2 shows the significance for both functions; the row labeled 2 shows the significance of the second function alone. Unfortunately, in this analysis, even the two functions together cannot significantly differentiate between the groups. This probably reflects the fact that function 1 accounts for such a high proportion of variance, and the data set is relatively small, so the analysis is lacking in power.

Table 12-17. Wilks's lambda

Test of function(s)	Wilks's lambda	Chi-square	df	Sig.
1 through 2	0.003	23.362	14	0.055
2	0.233	5.822	6	0.443

Table 12-18 shows the standardized canonical discriminant function coefficients. These are analogous to standardized regression coefficients and show the relationship between each of the measures of ability and the functions derived in this analysis.

Table 12-18. Standardized canonical discriminant function coefficients

	Function	
	1	2
Reading	−0.706	−0.141
Music	1.838	−0.368
Arithmetic	−0.364	−0.707
Verbal	3.686	1.409
Sports	−0.150	1.309
Spelling	−1.884	−2.030
Geometry	1.916	0.945

Table 12-19 shows the structure matrix; the values in the table are canonical variate correlation coefficients and can be interpreted as factor loadings, so they tell us the contribution of each variable to each variate. In this table, you can see loadings of reading and music on function 1 and loadings of spelling, verbal, arithmetic, geometry, and sports on function 2. These are slightly different from what you might have expected, say, from PCA or cluster analysis, but keep in mind that the algorithms used for each type of analysis have different computational goals, so it's not surprising that their results are not identical.

Table 12-19. Structure matrix

	Function	
	1	2
Reading	0.243	−0.140
Music	0.188	0.034
Arithmetic	0.115	−0.708
Verbal	0.379	0.433
Sports	−0.046	−0.331
Spelling	−0.055	−0.225
Geometry	−0.043	0.121

Finally, Table 12-20 shows the relationship between the two discriminant functions and the group centroids.

Table 12-20. Functions at group centroids

Cluster number of case	Function	
1	4.804	−0.169
2	−14.483	−3.465
3	−9.573	2.324

Exercises

Find some examples of professional articles in your field that use the techniques demonstrated in this chapter, and observe how the technique is used and the results explained. Here are a few examples to start you off:

- Depken, Craig A., and Darren Grant. 2011. "Product pricing in Major League Baseball: A principal components analysis." *Economic Inquiry* 49 (April): 474–488.

  Depken and Grant use principal components analysis to investigate factors affecting the pricing of concessions, tickets, and parking for Major League baseball games.

- Williamson, Hannah C., Thomas N. Bradbury, Thomas E. Trail, and Benjamin R. Karney. 2011. "Factor analysis of the Iowa Family Interaction rating scales." *Journal of Family Psychology* 25(6): 993–999.

  Williamson and colleagues use factor analysis to discover the factor structure of an instrument describing different types of verbal and nonverbal behaviors couples use to communicate; their innovation is using an instrument previously applied to white, middle-class couples to a sample of low-income, racially diverse couples.

- Tuma, Michael N., Reinhold Decker, and Sören W. Scholz. 2011. "A survey of the challenges and pitfalls of cluster analysis application in market segmentation." *International Journal of Market Research* 53(3): 391–414.

  Tuma, Decker, and Scholz look at some ways cluster analysis has been used in market segmentation work in the past 50 years and suggest best practices in this type of work.

- Kaye, Barbara K., and Thomas J. Johnson. 2011. "Hot diggity blog: A cluster analysis examining motivations and other factors for why people judge different types of blogs as credible." *Mass Communication and Society* 14(2): 236–263.

  Kaye and Johnson use cluster analysis to identify groups of people who judge different types of blogs (general information, media/journalism, war, military, political, corporate, and personal) as highly credible.

- Gonzalez, Richard. 2012. "Determination of sex from juvenile crania by means of discriminant function analysis." *Journal of Forensic Sciences* 57(1): 24–34.

  Gonzalez uses discriminant function analysis based on craniofacial measurements to differentiate between the skulls of males and females (from individuals of European descent, age 5–16 years); he achieves 78–89% accuracy in classification.

13

Nonparametric Statistics

The basis of inferential statistics is parameter estimation, estimation of the parameters of a population from information gained from a random sample presumed to have been drawn from that population. Many of the most common statistical techniques rely on the underlying distribution being of a particular type, such as the normal distribution, for inferences made from the relevant statistical tests to be valid; hence, these techniques are called *parametric statistics*. What about scenarios in which you know or suspect that the population does not meet the assumptions for a particular statistical test? In these cases, a different set of statistical techniques, known as *nonparametric statistics*, can be used. These techniques are often known as distribution-free statistics because they make few or no assumptions about the underlying distribution of the data; some prefer the term "distribution-free-er" because some nonparametric tests do require assumptions about the population distribution, although those assumptions are generally less stringent than those made by common parametric tests.

Nonparametric statistics are often applied to data sets in which data has been collected as ranks rather than as raw scores, or rank data is substituted for raw scores due to concerns about the distribution of the raw data. Rank data by definition is ordinal, as discussed in Chapter 1, and should not be analyzed using procedures meant for interval- and ratio-level data. Class rank is a familiar example: students in a school may be ranked by their grades, and although we can be sure of the order (student #1 has higher grades than student #2), we can't be sure of the interval between the ranks. (Students #1 and #2 might have almost the same grades, or there may be a large gap between them.)

If your research design suggests you should calculate a particular parametric statistic but your data do not meet the assumptions for that statistic, often you can use a nonparametric equivalent instead. There are many nonparametric statistics besides the few covered in this chapter, and William Conover's textbook, *Practical Nonparametric Statistics*, listed in Appendix C, includes a chart showing which nonparametric test to choose for which combination of data and statistical problem.

You can find this type of chart on the Internet as well; a reference for one such chart, from the Department of Health of the United Kingdom, is listed in Appendix C.

This chapter presents the median test, the Mann-Whitney U test, the Wilcoxon matched pairs signed rank test, the Kruskal-Wallis test, and the Friedman test. A few nonparametric tests are covered in Chapter 5, including the chi-square test, Fisher's Exact Test, McNemar's test, phi, Cramer's V, the Spearman correlation, Goodman and Kruskal's gamma, Kendall's tau, and Somers's d. The median and interquartile range, both of which are often used with nonnormal data, are discussed in Chapter 4.

Nonparametric techniques are more *robust* than their parametric counterparts, meaning they are less influenced by departures from model assumptions or unusual values (such as outliers) but are typically less powerful than their parametric equivalents. For this reason, if your data meets the assumptions for a parametric test, you should do the parametric test; if not, use the nonparametric equivalent (or transform your data as described in Chapter 3).

Between-Subjects Designs

This section reviews some commonly used nonparametric tests for between-subjects designs, generally based on the rank sum and mean rank measures.

The Wilcoxon Rank Sum Test

Two main descriptive statistics are used to characterize ordinal data: the *rank sum* and the *mean rank*. To illustrate how these statistics can be used, consider an example. An Olympic Games selection committee must choose a champion tae kwon do team from two states (California and Nevada) to represent the United States. Bacause there are both individual and group events for which the members have trained together, the teams can't be combined to produce a composite team of the most highly performing individuals; instead, one or the other of the teams must be selected as a whole. Each team member has been given an overall performance score, based on the number of bricks they managed to break during a five-minute testing session. The results are shown in Table 13-1.

Table 13-1. Performance scores for tae kwon do teams from two states

California	Nevada
4	2
5	3
6	3
6	4
7	4
8	5
9	10
9	10

California	Nevada
9	11
9	11

A higher score here indicates greater skill (more bricks broken). Trying to interpret the results just by visual inspection is difficult; the scores for the California team are more consistent and clustered in a smaller range, whereas the Nevada results have a greater range and clusters of both high and low scorers. Because the top four performers are from Nevada, you might be tempted to select this team, but the median score for Nevada is just 4.5, whereas the median for California is 7.5.

There is no reason to believe that the data in this sample is drawn from a normal distribution, and the sample size of 10 is not large enough to use the central limit theorem. Neither can we assume the data is equal-interval; whereas two bricks is certainly twice as many as one, we can't be sure that breaking two bricks indicates twice as much tae kwon do achievement as breaking one. (In fact, this interpretation is almost certainly *not* true.) We are much more comfortable with the interpretation that breaking two bricks represents *more* achievement than breaking one, without being sure how much more.

The most appropriate way to describe this data is by ranks rather than values. We will assign a rank to each case and then add all the ranks for each team. To assign ranks, the two teams are combined, and every team member from both teams is ranked from low to high. (A higher rank means more bricks broken.) Table 13-2 shows this process.

Table 13-2. Team rankings

California	Nevada	Rank
	2	1
	3	2
	3	3
4		4
	4	5
	4	6
5		7
	5	8
6		9
6		10
7		11
8		12
9		13
9		14
9		15

California	Nevada	Rank
9		16
	10	17
	10	18
	11	19
	11	20

What about the ties? Where tied ranks occur, the average rank is instead computed from the sum of the ranks concerned and divided by their number; for instance, a tie for second and third is given a rank of 2.5. Table 13-3 shows the new ranking, including tied ranks.

Table 13-3. Ranks for individual tae kwon do performance scores, including ties

California	Nevada	Rank
	2	1
	3	2.5
	3	2.5
4		5
	4	5
	4	5
5		7.5
	5	7.5
6		9.5
6		9.5
7		11
8		12
9		14.5
9		14.5
9		14.5
9		14.5
	10	17.5
	10	17.5
	11	19.5
	11	19.5

The rank sum is then calculated for each group by adding their respective ranks, as shown in Figure 13-1.

$$\sum_R (California) = 5 + 7.5 + 9.5 + 9.5 + 11 + 12 + 14.5 + 14.5 + 14.5 + 14.5 = 112.5$$

$$\sum_R (Nevada) = 1 + 2.5 + 2.5 + 5 + 5 + 7.5 + 17.5 + 17.5 + 19.5 + 19.5 = 97.5$$

Figure 13-1. Calculating the rank sums

If the two groups are approximately equal, we would expect the means of the rank sums to be approximately equal. This comparison is only fair when we have equal sample sizes, as we do in this example. We can also compute a mean rank, a better measure for groups of unequal sizes, as shown in Figure 13-2.

$$\overline{R}(California) = \frac{112.5}{10} = 11.25$$

$$\overline{R}(Nevada) = \frac{97.5}{10} = 9.75$$

Figure 13-2. Calculating the mean ranks

Comparing the mean ranks shows us that the California team ranks higher than the Nevada team. Thus, using rank-based methods, the selectors should select the California team because the mean rank is higher. What if we want to test whether the difference between the two teams is significant? We can calculate a Z-test to determine whether the difference between the two groups is statistically significant by using the standard alpha = 0.05. Under the null hypothesis, the two groups have equal mean ranks, so we can compute the expected sum of ranks as shown in Figure 13-3.

$$\mu_W = \frac{n_1(n_1 + n_2 + 1)}{2} = \frac{10(10 + 10 + 1)}{2} = 105$$

Figure 13-3. Calculating the expected sum of ranks

where n_1 and n_2 are the number of cases in the first and second groups. Note that this expected sum of ranks is based on the number of cases, not on their values; if you have two groups of 10 cases and the null hypothesis is that the two groups have the same average rank, the expected sum of ranks will always be 105. In the preceding example, you can see that one group (California) has a rank sum above the expected mean, and the other group (Nevada) has a rank sum below the expected mean. The Z-test can be computed from the mean and standard deviation of W, as shown in Figure 13-4.

$$z = \frac{W - \mu_W}{\sigma_W}$$

Figure 13-4. Formula for the Z-test for ranks

In this formula, W is the smaller of the two rank sums, μ_W is the expected sum of ranks as calculated previously, and σ_W is the standard error, calculated as shown in Figure 13-5.

$$\sigma_W = \sqrt{\frac{n_1 n_2 (n_1 + n_2 + 1)}{12}} = \sqrt{\frac{10(10)(10 + 10 + 1)}{12}} = 13.23$$

Figure 13-5. Calculating the estimated standard deviation for ranks

In this formula, n_1 and n_2 are the number of cases in the first and second groups, and 12 is a constant. Note that the standard deviation of ranks depends only on the number of cases, not on their value.

The value for the Z-test for this data is calculated as shown in Figure 13-6.

$$z = \frac{97.5 - 105}{13.23} = -0.57$$

Figure 13-6. Calculating the Z-test for ranks

Using the standard normal table (Figure D-3 in Appendix D), we find that this result has a *p*-value greater than 0.05; thus, we fail to reject the null hypothesis.

In this example, we used the normal approximation to the Wilcoxon rank sum test because both $n_1 \geq 10$ and $n_2 \geq 10$. For smaller samples, we calculate the sums of ranks for each group as previously and then compare the sums to a table of probabilities for different values of T. Such a table is available here (*http://bit.ly/TfKwoR*).

The Mann-Whitney U test, which will produce the same Z-score for a given set of data, is also used with this type of data. Either test can be used as a substitute for the two-sample *t*-test in which the normality assumption of the underlying data is questionable.

The Sign Test

The *sign test* is a nonparametric analog to the one-sample *t*-test and is used to test whether a sample has a hypothesized median. Often, the sign test uses ranks and the binomial distribution to test hypotheses about dichotomous data, meaning data with only two possible outcomes. Data values in the sample are classified as above (+) or below (–) the hypothesized median; the number of cases with a value above the median is $n+$, the number below the median is $n-$. Under the null hypothesis that the sample is drawn from a population with the specified median, these classifications have a binomial distribution with $\pi = 0.5$; each data point is considered a trial, and the result is either $+$ or $-$, either result having a probability of 0.5. Note that π (the Greek letter *pi*) is the designation for probability in a population, just as p is the designation for probability in a sample. The sign test uses the binomial probability distribution to find the probability of an observed result, assuming the null hypothesis to be true.

Suppose you are a medical researcher studying a new metabolic disorder, provisionally termed Type X diabetes. It seems to you that Type X diabetes has a later age of onset (the age at which the individual first has symptoms of a disease) than Type II diabetes; the median age of onset for the latter is 35.5 years. Your null hypothesis is that $\pi \leq 0.50$, that is, no more than 50% of the people with Type X diabetes have an age of onset of 35.5 or older; the alternative hypothesis is that $\pi > 0.50$, or more than 50% of people with Type X diabetes have age of onset of 35.5 or older. You find that in a clinical sample of 40 patients with Type X diabetes, 36 had an age of onset greater than 35.5 years: $n+ = 36$. Using an alpha level of 0.05, you use the normal approximation to the binomial with continuity adjustment to see how likely this result is and whether the null hypothesis is true. The calculations are shown in Figure 13-7.

$$Z = \frac{(X \pm 0.5) - np}{\sqrt{np(1-p)}} = \frac{(36 - 0.5) - [(40)(0.5)]}{\sqrt{40(0.5)(0.5)}} = \frac{15.5}{\sqrt{10}} = 4.90$$

Figure 13-7. Calculating the sign test

Here, X is the number of observed values greater than the median ($n+$),
0.5 is the continuity correction (negative in this case because our hypothesis is that $\pi > 0.5$),
np is the mean of the binomial distribution (the expected value for X if the null hypothesis is true),
$\sqrt{np(1-p)}$ is the standard deviation of the binomial distribution,
and n is the sample size.

Using the standard normal table (Figure D-3 in Appendix D), you see that the probability of a result at least this extreme is 0.00002, far less than your chosen alpha value of 0.05, so you reject the null hypothesis that the median age of onset for Type X diabetes is equal to or lower than that for Type II diabetes.

The Median Test

Further metabolic research in the lab subsequently suggests that there might be two subtypes of Type X diabetes—Type X_1 and X_2—and raises the question of whether subtypes are associated with age. You decide to look at another sample of 40 cases, 20 of which have been provisionally given Type X_1 diagnoses and 20 classified as Type X_2. You decide to use the median test, which classifies cases in two samples as being above or below the median for the pooled sample (the cases from both samples). In this data, the median of the pooled sample is 36.4 years; you decide to use the alpha level of 0.05 and to conduct a two-tailed test because you want to find an age difference in either direction.

For Type X_1, 12 cases were above the median age, and 8 cases were below the median age. For Type X_2, 9 cases were above the median age, and 11 cases were below the median age. The null hypothesis is that π is the same in both groups. If any of the cases had the exact median value, they would be removed from the sample. Table 13-4 shows the tabulated frequencies.

Table 13-4. Frequencies of age of occurrence for Type X_1 and Type X_2 diabetes

	Above median	Below median	Total
Type X1	12	8	20
Type X1	9	11	20
Total	21	19	40

A chi-square test for independence (discussed in Chapter 5) can be used to test these data for significance. You can use the fast computational formula for χ^2 analysis, in which the cells are described as in Table 13-5, and then find the probability of the resulting chi-square under the null hypothesis of independence (that individuals from either population are equally likely to have an age below the median).

Table 13-5. Cell labels for chi-square test for significance

Type	Above median	Below median	Row sums
X₁	a	b	$a+b$
X₂	c	d	$c+d$
Column sums	$a+c$	$b+d$	n

The calculations for this data are:

$$\chi^2 = \frac{n(ad-bc)^2}{(a+b)(c+d)(a+c)(b+d)} = \frac{40[(12 \times 11)-(8 \times 9)]^2}{(12+8)(9+11)(12+9)(8+11)} = 0.902$$

Using the chi-square table (Figure D-11 in Appendix D), with 1 degree of freedom, we find the probability results at least as extreme as ours ($\chi^{2Q} = 0.902$) to have a probability greater than 0.10. We therefore fail to reject the null hypothesis and conclude that our study has produced no evidence that the age of onset is different for Type X_1 diabetes and Type X_2 diabetes.

Kruskal-Wallis H Test

The *Kruskal-Wallis H* test is a nonparametric analogue to one-way ANOVA. It can also be thought of as an extension of the Wilcoxon Rank Sum test for more than two groups. The Kruskal-Wallis H test tests the hypothesis that several groups have the same median, and it does not require the samples to be the same size.

Suppose we are evaluating the performance of three sales teams, one with six members and two with five members. Our goal is to pick the best overall team judging by its recent sales performance. Their sales for the last quarter (in thousands of dollars) are shown in Table 13-6.

Table 13-6. Quarterly sales in thousands of dollars

Team A	Team B	Team C
10	8	6
10	8	8
12	9	10
13	9	14
14	14	15
15		

Our first step is to rank the individual sales totals, without regard to team membership, and assign the average rank to any ties, as shown in Table 13-7.

Table 13-7. Ranked quarterly sales

Team A	Team B	Team C	Rank
		6	1
		8	3
	8		3
	8		3
	9		5.5
	9		5.5
		10	8
10			8
10			8
12			10
13			11
14			13
	14		13
		14	13
		15	15.5
15			15.5

We will use the Kruskal-Wallis H test with alpha = 0.05 to see whether there is any statistically significant difference among the performance of these three teams. The Kruskal-Wallis H formula is shown in Figure 13-8.

$$H = \frac{12}{N(N+1)} \sum \frac{T_i^2}{n_i} - 3(N+1)$$

Figure 13-8. Formula for the Kruskal-Wallis H test

In this formula, N is the total sample size (all three samples added together),
n_i is the sample size for sample i,
T_i is the sum of ranks for sample i, and
12 and 3 are constants.

For this data:

$N = 6 + 5 + 5 = 16$

The T_i calculation for each team is shown in Figure 13-9.

$$\sum_A = 8 + 8 + 10 + 11 + 13 + 15.5 = 65.5$$

$$\sum_B = 3 + 3 + 5.5 + 5.5 + 13 = 30$$

$$\sum_C = 1 + 3 + 8 + 13 + 15.5 = 40.5$$

Figure 13-9. Calculating the sums of ranks

Plug these values into the Kruskal-Wallis H formula, as shown in Figure 13-10.

$$H = \frac{12}{16(16+1)} \left[\frac{65.5^2}{6} + \frac{30^2}{5} + \frac{40.5^2}{5} \right] - 3(16+1) = 2.96$$

Figure 13-10. Calculating the Kruskal-Wallis H test

To see whether our chi-square value of 2.96 is significant, we compare it to the chi-square value for 2 degrees of freedom (1 less than the number of groups) in Appendix D. Our chi-square value is less than the table value (5.991) for alpha = 0.05 and $df = 2$, so we fail to reject the null hypothesis that the three groups have the same median.

Within-Subjects Designs

This section reviews some commonly used nonparametric tests for within-subjects designs.

Wilcoxon Signed Ranks Test

The *Wilcoxon signed ranks* test can be used as a nonparametric substitute for the repeated measures *t*-test. It is appropriate when the data represent measurements on pairs of subjects, such as before and after scores for the same person or measurements on siblings or husband–wife pairs. The null hypothesis for this test is usually that the mean of the differences between the pairs is 0. The Wilcoxon signed ranks test does not assume normality but does assume at least a symmetric distribution, so it is not appropriate for data that are highly skewed.

Suppose we are interested in the effects of exercise on mental functioning and mood state. We have a sample of 40 sedentary adults who have volunteered to take part in an exercise program and to undergo a battery of psychological tests before beginning the program and again after completing it. The measurement of interest in this particular study is a 100-point measure of mood state in which 0 is very low affect, and 10 is very high affect. We administer the mood state instrument to the study sample before they begin the exercise program and again after they complete it. We will conduct a two-tailed test with the null hypothesis that exercise makes no difference in mood state and use an alpha value of 0.05.

In Table 13-8, we show an excerpt from the following data set to illustrate the process of calculating this test. (The process is mechanical and follows ranking procedures discussed previously.) For each pair of scores, we compute the difference score and then the absolute value of the difference score. We rank the absolute difference scores and then reaffix the sign to each rank. Any cases with difference scores of 0 are eliminated from the analysis, and tied ranks are given the average rank (so cases with a tie for ranks 3-4-5 are given a rank of 4).

Table 13-8. Exercise and mood state

Subject	Before exercise	After exercise	Difference (after − before)	Absolute difference	Rank of absolute difference	Signed rank
1	60	68	8	8	5	5
2	65	70	5	5	3	3
3	52	50	−2	2	1	−1
4	74	85	11	11	6	6
5	65	60	−5	5	3	−3
...	...	...	...	...	...	...
40	70	77	7	7	4	4

Five cases had difference scores of 0, so after eliminating them $n = 35$, giving us a sufficiently large sample (rule of thumb: $n \geq 25$) to use the large-sample approximation to the Wilcoxon signed ranks test, which gives us a Z-value whose probability we can determine using a standard normal table. The sum of the positive ranks was 380.

After removing tied pairs, we have 35 pairs, so we calculate the normal approximation to the Wilcoxon signed ranks test by using the formula shown in Figure 13-11.

$$z = \frac{T^+ - \dfrac{n(n+1)}{4}}{\sqrt{\dfrac{n(n+1)(2n+1)}{24}}}$$

Figure 13-11. Large-sample Wilcoxon signed ranks test

In this formula, T^+ is the sum of the positive ranks,
n is the number of pairs, and
4 and 24 are constants.

Note the similarity to the Z-statistic:

$$\frac{n(n+1)}{4}$$

is the expected value for the rank sum, and:

$$\sqrt{\frac{n(n+1)(2n+1)}{24}}$$

is the standard error, so this formula asks us to compare the value we found from our sample calculations with an expected value (analogous to population mean) and divides that difference by a measure of variability.

Using our values produces the result shown in Figure 13-12.

$$z = \frac{T^+ - \dfrac{n(n+1)}{4}}{\sqrt{\dfrac{n(n+1)(2n+1)}{24}}} = \frac{380 - \dfrac{35(35+1)}{4}}{\sqrt{\dfrac{35(35+1)(70+1)}{24}}} = 1.06$$

Figure 13-12. Large-sample Wilcoxon signed ranks test, with values

Using the standard normal table (Figure D-3 in Appendix D), we find the probability of a value at least this extreme to be 0.28914, far above our alpha value of 0.05, so we fail to reject the null hypothesis.

If we had had a smaller data set ($n < 25$), we would have used the small-sample form of the Wilcoxon signed ranks test. For the small-sample test, you assign signed ranks to your cases as for the large-sample case and then calculate the rank sums for both the positive ranks (T^+) and the negative ranks (T^-). You then compare these values to a table of critical values for the Wilcoxon signed ranks test. (Such a table is included in the Wilcoxon [1957] article cited in Appendix C, and versions are available in statistical reference books and on the Internet (*http://facultyweb.berry.edu/vbis sonnette/tables/wilcox_t.pdf*).) For a two-tailed test, you reject the null hypothesis if either T^+ or T^- is less than the critical value in the table for your sample size.

Friedman Test

The *Friedman test* is an extension of the matched pairs signed rank test for more than two related samples; you can also think of it as a nonparametric equivalent to repeated measures ANOVA. Suppose we have been brought in to evaluate the fitness level of a tae kwon do team. One of our concerns is that, because competitions might require multiple performances over several hours, we need to know whether these athletes can perform consistently over a long period. We conduct a mock competition and evaluate the sparring performance of each athlete on a 10-point scale (10 is the best performance, 0 the worst) after one hour of competition, two hours, and three hours. We believe this scale to be ordinal (a score of 9 represents better performance than a score of 8) but not equal-interval or ratio. (We don't know whether the difference between 8 and 9 is the same as the difference between 7 and 8, or whether 8 represents a performance twice as good as 4.) Therefore, we will conduct the Friedman test to examine changes in performance over the three time periods. Our null hypothesis is that performance will not change over the three time periods, so we are conducting a two-tailed test, and we will use an alpha level of 0.05.

The data from this trial is presented in Table 13-9.

Table 13-9. Sparring performance scores at three hourly time periods

Athlete	1 Hour	2 Hours	3 Hours
1	9	8	7
2	9	7	8
3	6	8	7
4	8	7	6
5	8	7	6
6	9	8	7
7	9	8	7
8	7	5	6

Our first step is to rank the performance within athletes; for example, for Athlete 1, the lowest score was at three hours, the second lowest at two hours, and the highest at one hour. These rankings are shown in Table 13-10. Note also the last row, which is the rank sum for each time period.

Table 13-10. Sparring performance ranks at three different hourly time periods

Athlete	1 Hour	2 Hours	3 Hours
1	3	2	1
2	3	1	2
3	1	3	2
4	3	2	1
5	3	2	1
6	3	2	1
7	3	2	1
8	3	1	2
Rank Sum	22	15	11

The formula for the Friedman test is given in Figure 13-13.

$$T = \frac{12\sum s_i^2}{bt(t+1)} - 3b(t+1)$$

Figure 13-13. Formula for the Friedman test

In this formula, b is the sample size,
t is the number of measurements on each subject,
s_i are the rank sums for each time period,
and 12 and 3 are constants.

In this example, $b = 8$, $t = 3$, and the values for s_i are 22, 15, and 11. Plugging these values into the equation gives us the result shown in Figure 13-14.

$$T = \frac{12(22^2 + 15^2 + 11^2)}{8(3)(3+1)} - 3(8)(3+1) = 7.75$$

Figure 13-14. Calculating the Friedman test

This statistic has a chi-square distribution with 2 degrees of freedom ($df = t - 1 = 2$). Using Figure D-11 in Appendix D, we see that the critical value for a chi-square with 2 df and alpha = 0.05 is 5.991; our test statistic exceeds this number, so we reject the null hypothesis that there is no difference in performance over the different time periods. Looking at the raw data, we can see that performance deteriorates over time for most of the athletes, suggesting that they need to invest more time in conditioning.

The use of the Friedman test is not limited to measures made over time but could also be used to evaluate the effect of drug treatments or any other experimental situation in which a nonparametric approach might be most appropriate.

Exercises

Here are some exercises to review the topics covered in this chapter.

Problem

Suppose you want to conduct the Friedman test but find that there are ties within the data. For instance, in the example of testing the performance of members of a tae kwon do team over three time periods, some of the athletes might have received a score more than once. In this case, you have tied ranks and need to use the values for the mid-rank. Table 13-11 presents the results on a scale evaluating the sparring performance of eight athletes; measurements were taken at one, two, and three hours into a simulated competition. Conduct the Friedman test for this data, using the null hypothesis that performance is the same at all three time periods and alpha = 0.05, and decide whether to accept or reject the null hypothesis. For tied ranks, assign the mid-rank score; that is, for scores of (6, 6, 5), the ranks would be (2.5, 2.5, 1).

Table 13-11. Sparring performance scores at three time periods (with ties)

Athlete	1 Hour	2 Hours	3 Hours
1	8	8	6
2	6	6	7
3	6	8	7
4	8	7	6
5	9	9	7
6	9	8	7
7	8	7	6
8	8	7	7

Solution

The scores are ranked and rank sums computed in Table 13-12.

Table 13-12. Sparring performance ranks at three hourly time periods (with ties)

Athlete	1 Hour	2 Hours	3 Hours
1	2.5	2.5	1
2	1.5	1.5	3
3	1	3	2
4	3	2	1

Athlete	1 Hour	2 Hours	3 Hours
5	2.5	2.5	1
6	3	2	1
7	3	2	1
8	3	1.5	1.5
Rank Sum	19.5	17	11.5

The Friedman test is calculated as shown in Figure 13-15.

$$T = \frac{12(19.5^2 + 17^2 + 11.5^2)}{8(3)(3+1)} - 3(8)(3+1) = 4.19$$

Figure 13-15. Calculating the Friedman test with tied ranks

There are 2 degrees of freedom ($df = t - 1$). From the chi-square table (Figure D-11 in Appendix D), we see that the critical value for alpha = 0.05 with 2 df is 5.991; our test statistic is lower than this value, so we fail to reject the null hypothesis.

Problem

A marketing professional is interested in collecting demographic information about fans of different football (soccer, for Americans) teams. Because specific marketing campaigns are often developed for different age groups, one question is the median age of supporters of different teams. You are the statistician on this project and draw a random sample of members of fan clubs devoted to two teams (A and B); you collect demographic data on them, including age, through a telephone survey. You determine that the overall median age (for both groups) is 27.5 and classify each supporter as above or below this median. Your data is presented in Table 13-13. If you are conducting a study whose null hypothesis is that there is no difference in median age between the two groups, with alpha = 0.01, what is your decision?

Table 13-13. Comparing age for fans of two football teams

Team	Above median	Below median	Row sums
A	30	70	100
B	60	40	100
Column sums	90	110	200

Solution

You decide to perform the median test and thus calculate the chi-square value for this data, testing the null hypothesis of independence (because if the medians are the same for fans of both teams, that means age is unrelated to which team you follow). You use the fast computational formula for χ^2 analysis, presented in the chapter section on the median test, and compare your results to the critical chi-square value.

The computations are shown in Figure 13-16.

$$\chi^2 = \frac{n(ad-bc)^2}{(a+b)(c+d)(a+c)(b+d)} = \frac{200[(30 \times 40) - (70 \times 60)]^2}{(100)(100)(90)(110)} = 18.18$$

Figure 13-16. Calculating chi-square for the median test

From the chi-square table (Figure D-11 in Appendix D), you see that the critical value for $df = 2$ and alpha = 0.01 is 9.210. Your test statistic is larger than this value, so you reject the null hypothesis of equal median age among followers of the two teams. The result is $\chi^2 = 18.18$, $p < 0.01$. Looking at the data table, you see that fans of team A tend to be younger than fans of team B because only 30% of team A fans are above the median age as compared to 60% of team B fans.

Business and Quality Improvement Statistics

Many of the statistics used in business and quality improvement applications are those within the common repertoire of basic statistics, including the chi-square test (covered in Chapter 5), *t*-tests (Chapter 6), and regression and ANOVA (Chapter 8 to Chapter 11). However, other techniques have been developed for the specific needs of business and quality improvement applications, and those techniques will be the subjects of this chapter.

Index Numbers

Index numbers are commonly used in business to measure the change in quantity or price over time for some good or combination of goods and services. One well-known example is the Consumer Price Index (CPI), which represents the average price of a quantity of consumer goods and services believed to be typical household purchases in the United States. The U.S. CPI is calculated monthly by the Bureau of Labor Statistics of the U.S. Department of Labor; it's used as a measure of inflation and to calculate cost of living adjustments for pensions and wages. Although many criticisms have been made of the CPI, it has proven highly useful as a summary measure of the average cost of living and allows comparison across historical periods and geographic areas. Other countries that calculate a CPI or similar index include Canada, China, Israel, New Zealand, Australia, and many countries in Europe.

Calculation of indexes can be very simple (when the index reflects the change in the price or quantity of a single commodity) or very complex (when the index reflects a weighted average of a number of goods and services, as is true for the CPI). A *simple index number* displays the change in time of the price or quantity of a single commodity such as the number of television sets sold or the price of an ounce of gold. To calculate a simple index, you must choose a *base period* to be used for comparison. The index will then represent the change in price or quantity relative to that base period. To calculate a simple index, three steps are required:

1. Obtain the price or quantity of the commodity for the time period of interest.

2. Select a base period, and obtain the price or quantity for that year.

3. Calculate the index number for each time period, using the formula in Figure 14-1.

$$I_t = \frac{Y_t}{Y_0} \times 100$$

Figure 14-1. Formula for a simple index

In this formula, I_t = the index at time t,
Y_t = the price or quantity at time t, and
Y_0 = the price or quantity in the base period.

Suppose we wanted to track the health of the automobile manufacturing industry in the United States over the past 20 years. As part of this research, we could create an index expressing the number of automobiles manufactured each year in terms of the first year of the time period. If we had data for 1986–2005, 1986 would be the base year, and the quantity of cars manufactured that year would be Y_0. Consider Table 14-1, which shows a small and entirely hypothetical data set to demonstrate calculation of a simple index.

Table 14-1. Data for simple index calculation

Year	Number of automobiles manufactured
1986 (base year)	5,000
2005	4,000

For this data, the index for 2005 is shown in Figure 14-2.

$$I_{2005} = \frac{4000}{5000} \times 100 = 80$$

Figure 14-2. Calculating a simple index

An index of 100 represents the same quantity or price as the base period. An index less than 100 indicates a decline in quantity or price, and an index greater than 100 indicates an increase in quantity or price compared to the base period. One of the great advantages of index numbers is that they put quantities measured on different scales and with different ranges of scores into a common metric. For instance, using indexes, we can easily compare the relative increase or decrease over time in production of automobiles, motorbikes, and bicycles.

A *composite index* combines information about the price or quantity of several types of goods or services. For instance, we might calculate the quantity of beer sold by the three largest breweries in Scotland by adding the quantity sold by each manufacturer. If we performed this calculation for a number of years and selected one year to use as the base period, we could calculate an index number for each year, as we did the simple index in the preceding example. This type of index is known as a *simple composite index* because it is calculated by combining information from several sources without using any type of weighting.

When some type of weighting is used to create the totals used to calculate the index number, this is known as a *weighted composite index*. Price indexes are often weighted by the quantity of goods sold, for instance. There are several ways to apply a weighting scheme because the quantities of items purchased can change from one time period to the next, and the choice of weights can have an important influence on the results of the index calculations. Once a scheme of weighting is selected, however, calculations are straightforward. The total price is calculated for each time period, and the index numbers for each time period are calculated using a procedure analogous to that used for the simple index.

A *Laspeyres index* uses the base period quantities as weights, so inflation or deflation is measured for a fixed basket of goods or services. The CPI is an example of a Laspeyres index; the quantities used for weighting are based on samples of purchases by more than 30,000 families from 1982 to 1984. The steps in calculating a Laspeyres index are:

1. Collect price information (P_{1t}, P_{2t}, ... P_{kt}) for each time period for each item (1 through k) to be included in the index.

2. Collect purchase quantity information (Q_{1t0}, Q_{2t0}, ...Q_{kt0}) for the base period for each item to be included in the index.

3. Select a base period (t_0).

4. Calculate the weighted totals for each time period, using the formula shown in Figure 14-3.

$$\sum_{i=1}^{k} Q_{it_0} P_{i_t}$$

Figure 14-3. Formula to calculate weighted totals for one time period

5. Calculate the Laspeyres index, I_t, by dividing the weighted total for each time period by the weighted total for the base period and multiplying by 100, as shown in Figure 14-4.

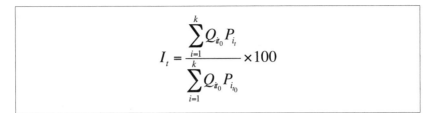

$$I_t = \frac{\sum\limits_{i=1}^{k} Q_{i_{t_0}} P_{i_t}}{\sum\limits_{i=1}^{k} Q_{i_{t_0}} P_{i_{t_0}}} \times 100$$

Figure 14-4. Formula for the Laspeyres index

Table 14-2 shows a simple example of the calculation of the Laspeyres index for a market basket containing only two types of goods.

Table 14-2. Laspeyres index example

Product	Base quantity (2000)	2000 price	2005 price
Bread	10	1.00	1.50
Milk	20	2.00	4.00

The weighted total for 2000 is:

$(10 \times 1.00) + (20 \times 2.00) = 50.00$

The 2005 weighted total is:

$(10 \times 1.50) + (20 \times 4.00) = 95.00$

The Laspeyres index for this basket of goods in 2005, using 2000 as the base year, is therefore shown in Figure 14-5.

$$I_{2005} = \frac{95}{50} = 190$$

Figure 14-5. Calculating the Laspeyres index

A *Paasche index* calculates weighted totals by using the quantities of items purchased in each time period. This has the advantage of adjusting for changes in consumer habits. For instance, if the price of a good rises, people tend to buy less of it and purchase less expensive substitutes. An example of substitution would be if the price of beef rose faster than the price of chicken, and people responded by buying more chicken and less beef. This change in consumer habits would not be reflected in the Laspeyres index but would be reflected in the Paasche index.

The steps to calculate a Paasche index are similar to those for a Laspeyres index. The main difference is that information about the quantities purchased in each time period must also be collected and used to calculate the weighted totals.

1. Collect price information (P_{1t}, P_{2t}, . . . P_{kt}) for each time period for each item (1 through k) to be included in the index.

2. Collect purchase quantity information (Q_{1t}, Q_{2t}, ...Q_{kt}) for each time period for each item to be included in the index.

3. Select a base period (t_0).

4. Calculate the weighted totals for each time period, using the formula shown in Figure 14-6.

$$\sum_{i=1}^{k} Q_{i_t} P_{i_t}$$

Figure 14-6. Formula to calculate the weighted totals for one time period with the Paasche index

5. Calculate the Paasche index, I_t, by dividing the weighted total for each time period by the weighted total for the base period and multiplying by 100, as shown in Figure 14-7.

$$I_t = \frac{\displaystyle\sum_{i=1}^{k} Q_{i_t} P_{i_t}}{\displaystyle\sum_{i=1}^{k} Q_{i_{t_0}} P_{i_t}} \times 100$$

Figure 14-7. Formula for the Paasche index

We will use the data in Table 14-3 to calculate a Paasche index.

Table 14-3. Calculating a Paasche index

Product	2000 quantity	2000 price	2005 quantity	2005 price
Bread	10	1.00	15	1.50
Milk	20	2.00	15	4.00

The 2000 weighted total is:

$(10 \times 1.00) + (20 \times 2.00) = 50.00$

The 2005 weighted index is:

$(15 \times 1.50) + 15 \times 4.00) = 82.50$

The Paasche index for this basket of goods in 2005, using 2000 as the base year, is shown in Figure 14-8:

$$I_{2005} = \frac{82.5}{50.0} \times 100 = 165.0$$

Figure 14-8. Calculating the Paasche index

Note that although the prices were the same in each example, the different methods of weighting resulted in substantial differences in the two index numbers (190 versus 165). The Paasche index has the advantage of comparing prices for a basket of goods at purchase levels appropriate to each time period. It has the disadvantage of requiring this information (quantities of each type of good purchased) to be collected for each time period, which might be prohibitively expensive. Another disadvantage of the Paasche index is that because both prices and quantities can change from one period to another, it is difficult to compare Paasche index numbers for any two periods when one of the periods is not the base period.

Criticisms of the U.S. Consumer Price Index (CPI)

The CPI is the principal measure of price changes in the United States and has been produced in some form by the Bureau of Labor Statistics since 1919. It is used for many purposes, including as a measurement of inflation and in calculating cost of living adjustments for negotiated wage packages and social security and civil service retirement benefits. Not surprisingly, an index used for so many purposes also comes under criticism from many quarters.

Among the principal criticisms, all of which tend to lead to the CPI overstating inflation, are the following:

Quality change and new product bias
> The CPI does not account for the improved quality of some items, such as electronics. A DVD player that sells for $150 in 2005 might be of a substantially higher quality and therefore worth more to the consumer than one that cost $100 in 2000, but this increase in quality is not reflected in the CPI. Similarly, because a fixed market basket of items is used, new items are not included in the index in a timely fashion. The result is that early declines in price (typical among new electronics products, for instance) are not captured in the index.

Substitution bias
> The use of a fixed basket of goods (weights are updated about once every 10 years) does not allow for changes in consumer purchasing patterns in response to changes in price. For instance, if the price of meat rises faster than that of other protein foods such as poultry or eggs, consumers can respond by purchasing more poultry and eggs and less meat, but this shift will not be reflected in the CPI.

> Because price information is gathered from traditional sales outlets such as department stores, newer outlets such as big-box discounters or Internet sales are not fully represented in the CPI surveys.

Time Series

Time series are used frequently in business statistics to chart the changes in some quantity over time. Strictly speaking, a time series is a sequence of measurements of some quantity taken at different times, often but not necessarily at equally spaced intervals. The previous example of the number of automobiles manufactured in each of the years from 1986 to 2005 would qualify, as would the measurements discussed later in this chapter in the section on control charts. Time series may be used for either descriptive or inferential purposes; the latter includes *forecasting*, or predicting values for time periods that have not yet occurred. The reader should bear in mind, however, that time series analysis is a complex topic with many specialized techniques and that this section can introduce only some of the terminology and a few simple examples. Anyone planning to work in this area should consult a textbook devoted to the subject, such as Robert S. Shumway's *Time Series and Its Applications: With R Examples* (Springer). Note also that some authors (e.g., Tabachnick and Fidell) specify that at least 50 data points are required for the appropriate use of time series techniques.

One characteristic of time series data is that data points in sequence are assumed to be not independent, as would be required for the standard general linear model and many other analytical techniques, but *autocorrelated*. This means that the value for a given time point is expected to be related to the points before and after it, and perhaps to points more distant in the series as well.

Time series data is assumed to be *stationary*, meaning that properties such as mean, variance, and autocorrelation structure are constant over the entire range of the data. Sometimes data has to be preprocessed by *differencing* to achieve stationarity; this means subtracting each data point from some previous point. The distance between the two points is called the *lag*. Techniques to test for the types of differencing required and to perform them automatically are included in software packages dedicated to time series analysis. Other transformations, such as taking the square root or logarithm of the data to stabilize the variance, may also be applied before the time series analysis begins.

Additive models are often used to describe the components of a time series, such as:

$$Y_t = T_t + C_t + S_t + R_t$$

The components of the trend Y_t in this model are:

T_t
    Secular or long-term trend, the overall trend over the time studied

C_t

The cyclical effect, fluctuations about the secular trend due to business or economic conditions, such as periods of general economic recession or expansion

S_t

The seasonal effect, fluctuations due to time of year such as the summer versus the winter months

R_t

The residual or error effect, what remains after the secular, cyclical, and seasonal effects have been accounted for; may include both random effects and effects due to rare events such as hurricanes or epidemics.

Much of time series analysis is devoted to resolving the variance observed over time in these components. The concept is similar to partitioning the variance in ANOVA models, although the mathematics involved is different.

Exact measurements plotted over time, also known as *raw time series*, will almost always show a great deal of minor variation that can obscure major trends that could help explain the pattern and make accurate future forecasts. Various types of *smoothing* have been devised to deal with this problem. They can be divided into two types: *moving average* or *rolling average* techniques, which involve taking some kind of average over a series of consecutive points and substituting this average for the raw values, and *exponential techniques*, in which an exponential series is used to weight the data points.

To calculate a *simple moving average* (SMA), take the unweighted mean of a specified number of data points (n) prior to the time point in question. The size of n is sometimes described as a *window* because the idea is that a window including n data points (a window of width n) is used to calculate the moving average. As you progress forward in time through the data, the window moves so you can see different data points each time, and the average is calculated using the points included in the window for each time point. For instance, a five-point SMA would be the average of a given value and the previous four data points.

The SMA for each new data point drops only one value and adds only one new value, reducing the fluctuation from point to point. This attribute gave rise to the term *rolling average* because the last value rolls off the series as the new value rolls on. This is similar to the methodology used to compute player standings on professional tennis tours, although in that case a total rather than an average is computed. Each player's total points in a given week is the sum of their points from the previous 52 weeks, and each week the total is recalculated as the oldest week's points are dropped from the total and the newest week's points added in.

The greater the size of the window used to calculate an SMA, the greater the smoothing because each new data point has less influence relative to the total. At some point, the data might become so smoothed that important information about the pattern is lost. In addition, the larger the window, the more data points that have to be discarded (because you need more points to calculate each average). This may be seen in the example in Figure 14-9 and Table 14-4.

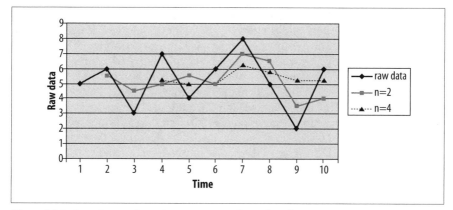

Figure 14-9. Raw data and moving averages with n = 2 and n = 4

Table 14-4. Simple moving average with different-sized windows

Time	1	2	3	4	5	6	7	8	9	10
Raw data	5	6	3	7	4	6	8	5	2	6
$n = 2$		5.5	4.5	5	5.5	5	7	6.5	3.5	4
$n = 4$				5.25	5	5	6.25	5.75	5.25	5.25

As would be expected, the largest fluctuations appear in the raw data, with less fluctuation with the width of size 2 and very little fluctuation when the window is increased to size 4.

When a window of size 2 is used, only one data point has to be dropped from the moving average (the first because it has no prior point to use in calculating the average). When a window of 4 is used, the first three points have to be dropped because none of them has three prior points to use in calculating the average. This is less of an issue when there are many data points but results in a serious loss of information with a data set containing only 10 observations.

The *central moving average* (CMA) is similar to the moving average but uses a window of size n with both past and future data used to calculate the average for each point. For a CMA of size 3, for instance, the value at time 2 would be 4.67 or (5 + 6 + 3)/3. Note that the future points are measured data, not forecasts; they are future only in that they are measured at a later time than the central data point for a given CMA. Table 14-5 shows an example.

Table 14-5. Central moving average (n = 3) for previous data

Time	1	2	3	4	5	6	7	8	9
Raw data	5	6	3	7	4	6	8	5	2
CMA ($n = 3$)		4.67	5.33	4.67	5.67	6.00	6.33	5.00	

The *weighted moving average* (WMA) uses values from a window of size n but assigns greater weight to the data points closer to the point in question. If not otherwise specified, arithmetic rather than exponential weights are used. A typical system assigns the weight n to the day whose weight is being calculated, where n is the number of days included in the weight. Every other day included in the WMA is weighted one less for each day it is removed from the day being weighted. Using this system, in a five-day WMA, the day being weighted would be given a weight of five, the previous day a weight of four, and so on down to four days previous, which would have a weight of one. This weighted sum is divided by the sum of the weight factors, which will be $[n(n-1)]/2$. The WMA makes intuitive sense in any situation when consecutive points can be assumed to be the most closely related, with the relationship lessening as the length of time between data points increases.

The *exponential moving average* (EMA) also applies more weight to closer measurements, but the weights allocated to data points farther from the point in question decrease exponentially rather than arithmetically. To calculate an EMA, an exponential smoothing constant α between 0 and 1 is selected. This constant is related to the number of time points included, n, by the equation presented in Figure 14-10.

$$\alpha = \frac{2}{n+1}$$

Figure 14-10. Formula to calculate the constant for an exponential moving average

Here, $\alpha = 0.2$ is equivalent to $n = 9$ because $(2/10 = 0.2)$. α is then applied as shown in Figure 14-11, which is continued until the terms become so small as to become negligible.

$$EMA = \frac{p_1 + (1-\alpha)p_2 + (1-\alpha)^2 p_3 + (1-\alpha)^3 p_4 + \dots}{1 + (1-\alpha) + (1-\alpha)^2 + \dots}$$

Figure 14-11. Formula for the exponential moving average

In the above formula, p_1 is the measure at the given time point for which the EMA is being calculated, p_2 is one time point removed, p_3 is two time points removed, and so on. The denominator approaches $1/\alpha$ as the number of points included increases, and 86% of the total weight in the calculation will be included in the first n time points. n is not the number of data points included in calculating the EMA because it is in the simple and weighted moving averages; the stopping point will be determined by the value chosen for α and by the researcher's decision about what constitutes a negligible value.

Decision Analysis

We all make decisions every day, but how do we go about making the best decision, particularly in a situation when a lot (for instance, a large amount of money) is at stake? *Decision analysis* is a body of professional practices, methodologies, and theories used to systematize the decision-making process in the service of improving the process. There are many schools of thought within decision theory, and each can be useful in a particular context. This section concentrates on several of the most common decision analysis methods, which will help introduce the student to the types of processes involved as well as provide concrete assistance in particular decision-making contexts. The decision-making process will be described in terms of financial costs and payoffs but can be used with other metrics as well (for instance, personal satisfaction or improved quality of life) if they can be quantified.

In decision analysis, the process of making a decision is usually conceived of as a series of steps that is not unlike the process involved in hypothesis testing. They are also not that different—except for the selection and application of a mathematical model in steps 5 and 6—from the ordinary type of decision-making process we engage in every day. Besides the potential to lead to better decisions, going through these steps (and justifying and documenting them) should make the reasons for a particular decision easier to explain and justify to someone who wasn't involved in the process. The basic steps are the following:

1. Define the situation or context, including *states of nature* (any situation in the real world that might influence the outcomes). States of nature must be stated as mutually exclusive and exhaustive alternatives, for instance, strong/medium/weak market or inadequate rainfall/adequate rainfall.

2. Identify the choices at hand, that is, the alternative decisions that could be made; these are known as *actions*.

3. Identify the possible *outcomes* or consequences.

4. Assign costs and profits associated with all possible combinations of choices and outcomes.

5. Select an appropriate mathematical model.

6. Apply the model using the information from steps 2 to 4.

7. Make a decision based on the best expected outcome, as predicted by the model.

Choice of a decision theory methodology depends in part on how much is known about a situation. There are three types of contexts in which one can apply decision theory:

- Decision making under certainty
- Decision making under uncertainty
- Decision making under risk

Decision making under certainty means that the future state of nature is known, so the decision-making process requires only stating the alternatives and payoffs to be able to pick the choices that will invariably lead to the best outcome. This situation will not be further discussed because no mathematical modeling is required and there is no uncertainty about what is the best choice.

Decision making under uncertainty is a more common situation; we don't know the probabilities of each state of nature and must make our decision based only on the gains or losses from different actions under each state. For instance, if we are choosing from several cities in which to open a restaurant, the success of the restaurant depends in part on the economic climate in each city when the restaurant opens, but we might not have good estimates of the future economic climates in these cities. Similarly, when choosing which crop or variety to plant, our success at harvest time depends partly on the amount of rainfall during the growing season, but we might feel that we don't have sufficient information to estimate this in advance.

In *decision making under risk*, we know the probabilities of each outcome (or have reasonable estimates of them) and can combine this information with information about expected payoffs to determine which decision is optimal.

Minimax, Maximax, and Maximin

The information needed to make a decision under uncertainty may be summarized in a payoff table, where each row represents a possible action taken and each column a state of nature. The numbers within the cells of the table represent the outcome expected under different combinations of actions and states of nature. Suppose we are considering whether to invest in staging an event in a large outdoor venue or a smaller indoor venue, with a third alternative not to invest in the event at all. Suppose also that the event is to be held in a climate where rainstorms are common during the season of the year when the event will take place, and we don't feel we can assign reasonable probabilities to the chance of rain on a particular day. It will cost us $50,000 to stage the event. The payoff table might look like Table 14-6.

Table 14-6. Payoff table for investing in an event

		Weather	
		Rain	No rain
Action	**Outdoor venue**	−$50,000	$500,000
	Indoor venue	$200,000	$200,000
	Do not invest	$0	$0

The outdoor venue is larger than the indoor, so if it doesn't rain that night, we stand to make a large profit (gain of $500,000). If it rains, the event will be canceled, and we will lose our investment and make no revenue (loss of $50,000). On the other hand, the indoor venue should return about the same profit ($200,000) whether it rains or not: less than the outdoor venue if the weather is good, more than the outdoor venue if it rains. Finally, we might decide that investing in staged events is too risky and choose to apply our money elsewhere.

We can create an *opportunity loss* table, which expresses the amount of money we lost the opportunity to make by choosing a particular course of action. For our hypothetical event-investment-in-rainy-country scheme, the opportunity loss table would look like Table 14-7.

Table 14-7. Opportunity loss table for investing in an event

		Weather	
		Rain	No rain
Action	**Outdoor venue**	$250,000	$0
	Indoor venue	$0	$300,000
	Do not invest	$200,000	$500,000

Note that there are no negative numbers in an opportunity loss table. The best action for a given state of nature has a loss of $0, whereas the others represent the amount of money lost by not choosing the best action for that state of nature.

Three procedures have been developed for decision making under uncertainty: *minimax*, *maximax*, and *maximin*. The *minimax* procedure involves choosing the action that will minimize opportunity loss. To make a minimax decision, we use the opportunity loss table to identify the maximum opportunity loss for each action and then choose the action with the lowest opportunity loss. In this example:

Maximum opportunity loss (outdoor venue) = $250,000
Maximum opportunity loss (indoor venue) = $300,000
Maximum opportunity loss (do not invest) = $500,000

Using the minimax procedure, we would decide to finance the event at the outdoor venue because it has the smallest maximum opportunity loss of the three choices.

The *maximin* strategy involves choosing the action that has the largest minimal outcome. This has been described as the strategy for pessimists because it chooses the alternative with the highest minimal gain or smallest loss— the best outcome under unfavorable conditions. In this example:

Minimum gain (outdoor venue) = -$50,000
Minimum gain (indoor venue) = $200,000
Minimum gain (do not invest) = $0

Using the maximin strategy, we would choose the indoor venue because the worst we could do is make $200,000, regardless of weather conditions.

The *maximax* strategy involves choosing the action that has the highest maximum outcome. For this reason, it might be called the strategy for optimists because it chooses the strategies that provide the best outcome under the most favorable state of nature. In this example:

Maximum gain (outdoor venue) = $500,000
Maximum gain (indoor venue) = $200,000
Maximum gain (do not invest) = $0

Using the maximax strategy, we would choose the outdoor venue because it offers the highest maximum outcome.

Decision Making under Risk

If the probabilities of different states of nature are known or can be reasonably estimated, we are in a decision making under risk situation. Let's say that in the previous example, we also had information about the probability of rain on the night when the event is scheduled. If the probability of rain is 0.6, that means the probability of no rain is 0.4 because they are mutually exhaustive states of nature. We add this information to Table 14-8.

Table 14-8. Expected payoff from various actions, given probabilities of different states of nature

		Rain	No rain	Expected payoff
	Probability	0.6	0.4	
Action	Outdoor venue	−$50,000	$500,000	$170,000
	Indoor venue	$200,000	$200,000	$200,000
	Do not invest	$0	$0	$0

The expected payoff is calculated by multiplying the payoff under each combination of actions and states of nature by the probability of the state of nature. For instance, for the outdoor venue option:

$$E(payoff) = (.6)(-50,000) + (.4)(500,000) = -30,000 + 200,000 = 170,000$$

We choose the option with the greatest expected payoff. In this case, we would choose to stage our event indoors. This method requires us to have reasonable estimates of the probability of the states of nature. If they were reversed in the preceding example, the highest expected payoff would come from the outdoor venue.

Decision Trees

If the probability of various outcomes given particular actions is known, a decision tree can be constructed that displays the actions and payoffs under different states of nature, and it can be used to clarify the outcomes of different combinations. The decision tree containing the same information as Table 14-8 is shown in Figure 14-12.

The purpose of a decision tree is to display decision-making information, including available actions, states of nature, and expected payoffs, in a clear and graphical manner. It does not include any rules for making decisions but can aid decision-making by presenting the relevant information in one graphical summary.

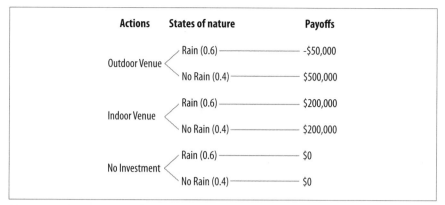

Actions	States of nature	Payoffs
Outdoor Venue	Rain (0.6)	-$50,000
	No Rain (0.4)	$500,000
Indoor Venue	Rain (0.6)	$200,000
	No Rain (0.4)	$200,000
No Investment	Rain (0.6)	$0
	No Rain (0.4)	$0

Figure 14-12. Decision tree for event venue example

Quality Improvement

The roots of quality improvement (QI) date back to the 1920s, when Walter Shewhart began developing a statistical approach to studying variation in manufacturing processes. Interest in QI got a major boost in the 1950s with the work of W. Edwards Deming, who developed a statistical approach to QI based on Shewhart's work. Ironically, Deming's approach was initially rejected in his native country (the United States) but enthusiastically embraced in Japan, where QI techniques were applied to manufacturing so successfully that Japanese companies were able to challenge and in some cases surpass the American supremacy in manufacturing. In response, American companies began adopting QI approaches in the 1980s; Motorola and General Electric are among the best-known early adopters.

There are multiple approaches to QI, including a popular program known as Six Sigma (6σ), which is part of a general approach known as Total Quality Management (TQM). This section concentrates on the basics of QI, which are common to many such programs, and avoids getting into the specifics of jargon and acronyms of any particular program. It also concentrates on the statistical methodology used in QI, although the reader should bear in mind that most QI programs are multifaceted and include psychological and organizational strategies as well as statistical measurement and analytic techniques.

Although QI began in the manufacturing sector, it is now applied in other areas, including health care and education. "Quality" may be the buzzword of the new century, so consideration of the basic aspects of quality measurement and improvement might prove useful to people working in widely disparate fields. Anywhere quality can be defined and measured, the QI field might provide useful tools.

The first step in measuring anything is defining it. *Quality* in the QI context is generally defined in terms of the customer; a high-quality product satisfies the needs and preferences of the customer. In manufacturing, this might mean machine parts with specified dimensions and durability. In health care, it might mean a doctor's visit that answers the patient's concerns and does not involve excessive waiting or other aversive experiences. The customer's needs and preferences must be translated into *product variables* that can be measured. Picking up on the healthcare example, "no excessive waiting time" might be operationalized into "waiting time of no more than 10 minutes." This would allow each visit to be evaluated as to whether the standard was met. Similarly, specific dimensions can be established for machine parts and specific parts evaluated as to whether they fall into the acceptable range as specified by the customer.

The language of QI is drawn from manufacturing and commonly refers to *products* that are created by *processes*, which are part of a larger *system*. For instance, a company might manufacture bolts (the *product*) through a series of *processes* (such as cutting, stamping, and polishing), which are part of a larger *system* that transforms *inputs* (such as metal) into *outputs* (the bolts). An inherent fact about any process is that it is *variable*. For instance, not every manufactured bolt will have exactly the same dimensions. QI largely is concerned with defining acceptable limits of variation, tracking variation within processes, and identifying causes and finding solutions when products are not within the acceptable range of variation.

Run Charts and Control Charts

Control charts, developed by Walter Shewhart in the 1920s, are a basic graphical technique used to monitor process variation. The control chart is a refinement of the basic run chart, which is simply a time series chart displaying some characteristic of the product in question on the *y*-axis and time or order of production on the *x*-axis. Often, the graphed data points represent a statistic such as the mean, calculated from small samples of product rather than individual values.

Graphing sample means allows us to invoke the central limit theorem and assume an underlying normal distribution for the data points without regard to the distribution of the individual values in the population. This is essential when using the decision rules that follow for determining when a process is going out of statistical control. If individual data points are represented in the control chart, these rules cannot be used unless the underlying process is normal, but graphing the points can still be useful as a graphical representation of the variation present in the process.

We expect to find variation in the output from any process but do not expect the distribution of the output to change, either in location (mean or median) or spread (standard deviation or range). If the distribution of output from a process is consistent over time, we say the process is *in statistical control* or simply *in control*. If it changes, the process is said to be *out of statistical control* or simply *out of control*. The process of monitoring and eliminating sources of variation for some process to bring it into or keep it in statistical control is called *statistical process control*.

There are two basic sources of the total variation of any process: *common causes* and *special* or *assignable causes*. *Common causes of variation* are those that are attributable to the design of a process and affect all output of the process. For a manufacturing process, common causes might include lighting in a factory, the quality of raw materials, and worker training. If the amount of variation due to common causes is too great, the process must be redesigned. Perhaps the lighting can be improved, workers can be given more training, the tasks can be broken down into smaller segments that can be performed more accurately, or a more consistent source can be found for the raw materials used in the manufacturing process. This type of correction is generally the responsibility of management and does not figure in the type of analysis discussed in this section.

For the purposes of this section, a process that has only common causes of variation is a process that is in control. Instead, we focus on *special causes of variation*, which are actions or events that are not part of the process design. Special causes are usually temporary and affect only small parts of the process. A worker might become fatigued and fail to execute his job accurately, or a machine might get out of adjustment and start producing products outside the range of acceptable values. Control charts are used to identify when processes are going out of statistical control and might aid in identifying special causes of variation.

Control charts usually include a *centerline* drawn at the process mean or median. The centerline acts as a reference point to evaluate the data points, for instance, to evaluate whether data points are close to or distant from a central value. The value of this centerline is usually specified in advance by the analyst and represents the expected value when the process is *in control* (running correctly, producing acceptable output) rather than the mean of the sample points. One other convention in control charts is the addition of lines connecting each consecutive point, which makes it easier to see the pattern across the sequence of measurements. Both features are displayed in the hypothetical run chart in Figure 14-13.

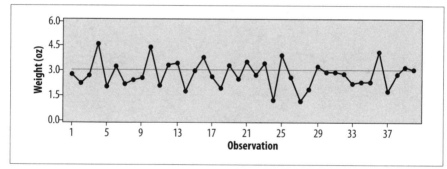

Figure 14-13. Control chart of weight in ounces for 40 screws (individual values) with a process mean of 3.0

This run chart displays the weight of 40 consecutively produced screws from a hypothetical manufacturing process. The y-axis displays the weight in ounces of each screw, the x-axis displays the order of observation, and the centerline displays the process mean of 3.0. We can observe therefore that the first three screws were slightly below the mean, the fourth was above, and so on. We can also see that the pattern is random and centered on the process mean and that the longest run (consecutive values in the same direction) involves four data points (values 29–32).

No particular pattern in the data is presented in Figure 14-14 (not surprising because it was created using a random number generator!), one of the indications that a process is in control. The charts in Figures 14-14 through 14-19 display some of the patterns that can be spotted by a run chart and might signal the need for further investigation.

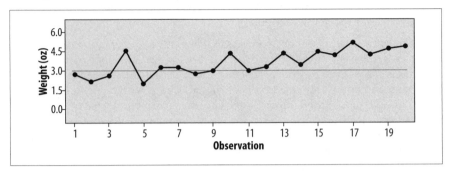

Figure 14-14. Control chart with an upward trend

Note that at this stage, because we are looking at individual data points, we are looking for general patterns rather than performing statistical tests. More formal rules are discussed shortly that may be used to determine when a data pattern cannot be attributed to random variation but should be investigated as evidence that a process is going out of control.

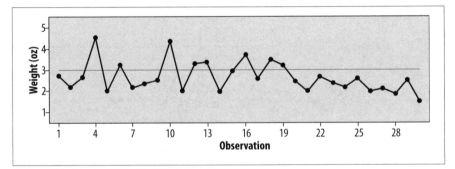

Figure 14-15. Control chart with a downward trend

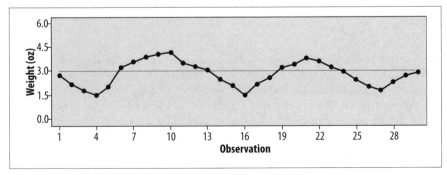

Figure 14-16. Control chart with a cyclical pattern

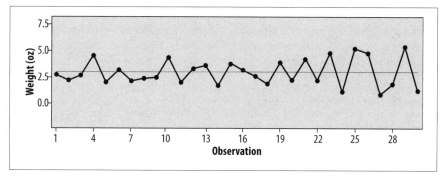

Figure 14-17. Control chart with increasing variability

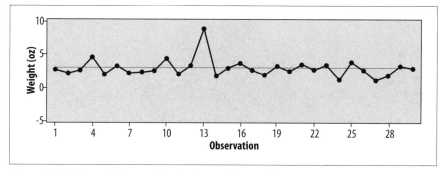

Figure 14-18. Control chart with a shock or outlier (single extreme value)

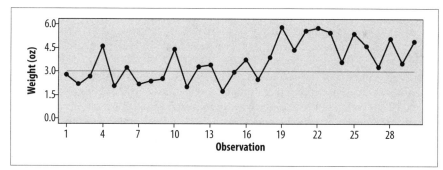

Figure 14-19. Control chart with a change of level (upward shift of mean)

When a control chart is based on sample means, thanks to the central limit theorem we can use the normal distribution to identify values or patterns that would be highly improbable for a process in statistical control. A number of rules have been determined that indicate a process is going out of control, based on the expected distribution of values of the data points if they were based on samples drawn from a normal distribution with mean and variance specified from the process when it is in control.

Use of the standard deviation to define acceptable ranges of values for the outputs from a process is the source of the name for the Six Sigma program because sigma (σ) is the symbol for standard deviation. The idea behind the Six Sigma program is to reduce variability sufficiently that output in the range of ±3σ will still be acceptable to the customer.

As discussed in Chapter 3, with normally distributed data, the probability of data points within particular ranges is known. The percentage of data from a normal distribution contained in different ranges, defined by standard deviations from the mean, is displayed in Figure 14-20.

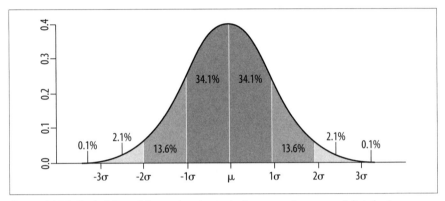

Figure 14-20. Probability of data points in particular ranges in a normal distribution

As discussed in Chapter 3, this figure tells us that, in a normal distribution, the probability of a data point within one standard deviation of the mean is about 68.2%. The probability of a data point in the range between one and two standard deviations above or below the mean is about 27.2%, the probability of a point between two and three standard deviations above or below the mean is about 4.2%, and the probability of a point beyond three standard deviations above or below the mean is about 0.2%. To look at it another way, in repeated samples from a normally distributed population, we would expect about 68% of the sample means to fall within one standard deviation of the mean, about 95% within two standard deviations, and about 99% to be within three standard deviations.

A control chart with the addition of control limits translates this information so the distribution of points is on the *y*-axis, whereas the *x*-axis displays the time or order of samples charted. The different ranges are often labeled as shown in Figure 14-21.

In this chart:

1. Zone A, or the three-sigma zone, is the area between two and three σ of the centerline.
2. Zone B, or the two-sigma zone, is the area between one and two σ of the centerline.
3. Zone C, or the one-sigma zone, is the area within one σ of the centerline.

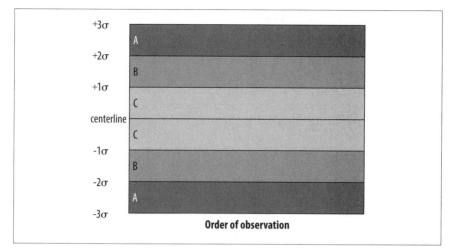

Figure 14-21. Control chart with sigma ranges

These zones are used in conjunction with a set of *pattern analysis rules* to determine when a process has gone out of control.

Because both the mean value and variability of the samples are important when determining whether a process is in control, control charts are usually produced in pairs, one representing mean values of the samples and one representing their variability. For continuous data, an *x-bar chart* (so called because $\bar{x}$, pronounced *x*-bar, is the statistical symbol for a sample mean) is used to track the mean value. Variability is represented with either an *s-chart* displaying the standard deviation of the samples or an *r-chart* representing the range of the samples.

The following pattern analysis rules are used to interpret data from the *x*-bar chart but could be applied to any of the various types of control charts. This list is an amalgam of several sets of rules, including the Western Electric rules developed at the Western Electric Company (now part of AT&T) and first published in 1956, and the Nelson rules developed by Lloyd S. Nelson and first published in 1984.

The circumstances under which a process is judged out of control under pattern analysis rules are:

1. If any point falls outside Zone A
2. If 9 consecutive points fall in Zone C or beyond (farther from the centerline) on the same side of the centerline
3. If 6 consecutive points fall in the same direction, that is, all increasing or all decreasing
4. If 14 consecutive points alternate up and down
5. If 2 out of 3 consecutive points fall in Zone A or beyond on the same side of the centerline
6. If 4 out of 5 consecutive points fall in Zone B or beyond on the same side of the centerline

7. If 15 points in a row fall in Zone C
8. If 11 consecutive points fall in Zone B or beyond

If data is binary rather than continuous (for instance, if items are simply classified as defective or acceptable), *p-charts* or *np-charts* based on the binomial distribution can be created in place of the *x*-bar chart. Note that binomial data is often referred to as *attribute data* within the field of quality control. If the interest is the *number of defects* rather than the *number of defective units* (if a unit can have more than one defect and the total count of defects is the variable of interest), *c-charts* and *u-charts* can be created in place of the *x*-bar chart. Because these charts are usually created using computer software (as are *x*-bar charts), they will not be discussed in detail here. The key point is that the principles of interpretation are the same as for *x*-bar charts. The following set of rules should help clarify which type of chart to use for each type of data:

1. Data points represent sample means from continuous data (*x*-bar chart).
2. Data points represent the number of defective items per sample, and all samples are the same size (*np*-chart).
3. Data points represent the proportion of defective items per sample, and samples are of different sizes (*p*-chart).
4. Data points represent the average number of defects per unit, and all samples are the same size (*c*-chart).
5. Data points represent the average number of defects per unit, and samples are of different sizes (*u*-chart).

W. Edwards Deming and Japan

Japan was not always the manufacturing powerhouse we know today. In the first half of the twentieth century, Japan was noted primarily for the manufacture of inexpensive products, and the industrial infrastructure of the country was severely damaged during the Second World War. However, after the war, the victorious Allied command assigned a group of engineers to help Japan rebuild its economy.

One aspect of this rebuilding was teaching Japanese manufacturers about statistical quality control methods. In 1950, W. Edwards Deming (1900–1993), a statistician who had studied with Walter Shewhart, was invited to present a series of lectures on statistical quality improvement under the auspices of the Japanese Union of Scientists and Engineers. During his visit, Deming also met with the top executives of many major Japanese companies.

Deming so impressed the Japanese industrial leaders that they established two annual awards in his name for achievements in the field of quality: the Deming Prize for Individuals (awarded to individuals who have made important contributions in the study, methodology, or dissemination of TQM) and the Deming Application Prize (awarded for outstanding performance improvement through application of TQM principles). Further information about these prizes is available from the Deming Institute website (*http://deming.org*).

Exercises

Here's a quick review of the topics covered in this chapter.

Problem

Calculate the simple index for 2000, using each of the other years (shown in Table 14-9) as a base year. What do the results tell you about the selection of the base period?

Table 14-9. Data for calculating an index using different base years

Year	Price
1970	1,000
1980	1,500
1990	2,000
2000	1,500

Solution

I_{2000} = 150 when 1970 is the base year, 100 when 1980 is the base year, and 75 when 1990 is the base year. This demonstrates the importance of the choice of the base year in index calculations and why it is important not to allow politics or other extraneous considerations to affect this choice.

Calculations when 1970 is the base year:

$$I_{2000} = (1,500/1,000) \times 100 = 150$$

When 1980 is the base year:

$$I_{2000} = (1,500/1,500) \times 100 = 100$$

When 1990 is the base year:

$$I_{2000} = (1,500/2,000) \times 100 = 75$$

Problem

Calculate the Laspeyres index and Paasche index for 2000 for the data in Table 14-10, using 1990 as the base year. Why do they differ?

Table 14-10. Data for comparing the Laspeyres index and Paasche index

Product	1990 quantity	1990 price	2000 quantity	2000 price
Beef	100 pounds	$3.00/pound	50 pounds	$5.00/pound
Chicken	100 pounds	$3.00/pound	150 pounds	$3.50/pound

Solution

The Laspeyres index is 141.67; the Paasche index is 87.50. The difference is due to different systems of weighting: the Laspeyres index uses the weights from the base year, but the Paasche index uses the weights from the index year. In this case, the same amount of meat was purchased in 1990 and 2000, but less beef and more chicken was purchased in 2000 relative to 1990. An inflation index based on the Laspeyres index would miss this change in consumer habits.

Figure 14-22 shows the calculations for the Laspeyres index.

$$\frac{(100 \times 5.00) + (100 \times 3.50)}{(100 \times 3.00) + (100 \times 3.00)} \times 100 = 141.67$$

Figure 14-22. Calculating the Laspeyres index

Figure 14-23 shows the calculations for the Paasche index.

$$\frac{(50 \times 5.00) + (150 \times 3.50)}{(100 \times 3.00) + (100 \times 3.00)} \times 100 = 129.17$$

Figure 14-23. Calculating the Paasche index

Problem

Calculate the SMA and CMA for $n = 3$ and $n = 5$ for the sixth time point for the data shown in Table 14-11.

Table 14-11. Data for calculating the SMA and CMA

Time	1	2	3	4	5	6	7	8	9
Raw data	3	5	2	7	6	4	8	7	9

Solution

SMA($n = 3$) = (7 + 6 + 4)/3 = 5.7
SMA($n = 5$) = (5 + 2 + 7 + 6 + 4)/5 = 4.8
CMA($n = 3$) = (6 + 4 + 8)/3 = 6.0
CMA($n = 5$) = (7 + 6 + 4 + 8 + 7)/5 = 6.4

Notice that because there is a general trend upward in this data, the CMA estimates are higher, particularly with the larger window.

Problem

Suppose you were considering whether to open a stationer's shop in a small or a large city. Greater potential profit is to be made in the large city but also greater potential loss (due to the greater expenses of setting up business there). The success of the shop will largely depend on the local business climate when you open. If other businesses in the city are expanding, you have a good chance to land some large orders, but if they are struggling, you might barely make enough sales to meet your expenses.

Table 14-12 contains data of payoffs under two states of nature. Calculate the minimax, maximax, and maximin decisions for this situation.

Table 14-12. Data for comparing investment prospects for a stationer's shop

		Business climate	
		Good	Poor
Location	Large city	$200,000	$10,000
	Small city	$100,000	$20,000

Solution

For the minimax solution, construct an opportunity loss table like Table 14-13.

Table 14-13. Opportunity loss table for the prospects for a stationer's shop

		Business climate	
		Good	Poor
Location	Large city	$0	$10,000
	Small city	$100,000	$0

The minimax solution is to choose the action that minimizes opportunity loss; in this case, we would choose to place our store in the large city.

The maximax solution is to select the action that has the highest maximum outcome, so in this case, we would place our store in the large city.

The maximin solution is to select the action that has the largest minimal outcome, so in this case, we would place our store in the small city.

Problem

What pattern analysis rules are violated in the control chart in Figure 14-24?

Note that for the in-control process for this example, the mean = 3 and the standard deviation = 0.5, so the centerline is at 3.0, the 3-sigma limits are at 1.5 and 4.5, the 2-sigma limits are at 4.0 and 2.0, and the 1-sigma limits are at 3.5 and 2.5.

Solution

The violations are identified in Figure 14-25 and listed after the figure.

1. Nine points in a row on the same side of the centerline (rule 2)
2. One point outside three-sigma range, that is, outside zone A (rule 1)
3. Six points in a row in the same direction (rule 3)
4. Four out of five consecutive points beyond the one-sigma range (zone B or beyond) on the same side of the centerline (rule 6)

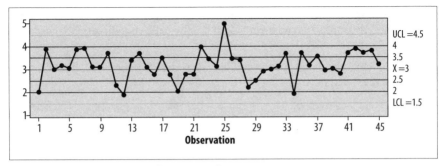

Figure 14-24. Control chart with pattern violations

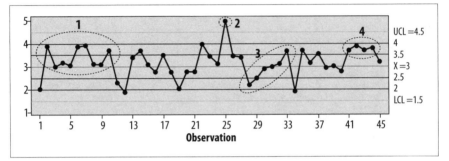

Figure 14-25. Control chart with pattern violations flagged

15

Medical and Epidemiological Statistics

Many of the statistics used in medicine and epidemiology are common to other fields; examples include the *t*-test (covered in Chapter 6), the correlation coefficient (covered in Chapter 7), and the various types of regression and ANOVA (covered in Chapters 8 through 11). However, some statistics have been developed specifically to meet the needs of medical and epidemiological research (such as the odds ratio), and others, although common to many fields, are used so frequently in medicine and epidemiology that they are covered in this chapter (for instance, power and sample size calculations).

Measures of Disease Frequency

Before getting into specific measures of disease frequency, it is worthwhile to discuss the meanings of several terms in common usage that are often confused. We can always report disease frequency in terms of the number of cases. For instance, there were 256 cases of tuberculosis (TB) in city A and 471 in city B last year. Raw numbers are useful for people who allocate current resources and plan future monetary and space allocation because they need to know how many cases of TB (and other diseases) to expect in the coming year so they can allocate resources accordingly. However, for research and planning at the national and international level, disease occurrence is more usefully described in terms of relative rather than absolute occurrence because we often want to look at trends over time or across different geographical areas with different population sizes. For instance, the preceding hypothetical raw numbers suggest that city B has a worse problem with TB than city A, but if city B has five times the population, the opposite would be true. Similarly, the number of cases of a disease can increase because the population is also increasing, so to make comparisons, we often need to translate counts into other metrics.

Ratio, Proportion, and Rate

Three related types of metrics are the ratio, the proportion, and the rate. A *ratio* expresses the magnitude of one quantity in relation to the magnitude of another quantity without making further assumptions about the two numbers and without requiring the two numbers to share a common unit. Ratios may be expressed as A:B, or A per B, and are often converted to standard metrics for easy comparison, such as 1:B or A per 10,000. We might be interested in the ratio of men to women living with AIDS in the United States. According to the Centers for Disease Control and Prevention (CDC), 769,635 men and 186,383 women were living with AIDS in the United States in 2005. The ratio of men to women living with AIDS is therefore 769,635:186,383, which can also be expressed as 4.13:1. The second formulation makes it clearer that just over four times as many men as women were living with AIDS in the United States in 2005.

Two types of ratios often used in epidemiology and public health are the risk ratio and the odds ratio, which this chapter further discusses. Ratios do not require the quantities compared to be measured in the same units; a common measure used to compare health care availability in different countries is the ratio of hospital beds to the population size. This is often expressed as the number of hospital beds per 10,000 people. According to the World Health Organization, the United Kingdom had 39.0 hospital beds per 10,000 people in 2005, whereas Sudan had 7.0 and Peru 11.0, suggesting that hospital care was more readily available in the United Kingdom than in the other two countries. This type of ratio is sometimes referred to as a *rate*, although it does not meet the strict definition of rate (discussed later) because the denominator does not include a measure of time.

A *proportion* is a particular type of ratio in which all cases included in the numerator are also included in the denominator. To return to the previous example, if we want to know what proportion of people living with AIDS in the United States were male, we would divide the number of males by the total number of cases (the number of cases in males plus the number of cases in females), as shown in Figure 15-1.

$$\frac{769,635}{769,635 + 186,383} = 0.805$$

Figure 15-1. Calculating a proportion

Proportions are often expressed as percents, which means, literally, *per cent* or *per 100*. (*Cent* is Latin for 100.) To translate proportions to percents, multiply by 100.

$0.805 \times 100 = 80.5\%$

The proportion of males among all people living with AIDS in the United States could also be expressed as 80.5 percent, or 80.5%.

A *rate*, strictly speaking, is a proportion in which the denominator includes a measure of time. For instance, we commonly measure a person's heart rate in terms of beats per minute and disease or injury occurrence in terms of the number of cases per week, month, or year. Morbidity and mortality (disease and death) statistics are often reported in terms of the rate per 1,000 or 100,000 per time unit because it is easier to interpret numbers such as 3.57 versus 12.9 annually per 100,000 population than 0.0000357 versus 0.0000129 annually per person.

Converting rates to standard quantities facilitates comparison across populations of different sizes. For instance, the CDC reports that the annual death rate in the United States in 2004 was 816.5 per 100,000 population compared to 1,076.4 per 100,000 in 1940 and 954.7 per 100,000 in 1960. There were more deaths in 2004 than in any of the comparison years (2,397,615 in 2004 versus 1,417,269 in 1940 and 1,711,982 in 1960), but because the population of the United States was also increasing, the annual death rate per 100,000 decreased.

A simple example using hypothetical data (Table 15-1) illustrates this.

Table 15-1. Population and death figures for several years

Year	Deaths	Population	Deaths per 100,000
1940	75	50,000	150.0
1950	95	60,000	158.3
1960	110	75,000	146.7
1970	125	90,000	138.9

We can see that although deaths increased each year, the population increased even faster, so the annual death rate per 100,000 population decreased in each year studied. To calculate the death rate per 100,000, use the formula in Figure 15-2.

$$\frac{deaths}{population} \times 100,000$$

Figure 15-2. Calculating the death rate per 100,000

So, given the data in Table 15-1, the death rate per 100,000 for 1940 is calculated as shown in Figure 15-3.

$$\frac{75}{50,000} \times 100,000$$

Figure 15-3. Calculating the death rate per 100,000 population for 1940

One issue in computing rates over a long period of time, such as a year, is to decide what number to use in the denominator because the population cannot be assumed to be constant over the period. One typical solution is to use the population at the midpoint of the period (e.g., the year).

There are several other issues involved in reporting disease incidence. One is whether the number of individuals with a condition or the number of conditions itself is being reported. For instance, if you were studying oral health, you might be interested in tooth decay, but a single person could have more than one cavity. Are you interested in the number of people who had at least one cavity or the total number of cavities?

A similar issue arises if you are studying a transient condition. For instance, if your topic is homelessness, are you interested in how many people had been homeless at least once over a time period, or would you count each instance of homelessness, with the understanding that some people might have been homeless more than once in the time period in question? These are problems of *unit of analysis*, meaning that you need to decide what entity you are studying (a *person* who might develop one or more cavities or a number of individual *teeth*, each of which can develop a cavity) and collect and analyze data with that definition in mind. Unit of analysis is discussed further in Chapter 3.

Prevalence and Incidence

In epidemiology and medicine, when we speak of the number of cases of a disease, we make a basic distinction about whether we are counting all the existing cases of a disease or only counting new cases. This might seem like hairsplitting to the average person, but it is important to people working in medicine and epidemiology because we often want to separate new cases of a disease from existing cases. We can determine whether a sanitation campaign is effective in preventing new cases of disease, for instance. We separate existing from new cases by measuring two types of disease frequency: prevalence and incidence.

Prevalence describes the number of cases that exist in a population at a particular point in time. Prevalence describes the disease burden on a population without differentiating between new versus existing cases; a diabetic diagnosed the day a survey is conducted is counted equally as having the condition as a diabetic who has been living with the condition for 20 years. Prevalence is particularly useful to people involved in resource allocation and planning because they need to know the disease burden in the population as well as what it will be like in the future. Prevalence has assumed increasing importance as the focus on epidemiology in the industrialized world has shifted from infectious to chronic diseases and conditions. This is because chronic diseases and conditions are often not curable but not rapidly fatal either, so a person can live for years with the disease or condition if appropriate medical care is provided.

Prevalence is defined as the proportion of individuals in a population who have the disease at a particular point in time and is calculated as shown in Figure 15-4.

$$P = \frac{\text{number of cases}}{\text{population}}$$

Figure 15-4. Calculating prevalence

If a survey of a city with a population of 150,000 people found that 671 were diabetics, the prevalence of diabetes at the time of the survey in that city would be 671 per 150,000 or 447.3 per 100,000. For ease of comparison, standards such as per 100,000 are commonly used when reporting data. Because prevalence tells you the disease status of a population at a particular point in time, it is sometimes called *point prevalence*. Note that the "point" can be a calendar time, such as a day, or a time in the life cycle or other course of events, such as the onset of menopause or the first day following surgery. Prevalence is sometimes referred to as *prevalence rate*, particularly when longer time intervals such as a year are used, although this is not strictly correct because there is no unit of time in the denominator.

Incidence is more complicated to calculate because it requires three elements to be defined. Incidence describes the number of *new cases* of a disease or condition that develop in a *population at risk* during a particular *time interval*. Population at risk means people who have the potential to develop the condition. Men are not at risk for pregnancy, for instance, so they would not be included in the population at risk. Similarly, after a person is infected with HIV (the virus that causes AIDS), that person cannot become infected again (or become uninfected, so far as we know), so the population at risk for HIV infection is restricted to those individuals who are not already HIV positive. Both incidence and prevalence are also used to describe health behaviors as well as diseases and conditions; for instance, we can refer to the prevalence of smoking in Mexico or the incidence of smoking onset in 2005 among teenagers at a particular school.

There are two types of incidence, cumulative incidence and incidence density. *Cumulative incidence* (CI) is the proportion of people who contract a disease during a specific time interval and is calculated as shown in Figure 15-5.

$$CI = \frac{\text{number of new cases}}{\text{population at risk}} \text{ for a specified period}$$

Figure 15-5. Cumulative incidence formula

CI is used to estimate the probability that an individual at risk will develop a disease or condition within a specified period, so it is important for the period to be identified. The CI of a woman developing breast cancer in a 1-year period following initial use of oral contraceptives will be different from the CI for a 10-year period.

The formula to calculate CI assumes the entire population at risk can be studied for the entire specified period; this means that, unless otherwise qualified, incidence is a proportion. If the population at risk changes over the period included in the incidence calculations, the *incidence density* (ID), also known as the *incidence rate* (IR), should be calculated instead. This would be necessary if people entered a study after it began or dropped out before it was completed. Calculation of the IR requires expressing the denominator in *person-time units*, which represent the amount of time each person was observed. The time of observation for each person is often referred to as the time each person *contributed* to the study.

The calculation of person-time units is demonstrated in Table 15-2. It represents hypothetical data on the annual rate of postsurgical infections at two hospitals. Because the hospitals serve different numbers of patients and patients are in the hospital for different lengths of time, we need to calculate the IR using person-time units in the denominator. Our statistic of comparison will be the number of complications per 100 patient-days. Each patient-day can be considered an opportunity for an infection to occur, so using patient-days in the denominator corrects for the different exposure to risk at the two hospitals.

Table 15-2. Data on postsurgical infections at two hospitals

Hospital	Patient ID	Days followed	Infection?
1	1	30	N
1	2	25	Y
1	3	15	N
Total for Hospital 1	70	1	
2	1	45	Y
2	2	30	N
2	3	50	N
2	4	75	Y
Total for Hospital 2	200	2	

The rate of infections per 100 patient-days is calculated as shown in Figure 15-6.

$$\frac{\text{number of infections}}{\text{patient - days studied}} \times 100$$

Figure 15-6. Calculating the infection rate per 100 patient-days

So for this example, the rates are shown in Figure 15-7 for hospital 1 and Figure 15-8 for hospital 2.

$$\frac{1}{70} \times 100 = 1.43 \text{ per } 100$$

Figure 15-7. The infection rate per 100 patient-days for hospital 1

$$\frac{2}{200} \times 100 = 1.00 \text{ per } 100$$

Figure 15-8. The infection rate per 100 patient-days for hospital 2

Even though hospital 1 had more postsurgical infections in the period studied, these occurred during proportionally more patient-days, so hospital 2 has a lower rate of postsurgical infections than hospital 1.

The relationship between incidence and prevalence for a particular disease depends largely on the duration of the disease. If a disease has short duration (such as the common cold), prevalence will be low relative to incidence. In contrast, if a disease has a long duration (typical of many chronic diseases such as diabetes), the prevalence will be high relative to incidence. Changes in prevalence across time periods can be due either to changes in incidence or in duration. For instance, incidence of a fatal disease might decrease, but prevalence might increase if new treatments are developed that allow people to live for longer periods with the disease without curing it (an increase in the duration of the average case of the disease). Similarly, the incidence of a disease can increase but the prevalence decrease if the duration of the disease is shortened through the development of new treatments that promote faster recovery.

Prevalence may be expressed mathematically as the product of incidence times average duration, as shown in Figure 15-9.

$$P = I \times \overline{D}$$

Figure 15-9. The relationship among prevalence, incidence, and duration

If two of these variables are known, the third can be calculated. For instance, if the incidence of a disease is 75 per 100,000 and the average annual prevalence is 45 per 100,000, the average duration can be calculated as shown in Figure 15-10.

$$\overline{D} = \frac{P}{I} = \frac{45/100,000}{75/100,000/year} = \frac{45}{75/year} = 0.6 \text{ years}$$

Figure 15-10. Calculating average duration, given prevalence and incidence

This assumes steady-state conditions for the time period under study and no major changes in disease incidence or duration. The formula can also be used to calculate how prevalence would change if either incidence or duration changes. For instance, if incidence of a particular disease remains steady at 125 per 100,000 but duration drops from 0.6 years to 0.1 years, prevalence will decrease from 75 per 100,000 per year to 12.5 per 100,000 per year. Similarly, if duration increases, prevalence will increase. If incidence of some disease remains steady at 200 per 100,000 per year but duration increases from 0.5 years to 2 years, prevalence will increase from 100 per 100,000 per year to 400 per 100,000 per year.

Crude, Category-Specific, and Standardized Rates

If not otherwise qualified, the term rate usually means the *crude rate*. The crude rate is the rate for the entire population under study with no weighting or adjustment.

Medical Statistics

A common example is the crude death rate. According to the CDC, the crude death rate for cancer in the United States in 2003 was 195.5 per 100,000. There is nothing wrong with crude rates, but sometimes we want more specific information, or we need to make adjustments in the rates to make comparisons more meaningful. For instance, the crude death rate for cancer in the United States in 2003 was not constant across ethnic groups, age groups, or genders, nor was it constant across types of cancer. Exploring these differences might be of interest to researchers, in which case they will want to look at the *category-specific* rates, in which both the numerator and denominator represent one population group or one type of disease. In the United States in 2003, the cancer mortality rate for men was 201.4/100,000; for women, it was 182.0/100,000. In the same year, the crude mortality rate for lung cancer was 76.9/100,000, whereas for skin melanomas, it was 2.7/100,000.

For white Americans in 2003, the crude cancer death rate was 203.8/100,000; for African-Americans, it was 164.3/100,000, a finding that might seem paradoxical until we consider that increased life expectancy is often associated with increased cancer mortality. Someone who dies as an infant is unlikely to have died of cancer, but someone who lives into her eighties has a much higher probability of a cancer-related death. This is true of general mortality as well. Under most circumstances, a person who is 90 years old has a much higher probability of dying in the next year than a person who is 12 years old. For this reason, death rates used to make comparisons across different populations or time periods are usually standardized by age and may also be standardized by categories such as ethnicity or gender.

The importance of age adjustment can be seen by comparing the crude and age-adjusted cancer mortality figures for the United States in 2003 in Table 15-3.

Table 15-3. Crude and age-adjusted cancer mortality rates (per 100,000) for the United States in 2003

	Crude	Age-adjusted
Overall	191.5	190.1
White	203.8	188.3
African American	164.3	234.5
Asian/Pacific Islander	79.4	114.3
American Indian/Alaska Native	69.3	121.0
Hispanic	60.3	127.4

This makes it clear that although the crude death rate from cancer is highest among white Americans, this result is due in part to a longer life expectancy. A longer life expectancy means that there are more white Americans in the older age categories, when mortality from cancer is higher. When age adjustment is considered, African-Americans have the highest death rate from cancer.

There are two types of standardization, *direct* and *indirect*. Both are used to compare morbidity and mortality in different populations while removing the influence of other population characteristics such as age or gender distribution. In *direct standardization*, a population is chosen to serve as the standard, and adjusted rates for the populations to be compared are calculated using weights from the standard population. Consider the hypothetical example of the prevalence of arthritis by employment status in Table 15-4.

Table 15-4. Arthritis prevalence by employment status

Employment status	Population	Persons with arthritis	Rate per 1,000
Employed	10,000	387	38.7
Unemployed	5,000	892	178.4

The rate (really the proportion) of arthritis is more than twice as high among persons not employed as among employed persons, according to this data. Could this be due to people being forced out of the labor market due to severe arthritis? Does working help keep arthritis at bay? Both are possible, but a more logical explanation is that people over the age of 65 are more likely not to be employed and more likely to have a diagnosis of arthritis. To test the hypothesis that age distribution is the reason for the observed differences in rate of arthritis diagnosis by employment status, we need to compute age-adjusted rates of arthritis by using a standard population. First, we need to calculate age-specific rates for employed and unemployed individuals, as in Table 15-5.

Table 15-5. Age-specific rates of arthritis diagnosis

	Employed			Unemployed		
Age group	Population	Persons with arthritis	Rate/1,000	Population	Persons with arthritis	Rate/1,000
18–44	5,000	127	25.4	1,000	32	32.0
45–64	4,500	260	57.7	1,500	100	66.7
65+	500	105	210.0	2,500	760	304.0
Total	10,000	387	38.7	5,000	892	178.4

Looking at the age distribution and age-specific rates for the employed versus unemployed populations, we see that for each age group, the rates of arthritis are somewhat higher in the unemployed group than in the employed group (the opposite pattern from that seen when data from all the age categories is combined). We also see, as we suspected, that a much higher proportion of the unemployed group (50%) as compared to the employed group (5%) is in the 65+ age category, where the rates of arthritis are highest.

Note that we used very broad age categories (corresponding to young working adult, older working adult, and retirement age) in this table for ease of calculation. Often, smaller categories are used, such as 10-year age ranges.

We can use age-specific rates to calculate the expected numbers of persons with arthritis in each age category for the two employment groups, using the age distribution from a hypothetical standard population. Usually, data from a recognized source is used for the standard population in these calculations, such as the U.S.

population in 2000 as determined by the U.S. Census Bureau. The expected number of persons with arthritis is shown in Table 15-6.

Table 15-6. Expected numbers of persons with arthritis by age category and employment category

Age group	Standard population Population	Employed Rate/1,000	Expected diagnoses	Unemployed Rate/1,000	Expected diagnoses
18–44	100,000	25.4	2,540.0	32.0	3,200.0
45–64	70,000	57.7	4,039.0	66.7	4,669.0
65+	30,000	210.0	6,300.0	304.0	9,120.0
Total	200,000		12,879		16,989

The expected cases within each age and employment category are calculated by applying the age-specific rates for each population to the number of people in that age category in the standard population. This may be considered a type of weighting and is equivalent to saying how many cases of diagnoses we would expect to see in each population if the age distribution were the same as in the standard population. For instance, for the 18–44 age group in the employed population, the calculation is shown in Figure 15-11.

$$E = \text{rate} \times \text{population} = \frac{25.4}{1000} \times 100{,}000 = 2{,}540$$

Figure 15-11. Expected cases for the employed, age 25–44, category

For the 65+ age category in the unemployed population, it would be as shown in Figure 15-12.

$$E = \text{rate} \times \text{population} = \frac{304}{1000} \times 30{,}000 = 9{,}120$$

Figure 15-12. Expected cases for the unemployed, age 65+ category

The total expected diagnoses for employed and unemployed people are found by adding up the expected diagnoses for each age group within the category. We can see that, if the two populations had the same age distribution, employed people would be expected to have fewer arthritis cases (12,879) than people who were unemployed (16,989). We can further refine this finding by calculating the age-adjusted arthritis rates for each population by dividing the number of expected cases by the total size of the reference population and then multiplying by 1,000 (to get the rate per 1,000). For employed people, this would be as shown in Figure 15-13.

$$\frac{12,879}{200,000} \times 1000 = 64.4 \text{ per } 1000$$

Figure 15-13. Age-adjusted arthritis rate for employed people

For unemployed people, the age-adjusted arthritis rate would be 84.9 per 1,000. Comparing these rates shows us that the rate of arthritis is higher in unemployed persons than in employed persons, but the difference is much less than the rates in Table 15-5. Note that the age-adjusted rates calculated through direct standardization do not represent the actual rates in any population; they represent what rate would be expected in one or more particular populations *if* they had the age distribution of some reference population.

Indirect standardization takes the reverse approach. It applies the category-specific rates from some standard population to the actual category distribution in two or more populations. Applying indirect standardization to our arthritis example, we will calculate the expected number of arthritis diagnoses if both populations had the same age-specific rate of diagnosis but kept their own specific population age distribution. The rates (which are hypothetical) are shown in Table 15-7.

Table 15-7. Indirect method of standardization

Age group	Standard rate/1,000	Employed population	Expected diagnoses	Unemployed population	Expected diagnoses
18–44	30.0	5,000	150	1,000	30
45–64	60.0	4,500	270	1,500	90
65+	200.0	500	100	2,500	500
Total		10,000	520	5,000	620

We can use these numbers to calculate the *standardized morbidity ratio* (morbidity means disease) by dividing the observed number of diagnoses (from Table 15-5) by the expected number of diagnoses (from Table 15-7). The standardized morbidity ratio for employed people is shown in Figure 15-14.

$$SMR = \frac{\text{observed diagnoses}}{\text{expected diagnoses}} = \frac{387}{520} = 0.744 \text{ or } 74.4\%$$

Figure 15-14. Standardized morbidity ratio for employed persons

For the unemployed group, the standardized morbidity ratio is 0.695 or 69.5%. If the standardized morbidity ratio is 1.0, we have the same number of observed and expected cases. In our example, the standardized morbidity ratio for both the employed and unemployed groups is less than 1.0, meaning that we observed fewer cases than expected. A standardized morbidity ratio higher than 1.0 means that more cases are observed than were expected.

If we were dealing with deaths rather than diagnoses of arthritis, we could use the same technique to calculate the standardized mortality ratio (SMR), a statistic commonly used to compare mortality (death) across populations; the difference is that we would be counting deaths rather than cases of disease.

The Risk Ratio

Many medical and epidemiological studies are concerned with the relationship between two dichotomous variables. A common example is the exposure to some risk factor (such as asbestos or tobacco smoke) and the development of some disease or condition (such as asbestosis or lung cancer). The exposure can be an inherent quality, such as gender or ethnicity, and need not be negative; for instance, engaging in regular physical activity is an exposure that has been shown to have a positive influence on health.

The relationship between two dichotomous variables is often presented in a *cross-tabulation* or *contingency table*, also called a 2×2 or two-by-two table because of its dimensions (two rows and two columns). Contingency tables are also discussed in Chapter 5, and the same principles apply here. However, there is a standard way to set up a contingency table for epidemiological studies, as illustrated in Table 15-8.

Table 15-8. The 2×2 table

		Disease		Total
		D+	D–	
Exposure	**E+**	a	b	$a+b$
	E–	c	d	$c+d$
Total		$a+c$	$b+d$	$a+b+c+d$

E+ means the person had the exposure, E– that he did not. D+ means the person has the disease, D– that he does not. The arrangement of categories (exposure as rows, disease as columns) and order of categories (so cell a is E+, D+) is common to many epidemiological studies, so it's a good idea to follow it unless you have a reason to do otherwise. Individuals in a study are classified by their exposure and disease status, and the cells labeled *a*, *b*, *c*, and *d* contain the frequencies for each combination of exposure and disease. For instance, cell *a* holds the frequency for people who have the exposure and have the disease, and cell *d* holds the frequency for people who have neither exposure nor disease.

The frequencies in the four cells *a*, *b*, *c*, and *d* are sometimes referred to as *joint frequencies* because the people in those cells are classified on both exposure and disease. On the margins of the table are the row and column totals, often referred to as *marginal* frequencies. For instance, *a* + *c* is the total number of people in the study with the disease regardless of exposure status, whereas *a* + *b* is the total number of people with the exposure regardless of disease status. The total number of people in the study is *a* + *b* + *c* + *d*.

The *risk ratio*, also called the *relative risk*, estimates the likelihood of developing the disease for people with the exposure relative to people without the exposure. It is the ratio of the proportion of the exposed that develop the disease to the proportion of the unexposed that develop it. The risk ratio is calculated as shown in Figure 15-15.

$$RR = \frac{a/(a+b)}{c/(c+d)}$$

Figure 15-15. Formula for the risk ratio

The risk ratio can also be thought of as the ratio of disease incidence in the exposed (I_e) versus unexposed (I_0) populations, as shown in Figure 15-16.

$$RR = \frac{\text{incidence in the exposed group}}{\text{incidence in the unexposed group}} = \frac{I_e}{I_0}$$

Figure 15-16. Risk ratio using disease incidence

For studies in which the denominator is person-time units, the calculation is analogous but uses the ratio of the incidence densities (ID) for the two populations, as shown in Figure 15-17.

$$RR = \frac{ID_e}{ID_0}$$

Figure 15-17. Risk ratio using incidence densities

Let's look at data from a hypothetical study designed to investigate whether there is a relationship between consumption of a high-fat diet (the exposure) and Type II diabetes (the disease). The data is presented in Table 15-9.

Table 15-9. Relationship between consumption of a high-fat diet and Type II diabetes

	D+	D–	Total
E+	350	1200	1550
E–	200	1900	2100
Total	550	3100	3650

The risk of Type II diabetes, given consumption of a high-fat diet, is calculated as shown in Figure 15-18.

$$\frac{a}{a+b} = \frac{350}{1550} = 0.226$$

Figure 15-18. Risk for the exposed group

This is the number of people with both exposure and disease (those who consumed a high-fat diet and had Type II diabetes) divided by all who had the exposure (those who consumed a high-fat diet, without regard to their disease state).

The risk of Type II diabetes for someone without the exposure—someone not consuming a high-fat diet—is shown in Figure 15-19.

$$\frac{c}{c+d} = \frac{200}{2100} = 0.095$$

Figure 15-19. Risk for the unexposed group

The relative risk of developing diabetes, comparing those consuming a high-fat diet with those who do not, is the ratio of these two risks (hence the term *risk ratio*), as shown in Figure 15-20.

$$RR = \frac{RR_{E+}}{RR_{E-}} = \frac{a/(a+b)}{c/(c+d)} = \frac{0.226}{0.095} = 2.38$$

Figure 15-20. Risk ratio for exposed versus unexposed

A relative risk greater than 1 indicates that the exposure increases the risk of the disease. If there is no relation between exposure and risk, the relative risk will be 1, but if the exposure is protective (associated with lower risk of disease), the risk ratio will be less than 1. In this case, we would say that people consuming a high-fat diet have 2.38 times the risk of Type II diabetes compared to people consuming a low-fat or normal diet.

Like many other statistics, risk ratios are usually reported along with their confidence interval (CI). Calculations of this confidence interval must take into account that the risk ratio is right skewed because it has a lower bound of 0 but no upper bound. To deal with this skew, we take the natural logarithm (*ln*) of the risk ratio, which transforms it to an approximately normal distribution. The procedure for calculating the CI for RR (risk ratio) requires taking the natural logarithm of RR, finding the confidence interval for this $ln(RR)$, and then taking the natural antilogarithm of the confidence interval limits to return to the original units. Note that in statistical notation, e^x is sometimes written as $\exp(x)$.

There are several ways to calculate the confidence interval for a risk ratio, the most common being to use statistical software. However, the calculation can also be done by hand, as shown in the formula in Figure 15-21 for the general case.

$$CI = (RR)\exp[\pm z\sqrt{Var(\ln(RR))}]$$

Figure 15-21. General formula for the CI for a risk ratio

In this formula, z is the value of the standard normal distribution associated with the desired confidence level; most often, this value is 1.96, resulting in a two-sided, 95% confidence interval. When the RR is estimated using the odds ratio (discussed later) from a case-control study, the CI may be calculated using values from the 2×2 table, using the formula in Figure 15-22.

$$CI = (ad/bc)\exp[\pm z\sqrt{(1/a + 1/b + 1/c + 1/d)}]$$

Figure 15-22. CI formula for a risk ratio estimated by an odds ratio

Using values from Table 15-9 and assuming a two-sided, 95% confidence interval, this translates to the result shown in Figure 15-23.

$$CI = \frac{350(1900)}{200(1200)}\exp(\pm1.96\sqrt{1/350 + 1/1200 + 1/200 + 1/1900})$$
$$= (2.77)\exp(\pm0.19)$$
$$= (2.30, 3.35)$$

Figure 15-23. Computing the CI, using the odds ratio formula

Because this CI does not include the null value of 1.0, we conclude that the relationship between consumption of a high-fat diet and diagnosis with Type II diabetes is significant.

The time period over which data is collected is important in interpreting relative risk. The risk of developing many chronic diseases increases with duration of exposure, for instance, so the risk of a high-fat diet for development of Type II diabetes would be expected to be higher in a 10-year study than in a 1-year study. This is particularly true for studies of mortality because if a study is continued long enough, the probability of mortality for all the subjects is 100%!

Attributable Risk, Attributable Risk Percentage, and Number Needed to Treat

Because there is often some risk of disease for people without the exposure being studied, epidemiology also uses the concept of *attributable risk* (AR). Attributable risk is the absolute effect of the exposure on disease occurrence, meaning the excess risk of disease in the exposed versus the unexposed group. AR is useful as a measure of the public health cost or benefit of some exposure because it subtracts from the exposed group the cases that would be assumed to have occurred anyway. AR can also be used to estimate the impact of a proposed intervention to remove an exposure by calculating how many cases of disease would be saved—would not occur—if the exposure were eliminated. Attributable risk is calculated by subtracting the incidence rate in the unexposed (I_0) from the rate in the exposed (I_e). In our example relating a high-fat diet and Type II diabetes, this would be as shown in Figure 15-24.

$$AR = I_e - I_0 = 0.226 - 0.095 = 0.131$$

Figure 15-24. Calculating attributable risk (AR)

Therefore, a high-fat diet accounts for about 131 excess cases of Type II diabetes per 1,000 people. If there is no relationship between exposure and disease, there would be no excess cases in the exposed group, and the AR would equal 0.

The *attributable risk percentage* (AR%, also called the *etiologic fraction*) is the proportion of cases in the exposed population that can be attributed to the exposure and would be prevented by eliminating the exposure. It is calculated, continuing with our example, as shown in Figure 15-25.

$$AR\% = \frac{AR}{I_e} \times 100 = \frac{I_e - I_0}{I_e} \times 100 = \frac{0.226 - 0.095}{0.226} \times 100 = 58.0\%$$

Figure 15-25. Calculating the attributable risk percentage (AR%)

We would interpret this by saying that 58.0% of the cases among the exposed groups are due to the exposure.

The AR% can also be calculated using RR as shown in Figure 15-26.

$$AR\% = \frac{RR - 1}{RR} \times 100 = \frac{2.38 - 1}{2.38} \times 100 = 58.0\%$$

Figure 15-26. Calculating the attributable risk percentage using the risk ratio

The *number needed to treat* (NNT) is the number of patients who would need to be treated with a procedure (as opposed to standard treatment or placebo) or to avoid an exposure to reduce the number of cases (people with the disease) in the population by 1. The NNT is useful in putting the expected gains from a new treatment into perspective and is calculated using the attributable risk (AR):

NNT = 1/AR

In our example, the AR was 0.131. The NNT would therefore be:

NNT = 1/0.131 = 7.6

NNT estimates are usually rounded up to the next whole number (no fractional patients, please!), so in this case, we would say that eight people need to avoid a high-fat diet for there to be one fewer case of Type II diabetes in the population.

The Odds Ratio

The odds ratio was developed for use in *case-control studies*, a methodology developed in epidemiology to facilitate research into diseases that are rare or slow to develop so that a conventional prospective study would be impractical. In the case-control study, individuals are selected based on their disease status; *cases* have the disease or condition under study; *controls* do not. The two groups are then compared on exposure status. Risk ratios cannot be calculated in case-control studies because risk ratios are sensitive to the number of controls (people without the disease), and this number is determined in case-control studies by the study design rather than the rate of disease in a population. As will be demonstrated, the odds ratio has the beneficial quality of being insensitive to the number of controls (persons without the disease), whereas the risk ratio does not share this property.

The *odds ratio* is the ratio of the odds of exposure for the case group to the odds of exposure for the control group. This is mathematically equivalent to the ratio of the odds of disease for the exposed group to the odds of disease for the unexposed group, so you might see odds ratios explained either way. In a 2×2 table, the odds of exposure given disease are a/c, and the odds of exposure given no disease are b/d. The odds ratio is calculated using the formula shown in Figure 15-27:

$$OR = \frac{\text{odds of exposure given disease}}{\text{odds of exposure given no disease}} = \frac{a/c}{b/d} = \frac{ad}{bc}$$

Figure 15-27. Formula for the odds ratio

Suppose we have a case-control study examining the effect of smoking on breast cancer. The hypothetical data is shown in Table 15-10.

Table 15-10. Relationship of smoking and breast cancer

	D+	D–	Total
E+	50	2,000	2,050
E–	25	1,900	1,925
Total	75	3,900	3,975

The odds ratio may be calculated as shown in Figure 15-28.

$$OR = \frac{50/25}{2000/1900} = 1.90$$

Figure 15-28. Calculating the odds ratio

Medical Statistics

Note that the risk ratio for this data is similar, as shown in Figure 15-29.

$$RR = \frac{50/2050}{25/1925} = 1.88$$

Figure 15-29. Calculating the risk ratio

If a disease or condition is rare (a rule of thumb is that disease incidence must be less than 10% in all exposure groups), the odds ratio provides a reasonable estimate of the risk ratio. The reason for the "rare disease" requirements is that as a disease becomes more common in a data set, the odds ratio diverges further from the risk ratio. This is demonstrated in the data presented in Table 15-11, which represents data from a hypothetical case-control study of smoking and lung cancer.

Table 15-11. Smoking and lung cancer

	D+	D–	Total
E+	50	50	100
E–	20	100	120
Total	70	125	195

The disease is common in both exposed and unexposed subjects; 50% of the exposed subjects have lung cancer, as do 16.7% of the unexposed. The odds ratio for this data is shown in Figure 15-30 and the risk ratio is shown in Figure 15-31.

$$OR = \frac{50(100)}{20(50)} = \frac{5000}{1000} = 5.0$$

Figure 15-30. Calculating the odds ratio

$$RR = \frac{50/100}{20/120} = 3.0$$

Figure 15-31. Calculating the risk ratio

The difference between 5.0 and 3.0 is substantial and can be attributed to the fact that the 10% standard was violated; in such data sets, the OR is not a good estimator of the RR for this data.

The RR is also sensitive to changes in the number of controls, whereas the OR is not. Suppose that because controls are easier to find than cases, we increased the number of controls ten-fold (unlikely, because diminishing returns set in for control-case ratios at about 4:1, but useful to demonstrate this point). This would give us the data shown in Table 15-12.

Table 15-12. Smoking and lung cancer, ten-fold increase in controls

	D+	D–	Total
E+	50	500	550
E–	20	1,000	1,020
Total	70	1,500	1,570

The odds ratio does not change from that calculated from the data in Table 15-11, as shown in Figure 15-32...but the RR does, as shown in Figure 15-33.

$$OR = \frac{50(1000)}{20(500)} = \frac{5000}{1000} = 5.0$$

Figure 15-32. The odds ratio is unaffected by an increased number of controls

$$RR = \frac{50/550}{20/1020} = 4.64$$

Figure 15-33. The risk ratio is affected by an increased number of controls

Confidence intervals for the OR may be calculated using the method described in the preceding section "The Risk Ratio" on page 362.

Odds

The odds ratio is an important statistic in medical and statistical research, yet it is based upon a concept that is not intuitive or familiar to most people: that of odds. The odds of an event are simply another way to express the likelihood of an event, similar to probability; the difference is that although probability is computed by dividing the number of events by the total number of trials, odds are calculated by dividing the number of events by the number of nonevents. To take an epidemiological example, the odds of a smoker contracting lung cancer is calculated by dividing the number of smokers with lung cancer by the number of smokers without lung cancer (a/b in a 2×2 table). The probability of lung cancer for smokers would be calculated by dividing the number of smokers with lung cancer by the total number of smokers (a/(a + b)).

Because both odds and probability are based on the same information, you can convert one to the other using the following formulas:

Odds = probability/(1 – probability), Probability = odds (1 + odds)

Suppose $P(A) = 0.5$ or 50%. The odds of A are therefore $0.5/1 - 0.5 = 1.0$. This should make intuitive sense: 50% probability means an even chance of an event happening or not happening, whereas odds of 1.0 also means an even chance of the event happening. Working in reverse, if odds = 1.0, probability = 1.0/(1.0+1.0) = 0.5.

The odds ratio is simply the ratio of two odds, for instance, the odds of lung cancer for smokers and the odds of lung cancer for nonsmokers (mathematically equal to the odds of smoking for those with lung cancer compared to the odds of smoking for those without lung cancer). The odds ratio can be computed using probabilities with this formula (where $odds_1$ and $odds_2$ are the odds of the outcome under the two conditions, and p_1 and p_2 are the probabilities of the outcome under the two conditions), as shown in Figure 15-34.

$$OR = \frac{odds_1}{odds_2} = \frac{p_1(1 - p_1)}{p_2(1 - p_2)}$$

Figure 15-34. Calculating the odds ratio using probabilities

Confounding, Stratified Analysis, and the Mantel-Haenszel Common Odds Ratio

Confounding is a condition in which an observed statistical association is due at least in part to differences in the study groups other than the exposure of interest in the study. Confounding is sometimes described as the "third variable" problem; the relationship between two variables, say exposure and disease, is mixed up or confounded with the influence of a third variable related to both of them. More than one variable can be involved in confounding, but for the sake of simplicity, we demonstrate methods to deal with a single confounding variable.

Researchers in epidemiology need to be alert to the potential for confounding in their data, particularly in observational studies when group membership is not under the control of the investigator. For instance, studies of the effects of smoking on health have to take into account the fact that smoking is a voluntary behavior (people choose to smoke or not to smoke) and people who smoke can differ in many other ways (such as alcohol consumption, diet, or level of education) from those who do not.

If possible, it is preferable to control for confounding in the study design. *Randomization* is the method of choice for intervention studies because it theoretically controls for all potential confounders at once. This is because, on average, random assignment to groups should result in approximately the same distribution of any potential confounder in each group, including confounders of which the researcher is not aware.

Two other methods that may be used in observational studies to control for known or suspected confounders are restriction and matching. Both have the disadvantage of implementing control only over the confounders included in the design. With *restriction*, the researcher studies only a subset of the population, selected based on their values on the potential confounder. For instance, medical studies are sometimes done only on men or only on women to remove the influence of gender on the relationship between the exposure and disease. This has the disadvantage of restricting the applicability of study results; if a relationship between alcohol consumption and psychopathology is found in a group of men, that does not immediately justify generalizing the conclusions to women because women were not included in the study.

Matching is another technique that attempts to control for known confounders by a different method. Matching includes all levels of the confounders but controls enrollment in the study or assignment to groups so that the confounders will be equally distributed across the groups. Matching is commonly used in case-control studies in which controls are selected to match the cases already enrolled in the study. There are different systems for matching, but the basic concept is that categories are constructed for the confounding variables, and assignment to groups is controlled so that the distribution of the confounders is the same in each group.

There are two ways to implement matching. In *direct matching*, individuals are matched on a one-to-one basis. In *frequency matching*, assignment to the groups is directed or monitored so that equal numbers of the confounders are present in each group. If the confounders are gender and age category, in direct matching a woman of age 60–70 years in the treatment group would be matched by a woman of age 60–70 years in the control group. In frequency matching, the project manager would monitor enrollment to see that an equal number of females and persons in the different age categories were included in the treatment and control groups. Frequency matching is sometimes called matching on cell because you can think of the different combinations of characteristics as forming different cells (for instance, men aged 20–29, men aged 30–39, etc.), and you want the groups to have equal numbers in each cell. Frequency matching is particularly common in case-control subjects because often, all the cases are enrolled and then controls selected to match them. Because you know the distribution of the cases' characteristics, you can then select controls with the same distribution of characteristics.

If it is not possible to control for confounding in the research design, it can be dealt with during the analysis. There are numerous statistical methods to control for confounding after the fact, including multivariable methods that can become quite complex. However, confounding is often treated more simply in epidemiology and regression, particularly in studies focused on a single exposure and disease. This presentation demonstrates one of the most common methods to evaluate and control for confounding: computation and comparison of the crude and Mantel-Haenszel common odds ratio.

There is no implication of causality in classifying a variable as a confounder; in fact, many of the most common confounders are assumed to be only correlates of another factor. To qualify as a confounder, a variable must meet three requirements:

1. It must be related to the exposure.
2. It must be related to the disease independent of its association with the exposure.
3. It must not be wholly intermediate in the causal pathway between exposure and disease.

A fourth requirement, which is practical rather than theoretical, is that to function as a confounder in a particular study, a variable must be unequally distributed among groups in the study. For instance, we know that age could serve as a confounder for mortality, but if in a particular study the age distribution is the same among all groups studied, then age cannot act as a confounder in that particular study.

Let's take as an example a study of the protective effect of voluntary leisure-time physical activity (exposure) on the occurrence of heart attack (myocardial infarction, or MI, the disease); we believe this relationship may be confounded by age. All three requirements are met:

1. Age is related to physical activity. (On average, young people exercise more than older people.)
2. Age is a risk factor for MI independent of physical activity. (On average, older people are more likely to have an MI.)
3. Age is not wholly intermediate in the causal pathway between physical activity and MI. (There is no way physical activity could affect a person's age, which would then affect her probability of MI.)

One method to control confounding is the use of stratified analysis, in which the groups to be studied are divided into *strata* or subgroups based on values of the confounding variable. Stratification by age category is a common example. As discussed in the section earlier on standardized rates, the populations of different countries have different age structures; some countries have relatively more young people, others relatively more older people. Age is related to mortality and many types of morbidity. For these reasons, comparison of morbidity and mortality between populations is often accomplished by stratifying by age category and then standardizing so the age distribution is comparable in the populations being compared.

An example demonstrates the need to evaluate confounding. In 2007, mortality rate in the United States was 8.26 deaths per 1,000, whereas in Ecuador, it was 4.21 per 1,000. Should this be interpreted as evidence that Ecuadorans lead more salubrious lifestyles than Americans? That's an intriguing possibility but is not supported by examination of detailed life tables, which show that Ecuadorans have higher death rates than Americans in each specific age category. For instance, for the 45–49 age group, the probability of death for Americans is 0.00341; for Ecuadorans, it is 0.00513.

The difference in mortality is due to the differences in age structure between the two populations. Ecuador, like most developing countries, has a higher percentage of its population in younger age groups. The United States, like most industrialized countries, has a higher percentage of people in the older age categories, when the risk of mortality increases. This distinction would be missed if only crude mortality rates were considered but becomes clear when a stratified analysis removes the influence of the confounding variable (age) from the outcome (mortality).

There is no absolute test for confounding, but there are ways to examine the effects of potential confounders on the relationship of interest and make a reasoned decision about whether confounding is present. The general steps to follow in assessing confounding are as follows:

1. Calculate the crude measure of association, ignoring the confounding variable.
2. Stratify the study population by the confounding variable, that is, divide the population into smaller subgroups based on values of the confounding variable.
3. Calculate an adjusted measure of association.
4. Compare the crude and adjusted measures; a difference of 10% or more is generally considered evidence of confounding.

The appropriate measure of association depends on the study design; we will demonstrate stratified analysis using the crude odds ratio and the Mantel-Haenszel adjusted odds ratio. Note that to use the Mantel-Haenszel method, two assumptions must be met: the overall sample size must be large, and the association between exposure and outcome should be in the range of approximately 0.5 to 2.5.

The Mantel-Haenszel (MH) estimator of the common odds ratio for stratified data allows information to be combined from a series of two or more 2×2 tables, using the formula shown in Figure 15-35.

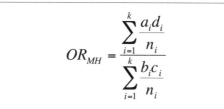

$$OR_{MH} = \frac{\sum\limits_{i=1}^{k} \frac{a_i d_i}{n_i}}{\sum\limits_{i=1}^{k} \frac{b_i c_i}{n_i}}$$

Figure 15-35. Formula for the Mantel-Haenszel odds ratio

In this formula, there are k individual tables,
i represents one of the tables (one strata of the population),
n is the sample size for that table,
and a_i, b_i, c_i, and d_i are the values of cells within that table.

Suppose we are interested in the relationship between smoking and liver disease. We know that people who smoke are also more likely to consume alcohol, alcohol consumption is an independent risk factor for liver disease, and alcohol consumption is not wholly intermediate in the hypothesized causal chain between smoking and liver disease. Alcohol consumption is therefore a potential confounder in this study, which we examine by stratifying our study population on alcohol consumption (as a dichotomy: those who drink alcohol versus those who don't) and examining the difference (if any) between the crude and adjusted odds ratios for our population.

The data looks like Table 15-13 before we consider the effect of alcohol consumption.

Table 15-13. Smoking/liver disease data before stratification

	D+	D−	Total
E+	50	100	150
E−	30	120	150
Total	800	220	300

The crude odds ratio for this data is shown in Figure 15-36.

$$OR = \frac{ad}{bc} = \frac{50(120)}{30(100)} = 2.00$$

Figure 15-36. Calculating the crude odds ratio

This is a strong positive OR and indicates that smoking is positively associated with liver disease: smokers are twice as likely to have liver disease as nonsmokers. To examine whether alcohol consumption is a confounding factor, we construct separate 2×2 tables for those who do and don't consume alcohol (Table 15-14 and Table 15-15).

Table 15-14. Smoking/liver disease for those who don't consume alcohol

	D+	D−	Total
E+	40	35	75
E−	30	45	75
Total	70	80	150

Table 15-15. Smoking/liver disease for those who do consume alcohol

	D+	D−	Total
E+	60	15	75
E−	50	25	75
Total	110	40	150

We can compute the MH common odds ratio for this data as shown in Figure 15-37.

$$OR_{MH} = \frac{\displaystyle\sum_{i=1}^{k} \frac{a_i d_i}{n_i}}{\displaystyle\sum_{i=1}^{k} \frac{b_i c_i}{n_i}} = \frac{(40 \times 45)/150}{(30 \times 35)/150} + \frac{(60 \times 25)/150}{(50 \times 15)/150}$$

Figure 15-37. Calculating the MH common odds ratio

Because this is more than 10% different from the crude odds ratio of 2.00, we conclude that alcohol consumption is a confounder in the relationship between smoking and liver disease and should be included as such in our analyses.

Power Analysis

This section deals with the theory of *power and sample size* and presents a few simple examples. Sample and power calculations are frequently simple, but they are also specific; every type of research design uses a different formula, and there's no point in listing them all when they are available in reference books. For those working in medicine and epidemiology, one particularly recommended source is the chapter on sample size calculation in the *Handbook of Epidemiology* (Springer). Many software packages, such as SAS and Minitab, include packaged routines to do power and sample size calculations, and there are various power and sample size calculators on the Web as well; a good collection of links to online calculators may be found here (*http://statpages.org*).

The practice of doing inferential statistics always includes the possibility of making a wrong decision because inferential statistics uses calculations on a sample to make conclusions about a population. As discussed in Chapter 3, there are two kinds of common errors in inferential statistics:

1. Type I error or α, when you incorrectly reject the null hypothesis
2. Type II error or β, when you fail to reject the null hypothesis when you should have rejected it

Another way to look at this is to say that Type I error is finding significance where none exists, while Type II error is failing to find significance when it does exist.

Power is $1 - \beta$ and is the probability of rejecting the null hypothesis when you should reject it. We'd all like to have high power all the time, but practical considerations, in particular the cost and availability of subjects, usually force us to compromise. A rule of thumb is that you should have at least 80% power, that is, 80% chance of finding significant results in your sample if they exist in the population. That means that 20% of the time, you won't find significance when you should. The standard of 90% power is regularly used as well.

The following four main factors affect power.

1. α level, that is, P(Type I error) (higher α increases power)
2. Difference in outcome between the populations (greater difference increases power)
3. Variability (reduced variability increases power)
4. Sample size (larger sample size increases power)

A change in any one of these factors while the others are held constant will change the power level for a given design. The α level is usually chosen to be 0.05 or less (for instance, 0.01); a larger value of α increases power. A greater difference in outcome between the populations increases power. Differences in outcome can be increased by improving the intervention so it has a stronger effect or by choosing study groups to increase the expected difference in outcomes between them. Reduced variability also increases power. Variability can sometimes be decreased by improving measurement or through selection of study subjects (such as restricting them to a particular age range or income level). However, our ability to exert control over these factors is usually minor in a given study.

We are then left with *sample size*, the one factor primarily under the control of the experimenter at the planning stages of his research project. All things being equal, more subjects = greater power. However, recruiting more subjects usually costs more money and requires more effort on the part of the research team. The goal of power analysis is to find a reasonable compromise in which you have acceptable power but are not going bankrupt or collecting more data than is necessary.

The concept of power, as well as of Type I and Type II error, may be clarified by considering Figure 15-38.

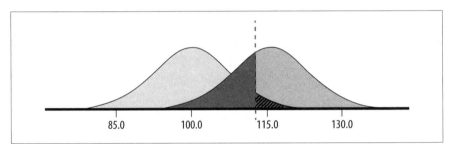

Figure 15-38. Power diagram for two normally distributed populations

Figure 15-38 illustrates aspects of a power calculation in which the null hypothesis is that the mean of the population is 100, whereas the alternative hypothesis is that the population mean is 115. Both populations are assumed to have an approximately normal distribution. In this figure, the leftmost (light gray) distribution is the null population, which represents the distribution of the population if the null hypothesis is true and the population mean is 100. The right-most (dark gray) distribution is the alternative population, which represents the data distribution if the alternative hypothesis is true and the population mean is 115.

Power calculations are always carried out with respect to a particular alternative hypothesis. In this case, the alternative is not simply that the population mean is greater than 100 but that the mean is 115. Note that hypothesis testing involves the location of population means, although the hypotheses are tested using means calculated from samples. For simplicity's sake in this example, both populations are assumed to have equal standard deviations of 15.

The hypothesis being tested is one-tailed, so a single cut point or critical value, represented by the dotted line, is established. If the sample mean is above this cut point, the null hypothesis will be rejected. If the sample mean is below this cut point, the null hypothesis will not be rejected. The location of the cut point, 112.5, was set with regard to the null population, which has a mean of 100 and a standard deviation of 15; it is the critical value for a significance test when $\alpha = 0.05$ because 95% of the null population lies to the left of 112.5 and 5% to the right.

The area of the null population above (to the right of) the cut point represents the P(Type I error) or the probability of rejecting the null hypothesis when it is true. In this example, P (Type I error) is 0.05.

The area of the alternative population below (to the left of) the cut point represents β or the P(Type II error) if the alternative hypothesis (population mean = 115) is true. This is the probability that if the true mean is 115, the sample value will be below the cut point of 112.5.

The area of the alternative population to the right of the cut point is the power of the test for this specific null hypothesis. This represents the probability that if the alternative hypothesis is true and the population mean is 115, the sample mean will be above the cut point of 112.5, and we will conclude that the population mean is significantly greater than 100.

Let's consider how each of the four factors cited could increase power in this example, assuming that only one factor can change at once:

1. If α were increased to 0.10, the cut point would be lower (farther to the left), and the power would increase, while P(Type II error) would decrease. The area below the cut point would decrease, representing a reduction in P(Type II error).

2. If the effect size were greater, for instance if the mean of the alternative population were 120 instead of 115, the distribution for the alternative population would be shifted up the number line. The result would be a decrease in P(Type II error) and an increase in power.

3. If the standard deviation were decreased, the two populations would have narrower distributions (more closely clustered around their mean) and thus would overlap less. This would result in a reduction in the probability of Type II error and an increase in power.

4. If sample size were increased, this would have a similar effect to decreasing standard deviation and would result in a reduction in the probability of Type II error and an increase in power.

One good way to become familiar with the influence of different factors on power is to experiment with a graphical power calculator; one example is the Statistical Power Applet created by Claremont Graduate University (*http://wise.cgu.edu/power_applet/power.asp*).

Sample Size Calculations

As mentioned before, each type of power or sample size calculation requires the appropriate formula to be used. However, if the principles of research design as well as power analysis are understood, finding the correct formula is not difficult. Two simple examples of sample size calculations are demonstrated here because they are a good illustration of the principles at work and are easily performed using only a hand calculator.

Confidence Interval for a Proportion

One common sample size problem is determining the sample size required to calculate a proportion with acceptable precision. For instance, you might be calculating agreement among different employees assigned to do medical chart reviews, and you want an estimate of the proportion in agreement, plus or minus five percentage points. Alternatively, you might be conducting a survey of the proportion of adults immunized against influenza in a population and want to estimate the proportion immunized plus or minus 10 percentage points. This is not a power calculation because no hypothesis is being tested, but it is a sample size calculation because you need to determine the minimum sample size required for a specified level of precision.

The formula used for a two-sided confidence interval is shown in Figure 15-39.

$$n = \left(\frac{Z_{1-\alpha/2}}{\omega}\right)^2 [\pi(1-\pi)]$$

Figure 15-39. Sample size formula for a two-sided confidence interval of specified precision for a proportion

In this formula, n is the required sample size,
π (Greek letter pi) is the hypothesized population proportion,
Z is the value from the standard normal distribution table corresponding to half the alpha level, and
ω (Greek letter omega) is the half-width of the desired confidence interval. The half-width is half the confidence interval. If we use a confidence interval of 10 percentage points, the half-width is 5 percentage points.

We want to calculate a two-sided confidence interval with $\alpha = 0.05$, so $Z = 1.96$. We believe π to be 0.8, and we want a confidence interval of 10 percentage points (0.10), so $\omega = 0.05$. Plugging these values into the equation gives us the result in Figure 15-40.

$$n = \left(\frac{1.96}{0.05}\right)^2 [0.8(0.2)] = 245.9$$

Figure 15-40. Calculating the sample size required for a confidence interval of specified precision for a proportion

We round this estimate up to 246 because there generally are no fractional subjects available! So we need 246 subjects, assuming our estimate of π is correct, to have an estimate with a 95% confidence interval of 0.10 (0.05 above and 0.05 below the estimate).

Power for the Test of the Difference between Two Sample Means (Independent Samples t-Test)

For an example of a simple power calculation, let's assume we want to calculate how many subjects per group we need to conduct a two-tailed independent samples t-test with acceptable power. The formula is given in Figure 15-41, where δ is effect size, calculated as shown in Figure 15-42.

$$n = \frac{2(Z_{1-\alpha/2} + Z_{1-\beta})^2}{\delta^2}$$

Figure 15-41. Sample size formula for an independent samples t-test

$$\delta = \frac{\mu_1 - \mu_2}{\sigma}$$

Figure 15-42. Effect size for an independent samples t-test

σ in this case is determined using whichever method of calculating the standard deviation for a t-test is appropriate for the data in question. (See Chapter 6 for details.) We need Z-values for both α and β to use this formula. We will stick with the 95% confidence interval for a two-tailed test used in the previous example, so the Z-value for $1 - \alpha/2$ will be 1.96. We will compute the sample size required for 80% power, so the Z-value for $1 - \beta$ will be 0.84. Note that if we were doing a one-tailed test, Z_α would be 1.645, and if we were calculating 90% power, $Z_{1-\beta}$ would be 1.28.

The effect size is the difference between the two populations divided by the appropriate measure of variance, as noted earlier. If $\mu_1 = 25$, $\mu_2 = 20$, and $\sigma = 10$, the effect size is 0.5. We can plug these numbers into the sample size formula as shown in Figure 15-43.

$$n = \frac{2(1.96 + 0.84)^2}{0.5^2} = 62.72$$

Figure 15-43. Calculating the sample size needed for an independent samples t-test

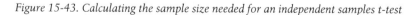

We round fractional results up to the next whole number, so we need at least 63 subjects per group to have an 80% probability of finding a significant difference between two groups when the effect size is 0.5.

How to Lie with Percentages

You can't work in statistics for very long before someone demonstrates her cleverness by quoting some form of the aphorism, attributed to the British politician Benjamin Disraeli and popularized in the United States by Mark Twain, that there are three kinds of lies: lies, damned lies, and statistics. There's even a popular book called *How to Lie with Statistics* by Darrell Huff (Norton), which is sometimes said to be the most-read statistics book in the world. One purpose of Huff's book, and this one as well, is not to teach you how to lie with statistics but to help you spot other people lying.

One of the easiest ways to lie (or mislead, if you prefer) with statistics is to quote percentages without reference to the raw numbers underlying them, a practice beloved of politicians but not exclusively practiced by them. For instance, if you heard that there was a 100% increase in cholera cases in the United States, you might find that cause for alarm, until you learned that the increase was from one case to two. Similarly, a 50% increase in cancer risk for some rare exposure (affecting, say, only 15 people nationally) might not have as much public health significance as a 5% increase of a common exposure (which might affect millions of people).

Another reason interpreting percentages can be confusing is that people often forget that percentage increases and decreases are not symmetrical. If you increase the number of college graduates by 10% one year, then decrease it by 10% the next year, you are not back to your original total. Say you have 100,000 college graduates to begin with. A 10% increase gives you 110,000. A 10% decrease of the new total gives you 99,000 (110,000 × 0.9), which is fewer than you started with.

Exercises

Here's a set of questions to help you review the topics covered in this chapter.

Problem

A classic example of the use of contingency tables in epidemiology is investigation of food poisoning outbreaks. When a number of people become ill after eating at a restaurant, the public health department will launch an investigation to identify the food or foods responsible. This effort is complicated by the fact that the people who got sick probably ate many foods, and some people who ate the same foods might not have gotten sick. One approach to sorting out this information is to interview the customers to ascertain what they ate and whether they got sick. The data is then arranged into a series of 2×2 tables, as in Tables 15-16 and 15-17, in which the exposure is the particular food in question and the disease is food poisoning. Calculate the risk ratios for the two foods listed and justify a decision about whether they are a likely cause of food poisoning.

Table 15-16. Contingency table for roast beef and food poisoning

	D+	D–
E+	15	85
E–	20	80

Table 15-17. Contingency table for chicken salad and food poisoning

	D+	D–
E+	80	20
E–	20	80

Solution

The RR for roast beef is calculated in Figure 15-44.

$$RR = \frac{a/(a+b)}{c/(c+d)} = \frac{15/100}{20/100} = 0.75$$

Figure 15-44. Calculating the risk ratio for roast beef and food poisoning

The RR for chicken salad is shown in Figure 15-45.

$$RR = \frac{a/(a+b)}{c/(c+d)} = \frac{80/100}{20/100} = 4.0$$

Figure 15-45. Calculating the risk ratio for chicken salad and food poisoning

Looking at just these two foods, it appears that the culprit is chicken salad because people who ate it had four times the risk of food poisoning compared to people who didn't eat it. Roast beef seems to have a slightly protective effect, perhaps because people who ate the roast beef were less likely to eat the chicken salad; people who ate roast beef had only three-quarters of the risk of food poisoning as compared to people who did not eat roast beef.

Problem

Compute the odds ratio and confidence interval for the data shown in Table 15-18 from a case-control study of oral contraceptive use and breast cancer. Do the data show a significant relationship between the two?

Table 15-18. Contingency table for oral contraceptive use and breast cancer

	D+	D–
E+	30	70
E–	20	80

Solution

The odds ratio is shown in Figure 15-46.

$$OR = \frac{ad}{bc} = \frac{30(80)}{20(70)} = 1.71$$

Figure 15-46. Calculating the odds ratio for oral contraceptive use and breast cancer

To see whether this is significantly different from the null value of 1.0, compute the 95% confidence interval as shown in Figure 15-47.

$$CI = \frac{ad}{bc}\exp\left(\pm Z\sqrt{\frac{1}{a}+\frac{1}{b}+\frac{1}{c}+\frac{1}{b}}\right)$$

$$= 1.71\exp\left(\pm 1.96\sqrt{\frac{1}{30}+\frac{1}{70}+\frac{1}{20}+\frac{1}{80}}\right)$$

$$(0.89, 3.28)$$

Figure 15-47. Calculating the 95% CI for the odds ratio for oral contraceptive use and breast cancer

The CI of (.89, 3.28) includes the null value of 1.0, so we conclude that this study does not demonstrate a significant relationship between oral contraceptive use and breast cancer.

Problem

Calculate and interpret the attributable risk, attributable risk percentage, and number needed to treat, given the following information:

Incidence of disease in exposed = 0.05
Incidence of disease in the unexposed = 0.02

Solution

The calculations required are shown in Figure 15-48.

$$AR = I_e - I_0 = 0.05 - 0.03 = 0.02$$

$$AR\% = \frac{0.02}{0.05} \times 100 = 0.40$$

$$NNT = \frac{1}{0.02} = 50$$

Figure 15-48. Calculating attributable risk, attributable risk percentage, and number needed to treat

The excess occurrence of disease due to the exposure is 0.02, or 20 per 1,000. Forty percent of disease among the exposed is attributable to the exposure, and it would be necessary to prevent 50 exposures to prevent one new case of disease in the population.

Problem

Calculate the sample size needed to estimate a proportion with a 95% confidence interval of plus or minus 10 percentage points when the hypothesized proportion is 0.70.

Solution

Use the sample size formula for a proportion, and plug in the numbers:

$Z_{1-\alpha/2} = 1.96$ $\omega = 0.10$ $\pi = 0.70$

These calculations (Figure 15-49) show that you need a sample size of 81.

$$n = \frac{\left(Z_{1-\alpha/2}\right)^2}{\omega}[\pi(1-\pi)] = \left(\frac{1.96}{0.10}\right)^2 [0.70(0.30)] = 80.7$$

Figure 15-49. Calculating the sample size needed to estimate a sample proportion

Problem

Calculate the sample size needed when testing for the difference in means, using an independent samples t-test with a one-tailed hypothesis, 90% power, and an effect size of 0.4.

Solution

Use the sample size formula presented, and plug in the numbers:

$Z_\alpha = 1.645$ $Z_{1-\beta} = 1.28$ $\delta = 0.4$

These calculations (Figure 15-50) show that you need 107 subjects per group.

$$n = \frac{2\left(Z_\alpha + Z_{1-\beta}\right)^2}{\delta^2} = \frac{2(1.645 + 1.28)^2}{0.16} = 106.9$$

Figure 15-50. Calculating the sample size needed for an independent samples t-test

16

Educational and Psychological Statistics

Many statistical techniques used in education and psychology are common to other fields of endeavor: these include the *t*-test (covered in Chapter 6), various regression and ANOVA models (covered in Chapters 8 through 11), and the chi-square test (covered in Chapter 5). The discussion of measurement in Chapter 1 will also prove useful because much of educational and psychological research involves constructs that cannot be observed directly and have no obvious units of measurement. Examples of such constructs include mechanical aptitude, self-efficacy, and resistance to change. This chapter concentrates on statistical procedures used in the field of *psychometrics*, which is concerned with the creation, validation, and use of tests and measurements applied to human intelligence, knowledge, abilities, and psychological characteristics such as personality traits.

The first question you may ask with regard to the use of statistics in education and psychology is why they are necessary at all. After all, isn't every person an individual, and isn't the point of both education and psychology to perceive each person in all his individual richness, not to reduce the individual to a set of numbers or place him in comparison with others?

This is a valid concern and underscores what anyone working in the human sciences knows already: doing research on human beings is in many ways much more difficult than doing research in the hard sciences or in manufacturing because people are infinitely more varied than chemical molecules or lug nuts. The diversity and individuality of people makes research in those fields particularly difficult. It's also true that although some educational and psychological research is aimed toward making general statements about groups of people, a great deal of it is focused on understanding and helping individuals, each of whom has her own specific social circumstances, family histories, and other contextual complexities, making direct comparisons between one person and another very difficult.

However, standard statistical procedures can be useful even in the most specific and individual therapeutic circumstances, such as when the goal of an encounter is to devise an appropriate educational plan for one student or therapeutic regimen for one patient. Making such decisions is difficult but would be even more difficult without the aid of formal educational and psychological tests that yield numeric values and can be compared to scores for other individuals. No one would suggest that only formal, standardized tests and questionnaires be used in these contexts; interviews and observational testing play an important role in educational and psychological evaluations as well. But the advantages of including formal testing procedures and standardized tests in clinical and educational evaluations include the following considerations:

1. Objective comparisons are facilitated by the use of a normative group. For instance, is this patient, recovering from trauma, experiencing more side effects than is common among others who have experienced the same injury? Are the reading skills of this pupil comparable to others of his age and grade level?

2. Standardized testing can yield results quickly; you needn't wait for the end of the school term to discover which pupils are struggling because of poor language proficiency, and you don't need a lengthy interview or practical examination to discover that a patient is suffering from serious memory deficits.

3. Standardized tests are presented in a regulated situation and under specified conditions and can be scored objectively, so the only issue being evaluated is the student's or patient's performance, not her appearance, sociability (unless that is germane to the context), or other irrelevant factors.

4. Most standardized tests do not require great skill to administer (unlike clinical interviews, for example) and can be given to groups of people at once, making the tests particularly useful as screening procedures.

Percentiles

In many countries, school-age children are evaluated by tests that report their results in *percentiles*, also known as *percentile ranks*; one student might score in the 70th percentile in reading and the 85th percentile in math, but another scores in the 80th percentile in reading and the 95th percentile in math. Percentiles are a form of *norm-referenced* scoring, so called because an individual score is placed in the context of a *norm group*, meaning people similar to the test-taker. For school-age children, the norm group is often other children in the same grade within their country. Norm-referenced scoring is used in all kinds of testing situations in which an individual's rank in relation to some comparison group is more important than his absolute score.

The percentile rank of an individual score refers to the percentage in the norm group that scored lower than that individual score, so a percentile score of 90 indicates that 90% of the norm group scored lower. Here's a brief example illustrating how to find percentile ranks for scores on an exam that was given to 100 students. (On national exams, the norm group would be much larger and the scores would reflect a greater range, but this example will illustrate the point.)

The first step in translating from raw scores to percentiles is to create a frequency table that includes a column for cumulative percentage, as illustrated in Table 16-1. To find the percentile rank for a particular score, use the cumulative percentage from the next-highest score, the row just above in the table. In this example, someone who scored 96 on the exam was in the 75th percentile rank (meaning 75% of the test-takers scored below 96), whereas someone who scored 85 was in the 25th percentile rank. There can be no 100th percentile rank because, logically speaking, 100% of the test-takers couldn't have scored below a score that is included in the table. You can have a 0th percentile, however; a person who scored 53 would be in the 0th percentile because no one achieved a lower score.

Table 16-1. Scores of 100 students on an exam

Score	Frequency	Percentage	Cumulative percentage
53	1	1.0%	1.0%
55	2	2.0%	3.0%
58	1	1.0%	4.0%
61	2	2.0%	6.0%
65	3	3.0%	9.0%
67	1	1.0%	10.0%
70	2	2.0%	12.0%
71	3	3.0%	15.0%
78	2	2.0%	17.0%
80	4	4.0%	21.0%
82	2	2.0%	23.0%
84	2	2.0%	25.0%
85	5	5.0%	30.0%
86	4	4.0%	34.0%
88	3	3.0%	37.0%
90	5	5.0%	42.0%
91	7	7.0%	49.0%
92	8	8.0%	57.0%
93	7	7.0%	64.0%
94	5	5.0%	69.0%
95	6	6.0%	75.0%
96	4	4.0%	79.0%
97	3	3.0%	82.0%
98	7	7.0%	89.0%
99	6	6.0%	95.0%
100	5	5.0%	100.0%

In situations such as standardized testing at the national level, the norm group used to map the scores to percentiles is much larger, and generally, calculation of percentiles for individual students is not necessary. Instead, the test manufacturer usually provides a chart that relates raw scores to percentile ranks.

Standardized Scores

The *standardized score*, also known as the *normal score* or the *Z-score*, transforms a raw score into units of standard deviation above or below the mean. This translates the scores so they can be evaluated in reference to the standard normal distribution, which is discussed in detail in Chapter 3. Standardized scores are frequently used in education and psychology because they place a score in the context of other scores and can therefore be considered a type of *norm-referenced* scoring. For frequently used scales such as the Wechsler Adult Intelligence Scale (WAIS), population means and standard deviations are known and may be used in the calculations; for the WAIS, the mean is 100, and the standard deviation is 15. To convert a raw score to a standardized score, use the formula shown in Figure 16-1.

$$Z = \frac{X - \mu}{\sigma}$$

Figure 16-1. Formula for the Z-score

In this formula, X is the raw score,
μ is the population mean,
and σ is the population standard deviation.

The conversion to Z-scores puts all scores on a common scale, that of the standard normal distribution with a mean of 0 and a variance of 1. In addition, Z-score probabilities are distributed with the known properties of the normal distribution. (For instance, about 66% of the scores will be within one standard deviation of the mean.) We can convert a raw score of 115 on the WAIS to a Z-score as shown in Figure 16-2.

$$Z = \frac{115 - 100}{15} = 1.00$$

Figure 16-2. Computing a Z-score

Using the table for the standard normal distribution (Z distribution) in Figure D-3 from Appendix D, we see that a Z-score of 1.00 means that 84.1% of individuals score at or below that individual's raw score. Standardized scores are particularly useful when comparing scores on tests with different scales. For example, let's say we also administer a test of mathematical aptitude that has a mean of 50 and a standard deviation of 5. If a person scores 105 on the WAIS (Figure 16-3) and 60 on the mechanical aptitude (Figure 16-4), we can easily compare those scores in terms of Z-scores.

$$Z = \frac{105 - 100}{15} = 0.33$$

Figure 16-3. Computing a Z-score (WAIS)

$$Z = \frac{60 - 50}{5} = 2.00$$

Figure 16-4. Computing a Z-score (mechanical aptitude test)

The Z-scores tell us that this person scored slightly above average in intelligence but far above average in mechanical aptitude.

Some people find standardized scores confusing, particularly because a person can have a Z-score that is 0 or negative (and in the standard normal distribution, half the scores are below average and therefore negative). For this reason, Z-scores are sometimes converted to *T-scores*, which use a more intuitive scale, with a mean of 50 and a standard deviation of 10. Z-scores may be converted to *T*-scores by using the following formula:

$T = Z(10) + 50$

If a person has a Z-score of 2.0 (meaning he or she scored two standard deviations above the mean), this can be converted to a *T*-score as follows:

$T = (2.0 \times 10) + 50 = 70.$

Similarly, someone with a Z-score of –2.0 would have a *T*-score of 30. Because hardly anyone ever scores five standard deviations or more below the mean, *T*-scores are almost always positive, which makes them easier for many people to understand. For instance, the clinical scales of the Minnesota Multiphase Personality Inventory-II (MMPI-II), commonly used to identify and evaluate psychiatric conditions, are reported as *T*-scores.

Stanines offer another method to translate raw scores into scores based on the standard normal distribution. The term "stanine" is an abbreviation of "standard nine" and refers to the fact that stanines divide scores into nine categories (1–9), each category or band representing half a standard deviation of the standard normal distribution. The mean of the stanine scale is 5, and this category includes scores that translate to standard scores from –0.25 to 0.25 (one quarter of a standard deviation below or above the mean). The primary advantage of using stanines instead of Z- or *T*-scores is that reporting categories rather than specific scores might get around the human tendency to obsess over small differences in reported scores.

Because scores near the central value of the normal distribution are more common than extreme scores, stanines near the central value of 5 are more common than scores near the extremes of 1 or 9. Note also that the distribution of stanine scores is symmetrical, as is the distribution of scores in the standard normal distribution, so a stanine of 1 is as common as a stanine of 9, a stanine of 2 is as common as a stanine of 8, and so on. The application of these two principles can be seen in Table 16-2, which shows the stanine score, the corresponding standard (Z) scores, and the percentage of the distribution contained within each stanine category.

Table 16-2. Stanines

Stanine	Z-score range	Percent of total
1	$Z < -1.75$	4%
2	$-1.75 < Z <= -1.25$	7%
3	$-1.25 < Z <= -0.75$	12%
4	$-0.75 < Z <= -0.25$	17%
5	$-0.25 < Z <= 0.25$	20%
6	$0.25 < Z <= 0.75$	17%
7	$0.75 < Z <= 1.25$	12%
8	$1.25 < Z <= 1.75$	7%
9	$Z > 1.75$	4%

Stanines may be calculated from Z-scores by using the following formula:

$$Stanine = (2 \times Z) + 5$$

Stanines are rounded to the nearest whole number; values with a decimal of 0.5 are rounded down. Suppose we have a Z-score of –1.60. This translates to a stanine of 2, as shown below:

$$Stanine = (2 \times -1.60) + 5 = 1.8$$

The nearest whole number is 2, and this corresponds to the stanine value for a Z-score of –1.60 given in Table 16-2.

A Z-score of 1.60 translates to a stanine of 8 because:

$$Stanine = 2(1.60) + 5 = 8.2.$$

The nearest whole number is 8, and this stanine value corresponds with the value indicated in Table 16-2 for a Z-score of 1.60.

Test Construction

Most tests in psychology and education are used for what is called *subject-centered measurement*, in which the purpose is to place individuals on a continuum with respect to particular characteristics such as language-learning ability or anxiety. Creating and validating a test is a huge amount of work. (When I was in graduate school, students were barred from writing a dissertation that required creating and validating a new test out of fear that they would never complete the process.) The burden is entirely on the test's creator to convince others working in the same field that the test scores are meaningful. Therefore, the first move for someone beginning to investigate a field is to check whether any existing, validated tests would be adequate. However, particularly if you are researching a new topic or dealing with a previously ignored population, no existing test might be adequate to your purpose, in which case the only option is to create and validate a new test.

Tests can be either *norm-referenced* or *criterion-referenced*. Norm-referenced tests have already been discussed; their purpose is to place an individual in the context of some group. In contrast, the purpose of a criterion-referenced test is to compare an individual to some absolute standard, say, to see whether he has obtained a defined minimum competency in an academic subject. In a criterion-referenced test, everyone taking the test could receive a high score, or everyone could receive a low score, because the individuals are evaluated with reference to a predetermined standard rather than in reference to each other. Although criterion-referenced tests can yield a continuous outcome (for instance, a score on a scale of 1–100), a *cut point* (single score) is often established as well so that everyone who achieves that score or above passes, and everyone with a score below it fails.

Most tests are composed of numerous individual *items*, often written questions, which are combined (often simply added together) to produce a *composite* test score. For instance, a test of language ability might be constructed of 100 items with each correct item scored as a 1 and each incorrect item as a 0. The composite score for an individual could then be determined by adding up the number of correct items. Many of the statistical procedures used in examining tests have to do with the relationship among individual items and the relation between individual items and the composite score.

Although composite test scores are commonly used, they can be misleading measures of ability or achievement. One difficulty is that typically all items are assigned the same weight toward the total score, although they might not all be of equal difficulty. The distinction between someone who misses some easy questions but gets more difficult questions correct versus someone who gets the easy questions correct but can't answer the difficult questions is lost when a composite score is formed by simply summing the scores of items of differing homogeneity.

The mean and variance of dichotomous items (those scored as either right or wrong) are calculated using the value for *item difficulty*, signified as p. Item difficulty is the proportion of examinees who answer a question correctly. If N people are in the group of examinees used to establish item difficulty, p is calculated for one item (j) as shown in Figure 16-5.

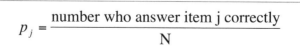

$$p_j = \frac{\text{number who answer item j correctly}}{N}$$

Figure 16-5. Formula for item difficulty

With dichotomous items scored 0 or 1 (0 for incorrect, 1 for correct), the mean is the same as the proportion answering the item correctly (Figure 16-6).

$$p_j = \mu_j = \frac{\sum_{j=1}^{n} X_j}{N}$$

Figure 16-6. Formula for item difficulty for dichotomous items

In this formula, X_j are the individual items,
and N is the number of examinees

Variance for an individual dichotomous item p_j may be calculated as shown in Figure 16-7.

$$\sigma_j^2 = p_j(1 - p_j)$$

Figure 16-7. Formula for dichotomous item variance

The correlation coefficient between two dichotomous items, also called the *phi coefficient*, is discussed in Chapter 5.

Computing the variance of a composite score requires knowing both the individual item variances and their covariances. Unless all pairs of variables are completely uncorrelated or are negatively correlated, the variance of a composite score will always be greater than the sum of the individual item variances. Although composite variance is usually computed using statistical software, the formula is useful to know because it outlines the relationship among the relevant quantities. The covariance for a pair of items j and k (whether the items are dichotomous or continuous) may be computed as shown in Figure 16-8.

$$\sigma_{jk} = \rho_{jk}\sigma_j\sigma_k$$

Figure 16-8. Formula for item covariance for a pair of items

In this formula, σ_{jk} is the covariance of the two items,
ρ_{jk} is the correlation between the two items,
and σ_j and σ_k are the individual item variances.

Often, we are interested in the variance of a composite such as a test score Y consisting of numerous items. Because there are two covariance pairs for each item pair (the covariance of j with k, and the covariance of k with j, which are identical), the covariance of a composite Y may be calculated as shown in Figure 16-9.

$$\sigma_Y^2 = \sum \sigma_i^2 + 2\sum_{i<j} \rho\sigma_i\sigma_j$$

Figure 16-9. Formula for covariance of a composite

The stipulation $i < j$ in the preceding formula stipulates that we compute only unique covariance terms. To get the right number of covariance terms, we then multiply each unique covariance by 2.

As items are added to a test, the number of covariance terms increases more quickly than the number of variance terms. For instance, if we add 5 items to a test that has 5 items to start with, the number of variance terms increases from 5 to 10, but the number of covariance items increases from 20 to 90. The number of unique covariance terms for n items is calculated as $n(n-1)$; therefore, a test with 5 items has $5(4) = 20$ covariance terms. A 10-item test would have $10(9) = 90$ covariance terms. The number of *unique* covariance terms is $[n (n-1)]/2$, so 5 items yield 10 unique covariance terms, and 10 items yield 45 unique covariance terms.

In most cases, adding items to a composite increases the variance of the composite because the variance of the composite is increased by the variance of the individual item plus its covariance with all the existing items on the test. The proportional increase is greater when items are added to a short test than to a long test and is greatest when items are highly correlated because that results in larger covariances among items. All else being equal, the greatest composite variance is produced by items of medium difficulty ($p = 0.5$ produces the largest covariance scores) that are highly correlated with one another.

Classical Test Theory: The True Score Model

In an ideal world, all tests would have perfect reliability, meaning that if the same individuals were tested repeatedly under the same conditions for some stable characteristic, they would receive identical scores each time, and there would be no systematic error (defined later) in the score. In this case, we would have no problem saying that a person's *observed score* on the test was the same as the person's *true score* and that the observed score was an accurate reflection of that person's score on whatever the test was designed to measure. In the real world, however, many factors can influence observed scores, and repeated tests on the same material taken by the same individual often yield different scores. For this reason, we must differentiate between the true score and the observed score. We do this by introducing the concept of *measurement error*, which is the part of the observed score that causes it to deviate from the true score.

Measurement error can be either random or systematic. *Random measurement error* is the result of chance circumstances such as room temperature, variance in administrative procedure, or fluctuation in the individual's mood or alertness. We do not expect random error to affect an individual's score consistently in one direction or the other. Random error makes measurement less precise but does not systematically bias results because it can be expected to have a positive effect on one occasion and a negative effect on another, thus canceling itself out over the long run. Because there are so many potential sources for random error, we have no expectation that it can be completely eliminated, but we desire to reduce it as much as possible to increase the precision of our measurements. *Systematic measurement error*, on the other hand, is error that consistently affects an individual's score in one direction but has nothing to do with the construct being tested. An example would be measurement error on a mathematics exam that is caused by poor language skills so that the examinee cannot read the directions to take the exam properly. Systematic measurement error is a source of bias and should be eliminated from testing whenever possible.

The psychologist Charles Spearman introduced the classic concepts of true and error scores in the early twentieth century. Spearman described the observed score X (the score actually received by an individual on a testing occasion), which is composed of a true component (T) and a random error component (E):

$$X = T + E$$

Over an infinite number of testing occasions, the random error component is assumed to cancel itself out, so the mean or *expected value* of the observed scores is the same as the true score. For individual j, this can be written as:

$$T_j = E(X_j) = \mu_{Xj}$$

where T_j is the true score for individual j, $E(X_j)$ is that individual's expected observed score over an infinite number of testing occasions, and μ_{Xj} is the mean observed score for that individual over the same occasions. Error is therefore the difference between an individual's observed score and her true score:

$$E_j = X_j - T_j$$

Over an infinite number of testing occasions, the expected value of the error for one individual is 0. Because "error" in this definition means random error only, true and error scores are assumed to have the following properties:

- Over a population of examinees, the mean of the error scores is 0.
- Over a population of examinees, the correlation between true and error scores is 0.
- The correlation between error scores by two randomly chosen examinees on two forms of the same test, or two testing occasions using the same form, is 0.

Reliability of a Composite Test

When we administer a test to an individual, one of our concerns is how well the observed score on that test represents the person's true score. In theoretical terms, what we seek is the *reliability index* for the test, which is the ratio of the standard deviation of the true scores to the standard deviation of the observed scores. The reliability index is calculated as shown in Figure 16-10.

$$\rho_{XT} = \frac{\sigma_T}{\sigma_X}$$

Figure 16-10. Formula for the reliability index

In this formula, σ_T is the standard deviation of the true scores for a population of examinees,
and σ_X is the standard deviation of their observed scores.

The reliability of a test is sometimes described as the proportion of total variation on the test scores that is explained by true variation (as opposed to error).

In practice, true scores are unknown, so the reliability index must be estimated using observed scores. One way to do this is to administer two parallel tests to the same group of examinees and use the correlation between their scores on the two forms, known as the *reliability coefficient*, as an estimate of the reliability index. Parallel tests must satisfy two conditions: *equal difficulty* and *equal variance*.

The reliability coefficient is an estimate of the ratio of true score variance to observed score variance and can be interpreted similarly to the coefficient of determination (r^2) in the general linear model. If a test reports a reliability coefficient of 0.88, we can interpret this as meaning that 88% of the observed score variance from administrations of this test is due to true score variance, whereas the remaining 0.12 or 12% is due to random error. To find the correlation between true and observed scores for this test, we take the square root of the reliability coefficient, so for this test, the correlation between true and observed scores is estimated as $\sqrt{0.88}$, or 0.938.

The reliability coefficient can be estimated using one of several methods. If we estimate the reliability coefficient by administering the same test to the same examinees on two occasions, this is called the *test-retest method*, and the correlation between test scores in this case is known as the *coefficient of stability*. We could also estimate the reliability coefficient by administering two equivalent forms of a test to the same examinees on the same occasion; this is the *alternate form method*, and the correlation between scores is the *coefficient of equivalence*. If both different forms and different occasions of testing are used, correlation between the scores under these conditions is called the *coefficient of stability and equivalence*. Because this coefficient has two sources of error, *forms* and *occasions*, it is generally expected to be lower than either the coefficient of stability or the coefficient of equivalence would be for a given group of examinees.

Measures of Internal Consistency

A different approach to estimating reliability is to use a measure of internal consistency that can be calculated from a single administration of a test to a single group of examinees. Consistency measurements are used to estimate reliability because a composite test is often conceived of as being composed of test items sampled from a large domain of potential items. An internal consistency estimate is a prediction of how similar an individual's score would be if a different subset of items from that domain had been chosen.

Consider the task of creating an exam to test student competence in high school algebra. The first steps in creating this test would be to decide what topics to cover. Then a pool of items would be written that evaluate student mastery of those topics. A subset of items would then be chosen to create the final test. The purpose of this type of exam is not merely to see how well the students score on the specific items included in the test they took but how well they mastered all the content considered to be within the domain of high school algebra. If the items used on the test are a fair selection from this content domain, the test score should be a reliable indicator of the students' mastery of the material. Item homogeneity is also a valued characteristic of this type of test because it is an indication that the items are testing the same content and do not have technical flaws such as misleading wording or incorrect scoring that would cause performance on an item to be unrelated to mastery of algebra.

Teaching to the Test

In some educational contexts, students are required to take a series of what have become known as high-stakes tests that are used to determine whether they may be allowed to progress through the school system (for instance, to move from 5th grade to 6th grade) or to graduate (for instance, from high school). Because administrators and teachers are understandably concerned for their students to do well on these exams, some schools allot part of the school day specifically to preparation for the exam. (Besides their concern for the students' educational progress, teachers and administrators can also be evaluated based on how well their students perform on the high-stakes exams.) When instruction is aimed toward improving performance on a specific exam rather than on increasing subject skill and knowledge, this is often referred to as teaching to the test. For instance, students might devote time to solving items in the exact format used by the upcoming test or confine their study to the known range of problems or information that will be covered on an upcoming test instead of studying a broad range of content and practicing applying their skills in many ways.

What is wrong with teaching to the test? The problem is that achievement tests are generally based on the assumption that items on a specific test represent a fair sample of its domain and that performance on the sample of items included in a particular test is a good indication of mastery of the domain as a whole. Under this assumption, if a different sample of items were selected, student performance should be similar. This assumption does not hold up if the students and teachers know in advance what the sample will be and prepare only for that sample; in that case, it is impossible to generalize from the performance on the sample to the mastery of the domain as a whole.

Suppose students are studying for a test to determine their competency in high school algebra. One of the topics included is geometric proofs; students should be able to construct a two-column proof demonstrating why a known algebraic theorem is true. If students are taught a general method for writing proofs, their knowledge should be equally applicable to any proof questions on the test, so their performance on the test should be a good indication of their general achievement in this aspect of algebra. However, if their instructor determines that only a few types of proofs appear on the exams from year to year, he could simply have the students memorize how to do those particular proofs. This is an example of teaching to the test. In this case, the students' ability to construct the types of proofs they have memorized bears no necessary relationship to their ability to construct other types of proofs. Thus, it is impossible to generalize from their performance on the test to their mastery of the domain of geometric proofs as a whole.

Split-Half Methods

Split-half methods to measure internal consistency require a test to be split into two parts or forms, usually two halves of equal length, which are intended to be parallel. All items on the full-length test are completed by each examinee. The split can be achieved by several methods, including alternate assignment (even-numbered items to one form, odd-numbered to the other), content matching, or random assignment. Whichever method is used, if the original test had 100 items, the two halves will each have 50 items. The correlation coefficient between examinee scores for the two forms is called the *coefficient of equivalence*. The coefficient of equivalence is an underestimate of the reliability for the full-length test because longer tests are usually more reliable than shorter tests. The *Spearman-Brown prophecy formula* can be used to estimate the reliability of the full-length test from the coefficient of equivalence for the two halves, using the formula shown in Figure 16-11.

$$\hat{\rho}_{XX'} = \frac{2\rho_{AB}}{1 + \rho_{AB}}$$

Figure 16-11. Spearman-Brown prophecy formula (for the coefficient of equivalence)

In this formula, $\rho_{XX'}$ is the estimated reliability of the full-length test, and ρ_{AB} is the observed correlation, that is, coefficient of equivalence, between the two half-tests.

For this formula to be accurate, the two half-tests must be strictly parallel. If the coefficient of equivalence for the two half-tests is 0.5, the estimated reliability of the full-length test is shown in Figure 16-12.

$$\hat{\rho}_{XX'} = \frac{2(0.5)}{1 + 0.5} = 0.67$$

Figure 16-12. Calculating the coefficient of equivalence

A second method to estimate reliability of a full-length test using the split-half method is to calculate the difference between scores on the two halves for each examinee. The variance of that difference score is an estimate of error variance of reliability, so the 1 minus the ratio of error variance to total variance may also be used as an estimate of reliability. Figure 16-13 presents the formula to use for the second method.

$$\hat{\rho}_{XX'} = 1 - \frac{\sigma_D^2}{\sigma_X^2}$$

Figure 16-13. An alternative formula for the coefficient of equivalence

In this formula, σ^2_D is the variance of the difference scores, and σ^2_X is the variance of the observed scores.

Estimates of reliability using either method will be identical when the variance of the two half-tests is identical. The more dissimilar the two variances, the larger the estimate using the Spearman-Brown formula will be relative to estimates using the difference-score method. Estimation of reliability by either method depends on how the items are chosen for the two halves because a different split will result in different correlations between the halves and a different set of difference scores.

Coefficient Alpha

Several methods of estimating reliability using item covariances avoid the problem of multiple split-half reliabilities; three of these methods follow. *Cronbach's alpha* may be used for either dichotomous or continuously scored items, whereas the two *Kuder-Richardson* formulas are used only for dichotomous items. The measure of internal consistency computed by any of these methods is commonly referred to as *coefficient alpha* and is equivalent to the mean of all possible split-half coefficients computed using the difference-score method. Coefficient alpha is, strictly speaking, not an estimate of the reliability coefficient but of its lower bound (sometimes called the *coefficient of precision*). This nicety is often ignored in interpretation, however, and coefficient alpha is usually reported without further interpretation.

Note that computing coefficient alpha for a test of any considerable length is tedious and therefore generally accomplished using computer software. Still, it is useful to know the formulas and work through a simple calculation to understand what factors affect coefficient alpha.

Cronbach's alpha is the most common method for calculating coefficient alpha and is the name often given for coefficient alpha in computer software packages designed for reliability analysis. It is computed using the formula shown in Figure 16-14.

$$\hat{\alpha} = \frac{k}{k-1}\left(1 - \frac{\sum \hat{\sigma}_i^2}{\hat{\sigma}_X^2}\right)$$

Figure 16-14. Formula for Cronbach's alpha

In this formula, k is the number of items,
$\hat{\sigma}_i^2$ is the variance of item i, and
$\hat{\sigma}_x^2$ is the total test variance.

Suppose we have a 5-item test, with a total test variance of 100 and individual item variances of 10, 5, 6.5, 7.5, and 13. Cronbach's alpha for this data set is shown in Figure 16-15.

$$\hat{\alpha} = \frac{5}{5-1}\left(1 - \frac{42}{100}\right) = 0.725$$

Figure 16-15. Calculating Cronbach's alpha

There are several Kuder-Richardson formulas to calculate coefficient alpha; two useful for dichotomous items are presented here. Note that KR-21 is a simplified version of the KR-20 formula; it assumes all items are of equal difficulty. KR-20 and KR-21 yield identical results if all items are of equal difficulty; if they are not, KR-21 yields lower results than KR-20. The KR-20 formula is shown in Figure 16-16.

$$KR_{20} = \frac{k}{k-1}\left(1 - \frac{\sum p_i(1 - p_i)}{\hat{\sigma}_X^2}\right)$$

Figure 16-16. Formula for KR-20

In this formula, k is the number of items,
p_i is the difficulty for a given item, and
$\hat{\sigma}_x^2$ is the total variance.

Note that the KR-20 formula is identical to the Cronbach's alpha formula with the exception that the item variance term has been restated to take advantage of the fact that KR-20 is used for dichotomous items.

The KR 20 formula can be simplified by assuming all items have equal difficulty, so it is not necessary to compute and sum the individual item variances. This simplification yields the KR 21 formula (Figure 16-17).

$$KR_{21} = \frac{k}{k-1}\left(1 - \frac{\hat{\mu}(k - \hat{\mu})}{k\hat{\sigma}_X^2}\right)$$

Figure 16-17. Formula for KR-21

In this formula, k is the number of items,
$\hat{\mu}$ is the overall mean for the test (usually estimated by $\bar{X}$), and
$\hat{\sigma}_x^2$ is the total variance for the test (usually estimated by s_x^2).

Item Analysis

Test construction often proceeds by creating a large pool of items, pilot-testing them on examinees similar to those for whom the test is intended, and selecting a subset for the final test that makes the greatest contributions to test validity and reliability. *Item analysis* is a set of procedures used to examine and describe examinees' responses to the items under consideration, including the distribution of responses to each item and the relationship between responses to each item and other criteria.

One of the first things usually computed in an item analysis is the mean and variance of each item. For dichotomous items, the mean is also the proportion of examinees who answered the item correctly and is called the *item difficulty* or p, as previously discussed. The total test score for one examinee is the sum of the item difficulties, which is the same as the sum of questions answered correctly. The average item difficulty is the sum of the item difficulties divided by the number of items, as shown in Figure 16-18.

$$\mu_p = \frac{\sum_{i=1}^{k} p_i}{k}$$

Figure 16-18. Formula for average item difficulty

In this formula p_i is the difficulty of item i,
and k is the total number of items.

Because item difficulty is a proportion, the variance for an individual item is:

$\sigma^2_i = p_i(1 - p_i)$

Often, items are selected to maximize variance to increase the test's efficiency in discriminating among individuals of different abilities. Variance is maximized when $p = 0.5$, a fact that you can confirm for yourself by calculating the variance for some other values of p:

If $p = 0.50$, $\sigma^2_i = 0.5(0.5) = 0.2500$
If $p = 0.49$, $\sigma^2_i = 0.49(0.51) = 0.2499$
If $p = 0.48$, $\sigma^2_i = 0.48(0.52) = 0.2496$
If $p = 0.40$, $\sigma^2_i = 0.40(0.60) = 0.2400$

Note that the variances for $p = 0.49$ and $p = 0.51$ are identical, as are the variances for $p = 0.48$ and $p = 0.52$, and so on.

In many common test formats, most obviously multiple choice, examinees might raise their scores by guessing if they don't know the correct answer. This means that the p value of an item will often be higher than the proportion of examinees who actually know the material tested by the item. To put it another way, the observed scores will be systematically higher than the true scores because the observed scores have been raised by successful guessing. For this reason, when an item format allows guessing (for instance, with multiple choice items that carry no penalty for incorrect answers), an additional step is necessary to calculate the observed difficulty of an item to maximize item variance. This is done by adding the quantity $0.5/m$ to the item difficulty, where m is the number of choices for an item. This formula assumes that the choices are equally likely to be selected if the examinee doesn't know the correct answer to the item. The observed difficulty p_0 of an item that is assumed to have a true difficulty of 0.5 (half the examinees know the correct answer without guessing) for different values of m would be as shown in Table 16-3.

Table 16-3. Item difficulties, corrected for guessing

Number of choices	p_0
2	$0.5 + 0.5/2 = 0.75$
3	$0.5 + 0.5/3 = 0.67$
4	$0.5 = 0.5/4 = 0.625$

Item discrimination refers to how well an item differentiates between examinees with high versus low amounts of the quality being tested, whether it is knowledge of geography, musical aptitude, or depression. Normally, the test creator selects items that have *positive discrimination*, meaning they have a high probability of being answered correctly or positively by those who have a large amount of the quality, and incorrectly or negatively by those who have a small amount. For instance, if you are measuring mathematical aptitude, questions with positive discrimination are much more likely to be answered correctly by students with high mathematical aptitude as opposed to those with low mathematical aptitude, who are unlikely to answer correctly. The reverse quality is *negative discrimination*; continuing with this example, an item with negative discrimination would be more likely to be answered correctly by a student with low aptitude than a student with high aptitude. Negative discrimination is usually grounds to eliminate an item from the pool unless it is being retained to catch people who are faking their answers (for instance, on a mental health inventory).

Four indices of item discrimination are discussed in this section, followed by an index of item discrimination that can be related either to total test score or to an external criterion. If all items are of moderate difficulty (which is typical of many testing situations), all five discrimination indices will produce similar results.

The *index of discrimination* is only applicable to dichotomously scored items; it compares the proportion of examinees in two groups that answered the item correctly. The two groups are often formed by examinee scores on the entire test; for instance, the upper 50% of examinees is often compared to the lower 50% or the upper 30% to the lower 30%. The formula to calculate the index of discrimination (D) is:

$$D = p_u - ftp_l$$

where p_u is the proportion in the upper group that got the item correct, and p_l is the proportion in the lower group that got it correct.

If 80% of the examinees in the upper group got an item correct, but only 30% of those in the lower group got it correct, the index of discrimination would be:

$$D = 0.8 - 0.3 = 0.5$$

The range of D is (-1, +1). $D = 1.0$ would mean that everyone in the upper group got the item correct and no one in the lower group did, so the item achieved perfect discrimination; $D = 0$ would mean that the same proportion in the upper and lower groups got the item correct, so the item did not discriminate between them at all. The index of discrimination is affected by how the upper and lower groups are formed; for instance, if the upper group were the top 20% and the lower group the bottom 20%, we would expect to find a larger index of discrimination than if the upper 50% and lower 50% were used.

There are no significance tests for the index of discrimination and no absolute rules about what constitutes an acceptable value. A rule of thumb suggested by Ebel (1965; full citation in Appendix C) is that $D > 0.4$ is satisfactory (items can be used), $D < 0.2$ is unsatisfactory (items can be discarded), and the range between suggests that the items should be revised to raise D above 0.4.

The *point-biserial correlation coefficient*, discussed in Chapter 5, is a measure of association between a dichotomous and a continuous variable; it may be used to measure the correlation between a single dichotomous item and the total test score (assuming the test contains enough items that the scores are continuous).

The *biserial correlation coefficient* may be calculated for dichotomous items if it is assumed that performance on the item is due to a latent quality that is normally distributed. The formula to calculate the biserial correlation coefficient is given in Figure 16-19.

$$p_{bis} = \left(\frac{\mu_+ - \mu_X}{\sigma_X} \right) \left(\frac{p}{Y} \right)$$

Figure 16-19. Formula for the biserial correlation coefficient

In this formula, μ_+ is the average total test score for examinees who answered the item correctly,
μ_X is the average total score for the entire examinee group,
σ_X is the standard deviation on the total score for the entire group,
p is the item difficulty, and
Y is the Y-coordinate (height of the curve) from the standard normal distribution for the item difficulty (e.g., from Figure D-3 in Appendix D).

Suppose for a given item, $\mu_+ = 80$, $\mu_X = 78$, $\sigma_X = 5$, and $p = 0.5$. The biserial correlation coefficient for this item is shown in Figure 16-20.

$$p_{bis} = \left(\frac{80 - 78}{5}\right)\left(\frac{0.5000}{0.3989}\right) = 0.5014$$

Figure 16-20. Calculating the biserial correlation coefficient

The value of the biserial correlation is systematically higher than the point-biserial correlation for the same data, and the difference increases sharply if $p < 0.25$ or $p > 0.75$. The biserial correlation coefficient is the preferred item difficulty statistic when a dichotomous item is assumed to reflect an underlying normal distribution, and the goal is to select items that are very easy or very difficult or if the test will be used with future groups of examinees with a wide range of ability.

The *phi coefficient*, discussed in Chapter 5, expresses the relationship between two dichotomous variables. If the variables are not true dichotomies but have been created by dichotomizing values from a continuous variable with an underlying normal distribution (such as a pass/fail score determined by establishing a single cut point for a continuous variable), the *tetrachoric correlation coefficient* is preferred over the phi coefficient because the range of phi is restricted when the item difficulties are not equal. Tetrachoric correlations are also used in factor analysis and structural equation modeling. The tetrachoric correlation coefficient is rarely computed by hand but is included in some of the standard statistical software packages, including SAS and R.

Item Response Theory

Although analyses based on classical test theory are still used in many fields, *item response theory* (IRT) offers an important alternative approach. Anyone working in psychometrics should be aware of IRT, and it is being used increasingly in other fields, from medicine to criminology. IRT will probably be used even more in the future because IRT capabilities are implemented into commonly used statistical packages. IRT is a complex topic and can be only briefly introduced here; those who wish to pursue it should consult a textbook such as Hambleton, Swaminathan, and Rogers (1991) or a similar introductory textbook. An inventory of computer packages for IRT is available from the Rasch SIG (*http://winsteps.com/rasch.htm*).

IRT addresses several failings of classic test theory chief among them is the fact that methods based on classic test theory cannot separate examinee characteristics from test characteristics. In classic theory, an examinee's ability is defined in terms of a particular test, and the difficulty of a particular test is defined in terms of a particular group of examinees. This is because the difficulty of a test item is defined in classic theory as the proportion of examinees getting it correct; with one group of examinees, an item might be classified as difficult because few got it correct, whereas for another group of examinees, it might be classified as easy because most got it correct. Similarly, on one test, an examinee might be rated as having high ability or having mastered a body of material because she got a high score on the test, whereas on another test ostensibly covering the same basic material, she might be rated as having low ability or mastery because she got a low score.

The fact that estimates of item difficulty and examinee ability are intertwined in classic test theory means that it is difficult to make an equivalent estimation of ability comparing examinees who take different tests or to rate the difficulty of items administered to different groups of examinees. Classic test theory has tried various procedures to deal with these issues, such as including a common body of items on different forms of a test, but the central problem remains:

- Performance of a given examinee on a given item can be explained by the examinee's ability on whatever the item is testing, and ability is considered to be a latent, unobservable trait.
- An item characteristic curve (ICC) can be drawn to express the relationship between the performance of a group of examinees on a given item and their ability.

Ability is usually represented by the Greek letter theta (θ), whereas item difficulty is expressed as a number from 0.0 to 1.00. The ICC is drawn as a smooth curve on a graph in which the vertical axis represents the probability of answering an item correctly, and the horizontal axis represents examinee ability on a scale in which θ has a mean of 0 and a standard deviation of 1. The ICC is a monotonically increasing function, so that examinees with higher ability (those with a higher value of θ) will always be predicted to have a higher probability of answering a given item correctly. This is shown in the theoretical ICC shown in Figure 16-21.

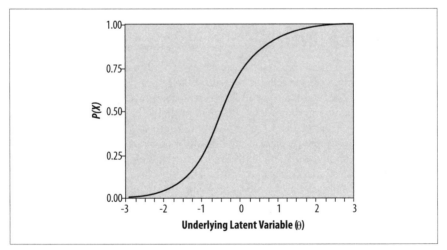

Figure 16-21. Theoretical ICC

IRT models, in relation to classic test theory models, have the following advantages:

1. IRT models are *falsifiable*; the fit of an IRT model can be evaluated and a determination made as to whether a particular model is appropriate for a particular set of data.

2. Estimates of examinee ability are not *test-dependent*; they are made in a common metric that allows comparison of examinees who took different tests.

3. Estimates of item difficulty are not *examinee-dependent*; item difficulty is expressed in a common metric that allows comparison of items administered to different groups.

4. IRT provides individual estimates of standard errors for examinees rather than assuming (as in classic test theory) that all examinees have the same standard error of measurement.

5. IRT takes item difficulty into account when estimating examinee ability, so two people with the same number of items correct on a test could have different estimates of ability if one answered more difficult questions correctly than did the other.

One consequence of points 2 and 3 is that in IRT, estimates of examinee ability and item difficulty are *invariant*. This means that, apart from measurement error, any two examinees with the same ability have the same probability of answering a given item correctly, and any two items of comparable difficulty have the same probability of being answered correctly by any examinee.

Note that although in this discussion we assume items are scored as right or wrong (hence, language such as "the probability of answering the item correctly"), IRT models can also be applied in contexts in which there is no right or wrong answer. For instance, in a psychological questionnaire measuring attitudes, the meaning of item difficulty could be described as "the probability of endorsing an item" and θ as the degree or amount of the quality being measured (such as favorable attitude toward civic expansion).

Several models are commonly used in IRT that differ according to the item characteristics they incorporate. Two assumptions are common to all IRT models.

Unidimensionality
   Items on a test measure only one ability; this is defined in practice by the requirement that performance on test items must be explicable with reference to one dominant factor.

Local independence
   If examinee ability is held constant, there is no relationship between examinee responses to different items; that is, responses to the items are independent.

The simplest IRT model includes only one characteristic of the item, item difficulty, signified by b_i. This is the one-parameter logistic model, also called the *Rasch model* because it was developed by the Danish mathematician Georg Rasch. The ICC for the one-parameter logistic model is computed using:

$$P_i(\theta) = \frac{e^{\theta - b_i}}{1 + e^{\theta - b_i}}$$

where $P_i(\theta)$ is the probability that an examinee with ability θ will answer item i correctly,

and b_i is the difficulty parameter for item i.

Item difficulty is defined as the point on the ability scale (x-axis) where the probability of an examinee getting the item correct is 0.5. For more-difficult items, greater examinee ability is required before half the examinees are predicted to get it right, whereas for easier items, a lower level of ability is required to reach that point. In the Rasch model, the ICCs for items of differing difficulties have the same shape and differ only in location. This is apparent in Figure 16-22, which displays ICCs for several items of equal discrimination that vary in difficulty.

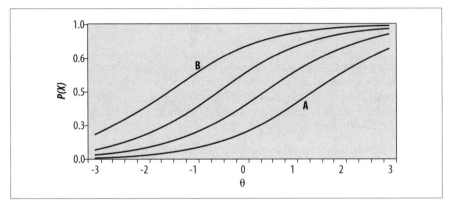

Figure 16-22. ICCs for several items of identical discrimination but varying difficulty: item A is the most difficult, item B the easiest

Bearing in mind that θ is a measure of examinee ability, one can see that a greater amount of ability is required to have a 50% probability of answering item A correctly compared with items further to the left. It is also clear that among the items graphed here, the least amount of θ is required to have a 50% chance of answering item B correctly. Therefore, we would say that item B is the easiest among these items and item A the most difficult. You can demonstrate this to yourself by drawing a horizontal line across the graph at $y = 0.5$ and then a vertical line down to the x-axis where the horizontal line intersects each curve. The point where each vertical line intersects the x-axis is the amount of θ required to have a 50% probability of answering the item correctly, and this quantity is clearly larger for item A than for item B.

The two-parameter IRT model includes an item discrimination factor, a_i. The item discrimination factor allows items to have different slopes. Items with steeper slopes are more effective in differentiating among examinees of similar abilities than are items with flatter slopes because the probability of success on an item changes more rapidly relative to changes in examinee ability.

Item difficulty is proportional to the slope at the point where $b_i = 0.5$, that is, where half the examinees would be expected to get the item correct. The usual range for a_i is (0, 2) because items with negative discrimination (those that an examinee with less ability has a greater probability to answer correctly) are usually discarded and because, in practice, item discrimination is rarely greater than 2. The two-parameter logistic model also includes a scaling parameter, D, which is added to make the logistic function as close as possible to the cumulative normal distribution.

The ICC for a two-level logistic model is computed using the following formula:

$$P_i(\theta) = \frac{e^{Da_i(\theta - b_i)}}{1 + e^{Da_i(\theta - b_i)}}$$

Two items that differ in both difficulty and discrimination are illustrated in Figure 16-23.

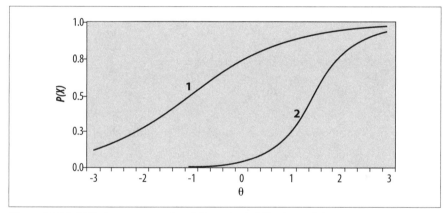

Figure 16-23. ICCs for two items that differ in both difficulty and discrimination

The three-level logistic model includes an additional parameter, c_i, which is technically called the *pseudo-chance-level parameter*. This parameter provides a lower asymptote for the ICC that represents the probability of examinees with low ability answering the item correctly by chance. This parameter is often called the guessing parameter because one way low-ability applicants could get a difficult question correct is by guessing the right answer. However, often c_i is lower than would be expected by random guessing because of the skill of test examiners in devising wrong answers that can seem correct to an examinee of low ability. The ICC for the three-parameter logistic model is calculated using this formula:

$$P_i(\theta) = c_i + (1 - c_i)\frac{e^{Da_i(\theta - b_i)}}{1 + e^{Da_i(\theta - b_i)}}$$

A three-parameter model is shown in Figure 16-24; it has a substantial guessing parameter, which can be seen from the fact that the curve intersects the x-axis around 0.20. This means that a person with very low θ would still have about a 20% chance of answering this item correctly.

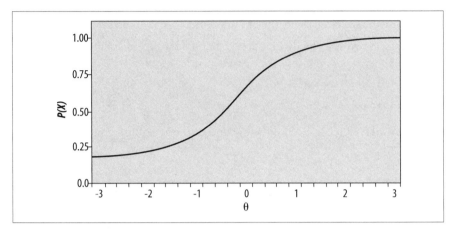

Figure 16-24. ICC for item with substantial guessing parameter

Exercises

Here is a set of questions to review the topics covered in this chapter.

Problem

Given the data distribution in Table 16-1:

1. What is the percentile rank for a score of 80?
2. What score corresponds to a score at the 75th percentile?

Solution

You find the percentile by looking at the cumulative probability for the score just above the score you are interested in. To find a score corresponding to a percentile rank, reverse the process.

1. A score of 80 is in the 17th percentile.
2. A score of 96 is in the 75th percentile.

Problem

Assume you are working with a published test whose mean is 100 and whose variance is 400. Convert the following individual scores to Z-scores, T-scores, and stanines.

1. 70
2. 105

Solution

1. For 70, $Z = -1.5$, $T = 35$, and stanine = 2.
2. For 105, $Z = 0.25$, $T = 52.5$, and stanine = 5.

The computations for a score of 70 are shown in Figure 16-25 and below.

$$Z = \frac{70 - 100}{20} = -1.5$$

Figure 16-25. Calculating the Z-score

$T = -1.5(10) + 50 = 35$
Stanine = $2(-1.5) + 5 = 2.0$

The computations for a score of 105 are shown in Figure 16-26 and below.

$$Z = \frac{105 - 100}{20} = 0.25$$

Figure 16-26. Calculating the Z-score

$T = 0.25(10) + 50 = 52.5$
Stanine = $0.25(2) + 5 = 5.5$; rounds down to 5.

17

Data Management

You might wonder what a chapter on data management is doing in a book about statistics. Here's the reason: the practice of statistics usually involves analyzing data, and the validity of the statistical results depends in large part on the validity of the data analyzed, so if you will be working with statistics, you need to know something about data management, whether you will be performing the management tasks yourself or delegating them to someone else.

Oddly enough, data management is often ignored in statistics classes, as well as in many offices and labs; some professors and project managers seem to believe that data will magically organize into a usable form without human intervention. However, people who work with data on a daily basis have quite a different view of the matter. Many describe the relationship of data management to statistical analysis by invoking the 80/20 rule, meaning that on average 80% of the time devoted to working with data is spent preparing the data for analysis, and only 20% of the time is spent actually analyzing the data. In my view, data management consists of both a general approach to the problem and the knowledge of how to perform a number of specific tasks. Both can be taught and learned, and although it's true that some people can pick up this knowledge on an informal basis (through the college of hard knocks, so to speak), there is no good reason to leave such matters up to chance. Instead, it makes more sense to treat data management as a skill that can be learned like any other, and there's no reason not to take advantage of the collective wisdom of those who have gone before you.

The quality of an analysis depends in part on the quality of the data, a fact enshrined in a phrase that originated in the world of computer programming: garbage in, garbage out, or GIGO. The same concept applies to statistics; the finest statistician cannot produce valid results if the data is a mess. The process of data collection by its nature is messy, and seldom does a data file arrive in perfect shape and ready for analysis. This means that at some point between data collection and data analysis, someone has to get her hands dirty working directly with the data file, cleaning, organizing, and otherwise getting it ready for analysis. There's usually no mystery about what needs to be done during this process, but it does require a systematic approach guided by knowledge of the data and the uses to which it will be put as well as an inquisitive attitude informed by common sense.

GIGO has another meaning that applies equally well to statistical analysis: garbage in, gospel out. This phrase refers to the distressing tendency of some people to believe that anything produced by a computer must be correct, which we can extend to the equally distressing belief that any analytic results produced using statistical procedures must be correct. Unfortunately, there's no getting around the need for human judgment in either case; computers and statistical procedures can both produce nonsense instead of valid results if the data provided to them is faulty. To take an elementary example, the fact that you can calculate the mean and variance of any set of numbers (even if they represent measurements on a nominal or ordinary scale, for instance) does not mean that those numbers are meaningful, let alone that they provide a reasonable summary of the data. The burden is on the analyst to provide correct data and to choose an appropriate procedure to analyze it because a statistical package simply performs the operations you request on the data you provide and cannot evaluate whether the data is accurate or the procedures appropriate and meaningful.

If your interest is restricted to learning statistical procedures, you might want to skip this chapter. Similarly, if you have no practical experience working with data, this chapter might seem entirely abstract, and you might want to skim or pass over it until you've actually handled some data. On the other hand, in either circumstance, you might still find having a basic understanding of what is involved in the process of data management useful and want to become aware of what can happen when it isn't done correctly. In addition, it's always good to know more than you need to for your immediate circumstances, particularly given that career change is a salient feature of modern life. You never know when a little knowledge of data management will give you an edge in a job interview, and reading this chapter should help you speak convincingly on topic, giving you an advantage over many other candidates. In addition, if in the future data management should become one of your responsibilities, the information in this chapter will start you off with a good understanding of why data management is important and how it is done.

An Approach, Not a Set of Recipes

Because many methods and computer programs are used to collect, store, and analyze data, it's impossible to write a chapter spelling out how to carry out data management procedures that will work in all circumstances. For that reason, this chapter focuses on a general *approach* to data management, including consideration of issues common to many situations as well as a generalized process for transforming raw data into a data set ready for analysis.

If I had to give one piece of advice concerning data management, it would be this: *assume nothing*. Don't assume that the data file supplied to you is the file you are actually supposed to analyze. Don't assume that all the variables transferred correctly when the file was translated from one program to another. (Volumes could be written on this subject alone, and every version of any software seems to include a new set of problems.) Don't assume that appropriate quality control was exercised during the data entry process or that anyone else has examined the data for out-of-range or otherwise impossible values. Don't assume that the person who gave you the project is aware that an important variable is missing for 50% of the cases or that another variable hasn't been coded in the way specified by the codebook. Data collection and data entry are activities performed by human beings who have been known to make mistakes now and then. A large part of the data management process involves discovering where those mistakes were made and either correcting them or thinking of ways to work around them so the data can be analyzed appropriately.

The Chain of Command

Without getting too carried away with the military metaphor, it is true that efficient data management for a large project requires establishing a structure or hierarchy of people who are responsible for different aspects of the process. Equally important, everyone involved in the project should know who is authorized to make what decisions so that when a problem arises, it can be resolved quickly and reasonably. This might sound like simple common sense, but in fact, it is not always exercised in practice. If the data entry clerk notices that data is coming in with lots of variables missing, for instance, he should know exactly who to report this problem to so it can be corrected while the project is still in the data collection phase. If an analyst finds out-of-range values during initial inspection of the data file, she should know who is authorized to make the decision about what to do with those values so they can be corrected or recoded before the main analysis begins. Make it difficult for such issues to be resolved, and the staff is likely to impose its own ad hoc solutions or give up trying to deal with them, leaving you with a data set of uncertain quality.

Codebooks

The codebook is a classic tool of research, and the principle of the codebook applies to any project that involves collecting and analyzing data. The codebook is simply a means to collect and organize important information about a project. Sometimes the codebook is a physical object such as a spiral notebook or a three-ring binder, and sometimes it is an electronic file (or a collection of files) stored on a computer. Some projects use a hybrid system in which most of the codebook information is stored electronically, but some or all of it is also printed and kept in a binder. The bottom line is that it doesn't matter what method you choose as long as the vital information about the project and the data set is reliably recorded and stored for future reference.

At a minimum, the codebook needs to include information in the following categories:

- The project itself and data collection procedures used
- Data entry procedures
- Decisions made about the data
- Coding procedures

Details about the project include its goals, timeline, funding, and some statement of the personnel involved (original plus any changes) and their duties. Information about data collection procedures includes when the data was collected, what procedures were used, whether any sort of quality control was used, and who actually collected the data. If a form like a questionnaire was used, a copy should be included in the codebook, as should any instructions given to the data collection team. Decisions made about the data include matters such as definitions of outliers (a case whose value is far different from others in the data set) or other unusual values, details about any cases that were excluded from analysis and why, and any imputation or other missing data procedures that were followed. Information about coding procedures include the meaning of variables and their values, how and why variables were recoded, and the codes and labels applied to them.

Recording information about data entry procedures is particularly important when data is collected in one medium, for instance by using paper questionnaires, and analyzed in another, such as in an electronic file. However, even if a CATI (computer-assisted telephone interviewing) system or other method of electronic data collection was used, the codebook should explain how the individual files were collected and transferred. Usually, electronic file transfer works smoothly, but not always, and every time a file is transferred it creates an opportunity for a data file to become corrupted. If the file for analysis is discovered to be corrupted, it might be necessary to trace backward through the transfer process to determine what happened and to develop a way to correct it. Information about the training of data entry personnel and any quality control methods used (such as double entry of a sample of the data) should also be recorded.

In my experience, companies whose data consists of the records of their day-to-day business operations do a better job of documentation than academics and others working on small projects with data collected specifically for each project. Several factors are involved here. One is that when data collection and storage processes are ongoing, it is relatively easy to establish a set of procedures and follow them. Another is that large companies that deal with data on a regular basis often have a staff of people assigned specifically to manage that data, and those people receive special training relevant to their job. In academia, the opposite situation is often the rule; a lab might be involved in a number of projects, each involving different data and each data set having its own set of quirks. Matters are often complicated by the fact that the responsibilities of collecting and organizing this data can be relegated to undergraduates with minimal experience or training or to PhDs or MDs who are subject matter experts but unfamiliar with (and possibly uninterested in) the day-to-day issues of data management.

The main reason you need a codebook or its equivalent is to create a repository of information about each project and its data, so that people who join a project or analyze the data long after the collection process has ceased know what the data is and how to interpret it. The existence of a reliable codebook is also helpful for people who have been involved in a project from the start because no one's memory is perfect, and it's easy to forget what decisions were made six months or two years ago. Having the codebook information easily accessible is also a great time-saver when it's time to write up your results or when you need to explain the project to a new analyst.

Seldom is data ready to be analyzed exactly as it has been collected. Before analysis begins, someone needs to examine the data file and make decisions about problems such as out-of-range values and missing data. All these decisions should be recorded, as well as the location of each version of the file. An archived version of the original data file should be stored somewhere it can't be changed in case you want to reverse a coding decision later or in case the edited file becomes corrupt and has to be recreated. It's also sensible to store versions of the file after each major round of editing in case you decide that decisions made in rounds 1, 2, 3, and 5 were valid but not those of round 4. Being able to go back to version 3 of the data file saves you from having to process the original version from scratch. The number of variables and cases in each version of the file, as well as the file layout, should also be recorded. Every time a file is transferred, you need to confirm that the right number of cases and variables appear in the new version, and the file layout is useful when you need to refer to variables by position rather than name (for instance, if the last variable in the file didn't survive a transfer). If any method such as imputation is used to deal with missing data, details on the method used and how this changed the data file should also be recorded.

Records of the coding procedures used for a project will probably occupy the largest part of your codebook. Information that should be recorded here includes the original variable names, labels added to variables and data values, definitions of missing value codes and how they were applied, and a list of any new variables and the process by which they were created (for instance, by transforming an existing variable or recoding a continuous variable into categories).

The Rectangular Data File

There are many ways to store data electronically, but the most common format is the rectangular data file. This format should be familiar to anyone who has used a spreadsheet program such as Microsoft Excel, and although statistical packages such as SAS and SPSS can read data stored in many formats, the rectangular data file is often used because it facilitates the exchange of data among different programs.

The most important aspect of a rectangular data file is the way it is laid out. For data prepared for statistical analysis, the usual convention is that each row represents a case and each column represents a variable. The definition of a case depends partly on the analysis planned and involves the concept known as the unit of analysis (discussed further in the sidebar "Unit of Analysis" on page 417). Because sometimes data about one case is recorded on multiple lines or data about multiple cases is recorded on a single line, some prefer to say that one line represents one record rather than one case.

Figure 17-1 displays an excerpt of data from the General Social Survey of 1993, a nationally representative survey that has been conducted by the National Opinion Research Center at the University of Chicago almost every year since 1972. Each line holds data collected from one individual, identified by the variable *id* in the first column. Each column represents data pertaining to a particular variable. For instance, the second column holds values for the variable *wrkstat*, which is the individual's response to a question about her work status, and the third column holds values for the variable *marital*, which is the individual's response to a question about her marital status.

	A	B	C	D	E	F	G	H	I	J	K
1	id	wrkstat	marital	agewed	sibs	childs	age	birthmo	zodiac	educ	degree
2	1	1	3	20	3	1	43	5	2	11	
3	2	1	5	0	2	0	44	8	6	16	
4	3	1	3	25	2	0	43	2	11	16	
5	4	2	5	0	4	0	45	99	99	15	
6	5	5	5	0	1	0	78	10	7	17	
7	6	5	1	25	2	2	83	3	12	11	
8	7	1	1	22	2	2	55	10	7	12	
9	8	5	1	24	3	2	75	11	9	12	
10	9	1	3	22	1	2	31	7	4	18	
11	10	2	5	0	1	0	54	3	12	18	
12	11	1	5	0	1	0	29	4	2	18	
13	12	1	5	0	0	0	23	10	8	15	
14	13	1	1	31	0	1	61	99	99	12	
15	14	5	4	24	3	4	63	3	1	4	
16	15	4	5	0	4	3	33	3	12	10	
17	16	1	5	0	0	1	36	11	8	14	
18	17	7	5	0	98	4	39	3	12	8	
19	18	1	1	22	9	0	55	1	10	15	
20	19	1	1	32	1	1	55	9	7	16	
21	20	1	1	24	2	2	34	4	2	16	
22	21	3	1	24	5	2	36	6	3	14	
23	22	2	1	23	0	3	44	8	5	18	
24	23	5	2	25	2	2	80	5	2	18	

Figure 17-1. Rectangular data file in Excel

Figure 17-2 shows the same excerpt from the same data file in SPSS. The chief difference is that in Excel, the first row stores variable names (*id*, *wrkstat*, etc.), whereas in SPSS, variable names are linked to the data but do not appear as a line in the data file. This difference in storage procedure means that when moving a data file from Excel to SPSS, there will appear to be one fewer case in SPSS than in Excel, but in fact, the difference is due to the row of data names used in Excel but not in SPSS. Transferring data from one program to another often involves this type of quirk, so it's good to know something about each system or program through which the data will pass.

Although other data arrangements are possible in spreadsheets, such as placing variables in rows and cases in columns, these methods are generally not used for data that will be imported into a statistical program. In addition, although spreadsheets allow for the inclusion of other types of information beyond data and variable names, such as titles and calculated fields, that information should be removed before the data is imported into a statistical program.

The main consideration when setting up a system of electronic data storage should be to facilitate whatever you plan to do with it. In particular, remember that whatever program or statistical package you intend to use to analyze this data (Minitab, SPSS, SAS, or R) has specific requirements, and it is your responsibility to provide the data in a form that your chosen program can use. Fortunately, many statistical analysis packages provide built-in routines to transform data files from one format to another, but it remains the responsibility of the data manager and/or statistical analyst

	id	wrkstat	marital	agewed	sibs	childs	age	birthmo	zodiac	educ	degree
1	1	1	3	20	3	1	43	5	2	11	
2	2	1	5	0	2	0	44	8	6	16	
3	3	1	3	25	2	0	43	2	11	16	
4	4	2	5	0	4	0	45	99	99	15	
5	5	5	5	0	1	0	78	10	7	17	
6	6	5	1	25	2	2	83	3	12	11	
7	7	1	1	22	2	2	55	10	7	12	
8	8	5	1	24	3	2	75	11	9	12	
9	9	1	3	22	1	2	31	7	4	18	
10	10	2	5	0	1	0	54	3	12	18	
11	11	1	5	0	1	0	29	4	2	18	
12	12	1	5	0	0	0	23	10	8	15	
13	13	1	1	31	0	1	61	99	99	12	
14	14	5	4	24	3	4	63	3	1	4	
15	15	4	5	0	4	3	33	3	12	10	
16	16	1	5	0	0	1	36	11	8	14	
17	17	7	5	0	98	4	39	3	12	8	
18	18	1	1	22	9	0	55	1	10	15	
19	19	1	1	32	1	1	55	9	7	16	
20	20	1	1	24	2	2	34	4	2	16	
21	21	3	1	24	5	2	36	6	3	14	
22	22	2	1	23	0	3	44	8	5	18	

Figure 17-2. Rectangular data file in SPSS

to determine which format is required for a particular procedure and to get the data into that format before beginning the analysis.

Unit of Analysis

The *unit of analysis* in a research project is the major entity that is the focus of interest for a particular analysis. For example, a study about school achievement could have as its unit of analysis the student, the classroom, the school, the neighborhood, or the city. A study of health care usage could use as its unit of analysis the visit, the patient, the physician, the unit, or the hospital. We refer to the unit of *analysis* because the same data could be analyzed using different units. For instance, one analysis of a data set might look at the academic achievement of individual schoolchildren, whereas another analysis of the same data could look at achievement levels among different schools, and a third could look at differences in academic achievement across a number of cities.

Data that is specific to one unit of analysis is often referred to as belonging to a particular level. In the school data example, the variables collected about individual schoolchildren (age, gender, etc.) would be called individual-level data, and the variables collected about schools (such as enrollment or type of funding) would be called school-level data. Although in some fields it is still acceptable to mix data from different levels in a conventional statistical analysis, this can produce misleading results. Instead, it is increasingly becoming the expectation that specialized techniques such as multilevel modeling will be used if data from different levels is combined in a single analysis.

Spreadsheets and Relational Databases

Even if a project's data will ultimately be analyzed using a specialized statistical analysis package, it is common to collect and/or enter the data by using a different program such as Excel, Microsoft Access, or FileMaker. These programs can be simpler to use for data entry than a statistical package, and many people have them installed on their computers anyway (particularly Excel), limiting the number of licenses of specialized statistical software that must be purchased. Excel is a spreadsheet, and Access and Filemaker are relational databases. All three can open electronic files from other programs and write files that can be opened by other programs, making them good choices if data will be transferred among programs. In addition, all three can also be used to inspect the data and compute elementary statistics.

For small projects with simple data sets, a spreadsheet can be completely adequate for data entry. The advantage of spreadsheets is their simplicity; you can create a new data file simply by opening a new spreadsheet and typing the data into the window, and the entire data set can be contained in a single document. Beginners find spreadsheets easy to use, and the spreadsheet format encourages entering data in the rectangular data file form, facilitating data sharing among programs.

Relational databases can be a better choice for larger or more complex projects. A relational database consists of a number of separate tables, each of which looks similar to a spreadsheet page. In a well-designed database, each table holds one particular type of data, and the tables are linked by key variables. This means that within the database, data for one case (for instance, for one person) might be contained in many separate, specialized tables. A student database might have one table for student home addresses, one for birth dates, one for enrollment dates, and so on. If data needs to be transferred to a different program for analysis, the relational database program can be used to write a rectangular data file that contains all the desired information in a single table. The chief advantage of a relational database is efficiency; data need never be entered more than once, and multiple records can draw on the same data. In the school example, this would mean that several siblings could draw on the same home address record, but in a spreadsheet, that information would have to be entered separately for each child, raising the possibility of typing or transcription errors.

Inspecting a New Data File

Let's assume you have just been sent a new data file to analyze. You have read the background information on the project and know what type of analysis you need to perform, but you need to confirm that the file is in good shape before you proceed. In most cases, you will need to answer the following questions (at least) before you begin to analyze the data. To answer these questions, you must open the data file and, in some cases, run some simple procedures such as creating frequency tables (discussed in Chapter 4). Some statistical packages have special procedures to aid in the process of inspecting a new data file, but almost any package allows you to perform most of the basic procedures required. However, you might also wish to consult one of the specialized manuals that explain the specific data inspection and cleaning techniques available with particular statistical packages; several such books are listed in Appendix C.

The following are some basic questions for a new data file:

1. How many cases are in the file?
2. How many variables are in the file?
3. Are there any (unintended) duplicate cases?
4. Did the variable values, names, and labels transfer correctly?
5. Is all the data within reasonable range?
6. How much data is missing and in what patterns?

You should know how many cases are expected to be in the data file you received. If that does not match up with the number actually in the file, perhaps you were sent the wrong file (which is not an uncommon occurrence), or the file was corrupted during the transfer process (also not uncommon). If the number of cases in your file does not match what you were expecting, you need to go back to the source and get the correct, uncorrupted file before continuing in your investigation.

Assuming the number of cases is correct, you also need to confirm that the correct number of variables is included in the file. Aside from being sent the wrong data file, missing variables can also be due to the file becoming corrupted during transfer. One thing in particular to be aware of is that some programs have restrictions on the number of variables they will handle; if so, you need to find another way to transfer the complete file. If this is not possible, another option is to create a subset of the variables you plan to include in your analysis (assuming you won't be using all the variables in the original file) and just transfer that smaller file instead. A third possibility is to transfer the file in sections and then recombine them.

Assuming you have a file with the correct number of cases and variables, you next want to see whether it contains any unintended duplicate cases. This requires communication with whoever is in charge of data collection on the project to find out what constitutes a duplicate case and whether the data includes a key variable (see the upcoming sidebar "Unique Identifiers" on page 420 if this term is unfamiliar) to identify unique cases. The definition of a duplicate case depends on the unit of analysis. For instance, if the unit of analysis is hospital visits, it would be appropriate for the same person to have multiple records in the file (because one person could have made multiple hospital visits). In a file of death records, on the other hand, you would expect only one record per individual. Different methods are available to identify duplicate records, depending on the software being used as well as the specifics of the data set. Sometimes it is as simple as confirming that no unique identifier (for instance, an ID number) appears more than once, whereas in other cases, you might need to search for multiple records that have the same values on several or all variables.

Unique Identifiers

The concept of the unique identifier is vital to data management and familiar to people who work with databases but might be a new concept to those who have never constructed a database or otherwise worked in data management. An identifier is a code, usually a number, that identifies cases in a data set. A unique identifier is a code that is unique for each case. The simplest way to assign a unique identifier to each case is to use numeric ID codes, assigned sequentially; in other cases, you might want to use preexisting ID codes (such as registration numbers for patients in a medical system). Even if unique, preexisting codes are available, however, a simple sequential ID code is often preferred because it reduces concerns about breaches of confidentiality.

Most data sets need at least one unique identifier for each potential unit of analysis. For instance, if data from a medical clinic could be analyzed at either the patient level or the visit level, one identifier is required that is unique for each patient but common to all the records for one patient, and a second identifier is required to identify all the records belonging to a specific patient visit (chart notes, blood tests, etc.). The unique identifier is useful to confirm that there are no duplicate records, to identify common records belonging to one unit (for instance, all the clinic visits for an individual), and to avoid confusing records for different individuals. Multiple Bill Smiths might be in a large file, for instance, and you wouldn't want to mix up their records. By the same principle, a particular Bill Smith might come to the clinic five times in a year; when looking at his health care history, you want to be able to identify easily all the records relating to him.

Checking that variable values, names, and labels are correct is the next step in inspecting a data file. Correct transfer of data values is the most important issue because names and labels can be recreated, but the data must be correct, and many unexpected things can happen to data in the file transfer process. Among the things you should check are correct variable type (sometimes numeric variables are unexpectedly translated to string variables or vice versa; see the following section on string and numeric variables), length of string variables (which are often truncated or padded during transfer), and correct values, particularly for date variables. Most statistical packages have a way to display the type, length, and labels associated with each character, and this should be used to see that everything transferred as expected.

Variable names can change unexpectedly during the file transfer process due to different programs having different rules about what is allowable in a variable name. For instance, Excel allows variable names to begin with a number, but SAS and SPSS do not. Some programs allow names up to 64 characters in length, whereas others truncate names at 8 characters, a process that can result in duplicate variable names or the substitution of generic names such as var1. Although data can usually be analyzed no matter how the individual variables are named, odd and nonmeaningful names impose an extra burden on the user and can make the analytical process less efficient. Some advance planning is in order if data will be shared among several programs. In particular, someone needs to confirm the naming conventions for each program whose use is anticipated and to create variable names that will be compatible with all the programs that will be used.

Variable and value labels are a great convenience when working with a data file but often create problems when files are moved from one program or platform to another. Variable labels are text phrases attached to a variable that provide one way to work around name length restrictions. For instance, the variable *wrkstat* in the GSS example could be assigned the label "Work status in the previous six months," which does a much better job of conveying what the variable actually measures. Value labels are assigned to variable labels but are assigned to the *values* of individual variables. Continuing with the previous example, for the variable *wrkstat*, we might assign the label "Full-time employment" to the value 1, "Part-time employment" to the value 2, and so on. Convenient as variable and value labels might be, they often don't transfer correctly from one program to another because each program stores this information differently. One solution, if you know that the data will be shared across several platforms and/or programs, is to use simple variable names such as *v1* and *v2* and simple numeric codes for values (0, 1, 2, etc.), and write a piece of code (a short computer program) to be run on each platform or program that assigns the variable and value labels.

The next step is to examine the actual values in the data set and see whether they seem reasonable. Some simple statistical procedures (such as calculating the mean and variance of numeric variables) can help confirm that the data values were transferred correctly (assuming you have the values for mean and variance for the data set before it was transferred). Date variables should be checked particularly carefully; they are a frequent source of trouble because of the different ways dates are stored in different programs. Generally, the value of a date is stored as a number reflecting the number of units of time (days or seconds) from a particular reference date. Unfortunately, each program seems to use a different reference date, and some use different time units as well, with the consequence that date values often do not transfer correctly from one program to another. If date values cannot transfer correctly, they can be translated to string variables, which can then be used to recreate the date values in the new program.

Even if you have confirmed that the file transferred correctly, there might still be problems with the data. One thing you have to check for is impossible or out-of-range variables, which is easily done by looking at frequencies (or the minimum and maximum values if a variable has many values) to see whether they make sense and match with the way the variable was coded. (Frequency tables are discussed in Chapter 4.) If a data file is small, it might also be feasible simply to sort each variable and look at the largest and smallest values. A third option, if you are using Excel, is to use the data filter option to identify all the values for a particular variable. Typical problems to watch out for include out-of-range data (someone with an age of 150 years), invalid values (3 entered in response to a question that has only two valid values, 0 and 1), and incongruous patterns (newborn infants reported as college graduates). If you find unusual values or obvious errors after confirming that the file transferred correctly, someone will have to make a judgment call about how to deal with these problems because once you begin statistical analysis, the program will treat all the data you supply for an analysis as valid.

Data Management

The final step before beginning an analysis is to examine the amount of missing data and its patterns. Your first goal is to discover the extent of the missing data, a task that can be accomplished using frequency procedures. The second is to examine the patterns of missing data across multiple variables. For instance, is data frequently missing on particular sets of variables? Are there some cases with lots of missing data, whereas others are entirely or primarily complete? Does the file include information about why data is missing (for instance, because a person declined to provide information versus because a question did not apply to her) and, if so, how is that information coded? Finally, you need to decide how you will deal with the missing data, a topic that is discussed later in this chapter.

String and Numeric Data

One distinction observed in most electronic data processing and statistical analysis systems is the difference between *string* and *numeric* variables, although they might use different names for the concepts. The values stored in string variables, which are also called character or alphanumeric variables, can include letters, numbers, blanks, and symbols such as #. (The specific characters allowed vary across different systems.) String variables are stored as a series of coded values; the coding systems most commonly used are EBCDIC (Extended Binary Coded Decimal Interchange Code) and ASCII (American Standard Code for Information Interchange). Because string variables are stored as a series of codes, each with a defined position within the variable, certain procedures are possible that refer to the position of the characters. For instance, many programming systems allow you to perform tasks such as selecting the first three characters of a string variable and storing it in a new string variable.

Numeric variables are stored as values rather than as the characters that are used to write those values. They may be used in mathematical and statistical procedures such as addition and subtraction, whereas string variables may not. In some systems, certain symbols such as the decimal point, comma, and dollar sign are also allowed within numeric variables. One point to be aware of is that the values of string variables coded with leading zeroes (0003) will lose those leading zeroes (3) if converted to numeric variables.

The specific method used to store the values of numeric variables differs across platforms and systems, as does the precision with which those values are stored. You should be aware that when transferring electronic files from one system to another, the variable type can change, or certain values that were read as valid in the first system might be recoded as missing in the second. This is a problem that must be handled on a file-by-file basis; the specific problems that occur when transferring files from Excel to SPSS, for instance, might be different from those that occur when transferring files from Access to SAS.

Missing Data

Missing data is a common problem in data analysis. Despite the ubiquity of missing data, however, there is not always a simple solution to deal with this problem. Instead, a variety of procedures and fixes is available, and analysts must decide what approach they will take and how many resources they can afford to dedicate to the problem of missing data. This discussion can only introduce the main concepts concerning missing data and suggest some practical fixes. For a more in-depth and academic discussion, see the classic text, *Statistical Analysis with Missing Data*, by Little and Rubin (Wiley) listed in Appendix C.

Data can be missing for many reasons, and it is useful if the reasons are recorded within the data set. Often, programs allow you to use specific data codes to differentiate among different types of missing data, using values such as negative numbers that cannot appear as true values for the variable in question. An individual completing a survey might refuse a particular question, might not have the information requested, or the question might simply not apply to him. These three types of responses could be assigned different codes (say, –7, –8, and –9) and the meaning of each code recorded in the codebook. Some systems also allow you to record the meaning of these codes by using value labels. The reason for differentiating among types of missingness is so you can use the information to perform further analyses. You might want to examine whether those who declined to answer a particular question differed in terms of gender or age from those who did not know the answer to the question.

Missing data poses two major problems. It reduces the number of cases available for analysis, thereby reducing statistical power (your ability to find true differences in the data, a topic discussed further in Chapter 15), and it can also introduce bias into the data. The first point is based on the fact that, all things being equal, statistical power is increased as the number of cases increases, so any loss of cases might result in a loss of power. To explain the second point requires an excursion into missing data theory.

Missing data is traditionally classified into three types: missing completely at random (MCAR), missing at random (MAR), and nonignorable. MCAR means that the fact that a piece of data is missing is not related to either its own value or the value of other variables in the data set. This is the easiest type of missing data to deal with because the complete cases can be considered to be a random sample drawn from the entire data set. Unfortunately, MCAR data rarely occurs in practice. MAR data is a missing piece of data that is not related to its own value but is related to the values of other variables in the analysis. Failure to complete a survey item about household income can be related to an individual's level of education. Nonignorable missing data is unfortunately the most common type and the type most likely to introduce bias into a statistical analysis. Nonignorable refers to data whose missingness is related to its own value. For instance, overweight people might refuse to supply information about how much they weigh, and people with low-prestige jobs might be less likely to complete an occupational survey.

This discussion might seem a bit theoretical; how can you tell which type of missing data you have when you, by definition, don't know the values of the data that is missing? The answer is that you have to make a judgment based on knowledge of the population surveyed and your experience in the field. Because the most common methods of statistical analysis assume you have complete, unbiased data, if a data set has a large quantity of missing data, you (or whoever is empowered to make such decisions) will have to decide how to deal with it. Implementing some of the following solutions suggested might require calling in a statistical consultant or using software designed specifically for dealing with missing data, so the departmental budget and availability of such experts and software will also play a role in the decision. Some potential solutions are listed here. The most preferable is the first, although this solution might not always be possible (and even if attempted might not be successful). Solution 3 is the second most preferable in most circumstances. Solutions 5 through 7 are seldom justified from a statistical point of view, although they are sometimes used in practice.

1. Make an extra effort to collect the missing data by following up with the source, which solves the problem by making the missing data no longer missing.

2. Consider a different analytical design, such as a multilevel model rather than a classic repeated-measures model.

3. Impute values for the missing data using maximum likelihood methods, such as those available in the SPSS MVA module, or use multiple imputation capabilities provided in programs such as SAS PROC MI to generate a distribution for the missing values. An imputation process provides substitute values for those that are missing based on the values that do exist in the data, creating a complete data set.

4. Include a dummy (0, 1) variable in your analysis that indicates that data is missing, along with an imputed value replacing the missing data.

5. Drop the cases or variables with large amounts of missing data from the analysis. (This is feasible only if the problem is confined to a small percentage of cases and/or variables that are not central to your analysis, and it can introduce bias if the data is not MCAR.)

6. Use conditional imputation by using available values to impute missing values (not recommended because it can result in an underestimate of variance).

7. Use simple imputation to substitute a value such as the population mean for the missing value (not recommended because it nearly always results in an extreme underestimate of variance).

18

Research Design

Often, one of the responsibilities of a statistician is to design research studies. To do this well, you must be familiar with the different types of research designs, know their strengths and weakness, and be able to draw on this knowledge to design studies to examine different types of questions. You also need to be familiar with the customs and practices of your profession, such as what type of study is generally used for a particular type of data or to answer a particular type of question. Research design is a larger subject than can be covered in a single chapter, so this chapter can only introduce the major issues in designing research studies and discuss some of the most common types of designs. Typically, designing a study involves compromise between what the researcher would ideally like to do and what is feasible, and the choice and execution of a design should be guided by consideration of what is most important to the research question and the traditions and standard practices in the relevant field of study. We'd all love to conduct research that is both perfectly controlled (meaning the experimenter can manipulate or otherwise control all the factors relevant to the research) and perfectly naturalistic (meaning that whatever is being measured is being so in a realistic, natural environment). However, the characteristics of control and naturalism are often in competition with each other, and learning to judge how much to emphasize one versus the other is one aspect of becoming a competent research designer. One influence on these decisions is the purpose of the research. Is it to determine some underlying cause for a phenomenon, as is common in science, or to optimize the yield or output for a specific process while minimizing expense and effort, as is often the case in business and technology research? Practical and ethical considerations also come into play—some research designs can simply be impossible to execute, prohibitively expensive, or considered unethical—and the researcher must be aware of community as well as scientific standards concerning the ethical conduct of research.

Basic Vocabulary

Research designs can be divided into three types: *experimental, quasi-experimental,* and *observational.* For a design to be experimental, subjects must be randomly assigned to groups or categories. The classic experimental design is the randomized controlled trial used in medicine, in which subjects are randomly assigned to experimental and control groups, administered some treatment, and the outcomes collected for both groups. The controlled experiment is considered the strongest type of research design as far as drawing conclusions from the results of research (in fact, some refer to the results from experimental controlled trials as the gold standard of evidence), but it is not always possible or practical to conduct this type of research. The next strongest design is the quasi-experimental, in which some sort of control or comparison group is used, but subjects are not randomly assigned to experimental and control groups. In an observational study, the researcher makes no assignments to groups or treatments but observes the relationships of different factors and outcomes as they exist in the real world. Although experimental designs are preferred for their ability to minimize *systematic error* or *bias* (a topic discussed in Chapter 1), quasi-experimental and observational designs have the advantage of minimizing experimenter intrusion into natural processes. This is important particularly in studies with human subjects because human behavior is highly situational, and a person's behavior in a lab situation where she knows she is being observed might be quite different from the way she behaves in her normal life. Again, decisions about which kind of design to use depend on what is most important for the research and possible from a practical and ethical point of view.

A *factor* is an independent variable (predictor variable) in a research design, that is, a variable that is believed to exert some influence on the value of a dependent variable (outcome variable) in the design. Often, experimental designs include multiple factors. If you were studying childhood obesity, factors that you might include in your study include parental obesity, poverty, diet, physical activity level, gender, and age. Some researchers call any research design that includes more than one factor a *factorial design*; others reserve this terms for designs in which all possible combinations of factors appear, also called a *fully crossed* or *fully factorial* design. You might be interested in the influence of each of these variables alone (*main effects*) and in their joint influence (*interaction effects*). You might believe that diet plays a role in child obesity (a main effect) but that the influence of diet is different depending on whether the study subject is male or female (an interaction effect).

Studies can also be classified by the relationship between the time when events occurred and the time when information about them is collected for use in the study. In a *prospective* study, data is collected from the starting point of the study into the future. A group that shares a common point of origin, such as their time of entry into the study or their year of birth, is called a *cohort*, so a *prospective cohort study* is one that follows a group of people (or other objects of study) forward in time, collecting information about them for analysis. In contrast, a *retrospective study* collects information about events that occurred at some time before the study began.

In terms of data types, researchers often differentiate between *primary* and *secondary* data. Primary data is collected and analyzed for a particular research project, whereas secondary data is collected for some purpose and later analyzed for some other purpose. There are tradeoffs between the two, and some researchers might work only with one or the other, other researchers with both. The greatest advantage of primary data is its specificity; because it is collected as part of the project that will also analyze it, it can answer the specific needs of a particular research project. In addition, the people analyzing primary data are most likely familiar with when and how it was collected. On the downside, because data collection is expensive, the scope of data collected by any one researcher or research team is limited. The greatest advantage of secondary data is scope. Because secondary data sets are often collected by governmental entities or major research institutions such as the National Opinion Research Center (located in Chicago), they are often national or international in scope and may be collected over many years, achieving a breadth of coverage of which individual researchers can only dream. The downsides with secondary data are that you have to take the data as it is, it might not correspond exactly to the purposes of your study, and there might be limitations on what data you are allowed to use. (For instance, confidentiality concerns can mean that individual-level data is not available.)

A final consideration is the unit of analysis in a study. The unit of analysis (further discussed in Chapter 17) is the primary focus of a research study. In human research, the unit of analysis is often the individual person, but it also can be collections or populations of individuals who are members of some larger unit such as a school, a factory, or a country. Studies in which the unit of analysis is populations rather than individuals are known as *ecological studies*. Although ecological studies can be useful in identifying potential areas of research (the relationship between a high-fat diet and heart disease) and are relatively cheap to carry out because they generally rely on secondary data, conclusions from ecological studies must be interpreted with caution because they are subject to the *ecological fallacy*. The ecological fallacy refers to the belief that relationships that exist at one level of aggregation (say, the country) also exist at a different level (the individual). In fact, the strength and/or direction of the relationship analyzed with one unit of analysis can be quite different when the same data is analyzed using a different unit of analysis. A classic paper by W.S. Robinson, listed in Appendix C, demonstrated the ecological fallacy with a series of analyses of the relationship between race and literacy in the United States at different levels of geographic aggregation.

Cook and Campbell Notation

Thomas D. Cook and Donald T. Campbell developed a style of notation for research design that has been adopted and adapted by many researchers. The key elements of this notation include O for an observation (collection of data), X for an intervention, R for randomization, a dashed line to indicate groups formed without randomization, and subscripts to indicate the order of observations or interventions. In Cook and Campbell notation, a randomized pretest-posttest design with an experimental and control group would be notated as shown in Figure 18-1.

$$R \quad O_1 \quad X \quad O_2$$
$$R \quad O_1 \quad\quad O_2$$

Figure 18-1. Randomized pretest-posttest design

This notation means that subjects are randomly assigned to treatment and control groups, initial measurements are taken from both groups, a treatment or intervention is delivered to the treatment but not to the control groups, and then measurements are taken again on both groups. This type of design is common in medical studies, in which the experimental intervention is a drug or other type of treatment; the control group does not receive this intervention but receives either standard treatment or no treatment. In the latter case, the members are sometimes called the *placebo* group.

In contrast, a quasi-experimental pretest-posttest design is notated as shown in Figure 18-2.

$$O_1 \quad X \quad O_2$$
$$O_1 \quad\quad O_2$$

Figure 18-2. Quasi-experimental pretest-posttest design

The difference in the quasi-experimental design is that subjects are not randomly assigned to groups. Often in this type of design, preexisting groups such as a classroom or a school are used rather than the random assignment of individuals to groups; the group that does not receive the intervention is called the *comparison* group.

Cook and Campbell's notation is simple and flexible, which explains its continued popularity. They also did a great deal to call attention to the use of weak designs in educational and social research and to point out the problems with trying to draw any conclusions from the data produced from poorly designed studies. Their catalogue of threats to validity and threats to reliability is an excellent reminder to researchers of the many factors that can bring into question the conclusions of even well-designed studies. Cook and Campbell's classic textbook on research design has been updated by William Shadish and is listed in Appendix C.

Observational Studies

Observational studies are generally conducted when it is not possible to conduct an experimental study or when collecting information from subjects in a natural environment is more valued than the control that is possible in an experimental setting. For an example of the first reason, consider research into the effects of cigarette smoking on human health. This research can be done only through observational studies because it would be unethical to assign some people to take up tobacco smoking, a practice known to be harmful to human health. Instead, we observe people who choose to smoke and compare their health outcomes with those of people who do not smoke. As an example of the second reason, consider research into disruptive behavior by primary school students. Because the disruptive behavior might be set off by specific triggers that occur in the school setting, researchers might choose to observe students in their usual classrooms rather than bringing them into the lab for observation.

One well-known type of observational study is the *case control design*, often used in medicine to study diseases that are rare or take a long time to develop. A prospective cohort study is an impractical design to study a rare or slow-developing disease because you would need to follow an extremely large cohort to have a chance of a sufficient number developing the disease in question, and the study might have to last 20 or 30 years (or longer) before members of the cohort begin to be diagnosed with the disease. A case control design circumvents these difficulties by beginning with people who have the disease (the *cases*), then collecting another group of people (the *controls*) who do not have the disease but who are similar in other ways to the cases. Case control research generally focuses on identifying factors (diet, exposure to occupational chemicals, smoking habits, use of prescription drugs) that differentiate the cases and controls in the hope of finding a key factor or factors that could explain why the cases have the disease and the controls do not. Some would classify the case control design as quasi-experimental because it includes a control group; however, the term "quasi-experimental" is more often used to describe prospective experiments in which groups are designated and then observed going forward in time.

The strength of a case-control study rests in large part on the quality of the match between cases and controls; ideally, the controls should be like the case in every way except that they do not have the disease. As a practical matter, matching is more often done on just a few variables that are considered important in terms of risk for the disease such as age, gender, presence of comorbidities (other diseases), and smoking habits. A recent method to improve matching is the use of a *propensity score*, which uses various factors to predict the probability of a given individual being a case or a control. Donald Rubin and Paul Rosenbaum first proposed use of the propensity score; see Appendix C for the citation to the article in which they introduced this approach.

A *cross-sectional design* involves a single time of observation; the most common example of cross-sectional design is a survey that collects data through a questionnaire or interview. The data collected by this type of design is like a snapshot that captures the state of the individuals surveyed at a particular moment. Although cross-sectional studies can be extremely helpful in tracking population trends and in collecting a wide variety of information from a large number of people, they are less useful in terms of establishing causality because of the lack of temporal sequence in the data. For instance, a cross-sectional survey might ask how many hours of television per week a person watches and about his height and weight. From this information, a researcher can calculate the BMI (body mass index, a measure of obesity) for all the individuals surveyed and investigate the *association* between television viewing habits and obesity. She could not, however, claim that excessive television watching *causes* obesity because all the data was collected at one point in time. In other words, even if the data shows that obese people on average watch more television than thin people, it can't tell you whether the obese people watched a lot of television and then became obese, or they became obese first and then took up watching television because more active pursuits became too difficult.

A cohort study can also be observational. A good example is the famous Framingham Heart Study, which began following a cohort of more than 5,000 men living in Framingham, Massachusetts (U.S.) in 1948 to identify factors associated with cardiovascular disease (heart disease). The men in the study, who were between the ages of 30 and 62 and without symptoms of cardiovascular disease at the start of the study, have returned every two years to allow researchers to gather data from them based on lab tests, a physical exam, and a medical history. The study is still active and has enrolled two subsequent cohorts, including spouses, children, and grandchildren of the original participants. The Framingham Heart Study (*http://www.fra minghamheartstudy.org/*) has made important contributions toward identifying major risk factors for heart disease (high blood pressure, smoking, diabetes, high cholesterol, and lack of physical activity) as well as the relationships between heart disease and factors such as age, blood triglycerides, and psychosocial factors.

The major criticism of observational studies is that it is difficult if not impossible to isolate the effects due to individual variables. For instance, some observational studies have noted that moderate wine consumption is associated with a higher level of health as compared to abstinence, but it is impossible to know whether this effect is due to the wine consumption or to other factors characteristic of people who drink wine. Perhaps wine drinkers eat better diets than people who don't drink at all, or perhaps they are able to drink wine *because* they are in better health. (Treatment for certain illnesses precludes alcohol consumption, for instance.) To try to eliminate these alternative explanations, researchers often collect data on a variety of factors other than the factor of primary interest and include the extra factors in the statistical model. Such variables, which are neither the outcome nor the main predictors of interest, are called *control variables* because they are included in the equation to control for their effect on the outcome. Variables such as age, gender, socioeconomic status, and race/ethnicity are often included in medical and social science studies, although they are not the variables of interest, because the researcher wants to know the effect of the main predictor variables on the outcome after the effects of these control variables have been accounted for. Such corrections after the fact are always imperfect, however, because you can never know all the variables that might affect your outcome, and there are practical limitations to how much data you can collect and how much you can include in any analysis.

Although observational studies are generally considered weaker in terms of statistical inference, they have one important characteristic: response variables (such as human behavior) can often be observed within the natural environment, enhancing their *ecological validity*, or the sense in which what is being observed has not been artificially constrained by engaging in a narrowly defined experimental paradigm. Going one step further, some observational studies use participant observation methods in which a researcher becomes involved in the activity under study. If this participation is hidden from the actual participants, ethical issues can arise around the use of deception, so safeguards must be built into the study to see that no inadvertent harm occurs because of the experimental procedures.

Quasi-Experimental Studies

Quasi-experimental studies are similar to experimental studies in that they use a control or comparison group but differ in that participants are not randomly assigned to those groups. Quasi-experimental designs are often used in field research (research in which data is collected in a natural setting as opposed to the laboratory or other obviously experimental setting) and are particularly popular in education and social science research, in situations in which experimental designs would often be impractical. For instance, if you want to study the effects of a new approach to teaching math, you can assign one preexisting classroom to use the new method and another to continue using the old method; at the end of the school year, you would then compare student achievement in the two classrooms. This is not an experimental design because students were not randomly assigned to the treatment (new method) and control (old method) groups, but a true experimental design would be impractical in a school setting. Instead, selecting a comparable group of students to use as a comparison group to the students who get the experimental treatment (the new method of teaching) is a compromise solution and is better than having no source of comparison at all.

The usefulness of a quasi-experimental design might be clearer if we contrast it with some weaker designs often used out of expediency. The terminology and notation used in this section was developed by Thomas D. Cook and Donald T. Campbell (see the sidebar "Cook and Campbell Notation" on page 427 and the Shadish reference in Appendix C), and it has become a widely used research design. Three particularly weak yet commonly used designs are the one-group posttest-only design, the posttest-only nonequivalent groups design, and the one-group pretest–posttest design. As Cook and Campbell note, results from studies using these designs could be due to so many reasons other than the factor of interest that it is difficult to draw any conclusions from them.

In the *posttest only design*, an experimental treatment is delivered and data collected on the group that received the treatment, as shown in Figure 18-3.

X 0

Figure 18-3. One-group posttest-only design

This design is as simple as it looks; you deliver an intervention to a single group and then observe the members once. It can be useful if information is available from other sources about the condition of the experimental group before the intervention, and it may be used in the very early stages of research to gather descriptive information used to create a stronger design for the main study. However, without contextual information, this design amounts to little more than "We did some stuff, and then we took some measurements." True enough, but what is the value of the

Research Design

resulting data? It is difficult, if not impossible, to justify drawing causal inferences from this design because so many factors other than the intervention could also be responsible for any observed outcomes. Without knowing the precise state of the group before the intervention, it is difficult to say anything about how the members might have changed, and without a control group, it is not possible to say that the changes were due to the intervention. Other possible explanations include chance, the influence of events outside the study, maturation (normal growth processes; this cause is particularly relevant when studying children and adolescents), and the effects of being studied.

The *posttest-only non-equivalent groups design* adds one refinement to the one-group posttest-only design: a comparison group that does not receive the intervention but is measured or observed at the same time as the intervention group, as shown in Figure 18-4.

$$X \quad 0$$
$$\text{- - - - -}$$
$$0$$

Figure 18-4. Non-equivalent groups posttest-only design

This design might provide useful preliminary descriptive data if information is available from another source about the state of the two groups before administration of the intervention, and the use of a comparison group (ideally one as similar as possible to the experimental group, such as a comparable class within the same school) provides some information that might help place the measurements in context. Data from the comparison group also helps eliminate some alternative explanations such as maturation (assuming the two groups are the same age and have roughly comparable experience regarding whatever is being measured). However, differences observed between the experimental and comparison groups could be due to initial differences between the two groups rather than to the intervention, and the lack of random assignment, as well as the lack of pretest information, makes it difficult to eliminate this explanation for any differences observed between the groups.

The *one-group pretest-posttest design* uses only one group but adds an observation (the pretest) before the intervention is delivered, as shown in Figure 18-5.

$$0_1 \quad X \quad 0_2$$

Figure 18-5. One-group pretest-posttest design

Although the collection of information about the experimental group prior to the intervention is certainly useful, it is still not possible to draw causal inference from data collected from this type of design. The reason is that so many other explanations for the observed results are possible. Besides obvious issues such as maturation and influence from outside events, *statistical regression* (also called *regression to the mean*) must always be considered with this type of design, particularly if the experimental group was selected due to high or low performance on some measure related to the purpose of the study. Suppose a group of children who score poorly on a reading test (the pretest) is given extra coaching in reading (the intervention) and then tested again (the posttest). They may well perform better on the posttest than on the pretest, but ascribing this change to the influence of the intervention requires a leap of faith not supported by the study design because all measurements include a component of random error. (This is further discussed in Chapter 16.) For instance, each student in our hypothetical study has a true level of competency in reading, but any particular measurement of his reading competency (the score on a reading test) includes some error of measurement that might make his observed score higher or lower than the true score that reflects actual competency. Thus, a student who scores low on one reading exam might score higher on the next simply due to this random error of measurement, without her level of competency in reading having changed at all. If a study group is selected for its extreme scores (for instance, children who perform poorly on a reading test), the probability of regression to the mean resulting in higher scores on a second testing occasion is increased.

Cook and Campbell present many quasi-experimental designs that are preferable to these three (see the Shadish, Cook, and Campbell book in Appendix C for more on this topic); all represent attempts to improve control in situations when it is not possible to include random assignment to groups. One simple example is the *pretest-posttest design with comparison group* illustrated in Figure 18-6.

$$O_1 \quad X \quad O_2$$
$$\text{-----}$$
$$O_1 \qquad O_2$$

Figure 18-6. Pretest-posttest design with comparison group

In this design, a comparison group is selected that is similar to the experimental group, but subjects are not assigned randomly to either group; instead, most often, preexisting groups are used. Measurements are taken on both groups, the intervention is delivered to the experimental group, and measurements are taken on both groups again. The main downside to this design is that without random assignment, the experimental and comparison groups might not be truly comparable; administering the pretest to both groups helps but does not completely overcome this difficulty. Other threats to this design include the fact that simply receiving any intervention can cause a difference in the outcome (for this reason, the control group sometimes receives a different intervention not believed to affect the outcome) and that the two groups might have different experiences outside of the experimental context. Different classrooms have different teachers; different towns might have different economic conditions, and so on.

The *interrupted time series* shown in Figures 18-7 and 18-8 is a quasi-experimental design that might or might not include a comparison group.

Research Design

$$O_1 \; O_2 \; O_3 \; O_4 \; O_5 \quad X \quad O_6 \; O_7 \; O_8 \; O_9$$
$$\text{-----------------------}$$
$$O_1 \; O_2 \; O_3 \; O_4 \; O_5 \quad\quad O_6 \; O_7 \; O_8 \; O_9$$

Figure 18-7. Interrupted time series with a comparison group

$$O_1 \; O_2 \; O_3 \; O_4 \; O_5 \quad X \quad O_6 \; O_7 \; O_8 \; O_9$$

Figure 18-8. Interrupted time series (no comparison group)

The number of observations can vary from one study to another, but the basic idea is that a series of measurements is recorded over a period of time, an intervention is delivered, and then another series of measurements is recorded over a period of time. This design is often used to judge the effects of legal or social policies that affect large groups of people, such as passage of a law requiring drivers to wear seat belts or increasing fees charged to households for waste disposal. Multiple measurements are taken over a period of time before the intervention to establish a baseline level and again after the intervention to establish the new level. Multiple measurements are necessary to control for the natural fluctuation of the phenomena. For instance, even without any change in laws, the number of traffic accidents varies from one month to the next. Ideally, the baseline measurements should be stable around a particular value, and the post-intervention measurements should be stable but around a different value and in the expected direction. The addition of a comparison group to this design strengthens the researcher's ability to draw conclusions because it can help control for nonintervention influences that might influence the results. (A public conservation campaign might influence people to begin recycling and composting, independent of the effect of the higher trash disposal fees.)

Suppose a state is concerned about the number of traffic deaths and decides to lower the speed limit on highways, a decision that it believes will result in fewer deaths. Because the reduced speed limit will affect everyone who drives in that state, it is not possible to have a control group; instead, a neighboring state with similar demographics and traffic death rates is chosen to serve as a comparison group. Data from this study is presented in Figure 18-9.

The black line represents traffic deaths in the intervention state, the grey line those in the comparison state; the vertical dotted line shows when the intervention (the new speed limit) took effect. As can be seen, the two states had comparable rates of traffic deaths in the five months preceding the intervention; then deaths dropped in the intervention state and became stable around a new, lower level, as would be expected if the law were effective in reducing traffic deaths. No such change was observed in the comparison state (in fact, the rate of traffic deaths might have risen slightly), giving support to the contention that the law, rather than any other influence, was responsible for the observed decline in traffic deaths. Of course, we would also conduct a statistical investigation to see whether the change was significant, but the graph suggests that the intervention did have an effect in the desired direction.

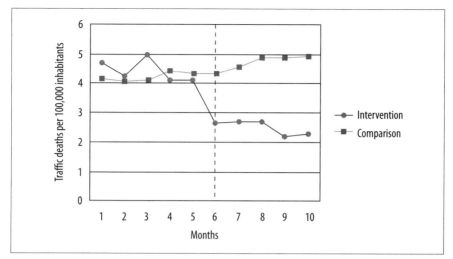

Figure 18-9. Effects of a speed limit on traffic deaths

Is the Sports Illustrated Jinx Real?

You might have heard of the *Sports Illustrated* jinx: the claim that athletes who appear on the cover of the weekly magazine *Sports Illustrated* (*SI*) are subject to a curse of sorts, reflected in poor subsequent performance in their sport or other misfortune. Believers can cite many instances to support their claim. Golfer Ben Hogan, one of the greatest players of his era, was featured on the cover of *SI* on January 10, 1949, only to suffer career-threatening injuries in an automobile accident a few weeks later. Belarus gymnast Ivan Ivankov, lauded on a September 2000 *SI* cover as the best gymnast in the world, was shut out of the medals at the 2000 Olympic Summer Games shortly thereafter.

Everyone knows that anecdotes are not the same thing as evidence, so three *SI* writers sat down to tally up the fates of the athletes appearing on almost 2,000 *SI* covers. Their conclusion: more than one third (37.2 percent) of the cover subjects suffered a misfortune relatively soon after appearing on the cover, with misfortune defined to include anything from a decline in individual or team performance to injury or death. Of course, to test this result for statistical significance, we'd need a lot more information, including the frequency of misfortune for each athlete through his career; because collecting such data would be extremely time-consuming, if not impossible, the issue will probably never be settled.

A much simpler explanation is at hand, however: *regression to the mean*. Athletes selected to appear on the *SI* cover are usually top performers in their sport at the time of their selection, and because everyone's performance fluctuates, it's understandable that their performance would not always remain at that high level. For the superstitious, this decline in performance could easily be interpreted as a jinx rather than as natural variability. For more on this subject, see the article by Alexander Wolff and colleagues cited in Appendix C.

Experimental Studies

Experimental studies provide the strongest evidence for causal inference because a well-designed experiment can control or eliminate the influence of many sources of variation, making us more comfortable in asserting that observed effects are due to the experimental intervention rather than to any other cause. There are three elements to an experimental study, and the configuration of the design can range from the very simple to the very complex:

Experimental units
> The objects under examination. In human experiments, units are generally referred to as participants, given their active engagement in the experimental process.

Treatments
> The interventions applied to each unit in the experimental setting.

Responses
> The data collected after the treatment has been delivered that form the basis for evaluating the effects of the treatment.

Besides the treatments that are the focus of the study, other variables might be believed to affect the responses. Some of these are characteristics of the experimental subjects; in the case of human subjects, they might include qualities such as age and gender. These characteristics can be of interest to the researchers (it might be hypothesized that a treatment is more successful with males than with females), or they might simply be *nuisance variables* or *control variables* that might obscure the relationship between treatment and response. For nuisance or control variables, you want to neutralize their effects on the response variables, and normally this is done by having approximately equal representation on the important nuisance or control variables in the experimental and control groups. Normally, random assignment will make the distribution of characteristics such as gender and age approximately equal in each group; if this is not sufficient, matching or blocking procedures can also be used, as described later.

In some experimental designs, a comparison is made between a *baseline* measurement for each unit before treatment and the measurement for the unit after treatment (also known as pretest and posttest responses). This type of design is known as a *within-subjects* design and provides a high degree of experimental control because measurements on units are made only with themselves; participants act as their own controls. The example of the matched-pairs *t*-test discussed in Chapter 6 is an example of a within-subjects design. In a *between-subjects* design, comparisons are made between different units, and frequently, the units are matched on one or more characteristics to ensure the least confounded comparison of the treatment on units from the control and experimental groups.

Ingredients of a Good Design

The goal of an experiment is to determine the effect of an experimental treatment; this is often measured by the differences in response values for members of the treatment and control groups. It's important to use good procedures for allocating experimental units to treatment and control; indeed, the method of allocating units is a basic difference that separates experimental from observational studies. A major goal of any experimental design is to minimize or preferably eliminate *systematic errors* or *biases* in the data collected.

For many reasons—including ethical and resource considerations—the amount of data collected should be minimally sufficient to answer a particular research question. The use of effective sampling and *power* calculations (discussed in Chapter 16) ensures that the smallest number of experimental units is subjected to experimentation and that a result can be achieved with the least cost and effort.

An effective research design makes analysis much easier later on. For example, if you design your experiment so that you will not encounter missing observations, you will not need to worry about coding missing data and any limitations of interpretation from your results that might subsequently arise (a topic explored further in Chapter 17), including the biases that might accompany nonrandom missing data.

Statistical theory is flexible to the extent that many sophisticated types of designs are mathematically possible, but in practice, most statistics (and therefore designs) are structured according to the requirements of the *general linear model*. This simplifies analysis because many techniques, such as correlation and regression, are based on this model. However, to make valid use of the general linear model, an experiment must be designed with several important factors in mind, including balance and orthogonality.

Balance means that treatments are administered in equal numbers within each experimental block; they will occur with the same frequency. Balanced designs are more powerful than unbalanced designs for the same number of subjects, and an unbalanced design can also reflect a failure in the process of subject allocation. Randomization of group allocation, blinding, and identifying biases are all mechanisms for ensuring that balance is maintained; these are discussed later in this chapter.

Orthogonality means that the effects of different treatments can be independently estimated without interfering with each other. For example, if you have two treatments in an experiment and you build up a statistical model that measures their effects on experimental units, you should be able to remove either treatment from the model and get the same answer for the remaining treatment.

None of this is as complicated as it first sounds, and if you stick to some well-known recipes and templates for factorial design, you won't need to worry about more exotic exceptional cases.

Gathering Experimental Data

So, you want to run an experiment, but where do you start? This section offers a general outline of the research process, roughly in the order in which you need to carry out each step, but your plan should also consider how experiments such as the one you are planning are usually approached in your field of work or study. Stand on the shoulders of giants, in other words; if you are running experiments in a scientific discipline, look at some articles from research journals in that specific discipline, and ensure that the designs and analyses that you carry out are consistent with what others are using in the field. The process of peer review, although not flawless, ensures that the methodology used in an article has been vetted by at least two experts. If you have an advisor or supervisor, you might ask her for advice as well—there's no sense in reinventing the wheel. In industry or manufacturing, it might be harder to find guidance, but company technical reports and previous analyses should provide some previous examples—even if they have not been peer-reviewed—that you might find instructive.

Having said that, you will be surprised at just how much variation and urban mythology surrounds experimental design, so let's walk through the steps one by one:

1. Identify the experimental units you want to measure.
2. Identify the treatments you want to administer and the control variables you will use.
3. Specify treatment levels.
4. Identify the response variables you will measure from the experimental units.
5. Generate a testable hypothesis that predicts what effect the treatment will have on the response variables.
6. Run the experiment.
7. Analyze the results.

Design (steps 1–5) can seem easy when looked at in the abstract, but let's look at each step in detail to see what's really involved.

Identifying Experimental Units

Recall that statistics are calculated on samples and are estimates of the parameters of the populations from which the samples were drawn. To ensure that these estimates are accurate estimates, most statistical procedures rely on the assumption that you have selected these units randomly from the population. (Matched and paired designs are an obvious exception to this rule.) *Bias* can very easily creep into a design at this first stage, and yet, circumstances might dictate that bias cannot be easily avoided.

For example, many research studies in psychology use undergraduate psychology students as participants. This serves two purposes. First, as part of their coursework, students are exposed to a wide variety of experimental designs and get to experience firsthand what is involved in running an experiment; second, the participant group is easily accessible for psychological researchers. In some ways, the homogeneity of the students serving as the subject pool provides a type of control because the participants can be of similar ages, have an even split in terms of sex, come from the same geographical area, have similar cultural tastes, and so on. However, they are not a random sample from the general population, and this might limit the inferences you can make from your data. Research papers based on a sample of college students might tell us more about the behavior of college students than about the population at large; whether this is an important distinction depends on the type of research being conducted. This problem is not limited to psychology; despite the expectations of random selection of subjects, in practice, researchers in many fields select their subjects nonrandomly. For instance, medical research is often performed with patients in a particular hospital or people who receive care at particular clinic, yet the results are generalized to a much larger population; the justification is that biological processes are not dependent on matters of geography, so results from one set of patients should generalize to similar patients as well. It's important to know what the expectations and standard practices are in your particular field in terms of sample selection and the validity of generalizing results from a sample because no one rule applies in all fields of study.

What is meant by random selection in this context? Imagine a lottery in which every citizen of a country receives a ticket. All the tickets are deposited in a large box, which is mixed by rotation through many angles. An assistant is then asked to pick one ticket by placing his hand in the box and selecting the first ticket he touches. In this case, every ticket has an equal chance of being selected. If you needed 100 members for a control and 100 for an experimental group, you could select them using a similar process by which the first 100 selections are allocated to a control group, and the next 100 are allocated to the treatment group. Of course, you could alternate selection by allocating the first ticket to a control group, the next to the treatment group, the third to a control group, and so on. If the sampling is truly random, the two techniques will be equivalent. It's important, for the selection to be random, that the allocation of any particular individual must be truly independent of the selection of any of the others.

Different procedures for drawing samples are discussed further in Chapter 3. The main point to remember here is that in real-world situations, it is often not possible or feasible to draw a random sample from the population, and practical concerns dictate that your sample will be drawn from a population smaller than the one you wish to generalize to (that you believe your results apply to). This is not a problem if you are clear about where and how your sample was obtained. Imagine that you are a microbiologist interested in examining bacteria present in hospitals. If you use a filter with pores of diameter one μm (micrometer), any bacteria smaller than this will not be part of the population that you are observing. This sampling limitation will introduce systematic bias into the study; however, as long as you are clear that the population about which you can make inferences is bacteria of diameter greater than one μm, and nothing else, your results will be valid. In reality, we often want to generalize to a larger population than we sampled from, and whether we can do this depends on a number of factors.

In terms of medical or biological research, generalizing far beyond the population that was sampled is common because of the belief that basic biological processes are common to all people. For this reason, the results of medical research conducted with patients in one hospital can be assumed to apply to patients all over the world. (Of course, not every medical result generalizes this easily.) Another fact to keep in mind is that by explicitly stating the limitations of your sample, you can produce valid results that add to the general body of knowledge. Because many such studies are conducted in a particular field, it might be possible to generalize the results to the general population as well. For example, carrying out tests of reaction time to

English words might be used to make inferences about the perceptual and cognitive processing performance of English-speaking people. Subsequent experiments aimed at increasing the possibility of generalizing the finding might include the same experiment but with German words displayed to German speakers, French words to French speakers, and so on. Indeed, this is the way more general results are built up in science.

Identifying Treatments and Controls

Treatments are the manipulations or interventions that you want to perform to demonstrate an experimental effect. Suppose a pharmaceutical company has spent millions of dollars developing on a new smart drug, and after many years of testing in the lab, it now wants to see whether it works in practice. It sets up a clinical trial in which it selects 1,000 participants by randomly selecting names from a national phone book, giving a truly representative sample of the population on significant parameters such as age, gender, and so on. Luckily, it has a 100% success rate in recruiting participants for the study (everyone wants to be smarter, right?), so they don't have to worry about the selected sample refusing to participate or dropping out (both of which could introduce bias into the sample). All participants will be tested on the same day in identical experimental conditions (exactly the same location, temperature, lighting, chair, desk, etc.). At nine in the morning, participants are administered an intelligence test by computer; at noon, they are given an oral dose of the smart drug with water; and at 3 P.M., they sit for the same intelligence test again. The results show an average increase in intelligence of 15%! The company is ecstatic, and it releases the results of the test to the stock exchange, resulting in a large increase in the company's share price. Nevertheless, what's wrong with the treatments administered in this experiment?

First, because everyone was tested in exactly the same place and under exactly the same experimental conditions, the result cannot be automatically assumed to apply to other locations and environments. If the test were administered under a different temperature, the results might be different. In addition, some aspect of the testing facility might have biased the result—say, the chair or desk used or building oxygen levels—and it's difficult to rule these confounding influences out.

Second, the fact that the baseline and experimental conditions were always carried out in the same order and the same test used twice will almost certainly have been a contributing factor in the 15% increase in intelligence. It's not unreasonable to assume that there will have been a learning effect from the first time that participants undertook the test to the second time, given that the questions were exactly the same (or even if they were of the same general form).

Third, there is no way the researchers can be sure that some other confounding variable was not responsible for the result because there was no experimental control in the overall process; for example, there could be some physiological response to drinking water at noon (in this paradigm) that increases intelligence levels in the afternoon.

Finally, participants could be experiencing the placebo effect by which they expect that having taken the drug, their performance will improve. This is a well-known phenomenon in psychology and requires the creation of an additional control group to be tested under similar circumstances but with an inert rather than active substance being administered.

There are numerous such objections that could be made to the design as it stands, but fortunately, there are well-defined ways in which the design can be strengthened by using experimental controls. For example, if half of the randomly selected sample was then randomly allocated to a control group and the remaining half allocated to an experimental group, an inert control tablet could be administered to the control group and the smart drug to the experimental group. In this case, the learning effect from taking the test twice, as well as the effects of being part of an experiment, can be estimated from the control group, and any performance differences between the two groups can be determined statistically after the treatment has been applied.

Of course, in real clinical drug trials, the research designs are structured quite differently, and investigations are staged in phased trials that have explicit goals at each step, starting with broad dose-response relationships, investigations of toxicity, and so on, with controls being tightened at each stage until an optimal and safe dosage can be identified that produces the desired clinical outcome. Participants are virtually never a random selection of the population but instead are required to meet a particular set of restrictions (age, health, etc.). However, after the study sample is selected, subjects are generally *randomly assigned* to treatment and control groups, an important point in experimental design that helps control bias by making the treatment and control groups as equal as possible.

Specifying Treatment Levels

In practice, you might not be specifically interested in determining whether some factors are influencing the experimental result—you might simply wish to cancel out any systematic errors that can arise. This can often be achieved by balancing the design to ensure that equal numbers of participants are tested in different levels of the treatment. For example, if you are interested in whether the smart drug increases intelligence in general, your sampling should ensure that there is an equal number of male and female participants, a spread of testing times, and so on. However, if you are interested in determining whether sex or time of drug administration influences the performance of the drug treatment, these variables would need to be explicitly recognized as experimental factors and their levels specified in the design. For categorical variables such as sex, the levels or categories (male and female) are easy to specify. However, for continuous variables (such as time of day), it might be easier to collapse the levels to hourly times (in which case there will be 24 levels, assuming equal dosage across the 24-hour day) or simply to morning, afternoon, and evening (3 levels). The research question guides the selection of levels and the experimental effects that you are interested in. Otherwise, counterbalancing and randomization can be used to mitigate error arising from bias. Indeed, replication of the results while extending or being able to generalize across spatial and temporal scales is important for establishing the possibility of generalizing the result.

After treatment levels have been determined, researchers generally refer to the treatments and their levels as a formal *factorial* design, in the form of $A_1 (n_1) \times A_2 (n_2) \times \ldots A_x (n_x)$, where $A^{1..A}x$ are the treatments and $n_1 \ldots n_x$ are the levels within each treatment. For example, if you wanted to determine the effect of sex and time of drug administration on intelligence, and you had a control and experimental group, there would be three treatments, with their levels as follows:

SEX: male/female
TIME: morning/afternoon/evening
DRUG: smart/placebo

Thus, the design can be expressed as SEX (2) × TIME (3) × DRUG (2), which can be read as a 2 by 3 by 2 design. The analysis of *main effects* within and *interactions* between these treatments is discussed in the analysis chapters.

Treatments or Characteristics?

One important difference arises between the physical and social sciences in the definition of treatments. The word "treatment" implies an active process of applying a process that is transformative, such as administering a drug to improve intelligence. However, in the social sciences, treatments are quite often made up of fixed characteristics such as gender. Should such characteristics be regarded as treatments because no transformation takes place? Are designs that use such treatments experimental, quasi-experimental, or actually observational? The issue is fundamental to demonstrating a causal relationship between treatments and responses because using a nontransformative treatment leaves open the question of which characteristic of the experimental units is actually responsible for any differences in responses observed between treatment levels. For this reason, some researchers prefer to call characteristics such as gender independent variables rather than treatments, and some use the term "independent variables" for all variables believed to influence outcomes. Ultimately, the type of inferences that can be made from any research design are limited by such considerations. In technological research, an experiment can have a more explicit optimization goal, such as the estimation of an effect size, to determine the optimal combination and proportions of different treatments and levels that will maximize the value of the response variable.

Specifying Response Variables

Sometimes the response variable will be obvious, but in other cases, more than one response variable might need to be measured, depending on how precisely the variable can be operationalized from some abstract concept. Intelligence is a good example; the abstract concept might appear to be straightforward to the layperson, yet there is no single test to measure intelligence directly. Instead, many measures of general ability across different skills (numerical, analytical, etc.) are measured as response variables and might be combined to form a single number (an intelligence quotient, or IQ), representing some latent structure among the correlated responses. There are advanced techniques (covered in Chapter 12) that describe how to combine and reduce the number of response variables to a smaller, more meaningful (in the sense of interpretation) set.

The safest bet when working with a complex and problematic concept such as intelligence might be to use a number of instruments to obtain response variables and then determine how much they agree with one another. Indeed, techniques for determining the mutual consistency of response variables play an important role in validating experimental designs.

There are three main types of response variables: *baseline*, *response*, and *intermediate*. In the previous section, we saw how a baseline measure of intelligence was used to estimate a direct experimental effect on a response variable (intelligence). An intermediate variable is used to explain the relationship between the treatment and response variable when this relationship is indirect but controllable. If you're interested in establishing a causal relationship as part of an explanatory model, you clearly want to be aware of all the variables involved in a process.

In some designs, the distinction between treatments and intermediate variables might not be important. For instance, if you are a chemist and you are interested in the chemical properties of water, you might be happy to work at the level of atomic particles (protons, neutrons, electrons) rather than at the subatomic level in your analysis. In psychology studies, in contrast, intermediate variables often receive much more attention, particularly if the goal of the research is to specify how some psychological process operates.

In very complicated systems, unanticipated interventions (or unobservable intermediate variables) can influence the result, especially if such variables are highly correlated with a treatment or the act of performing the experiment changes the behavior of what is being observed. Thus, it might be hard to draw out causally whether a treatment is specifically responsible for a change in response. Another general principle is that the longer the delay between a treatment being administered and a response being observed, the greater the likelihood of some intermediate variable affecting the result and possibly leading to spurious conclusions. For example, seasonal factors, such as temperature, humidity, and so on, exert a very strong influence on the outcomes of agricultural production, and these influences can be greater than that of the intervention (a new type of fertilizer) that is the focus of the study.

Hypothesis Testing Versus Data Mining

Given that a statistical significance level $p < 0.05$ implies that 1 in 20 experiments will result in a Type I error, the onus is on the researcher to construct experiments that are consistent with, or attempt to explain, phenomena based on a model or theory. However, it is the practice of some researchers to collect a large amount of data on many response variables and try to relate these to explicit treatments of known characteristics of the sample. When undertaken on a large scale, this approach is known as *data mining*. Data mining—as a form of secondary analysis —is incredibly useful in exploring large data sets, often collected through observation or aggregated from different sources. At its simplest, the purpose of data mining is to determine correlations between many variables, which might later form the basis for constructing experimental designs. Alternatively, in industrial settings, data mining may be used to create decision rules in production systems based on relationships observed in the data. For example, a financial database might reveal that bank customers with income greater than $100,000 and living at an address longer than 3 years have never defaulted on a home loan. Thus, the bank might decide to offer loans to customers who meet these requirements and who currently do not have a loan. However, generally no causal inference is made; the decision rules are pragmatic in nature.

The data mining approach is less defensible in traditional experimental contexts; it's not considered appropriate to forgo the process of stating hypotheses and testing them, and conducting many statistical tests in the hope that *something* will come up significant is called fishing (as in fishing for results). The reason is that *p*-values are valid for a single test, not for a variety of similar tests on the same data. When multiple tests are performed, the *experiment-wise* Type I error rate is almost certainly higher than the *p*-value for a single experiment. (The exception is if all the tests are completely independent.) Several statistical procedures have been established to adjust *p*-values for multiple testing, including the Greenhouse-Geisser correction and the Bonferroni correction.

Blinding

You might have heard of the so-called *placebo effect*, in which participants in an experiment who have been allocated to a control group appear to exhibit some of the effects of the treatment. This effect arises from many sources, including an expectancy effect (because in drug trials, for example, the experimental substance and its known effects and risks would be disclosed to participants) as well as bias introduced by the behavior of the treatment allocators or response gatherers in an experiment. For example, if a treatment allocator knows that a participant will receive the treatment, she might act more cautiously toward the participant than if the allocator were administering a control. Conversely, the response gatherer (the person responsible for observing and measuring data in an experiment) might also be influenced by membership knowledge of the treatment and control groups.

Using single-, double-, or triple-blind experimental methods can effectively control these sources of error.

Single-blind
> The participant does not know whether he has been allocated to a treatment or control group.

Double-blind
> Neither the participant nor the treatment allocator knows whether the participant has been allocated to a treatment or control group.

Triple-blind
> The participant, the treatment allocator, nor the response gatherer knows whether the participant has been allocated to a treatment or control group.

In small laboratories, the roles of treatment allocator and response gatherer can be carried out by the same individual; thus, triple-blind status can often be as easily achieved as double-blind status. Although blinding is highly desirable, it might not always be possible to achieve at one or more of the levels. For example, most adults are familiar with the physiological effects of drinking alcohol, so coming up with a placebo that simulated the effects of alcohol yet did not affect reaction time in an experiment on the effects of alcohol consumption on reaction time would be difficult. (If it affected reaction time, it would no longer be an effective control.) In other cases, it might be possible to create an effective placebo so that the participants will not know which group they were assigned to. The principle is that experiments should use blinding when possible; this is part of the general effort to restrict effects on the treatment group to those caused by the intervention and to prevent extraneous factors from confusing the picture.

Retrospective Adjustment

The previous section mentioned the potential bias of the response gatherer arising from not being blind to the treatment status of participants. Another potential source of bias arises when there are multiple response gatherers or when different instruments are used to gather response data, making essentially independent judgments of responses in either control or experimental treatments. Good training of the judges or response gatherers can help limit this source of bias, and other ways can be used to reduce it. For instance, responses from multiple judges could be averaged to reach a consensus value. Another possibility is to examine the overall set of decisions made by each judge and attempt a retrospective adjustment for perceived bias.

Blocking and the Latin Square

The purpose of *blocking* is to set up experiments in such a way that comparable (and preferably identical) responses can be elicited from the same treatment. The idea is to use as much *a priori* information as possible about experimental units to allocate them to experimental blocks so that all units in a specific block give the same response to a treatment. Perhaps the most famous example of blocking is the use of identical twins in psychological research to examine the effect of nature versus nurture because identical twins have exactly the same genetic makeup. When the twins have been separated at birth, for example, or sent to different schools, the impact of differences in environment can be determined while controlling for genetic factors. The advantage of blocking with identical twins is that variation due to one factor (genetics) can be tightly controlled; the disadvantage is that the subject pool is limited, and the numbers of separated identical twins are even fewer.

Matching can be used to limit the influences extraneous factors exert in experimental design. The differences in responses *between subjects* can be controlled by matching on as many potentially confounding (or unit treatment–correlated) factors as possible. In psychological research, this typically means matching on factors such as age, sex, and IQ but can also include quite specific controls such as visual acuity or color blindness in perceptual experiments.

It might not be possible to match participants on all possible sources of influence extraneous to the research question, but most scientific fields have a set of well-known criteria on which matching has been shown to be effective. The advantage of matched designs is that, on a per-unit basis, you can establish more confidence that an experimental effect genuinely occurs for all units rather than hoping randomization will iron out any differences. A further refinement is to use a *randomized block design*, which allows the researcher to allocate treatments to matched units in a random way, thus gaining the control of matching while also preserving the reduction in bias achieved through randomization.

 A rule of thumb in research design is to block wherever possible and when you can't block, to randomize.

Recall that a matched-pair design attempts to control extraneous factors by matching experimental and control treatment units on important variables. Further control can be achieved by allowing units to act as their own controls in a *within-subjects design* (as in the examples discussed in the paired-samples *t*-test section in Chapter 6), although it might not always be physically possible or practical to do this. Within-subjects designs are used extensively in psychology; however, because many of the experiments involve some modification of behavior or cognition, you might wonder whether there isn't a possible confounding learning effect. If all units were given the control treatment first and then administered the experimental treatment (or vice versa), there certainly would be potential for a learning effect (or *maturation bias*) to influence the results.

However, randomization again provides an antidote in the form of a *Latin square*, which provides an unbiased way to randomize the allocation of participants to treatments. In any design in which y conditions are presented to each participant $(T_1, T_2, \ldots, T_y)$, the trials for each participant are grouped together and randomized, using a Latin square to ensure that no sequence is ever repeated for different subjects. For example, if the reaction time to five objects is measured with trials T_1, T_2, T_3, T_4, and T_5, so $y = 5$, and there are five participants, a randomized Latin square would produce the design shown in the following table, governing the order of stimulus presentation.

T_1	T_5	T_2	T_3	T_4
T_3	T_2	T_4	T_5	T_1
T_4	T_3	T_5	T_1	T_2
T_5	T_4	T_1	T_2	T_3
T_2	T_1	T_3	T_4	T_5

Using a Latin square in this way ensures that any between-subjects variation affects all treatments in an equal way. Note that there are 161,279 other possible randomizations of the 5×5 Latin square that would retain their characteristic property of no orthogonal (row or column) having the same number more than once. If your design required at least one instance of the ordinal presentation of treatments (T_1, T_2, T_3, T_4, and T_5), the reduced form could be used—because the first row and column would preserve ordinality—but would yield only 55 possible randomizations. Latin squares for a few conditions are easily constructed by hand, but you can find a table of Latin squares online (*http://statpages.org/latinsq.html*) as well as a simple algorithm to construct them (*http://rintintin.colorado.edu/~chathach/balancedlatins quares.html*).

Example Experimental Design

This section reviews a real example of an experiment and discusses the design decisions made, comparing how it could have been conducted using two common experimental designs, and provides examples that highlight the relative strengths and/or weaknesses of each strategy.

Frances H. Martin and David A.T. Siddle (2003; full citation is included in Appendix C) set out to investigate the main effects of alcohol and tranquilizers on reaction time, P300 amplitude and P300 latency, as well as their interaction. P300 amplitude and latency are measures derived from event-related potentials in the brain at 300ms. All three responses are related to different information-processing mechanisms in the brain.

The research question was based on previous studies that had independently demonstrated the impact of alcohol or tranquilizers on these response variables but not their interaction. In addition, studies investigating the effect of alcohol on the response variables tended to use large doses, and studies looking at tranquilizers focused on strong tranquilizers, whereas in this study, a mild tranquilizer, Temazepam, was selected. Thus, three questions were posed:

1. Does alcohol have a significant main effect on any of the response variables?
2. Does Temazepam have a significant main effect on any of the response variables?
3. Do alcohol and Temazepam interact?

The experiment used a within-subjects design so that participants acted as their own controls. The factorial design was 2 (alcohol, control) ×2 (tranquilizer, control); thus, every participant performed the same experiment four times with the following conditions:

- No alcohol and no Temazepam
- Alcohol only
- Temazepam only
- Both alcohol and Temazepam

The results indicated a significant main effect for Temazepam on P300 amplitude (that is, this effect was present with or without alcohol) and significant main effect for alcohol on P300 latency and reaction time. However, there was no significant interaction between the two factors. Given that alcohol and Temazepam have different main effects, and because they don't interact, the study supports the idea that alcohol and Temazepam independently affect different information-processing mechanisms in the brain.

If you were designing this experiment, what would you have done? Would you have selected a matched-pair design instead of a within-subjects design? This would have reduced the number of trials that each participant had to complete, but in this instance, using a within-subjects design also allowed for smaller participant numbers to be used ($N = 24$), whereas a larger sample might have been needed to demonstrate an effect between subjects. No doubt you would have randomized the selection of participants, perhaps by selecting names from a phone book, using page numbers and columns generated by a random number generator. Content validity would not be a concern because the response variables used are widely accepted in the field as reflecting information-processing characteristics of the brain. You would also have ensured blinding of the researcher administering the alcohol or Temazepam, ensuring that the control for each was physically the same in appearance. Would you have chosen to increase the number of factors rather than having a 2×2? For example, perhaps there would only be an interaction between alcohol and Temazepam at high respective dosages, so perhaps a 3×3 design would have been more appropriate. The question here is not necessarily experimental but ethical; you want to limit the amount of tranquilizer being administered to each participant, and in the absence of a compelling theoretical reason (or clinical evidence or observation) to suspect otherwise, the choice of a 2×2 study makes sense.

19

Communicating with Statistics

If you've been called upon to do statistics as part of your job or coursework, chances are your responsibility does not end with the calculations—you probably need to communicate the results of those calculations and the conclusions drawn from them to someone else. That someone else might be your boss, a group of coworkers, an audience of professional statisticians, a journalist, your instructors, your classmates...the possibilities are as diverse as the contexts in which statistics are used today.

The key to successful communication is to consider the audience and to shape your writing or presentation appropriately. Sometimes, the expectations are explicit. If you are writing an article for a professional journal, the expected format (covering everything from the parts of the article to the way references should be cited) is probably clearly specified, and you can consult articles previously published in the same journal for further guidance. Writing for a more general audience—as for a daily newspaper or popular magazine—poses a different set of challenges because you need to communicate your major points without confusing your readers (or worse, causing them to simply give up and move on) with a lot of technical jargon. Writing an article or presentation for your workplace poses yet another set of challenges because you must often communicate simultaneously with people of vastly different levels of statistical understanding.

The emphasis in this chapter is on writing, but much of the advice also applies to oral presentations, including talks at professional conferences. There are many good resources discussing matters such as organizing a useful slide presentation, and a few of these are listed in Appendix C.

General Notes

Unless you are writing a technical article for an audience of professional statisticians, the statistics themselves will probably play a supporting role, and the topic of your presentation or article will take the lead. For this reason, it's a good general practice to state your conclusion first and then follow with the statistic(s) that support the conclusion. "Participants in the exercise and diet condition lost an average of 20 pounds over the six-month course of the study compared to those in the diet-only condition, who lost an average of 15 pounds; this difference was statistically significant ($t = 2.75$, $p = .0071$)" communicates your point about the relative effectiveness of two weight-loss plans more effectively than "We found a t-statistic of 2.75, indicating a significant difference in the groups." Some refer to this as the BLUF approach, for Bottom Line Up Front.

Consider what level of precision is appropriate to your purpose, and round your numbers accordingly. Just because a statistical program gives results to eight decimal places does not mean you need to report all of them, and doing so might actually obscure your message. Particularly in tables of data, it's tedious to read numbers with lots of decimal places and difficult to compare them, so you are generally better served by rounding 10.77953201 to 10.8 or 10.78. If you are dealing with very large or very small numbers, scientific notation (2.38×10^{-5} instead of 0.0000238) data or the "x per y" convention can make the values clearer; the latter is commonly used when reporting population statistics, such as hospital beds per 1,000 population or murders per 100,000 population.

Remember that your audience is not as familiar with your analysis as you are and thus must do more work to grasp the meaning of your results. Don't be afraid to make the same point in more than one way—for instance, once in the text and a second time with a table or chart. This principle is particularly important when dealing with general audiences, who might have little understanding of statistics (or simply skip paragraphs with numbers in them) but who can easily grasp concepts presented in a well-designed graphic.

It's always wise to specify the source of the data you analyzed, particularly if you did not collect the data yourself. "Data from the Quarterly Census of Employment & Wages released yesterday by the U.S. Bureau of Labor Statistics indicates . . ." lets your readers know that you used a standard source and aids them in interpreting your results because every data set has its own set of limitations and peculiarities. This rule particularly applies when the data comes from a source that has an interest in the results of the analysis, so if your story on the health benefits of tangerine consumption is based on data collected by the Tangerine Growers Association, you owe it to your readers to state that information up front.

It is often material to include some information about the sample and techniques of data collection. In a professional article, you will discuss these matters in detail, but even in a general-audience article, you probably want to include information about the sample size, method of sample selection, and method of data collection. If your statistics were calculated on a convenience sample of 20, and if the data is entirely self-reported (meaning that you asked people about their behavior rather than measuring it yourself), this should be clearly stated so that the reader can use this information in weighing the meaningfulness of your results.

If you cite percentages, you need to include the base as well; saying that violent crime has doubled in Town A can mean an increase from 1 to 2 violent crimes or an increase from 500 to 1,000. Both examples represent a doubling, but the implications are quite different. The possibilities for deception or misunderstanding using percentages are even greater when comparing two or more entities of different sizes. If you will report that Town B had a 25% increase in violent crime, but Town C had only a 15% increase, you should also let your readers know that Town B has a population of 300 and experienced only four violent crimes the previous year, whereas Town C has a population of 3,000,000 and experienced 50,000 crimes the previous year.

If you are reporting data from a survey, including the actual text of some or all the questions might be advisable, particularly if the survey is short. Even if it is not possible to include the actual wording of the questions, you should at least make their scope clear. For instance, a survey about drug use might frame questions in terms of lifetime use, use in the past year, use in the past 30 days, or habitual use, and each of these frames would result in a different percentage of people classified as drug users.

Writing for a Professional Journal

In one sense, this is the easiest type of statistical writing to do because the expectations regarding audience and format are the clearest. Professionals quickly learn to write in the style of the journals most central to their field, so these remarks are oriented more toward students and young researchers working on their first articles. It goes without saying that you should use all the resources available to you, including your professors or supervisors, the published guidelines of your target journal (the one you want to publish in), the journal's guidelines for reviewers (these are sometimes available on the journal's website), and articles published in previous editions of the journal. Often, universities and research organizations have a journal club to help their students and employees keep in touch with current research, and you should certainly take advantage of such an opportunity if available. (Journal club presentations are discussed in Chapter 20.) Getting oriented to the major journals in your field, and to the types of research they publish, is one of the most useful things young scholars can do to further their careers.

The obvious beginning point in writing an article is that you have something to say—something important that will interest others in your field. To have something to say, you need to be familiar with the major issues and discussions in your field, a knowledge gained from your own reading as well as interactions with your peers and supervisors. This knowledge will also help you choose a target journal. Many fields have a clear hierarchy of journals, and you need to choose one that publishes articles of the type you are writing. In addition, you might be concerned about the quality or relative standard of the journal because, presumably, you want your work to be widely read and to have an impact in your field. One thing you might consider is the impact factor of the journal, the average number of citations an article in a particular journal has received in the preceding two years; a higher impact factor indicates more citations. Choosing an appropriate target journal can require some insider knowledge that a more experienced colleague can provide and is something of a trial-and-error process. It's not uncommon to send an article to a journal that might be a bit of a stretch, with plans to resubmit to other journals if the more prestigious journal rejects your article.

Writing the Article

Of course, writing an article for a particular journal means that you will follow the format of that journal. Fortunately, although there is a multitude of scholarly journals, there are many commonalities among the styles they use because the style is meant to serve a common goal, to facilitate communicating your findings to a specialist audience.

Many journals have a strict format for their articles that specifies the major parts of the article and their order and the format to be used for references (footnotes or endnotes). Many scholars use software (e.g., EndNote) to organize their references because this makes changing the citation format easy if an article is rejected by one journal and needs to be submitted to a different one using a different format. Journals generally have rules about the style used in the text also. For instance, when should numbers be spelled out ("one") and when should numerals be used ("1")? Use active or passive voice? Most adhere to one of several common styles that dictate these matters. Common styles include APA (American Psychological Association), ASA (American Sociological Association), Chicago/Turabian, AP (Associated Press), and ICJME (International Committee of Medical Journal Editors); the point is to find out the style used by the journal to which you are submitting and to follow it.

Professional research articles generally include the following sections (although they can be called by different names):

Abstract
> This is a summary of your research and conclusions and is generally limited in length (e.g., 250 words) as specified by the journal. The abstract is the section of your article most likely to be read, so it is important for it to be compelling as well as concise; you want the abstract to communicate that you have something important to say and to deliver a summary of your major results.

Background/literature review
> This section reviews the current state of knowledge in the field and sets the stage for your original contribution. It's easy to get hung up with trying to read everything ever written on your topic and never publish any research of your own. Your advisor, supervisor, and/or colleagues can help you find a happy medium.

Methods
> This section explains what you did in your research, including details such as the sample studied and any instruments used. Readers often skim this section if the abstract suggested that your article might be of interest to them, so be sure that this section answers all the important questions about how you conducted your research.

Results
> This section presents what you found in your study, including the results of any statistical tests you conducted. Next to the abstract, this is probably the most important section in terms of communicating the importance of your research. Basically, people want to know what you found out in your work.

Discussion

> In this section, you interpret your results, place them in the context of other studies, discuss limitations (perhaps your sample came from only one geographic area or was limited to people whose first language is English), and suggest further directions for research.

Usually, each section, and possibly subsections within them, is identified by headings. This standardized style is meant to facilitate the communication of information, bearing in mind that the typical reader does not read most professional articles from beginning to end. Instead, it is more typical to skim a number of articles (or perhaps simply their abstracts), looking at more sections (perhaps the methods and results) of those that seem most relevant and finally selecting a few to read in their entirety.

Tables and graphs (discussed in Chapter 4) play a key role in many scientific papers. Just as readers often skim the abstract, methods, and results to see whether an article is of interest, so they might glance at the tables and graphs to see whether anything is worth a further investment of their time. For this reason, if you are going to use tables and graphs, it is important to use them well. Each table or graph should tell a story and should be labeled so it is self-sufficient (meaning the reader doesn't need to search through the text to figure out what a graph or table means).

The Peer Review Process

Every journal has its own process for working with authors, and if you are lucky, this process will be explained on its website. In general, the peer review process runs something like this:

1. You submit your article (these days, this is usually done electronically).
2. The editor and/or several volunteer reviewers (generally peers—other professionals in the field) read your article and reply with one of several decisions:
   - Accept (with or without minor revisions)
   - Revise and resubmit
   - Reject
3. You respond accordingly. Celebrate the acceptance, make the requested revisions and resubmit, or submit to a different journal.

In most journals, it's rare for an article to be accepted as submitted, meaning that no changes are required. Some accepted articles require minor revisions, and even if your article is accepted as is, you should still expect to have a back-and-forth conversation (most often through email these days) about editorial matters.

The revise-and-resubmit response is common and should not be cause for discouragement; it means that the reviewers like your article enough that they are willing to work with you to make it acceptable for publication in the journal. Most reviewers sincerely want to help you make your article better, so you should consider their suggestions seriously. Usually, after you have revised your article, you send it back along with a cover letter specifying how you have responded to each of the reviewers' suggestions for revision. If you have a reason for not following one of the suggestions, be sure to explain this in your response. You don't necessarily have to do what the reviewers say, but failing to respond suggests you either didn't read the comments carefully or that you are ignoring them; neither is a good attitude to take if you want the journal to publish your article.

Sometimes, your article never makes it to peer review; many journals use a system in which an editor decides whether the article is within the journal's scope and does a rough estimation of quality before deciding whether it is worth sending out to review. If the editor decides to send an article on to peer review, those reviewers might still recommend that the article should be rejected. If your article is rejected entirely, you may submit it to another journal (unless you conclude, based on the reviewers' comments or other feedback, that it is simply not worth publishing). Even if you are submitting to a different journal, you should consider the reviewers' comments; you might want to act on some of them to strengthen your article before sending it elsewhere. On the other hand, you don't want to get too wrapped up in pleasing one set of reviewers before sending your paper to a different set. There's no guarantee that the second set will agree with the first, and you can end up undoing some of your changes for the second journal.

The review process can be frustrating, particularly when you are dealing with it for the first time. There's no denying that some reviewers can be difficult or unfair and that politics are sometimes involved, but if you want a career in research, you need to learn to deal with the peer review process. My best advice is not to take the process personally but to learn how to work with it and use the wisdom of your senior colleagues as you find your way through the process.

Historically, peer review has been an anonymous process. However, some journals, such as those published in the Public Library of Science (PLoS), encourage open peer review (in which case, the author of an article knows the names of the reviewers for his article). There have been discussions within the scientific community of alternatives to peer review or of ways to reform the process to make it more cooperative and less contentious, but for most journals, any such changes lie in the future if they take place at all.

Writing for the General Public

Writing for a broader public, such as writing an article in a newspaper or general-interest magazine, poses a different set of challenges. You can't assume that most of your audience knows or cares much about statistics; instead, they are probably interested in the subject matter (health, ecology, education, etc.) of your article. This means that you need to highlight your results and their practical implications, rather than the details of your methodology, and that you must clearly explain any essential statistical concepts in everyday language.

Just as with professional journals, popular publications generally have style guidelines and intended audiences, and you should be knowledgeable about both before you begin to write. Few things are more irritating to editors than receiving articles that are clearly not appropriate to their publication. There's an art to matching an article with a publication outlet, but it's your problem to solve, not the concern of the editor, and sending out-of-scope articles simply suggests you haven't bothered to read the publication before submitting your article.

People who write on technical issues for the public often describe their job as story-telling. You don't want just to throw a lot of information and numbers at your audience; you want to organize it into a narrative that includes a context and communicates the importance of whatever you have to say. You also need to be selective. You can't include all the information in a 100-page government study in a single news story, so pick out one or a few main points and organize them into a story line.

Popular science writing often relies on images to convey abstractions (either real pictures or word pictures) and should be written in everyday language with any technical terms paraphrased so they can be easily understood by a layperson. You also need to tell the reader why your topic is important; perhaps a new study has identified an important risk to health or found that there's no evidence that a supposed societal trend is really happening at all. As a general rule, the more surprising or contrarian the result, the more time you need to spend explaining it and the more evidence you need to present to support the result.

Writing for Your Workplace

Writing a report or presentation for your workplace can be tricky because you might be addressing quite a mixed audience in terms of technical and statistical understanding, and the stakes might be high as well; company policy might be guided by the results you present, for instance. It's helpful in such instances to think in terms of providing both an executive summary and a detailed report. In a written document, the executive summary is a short, concise summary of a longer and more detailed report, written in nontechnical language and including the key points of the longer report. As the title implies, the executive summary is written for upper-level managers who might have neither the time nor the technical expertise to tackle the entire report yet need to know something about the information contained within it. Often, executive summaries offer not only information but also recommendations or a set of choices for the person who will be the decision maker.

When writing a statistical report, it's easy to see how the concept of the executive summary applies; you write the full report, including all the details necessary, and then write a shorter summary of it, including your main points and recommendations, using plain language. You should be able to back up any statements in the executive summary with data and analysis from the detailed report. The executive summary is not the place to introduce new points or to throw in unsupported opinions but is meant as a summary of the findings reported in the longer document.

How does the executive summary concept apply to presentations? If called upon to present to an audience including both technical and nontechnical staff, you need to think in terms of preparing two presentations: an executive presentation that can be followed with little or no technical knowledge and a more detailed presentation with all the details and calculations that the technical staff will want to see. The trick is that you might need to do both presentations simultaneously. In this case, your slides should tell the executive version of the story, perhaps with a bit of detail included as asides or footnotes. You can also prepare a handout keyed to your presentation, with all the details that the technical staff will be interested in. Because you, as the statistical expert, already know the technical details of your project, preparing a two-level presentation in this way shouldn't require a lot of extra work.

When addressing a mixed audience, you should also be prepared to explain the technical details of any major issue (why did you use that particular statistical test, and what do the results mean?) in a way that your nontechnical listeners as well as the technical staff can understand. Nothing destroys your credibility faster than being unable to explain an important point in your presentation, and your audience will quickly become impatient if you can't address them at their own level.

20

Critiquing Statistics Presented by Others

This chapter explains how to read and critique statistics presented by someone else, including statistics contained in published research articles and workplace presentations. It begins with a general outline of how to critique a research article and then focuses more narrowly on ways to critique the statistics chosen and their presentation, as well as common ways authors and presenters try to cover up weaknesses in their data. In some ways, this chapter has the broadest applicability of any in this book because even if you never plan or carry out a statistical analysis yourself, there's a good chance you will consume statistics presented by others, whether in your workplace, at school, or simply in your daily life as an informed citizen.

Evaluating the Whole Article

Often, you are called upon to evaluate not just the statistics used in an article but the entire article. This can be intimidating, particularly the first time you face such a task, but following a systematic process can make it easier. If you are reviewing an article for a specific journal, there may be a checklist or other form provided to guide your evaluation process. If not, check other publications in the field to see whether they have a checklist or set of guidelines that might be useful to you. For instance, *Preventing Chronic Disease*, a journal published by the Centers for Disease Control and Prevention, has a peer review checklist (*http://www.cdc.gov/pcd/for_reviewers/ reviewer_checklists.htm*). This chapter presents a basic outline that provides some guidance, in the form of questions you should ask of each part of the article, for reviewing any research article.

Abstract

Is the research question interesting and relevant? Is enough detail included in the abstract (including statistical results) to give you confidence that the article is based on original and significant research? Are the claims in the abstract supported by the results presented in the article? (Surprisingly often, they aren't.)

Literature review

Does the literature review set up the research question, convincing you that it is important and necessary? Is the research cited current? It's fine for a review to include older articles, particularly if they are classics in the field, but the absence of any articles from the past several years suggests a paper that has been in the trunk (written but not published) for several years. The lack of current articles can also indicate a research question that is no longer central to its field or a literature review drawn from an older article and not updated.

Study design

What kind of a design is used—observational, quasi-experimental, experimental? What specific statistical methods are used—ANOVA, linear regression, factor analysis, and so on? Is the design appropriate to the data and the research question? Could a stronger design have been used instead? Are the hypotheses presented clearly?

Data

Is it clear how the sample was obtained and why? Is the process of data collection, processing, and analysis described in detail? Do you see any red flags, such as different sample sizes in different analyses, that are not discussed adequately? Are the data appropriate to the research question?

Results and conclusions

Are the results presented clearly and related to the hypotheses? Do the results support the conclusions drawn? Is sufficient information presented in tables and figures (not just the results from statistical tests) so that you have a good feel for the study and its results? Are there any potential biases or flaws in the study that should have been discussed but weren't? Do the results have practical as well as statistical significance? Are the limitations of the study presented clearly?

It's possible to be too critical—all studies are performed in the real world, after all, and we don't want the perfect to be the enemy of the good, let alone the excellent. When evaluating research, it is important to know what standards are expected in the professional field in question, and more experienced colleagues can provide excellent guidance in this regard.

The Misuse of Statistics

Broadly, the misuse of statistics falls into two very distinct categories: ignorance and intention. The ignorant use of statistics arises when a person attempts to use descriptive or inferential statistics to support an argument, yet the technique, test, or methodology is inappropriate. The intentional misuse of statistics arises when a person attempts to conceal, obfuscate, or overinterpret results that have been obtained. Intuitively, you might think that ignorance arises mostly with complex statistical procedures such as multivariate analysis—and it certainly does—but even basic descriptive statistical procedures are routinely misused. Intentional misuse is rife in descriptive statistics as well, from using misleading scales on graphs to ignoring the assumptions of inferential testing necessary to make their results valid.

In this chapter, some examples are drawn from the contemporary debate surrounding climate change and global warming because the public mood in most countries has clearly changed over the past few years. The purpose is not to convince you one way or the other regarding climate change but simply to provide real-world examples of some of the difficulties of doing statistical research and interpreting and communicating the results.

Common Problems

If you are presented with a set of dazzling statistics that are meant to prove or support some argument, theory, or proposition, begin with the following checklist to start asking the tough questions:

Representative sampling

If the investigator is attempting to make inferences about a population by using a sample, how was the sample selected? Was it truly randomly selected? Were there any biases in the selection process? The results of any inferential tests will be valid only if the sample is truly representative of the population that the investigator wants to make inferences about. In some cases, samples can be maliciously constructed to prove a particular fallacious argument. Alternatively, a volunteer bias might arise when some members of a population respond to a sampling request while others do not. For inferences about a population to be valid, the sample must be truly representative with all sources of bias removed.

Response bias

If the data was obtained through interviews or surveys, how were the questions worded and the responses collected? Be cognizant not only of push-polling (a poll whose true purpose is not to gain information but to influence public opinion) but also of social desirability bias (the tendency for respondents to give the response they think the data collector wants to hear and/or that makes them look like a better person).

Conscious bias

Are arguments presented in a disinterested, objective fashion or is there a clear intention to report a result at any cost?

Missing data and refusals

How is missing data treated in the analysis? If participants were selected randomly but some refused to participate, how were they counted in the analysis? How is attrition (loss of subjects after a study begins) handled?

Sample size

Were the sample sizes selected large enough for a null hypothesis to be rejected? Were the sample sizes so large that almost any null hypothesis would be rejected? Was the sample size selected on the basis of a power calculation?

Effect sizes

If a result is statistically significant, was an effect size reported? If not, how was the importance of the result established? Was it meaningful in the context of the phenomenon under investigation?

Parametric tests
    Was the data analyzed using a parametric test when a nonparametric test might have been more appropriate?

Test selection
    Was the correct inferential test used for the scale of variable? Different techniques are used for different DV (dependent variable) and IV (independent variable) combinations of categorical, ordinal, interval, or ratio data.

Association and causality
    Is the only evidence for a causal relationship between two variables a measure of association, such as correlation? In this situation, it is incorrect to assert a causal relationship, even if one variable is labeled dependent on an independent variable.

Training and test data
    Has a model been developed using one data set and then tested using the same data set? If so, is there any acknowledgement that the model might not work as well on a different data set? This problem occurs frequently in pattern recognition applications.

Operationalization
    Is the variable selected to measure some particular phenomenon actually measuring it? If not, does the operationalization of the phenomenon seem reasonable? This is a common problem in psychology, when latent variables (such as intelligence) are measured indirectly by performance on different cognitive tasks.

Assumptions
    Have the assumptions that underlie the validity of the test been met? How has the investigator ensured that they have been met? For example, if a test assumes that a population is normally distributed and it is in fact bimodal, the results of the test will be meaningless.

Testing the null hypothesis
    To determine whether two groups are drawn from the same or different populations, it is common practice to test the null hypothesis that they are drawn from the same population. This derives from basic scientific methodology in which theories are supported by numerous and reliable sets of tests of null hypotheses that are rejected rather than the (apparently) more straightforward approach of testing the hypothesis directly. Beware of any piece of research that attempts to prove a theory by a single experiment.

Blinding
    Was the study single-, double-, or triple-blinded? For example, could the participants or investigators have introduced some bias by having knowledge of the treatment or control conditions in an experiment?

Controls
    If the effect of a treatment is demonstrated in a pre-treatment or post-treatment model, are matched controls receiving a placebo within the same experimental paradigm to control for the placebo effect? *A designed experiment is the best (some would say the only) way to draw causal inferences reliably from data.*

Quick Checklist

Investigations supported by statistics follow a surprisingly standard life cycle. If you are reviewing a piece of work, try to determine what the sequence of events was during the investigation. Did the investigators start with one hypothesis and change their minds after the results were in? Did they try numerous tests with various post hoc adjustments to make sure that they could report a significance-test result? Have they split results from one study into several articles to increase the length of their CVs? Asking searching questions about the research process is like a detective asking questions about movements at a certain date and time—inconsistencies and story-changing can be very revealing!

Investigations based on statistics should proceed along the following general lines:

- Assuming that a period of observation and exploration has preceded the start of an investigation, research questions should be stated up front. Investigators must have formulated hypotheses (and the corresponding null hypotheses) well before they begin to collect data. Otherwise, the use of hypothesis testing is invalid, and the investigation can take on the flavor of a fishing expedition. Given that a $p = 0.05$ result represents a 1 in 20 chance of making a Type I error, and because thousands of studies are published each year in the scientific literature alone, many "facts" must surely be open to question. This is when independent repeatability and reliability are critical to the integrity of the scientific method.

- The relationship between the population of interest and the sample obtained must be clearly understood. It's not acceptable to make inferences about the entire human population based on a sample of highly educated, healthy, middle-class college students from one college.

- Hypotheses must relate to the effect of specific independent (predictor) variables on dependent (outcome) variables. Thus, it's critical to know as much about the dependent variables as possible, especially every source of variation in them. This is particularly important when dependent variables are thought or known to be highly correlated (i.e., multicollinearity). The dependent variables must be measurable and must operationalize underlying concepts completely.

- In complex designs, when there are both main effects and interactions to consider, all the possible combinations of main effects and interactions and their possible interpretations must be noted.

- Procedures for random sampling and handling missing data or refusals must be formalized early on to prevent bias from arising. Remember that a truly representative sample must be randomly selected. Where purely random sampling is not feasible, it might be possible to identify particular strata within the population and sample those in proportion to their occurrence within the population. If random sampling will not be used (and frequently it is not), this limitation must be acknowledged and addressed.

- The simplest test that is adequate to the purpose should be selected, that is, the simplest test that will allow you to explore the inferences that you need to examine. Multivariate techniques are incredibly important, but if you need to make only simple comparisons, they might be inappropriate.

- Tests should be selected based on known or expected characteristics of the data.

- Ideally, every result should be reported, even if the study did not find statistical significance. Failure to do so leads to *publication bias*, in which only significant results are published, creating a misleading picture of our state of knowledge. Don't be afraid to report deviations, nonsignificant test results, and failure to reject null hypotheses—not every experiment can or should result in a major scientific result!

Publication Bias and the Funnel Plot

It's easy to fall into the naïve belief that the published research literature presents a fair picture of our collective knowledge in any research field. If you do a proper literature search and find four research articles demonstrating the effectiveness of a particular drug and no articles saying it is ineffective, that's pretty good evidence that the drug works, right? Unfortunately, not always. The reason is publication bias (also known as the file drawer problem), the tendency for articles presenting statistically significant results to be published and articles without such results to remain unpublished (and in the file drawer). Other biases also influence the picture we obtain from public research. For instance, research published in English might be more readily available than equally good or better research published in other languages and thus more likely to be cited repeatedly by other articles. (The number of citations is sometimes used as a measure of an article's importance or influence.)

One way to evaluate publication bias on a topic is to create a funnel plot, a graph in which each data point represents a published study, with the log odds ratio of the study on the horizontal axis and the standard error of the study on the vertical axis. If there is no publication bias, we expect to see a pattern similar to an inverted funnel, as in Figure 20-1.

Note that in studies with a larger standard error (less precise studies), there is a greater variability of results (a wider range of values for the log odds ratio), whereas for more precise studies, the log odds ratio clusters more closely around a single value. Note also that this plot is basically symmetrical, indicating that a range of studies with positive, negative, and nonsignificant results has been published. A funnel plot with the general shape shown in Figure 20-1 suggests that publication bias is not a large concern in this particular area of research.

A funnel plot that looks more like Figure 20-2 does suggest publication bias; about half of the funnel is missing because few studies have been published with a neutral or negative result. The plot alone does not prove publication bias (several other possibilities are discussed in the Cochrane Collaboration document listed in Appendix C), but it does suggest it as a possibility.

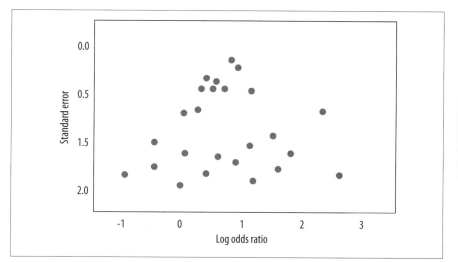

Figure 20-1. A funnel plot suggesting little to no publication bias

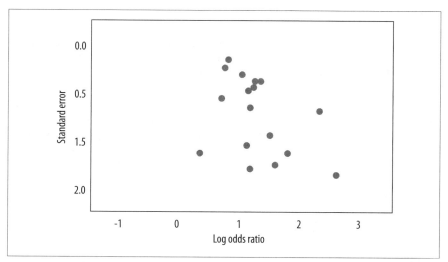

Figure 20-2. A funnel plot suggesting publication bias

Issues in Research Design

Generally, the design of an investigation of a question of interest needs to follow the guidelines presented in Chapter 18 if meaningful inferences are eventually to be made. However, many investigations do not follow these types of guidelines at all—especially if the investigation was produced for a publication that relies on sensational headlines to grab the attention of an inattentive reader or viewer. Even if a study followed appropriate procedures and produced valid results, the meaning of those results can be distorted if a news report extrapolates from the results of a single study to indicate a fundamental shift in knowledge.

Variation

Understanding *variation* is critical to all systems. Variation can arise from legitimate sources (true variation in the population) but also from measurement error. Variation can be cyclical, so cross-sectional designs might not always correctly identify that local minima might be perfectly acceptable in the life cycle of a system. In climatic systems, for example, variation in temperature occurred prior to the Industrial Revolution and the consequent increase in the release of greenhouse gases; how do you separate the variation expected due to normal cyclical effects from that which can be directly attributed to human activity? This is one of the critical issues facing environmental science because the atmosphere definitely warmed since the last ice age, without any human interference, until the Industrial Revolution. The point is that a scientific article should discuss the issue of variation and place its results in the context of expected natural variation.

Population

Scope in defining a population is critical in accurately specifying the limits of inference that can be made from a particular study. If all members of a population are measured in some way, and there is no missing data or refusals, you don't need statistics at all because you can calculate parameters of interest directly. However, this situation rarely arises in research. Part of the problem in defining a population is when there is some fundamental misunderstanding of the population in question. Imagine that a survey of attitudes is undertaken in Utah and the results extrapolated to apply to the population of California, or a survey taken in Italy but applied to the population of Denmark. This might not seem farfetched because in the first example, both states are not only located in the same country but are relatively close in geographical terms, and in the second they are both part of Western Europe. However, in either case, there are also many differences: the size and diversification of the economies, the ethnic and racial makeup of the populations, and so on, and the burden is always on the researcher to make the case that such a generalization is appropriate.

Sampling

There are two key aspects of sampling: size and randomness. A truly representative sample must be both large enough and randomly selected to give an accurate estimate (statistic) of any population parameters. Being sufficiently large to represent the population is a difficult problem—calculations of statistical power (discussed in Chapter 15) certainly provide a basis for this in terms of inferential testing—but more sophisticated sampling schemes will attempt to identify all sources of variation in the population that might introduce bias and sample within those appropriately. Drawing a random sample is the best way to avoid many types of bias in selection, but it is not always possible. An article should always report how the study sample was selected and discuss any consequences from the use of nonrandom sampling.

Controls

A recent study indicated that the administration of antidepressant medication to a large number of participants in a clinical study was no more effective than a placebo. Thus, the expectation of receiving a cure resulted in the same improvement in depressive symptoms as receiving a tablet with the active ingredient. The placebo effect is very powerful in humans, and most studies should provide some type of explicit control when the effect of a treatment is intended to be demonstrated. In clinical and pharmaceutical sciences, the methods and processes for controls are well established. When control groups are not possible (for instance, in climate modeling), the article should provide some other sort of context, if possible, for the results presented (historical data, results from other studies).

The Power of Coincidence

When statistical significance is measured at the $p = 0.01$ or 0.05 level, this means that there is a 1 in 100 or a 1 in 20 chance, respectively, of a Type I error being committed. Thus, in the case of $p = 0.05$, a repetition of the experiment would lead to 19 out of 20 cases being significant and 1 out of 20 being insignificant. This is why independent replication and repeatability are so important. In addition, the world is full of coincidences, and experiments are subject to measurement error. The interaction of coincidence and measurement error can lead to some downright wacky and unexpectedly "significant" findings to which no actual significance should be attached. Imagine that there are 20 earths surrounding the sun, and you choose one to examine the effects of global warming. You find a correlation between increases in industrial activity and temperatures for the past 200 years. Because you know that there is a 1 in 20 chance of committing a Type I error, you would check out at least some of the other planets or perform an experiment on them all, with half acting as matched controls for the others.

You can see the difficulty here in understanding the causal sources of global warming. There are no other 19 planets that you can experiment with or verify your model against—but at the same time, you know there is a strong possibility of committing a Type I error. A similar problem arises in the case of disease clusters; some geographical areas seem to have unusually high rates of a particular disease, leading residents to suspect an environmental cause. However, this type of reasoning is also susceptible to the shooting-the-barn fallacy in which you shoot first and then draw the target around where the bullets struck; the implication is that the geographical area was defined after the disease cluster was noted. In addition, purely by chance, some towns, counties, and so on will have unusually high rates of disease just as when, flipping a coin many times, you can expect long runs of heads and tails from time to time. The point is that you should always be alert for studies that appear to capitalize on chance, particularly if their results contradict what is known from other, better-designed studies.

Descriptive Statistics

The issues surrounding the appropriate interpretation of inferential tests are complex and prone to error. However, the use of descriptive statistics also has enormous potential to introduce errors in reasoning and understanding. Some of these errors are deliberate attempts to misguide and mislead. Others are simply poor choices. In this section, you learn about some common problems associated with descriptive statistics, especially measures of central tendency and graphing.

Measures of Central Tendency

The issue of selecting an appropriate measure of central tendency applies whenever data is not normally distributed, and the more extreme the departure (particularly due to the presence of outliers), the more important the choice. In right-skewed populations (those with a relatively small number of high values), the mean will be higher than the median, and if the high values are far removed from the rest of the population, the mean can be quite misleading as an average value. This is the reason information such as income and home values are usually reported as medians rather than means—the presence of a few very rich people or a few very expensive homes within a population can distort the mean while having minimal effect on the median.

Measures of central tendency can also be misleading when the sample and/or the population changes from measurement to measurement. Average house prices are a classic example; these are based solely on sales in a particular period such as one year. From year to year, the sample from which the average is calculated will almost certainly change, unless all houses sold in one year are resold the next and no other houses are sold. This would surely be a very unlikely event. And yet eager homeowners often take a 10% average rise in house prices to mean that their home's value has increased by the same proportion. Where the population itself changes—such as where many new homes are built and sold in one year—the median will almost certainly rise. And yet existing houses might sell for exactly the same price (or less) than the year before. A more valid method of determining the average house price would be to sample among the population so that each house has an equal chance of being valued and added to the sample. Furthermore, because the proportion of existing houses to new builds is known, the sample could be further stratified so that average prices for both types of houses could be reported and/or aggregated afterward.

To avoid undue influence from extreme cases, sometimes a rule is followed to remove them from analysis. For instance, cases that lie two standard deviations above or below the mean might be dropped before the statistics are calculated, or a certain proportion, such as the upper and lower 10%, might be dropped, a practice known as *trimming*. Removing extreme cases from analysis also helps minimize measurement error effects; in reaction time experiments, for example, it's not uncommon for participants to become incredibly bored and miss a stimulus. If the computer program waits only for two seconds to accept a response, but the stimulus is missed, a reaction time that is usually on the order of 20–80ms is now recorded as 2,000ms, which is up to two orders of magnitude greater. If this case is not culled, the mean would be greatly overestimated.

The key point in this example is that although there are legitimate reasons to remove outliers, any alteration of the study sample after it is collected must be reported and strongly justified by the authors. A good practice when reviewing articles is to check reported sample sizes throughout the article: if the sample drawn is not the sample analyzed, is a reasonable explanation provided? If cases were removed by the researcher, is this clearly explained and justified? Was a *sensitivity analysis* performed; that is, was the data analyzed twice, once with all the cases and once with the outliers removed, to examine the effect of removing the outliers?

Standard Error and Confidence Intervals

The standard error for the data should be reported, especially if the article is comparing the means of two groups. The standard error is an estimate of the standard deviation of a sampling distribution (such as the sampling distribution of the sample mean) and is thus an estimate of the variability of a reported statistic.

Normally, the standard error is estimated by using the standard deviation divided by the square root of n; thus, all else held equal, as the sample size increases, the standard error decreases, and the parameter estimates become more reliable. In most fields, the confidence interval for any point estimate (e.g., the mean) is generally reported as well as the standard error of the mean. The confidence interval provides a measure of the precision of the point estimate, and the article should not only provide confidence intervals but discuss their meaning. If the confidence intervals are wide, this should be discussed in terms not only of the precision of the particular study but in terms of generalizing the results.

Graphical Presentation of Data

Graphs provide an accessible way to communicate numerical information. However, graphs can be misused in a number of ways; for example, axes might be unlabeled, meaning that they cannot be correctly interpreted, or manipulated to obscure or enhance the real relationship between variables. In scientific work, actual data values should be presented as well as graphical displays; often in the popular media, only graphs are presented, which heightens the possibilities for deception.

The old adage "A picture tells a thousand words" is certainly true, but the thousand words can change dramatically depending on the choice of scale. Figure 20-3 shows a fictional set of temperatures that increase, ranging from 70–77 degrees Fahrenheit, over a 100-year time span. The rise in temperature is almost perfectly correlated with the year ($r = 0.94$); this fact can be either illuminated or obscured in the graphic presentation. Figure 20-3 certainly shows a strongly linear rise.

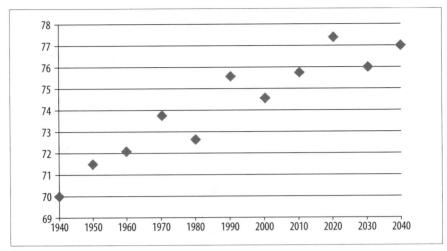

Figure 20-3. Temperature rise over 100 years

However, by stretching out the horizontal axis, suddenly the visual effect is of an overall slower rise in temperature, as shown in Figure 20-4.

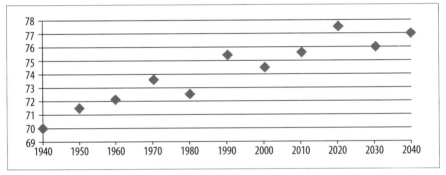

Figure 20-4. Stretched horizontal axis

Note that if the temperature scale is now adjusted to start at 0 degrees rather than 68, the relationship is even further flattened, and the two variables visually appear to be uncorrelated, as shown in Figure 20-5.

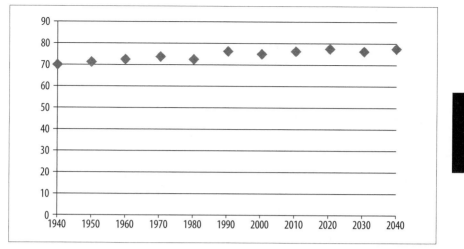

Figure 20-5. Increased range vertical axis

Of course, if you took the opposite view, you could always stretch the temperature axis vertically and make the temperature rise look even more dramatic, as shown in Figure 20-6.

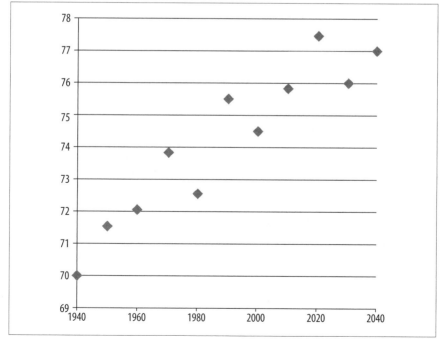

Figure 20-6. Stretched vertical axis

You will seldom find graphs as misleading as Figure 20-5 and Figure 20-6 in scholarly publications, but it's always wise to be alert to attempts to mislead the reader through peculiar choices of ranges, axes, or other tricks. Unfortunately, such deceptions are more common in the popular press, so you need to be particularly alert when interpreting graphical information from publications meant for a general readership.

Extrapolation and Trends

A common tool used in marketing is *extrapolation* of a known relationship between two variables, outside a measured range, to forecast a trend. For example, if the S&P 500 index has increased by 10 points for each of the past 10 weeks, a gambler might feel some confidence in betting that the index might increase by 10 points during the following week. In this case, using simple linear extrapolation provides the best estimate possible—but because the stock market is subject to a lot of random variation, the index will not always rise in accordance with previous experience. If the system is not a linear one, linear extrapolation is not appropriate.

Looking at trends can be useful, and it's a common practice in many fields. However, when the system under study is not deterministic, subject to random error, or chaotic, the usefulness of trending is limited and can give wildly inaccurate and potentially misleading results. Any forecasts reported in an article should be clearly identified as such and justified, as should any extrapolations beyond the range of the measured data.

Inferential Statistics

So far, you have learned about key problems in research design and descriptive statistics that are often present in reports of statistical work performed. In some cases, deception might be behind the incorrect presentation of an analysis, and the omission of key statistics should raise your suspicions. With inferential statistics, you must also be alert to the incorrect or inappropriate use of some tests. The most significant problem is that the assumptions of multivariate tests are routinely ignored, and yet the results of these tests are extremely sensitive to any violation of the assumptions. A research article should explain clearly how the appropriate assumptions were tested and what remedies, if any, were taken before the data was analyzed.

Assumptions of Statistical Tests

Here are some typical violations of the assumptions of common statistical tests and mechanisms to test whether the assumptions are violated. If an article does not discuss how the appropriate assumptions were tested, you should be suspicious of the results.

t-tests

Two-sample t-tests assume that the samples are unrelated; if they are related, then a paired t-test should be used. (t-tests are discussed further in Chapter 6.) Unrelated in this context means independent—you can test for linear independence by using the correlation coefficient. Serial correlation might become an issue if data is collected over a period of time.

t-tests are also influenced by outliers, so the article should mention whether the data was screened for outliers and, if any were found, what was done with them. Note that discarding outliers on the basis of sound statistical measures is an entirely separate activity from discarding data that happens to be unfavorable, and removing cases simply to strengthen the results is unethical.

t-tests assume that the underlying population variances of the two groups are equal (because the variances are pooled as part of the test), so the article should state that one of the tests for homogeneity of variance was used, what corrective measures were taken if necessary, or whether a test that does not rely on homogeneous variance (e.g., Welch's t-test or a nonparametric test) was used instead of the standard t-test.

Normality of the distributions of both variables is another assumption of the t-test, unless sample sizes are large enough to apply the central limit theorem. Again, there should be some mention of how this assumption was tested and, if necessary, what corrective measures were taken.

ANOVA

ANOVA has a large number of assumptions that need to be met, which usually requires directly determining whether the assumption is met (rather than hoping that it is met or ignoring it). ANOVA (discussed further in Chapter 8 and Chapter 9) assumes independence and normality, but the *most important* assumption from a practitioner's perspective is the equality of variances.

ANOVA is most reliable when the study is balanced (when the sample sizes are approximately equal) and when the population variances are equal; skewed distributions and unequal variances can make the interpretation of the F-test unreliable. An article using ANOVA should report how all these assumptions were tested and what remedial measures or adjustments, if any, were taken.

Linear regression

Linear regression (further discussed in Chapter 8 and Chapter 10) assumes the independence of errors in the independent and dependent variables. This assumption might not be met if there is a seasonal effect, for instance. (Sales of ice cream tend to be higher in hot months.) The article should describe how this assumption was tested (generally, through a residuals analysis) and what was done if nonindependence in the error terms was discovered (e.g., the use of time series analysis instead of linear regression).

Presenting at the Journal Club

Many research institutions and academic departments have a journal club, which is a group of people who meet on a regular basis to discuss published work in their field. Pizza or other culinary enticements might be included to improve attendance. Often, meetings are structured around one or two articles, each of which is presented and critiqued by a member of the club. How can you prepare best when it's your turn to present? Of course, you will follow any guidelines provided by the club, but here are a few suggestions that might be helpful:

1. Pick an article worth presenting. Remember that you might be taking up an hour or more of your colleagues' time, and they're just as busy as you are. Start by reading abstracts from a number of articles, and then pick one with an interesting hypothesis, a good research design, and adequate data to draw a firm conclusion.

2. Read the article at least three times, once to get a general feel for the author's arguments, the second time critically, and the third time to pick out the *major points* you want to emphasize in your presentation. Avoid the temptation simply to run through the material in the order presented in the article. Before you present, you should know this material so well that you can speak to issues rather than simply regurgitating the information on the page.

3. Put the article in context. Who else is working on this topic, how does the theory in this article relate to others in the field, and so on? In addition, who funded the research, and is there any obvious conflict of interest?

4. Briefly define any essential technical terms or statistical techniques that might be unfamiliar to your audience, including, if appropriate, a critique of how they are used in the article.

5. Outline your presentation, keeping in mind the time allotted to you and allowing time for questions and discussion. Many people find it helpful to sketch out the different sections of their presentation (sort of a crude outline) to be sure they're allotting adequate time to the most important issues.

6. It's not cheating to plant several questions with colleagues you know will be in attendance in case the discussion is slow to get started. This is also a good way to direct the discussion toward points on which you would like to elaborate further.

A

Review of Basic Mathematics

You don't need to be an ace in mathematics to learn statistics, and nowadays pocket calculators and computer programs can do much of the calculation drudgery for you. However, a good understanding of how numbers work, including the basic laws of arithmetic and algebra, is a prerequisite to being able to reason statistically. Although anyone can learn to churn out calculations, if you don't understand the meaning of the numbers thus produced, your efforts can be useless or counterproductive. Besides, it's always more fun to understand what you are doing, and if you truly understand numbers and can explain them to others, you'll find you have a great advantage over other candidates, whether in school or at work.

If the math you learned in school has faded to a distant memory, don't worry; you have lots of company! Even if you did well in high school algebra, a brief review of the basic concepts can ease your path into statistics, and working through some elementary problems will help sharpen your mind before you take on more complex calculations. Running through simple calculations is also a good way to get acquainted with a new calculator or a new software program. Start by working with calculations in which you know the right answer, and you'll be much more confident in using the technology to tackle new problems.

I had a calculus teacher in college who told us that most of the errors students made in their homework were errors in algebra. Not only was he right, but many of our mistakes came in when using principles we had learned in junior high school! The same principle applies in statistics; nothing is complicated about the math you need, at least at the beginning level, but you need to be very comfortable with the material, and you need it fresh in your mind. To that end, this appendix offers a friendly review of some basic mathematics, which I hope will reduce the anxiety and refresh the memories of those who don't quite remember the last time they multiplied exponents or plotted Cartesian coordinates.

If you want to see how much you remember, you can go straight to the quiz at the end of the chapter; if you do well on all the topics, you can safely skip this appendix. On the other hand, if you do poorly on the quiz, you might want to supplement the materials in this appendix by working through an algebra review text aimed at the high school or college freshman market. If you discover that you like statistics so much you want to major in it, you will eventually need to take several semesters of calculus as well as calculus-based statistics courses, but that level of math is far beyond anything you will need for the techniques presented in this book.

Laws of Arithmetic

It's often helpful to think of numbers as points along a number line, in which lower numbers are to the left and higher numbers are to the right. You might remember the number line from primary school (Figure A-1).

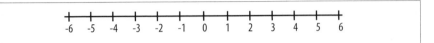

Figure A-1. Number line

The concept of the number line is useful in statistics because we often refer to a value in a distribution as being "farther to the right" when what we really mean is "of a higher value." The statement that a value is "at least as extreme" or "at least as far from the mean" as another value, which you will frequently encounter in hypothesis testing, also refers to the number line. Distributions such as the normal distribution are symmetrical and have a single most common central value; as values get farther from that central value (farther to the left or right), they become less likely.

Numbers may be written with either a positive or negative sign; if no sign is included, positive value is assumed. The *absolute value* of a is written $|a|$ and means the distance a occupies on the number line, whether in a negative or positive direction. This means that if $a = -5$ and $b = 5$, the absolute values of a and b are identical: $|a| = |b| = 5$. Another way to look at it is that the absolute value of the number is the same as the value of the number after any negative sign is removed. By this rule, $|-5|$ is larger than $|4|$, even though 4 is larger (farther to the right) than -5, because 5 (the absolute value of $|-5|$) is larger than 4 (the absolute value of $|4|$)

To add numbers with like signs, add their absolute values and keep the sign:

$3 + 5 + 8 - 3 + -5 = -8$

To add two numbers with different signs, subtract their absolute values and keep the sign of the number with the larger absolute value:

$-3 + 5 = 2 \ 3 + -5 = -2$

To add more than two numbers with different signs, group them by signs, add the absolute values of each set, and then subtract the negatives from the positives:

$$-3 + 5 + -2 + 4 = (5 + 4) - (3 + 2) = 4$$

As you can see, adding a negative number is the same as subtracting a positive number. This is formalized in the following law:

$$a - b = a + -b$$

So:

$$2 - 5 = 2 + (-5) = -3$$

To multiply numbers with like signs, multiply their absolute values. If all values are positive, the result is positive. If all are negative, count the number of negative signs. If there is an even number of negative signs, the result is positive; if an odd number, the result is negative:

$$4(2) = 8, -4(-2) = 8, -4(-2)(-3) = -24$$

To multiply numbers with unlike signs, multiply the absolute values and then count the number of negative signs; if even, the result is positive; if odd, the result is negative:

$$-4(2)(-3) = 24, -4(2)(3) = -24$$

To divide numbers with like signs, divide the absolute values and make the result positive. To divide numbers with unlike signs, divide the absolute values and make the result negative:

$$10/5 = 2, -10/-5 = 2, 10/-5 = -2$$

Order of Operations

In general, we solve arithmetic expressions from left to right but perform arithmetic operations within an expression in the following order:

1. Anything in parentheses
2. Exponents and roots
3. Multiplication and division
4. Addition and subtraction

Legions of schoolchildren have learned this by the mnemonic "Please excuse my dear aunt Sally": parentheses, exponents and roots, multiply and divide, add and subtract. If there are multiple layers of parentheses, you solve the expressions in order, beginning with the innermost parentheses. Table A-1 shows some examples.

Table A-1. Order of operations examples

Expression	Rule	Result
$2 + 5 \times 10$	Multiplication before addition	52
$(2 + 5) \times 10$	Expressions in parentheses first	70
10×2^2	Exponents before multiplication	40
$(10 \times 2)^2 + 5$	Expressions in parentheses first, then exponents, then addition	405
$10 - 4/(2 + 2)$	Expressions in parentheses first, then division, then subtraction	9
$[5 + 3(4 + 6)]/(3 + 2)$	Innermost parentheses first and multiplication before addition	7

Properties of Real Numbers

Real numbers are the types of numbers familiar from everyday life and that are used most often in math and statistics. They can be written using decimals and therefore include rational numbers, such as 4 and 7/5, and irrational numbers, such as π (3.1415 . . .) and the square root of 2 (1.4142 . . .), but not imaginary or complex numbers (numbers that are negative when squared). Unless otherwise specified, real numbers are assumed throughout this review. Some properties of real numbers include:

- The *associative property* for addition and multiplication:

  $(a + b) + c = a + (b + c)$ so $(1 + 2) + 3 = 1 + (2 + 3) = 6$, $a (b \times c) = (a \times b) c$ so $2 \times (3 \times 4) = (2 \times 3) \times 4 = 24$

- The *commutative property* for addition and multiplication:

  $a + b = b + a$ so $5 + 4 = 4 + 5 = 9$, $a \times b = b \times a$ so $2 \times 3 = 3 \times 2 = 6$

- The *distributive property* of multiplication:

  $a (b + c) = ab + ac$ so $5(2 + 3) = 5(2) + 5(3) = 10 + 15 = 5(5) = 25$

- The *additive identity* of 0: any number plus 0 = the number itself:

  $a + 0 = a$ so $5 + 0 = 5$

- The *multiplicative identity* of 0: any number times 0 = 0:

  $a \times 0 = 0$ so $5(0) = 0$

- The *multiplicative identity* of 1: any number times 1 = the number itself:

  $a(1) = a$ so $5(1) = 5$

- The *inverse property of addition*: the sum of any number and its inverse is 0:

  $a + -a = 0$ and $-a + a = 0$ so $5 + -5 = 0$ and $-5 + 5 = 0$

- The *rule of double negatives*: pairs of negatives cancel each other out:

  $-(-a) = a$ so $-(-5) = 5$

- The *inverse property of multiplication*:

  $a \times (1/a) = 1$ if $a \neq 0$ (because division by 0 is undefined) so $5 \times (1/5) = 1$

Exponents and Roots

An exponent tells you to multiply the base number by itself as many times as the exponent says:

- $a^n = a \times a \times a \ldots n$ times, where a is the *base* and n the *exponent*, so $2^4 = 2 \times 2 \times 2 \times 2 = 16$
- a^2 is often referred to as "a squared" and a^3 as "a cubed"; they can also be read as "a to the second power" or "a to the second" and so on, and this system is used for powers above 3 (a^7 would be read as "a to the seventh power").
- *Multiplying exponential numbers* with a common base: add the exponents and keep the base:

  $a^m \times a^n = a^{m+n}$ so $3^2 \times 3^3 = 3^{2+3} = 3^5 = 243$
- *Power rules* for exponents:

  $(a^m)^n = a^{mn}$ so $(2^2)^3 = 2^6 = 64$, $(ab)^n = a^n b^n$ so $(5 \times 4)^2 = 5^2 \times 4^2 = 400 = 25 \times 16$ $(a/b)^n = a^n/b^n$ so $(3/4)^2 = 3^2/4^2 = 9/16$, assuming $y \neq 0$
- *Zero exponent*: any number other than 0, with an exponent of 0, $= 1$:

  $a^0 = 1$ so $245^0 = 1$ and $-8^0 = 1$ (0^0 is undefined)
- A *negative exponent* is the same as dividing by the base raised to the power of the exponent:

  $a^{-1} = 1/a$ and $a^{-2} = 1/a^2$ so $2^{-1} = 1/2$ and $2^{-2} = 1/2^2 = 1/4$, $(a/b)^{-n} = (b/a)^n$ so $(5/3)^{-2} = (3/5)^2 = 9/25$
- When *dividing exponential numbers* with a common base, subtract the exponents:

  $a^m/a^n = a^{m-n}$ (assuming $a \neq 0$) so $3^5/3^2 = 3^{5-2} = 3^3 = 27$

Taking the root of a number is the inverse of raising it to an exponential value: the *n*th root of x is the number a such that $a^n = x$. This might be easier to understand if we consider the *square root*, which is the second root of a number. The square root of 9 is 3 because $3^2 = 9$. Technically speaking, 3 is the *principal* square root of 9 (–3 is also a square root of 9), but this distinction is often ignored in practice. Similarly, the third root of 125 is 5 because $5^3 = 125$. The third root is also called the *cube root*; beyond 3, the usual terminology is fourth root, fifth root, and so on.

Properties of Roots

Figures A-2 to A-4 show several important rules for working with roots.

$$\sqrt[n]{ab} = \sqrt[n]{a}\sqrt[n]{b} \text{ when both a and b } \geq 0$$

Figure A-2. Rule for multiplying roots

$$\sqrt[n]{\frac{a}{b}} = \frac{\sqrt[n]{a}}{\sqrt[n]{b}} \text{ when both a and b } \geq 0$$

Figure A-3. Rule for dividing roots

$$\sqrt[n]{a^m} = \left(\sqrt[n]{a}\right)^m = a^{\frac{m}{n}} \text{ when both a and b } \geq 0$$

Figure A-4. Rule for roots with exponents

You can demonstrate these rules for yourself by using your calculator, as shown in Figure A-5.

$$\sqrt{4 \times 16} = \sqrt{4}\sqrt{16} = 2 \times 4 = 8$$

$$\sqrt[3]{\frac{27}{64}} = \frac{\sqrt[3]{27}}{\sqrt[3]{64}} = \frac{3}{4} = 0.75$$

$$\sqrt[3]{8^2} = \left(\sqrt[3]{8}\right)^2 = 8^{\frac{2}{3}} = 4$$

Figure A-5. Applying the rules for roots

A *logarithm* (often abbreviated *log*) is the power to which you need to raise a given base to produce a particular number. Using a base of 10, log_{10} 100 = 2 because $10^2 = 100$. Although any number can serve as a base, in statistics we often work with base-e exponential functions. These are also called *natural logarithms* or *Naperian logarithms* and are written ln x which means $\log_e x$. The base e is the irrational number 2.718 . . . and is useful to describe many processes in the natural sciences, hence the name "natural log." Scientific calculators usually have an LN key to calculate natural logs, and many computer programs have built-in functions for the same purpose. Be forewarned, however; sometimes, the function to compute a natural log is abbreviated LOG rather than LN, so you need to determine the correct symbol for the calculator or computer program you are using.

The equation ln x = 1.5 is equivalent to writing $e^{1.5} = x$. In this case, x = 4.48 (rounded) because $e^{1.5}$ = 4.48, and we can say that the natural log of 4.48 is 1.5. The following principles hold for logarithms of whatever base (the base is signified by b in these examples):

- $\log_b 1 = 0$ because $b^0 = 1$ (because any number to the 0th power = 1)
- $\log_b b = 1$ because $b^1 = b$ (because any number to the first power equals itself)
- $\log_b b^x = x$ (because by definition the log of b^x is x if the base is b)
- $b^{\log_b x}$ where $x > 0$ (because $\log_b x$ is the exponent to which you raise b to get x)

The following properties of logarithms are also useful in statistics:

- $\log_b MN = \log_b M + \log_b N$ (The logarithm of a product is the sum of the logarithms.)
- $\log_b M/N = \log_b M - \log_b N$ (The logarithm of a quotient is the difference of the logarithms.)
- $\log_b M^p = p \log_b M$

You can demonstrate these principles to yourself by using a pocket calculator. For instance, using natural logs:

$\ln (2 \times 4) = \ln 2 + \ln 4 = 0.693 + 1.386 = 2.079$
$\ln (2/5) = \ln 2 - \ln 5 = 0.693 - 1.609 = -0.916$
$\ln 2^3 = 3 \ln 2 = 3(0.693) = 2.079$

Note that logarithms for numbers between 0 and 1 are negative, and logarithms for numbers less than 0 are undefined. (You'll get an error message on your calculator if you try to find $\ln - 1$.)

Solving Equations

The following *properties of equality* will help you solve equations:

- If $a = b$, then $a + c = b + c$ (Adding a constant to both sides of an equality does not change the equality.)
- If $a = b$, then $a - c = b - c$ (Subtracting a constant from both sides of an equality does not change the equality.)
- If $a = b$, then $ac = bc$ (Multiplying both sides of an equality by a constant does not change the equality.)
- If $c \neq 0$, then $a/c = b/c$ (Dividing both sides of an equality by a nonzero constant does not change the equality.)

These properties come in handy, as do the properties of the preceding real numbers, when solving linear equations. For instance, to solve:

$5(x - 4) = 40$

Multiply out the left side:

$5x - 20 = 40$

Then isolate x by adding 20 to both sides:

$5x = 60$

Then divide both sides by 5:

$$x = 12$$

To check the solution, we substitute 12 back into the original equation:

$$5(12 - 4) = 5(8) = 40, \text{ which is correct.}$$

For more complex problems, we need to *combine like terms* as follows:

$2(3x + 1) = 5(x + 2)$	
$6x + 2 = 5x + 10$	Multiply out both sides.
$x + 2 = 10$	Subtract 5x from both sides.
$x = 8$	Subtract 2 from both sides.
$2(24 + 1) = 5(8 + 2) = 50$	Check: substitute 8 for x in the original equation.

Logarithms are useful for solving equations that include exponents; you can take the log of both sides and then use the properties of logarithms to solve for the unknown. For instance, using a base of 10:

$5^x = 3$	
$\log 5^x = \log 3$	Take the log of both sides.
$x \log 5 = \log 3$	Use the law of exponents and logs.
$x = \log 3 / \log 5 = 0.683$	Divide both sides by log 5.
$5^{0.683} = 3$	Check: substitute 0.683 for x in the original equation.

Systems of Equations

A *system of equations*, also known as a *system of simultaneous equations*, is a set of algebraic equations with common variables. Solving a system of equations means finding a common solution, values for the variables that will be correct for all equations in the system. If there is a common solution (which is the case with all the systems presented here), the system is called *consistent*; if not, the system is called *inconsistent*. Systems of equations can be solved by graphing (by drawing the lines represented by the equations; the solution is the point of intersection) or by using algebra. We will present only the latter method here.

Solving some system of equations problems is a good review of algebra and logical reasoning. A simple approach to solving systems of equations, which will work for the examples presented here, is to simplify each equation as much as possible and then use either the method of substitution or the method of addition and subtraction to solve the system. We'll demonstrate with systems of *two equations in two unknowns*, although the same principles can be used to solve larger systems such as three equations in three unknowns. That's about the point, however, when it becomes more convenient to solve more complex problems using matrices, a topic that is beyond this basic review.

Here is a demonstration of the *method of substitution* used to solve a system of two equations in two unknowns (the unknowns are x and y):

$2x + y = 6$, $3x - 2y = 16$

Solve the first equation for y:

$y = 6 - 2x$

Substitute this value for y into the second equation:

$3x - 2(6 - 2x) = 16$

Solve the second equation for x:

$3x - 12 + 4x = 16$, $7x = 28$, $x = 4$

Substitute this value into the first equation to solve for y:

$y = 6 - (2 \times 4) = -2$

So the solution is $(4, -2)$. That is, $x = 4$, $y = -2$. Check by substituting these values into the equations:

$2(4) + (-2) = 6$, $3(4) - (2 \times -2) = 16$

To use the *method of addition* (or subtraction) to solve the same system of equations, you add or subtract the like terms from the two equations so that one of the variables drops out and then solve for the other variable. An additional step is often necessary, which is to multiply one or both equations by a constant so that one of the variables (x or y) will drop out when the systems are added or subtracted. In this case, we multiply the first equation by 2:

$2[2x + y = 6]$ becomes $4x + 2y = 12$

We then substitute this equation (which is equivalent to the original expression because all we have done is multiply both sides by a constant) in the system and add it to the second equation. Figure A-6 continues with the system of equations from the previous example.

$$+\frac{4x + 2y = 12}{3x - 2y = 16}$$

$$= 7x + 0y = 28 \text{ so } x = \frac{28}{7} = 4$$

Figure A-6. Solving a system of equations using the method of addition

We can then use this value to solve either equation for y.

$2(4) + y = 6$ so $y = -2$, $3(4) - 2y = 16$ so $y = -2$

This gives us the same solution as with the substitution method: $(4, -2)$.

Graphing Equations

Points in multidimensional space are often described using *Cartesian coordinates*, also called *rectangular coordinates*, which are simply the values on each dimension in a system that locate a particular point. We will demonstrate this system by using two dimensions because that is easier to display on a printed page, but the same concepts can be applied to higher numbers of dimensions.

Identifying the location of points in two-dimensional space is done using a plane with two axes, x (horizontal) and y (vertical), as in Figure A-7. Each point in this plane is identified by two numbers, the x-coordinate and the y-coordinate, always listed in that order. For instance, the point $(2, 3)$ has an x-coordinate of 2 and a y-coordinate of 3; the point $(-1.5, -2.5)$ has an x-coordinate of -1.5 and a y-coordinate of -2.5.

Linear equations may be written in the form $y = mx + b$, where m is called the *slope* and b is the *y-intercept*; this method of notation is called the *slope-intercept form* of a line. Lines can also be written using an alternative notation, $y = ax + b$, in which case a is the *slope* and b is the *y-intercept*. Either method of notation gives you the *slope-intercept* form of a line. To plot a linear equation (one that does not include squares or higher-order terms) using Cartesian coordinates, find two or more pairs of coordinates that satisfy the equation and then draw a straight line connecting them. Here's a simple example:

$y = 2x + 4$

Here are some possible solutions. (Note: There are an infinite number of solutions to this equation.)

$x = 0, y = 4; x = 1, y = 6; x = -2, y = 0$

Graphing these solutions can be done as in Figure A-8.

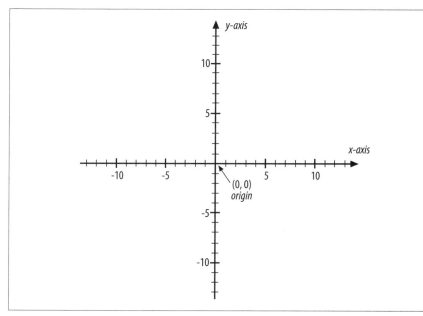

Figure A-7. The Cartesian coordinate system

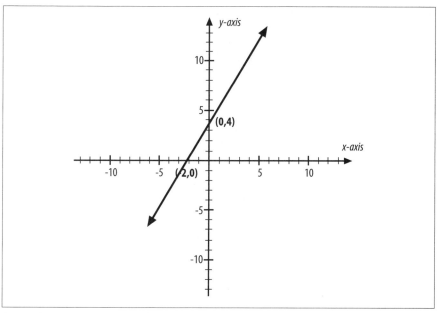

Figure A-8. Line representing the equation y = 2x + 4

The interpretation of the line's components:

Slope

The amount of increase in *y* for a one-unit increase in *x*

Intercept
    The value of *y* when *x* = 0, that is, the value when the line crosses the *y*-axis

Even without drawing a graph, you can interpret the equation and predict new values of *y* given *x*. Look at the following equation:

$$y = -3x + 6$$

Because the slope of this line is negative, we know that the line will run from the upper left to lower right of the graph (the opposite of the line graphed in Figure A-8, which has a positive slope). We also know that as *x* increases, *y* decreases, and vice versa. The intercept (6) also tells us that the line will cross the *y*-axis at 6. We can calculate some points on the line as follows (it's often easier to find the *x* - intercept and *y*-intercept immediately). Table A-2 shows some possible values.

Table A-2. Some values for the line y = –3x + 6

x	y
2	0
0	6
1	3

The graph of this equation is shown in Figure A-9.

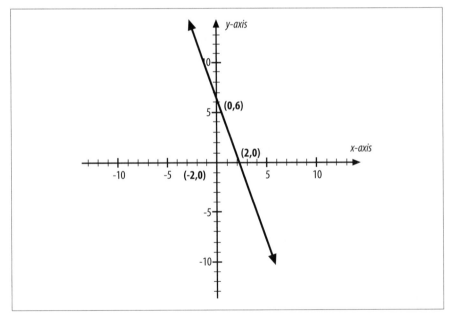

Figure A-9. Graph of the equation y = –3x + 6

Another way to write the equation of a straight line is by using what is called the *point-slope* form. This format relies on the fact that if we know the slope of the line and one point on it, we can draw the line and calculate the coordinates of any point on the line. Similarly, if we know two points on the line, we can calculate the slope. To put it another way, a straight line can be uniquely identified by two points or by one point plus its slope. The point-slope form of a line is written as:

$$y - y_1 = m(x - x_1)$$

where m is the slope of the line and (x, y) and (x_1, y_1) are two points on the line. We can find the slope, given two points on the line, using the formula shown in Figure A-10.

$$m = \frac{y - y_1}{x - x_1}$$

Figure A-10. Formula for the slope of a line

You might remember this as "slope = rise over run," where *rise* is the change in *y*-values (the change on the vertical axis), and *run* is the change in *x*-values (the change on the horizontal axis) between the two points. If we have the points (0, 6) and (2, 0), the slope of the line that contains them is shown in Figure A-11.

$$m = \frac{6 - 0}{0 - 2} = \frac{6}{-2} = 3$$

Figure A-11. Finding the slope of a line

This corresponds to the slope that we found in the previous example. If instead our line included the points (6, 6) and (4, 2), its slope would be as shown in Figure A-12.

$$m = \frac{6 - 2}{6 - 4} = \frac{4}{2} = 2$$

Figure A-12. Finding the slope of a line

Continuing with this example, if we know that a line with slope 2 runs through the point (6, 6), we can find the *y*-coordinate for 4 using the point-slope equation:

$$y - y_1 = m(x - x_1), 6 - y_1 = 2(6 - 4), -y_1 = 4 - 6 = -2 \ y_1 = 2$$

Linear Inequalities

An *equation* connects two expressions with an *equals sign*; for instance, $y = mx + b$ is the equation of a line. Often, we want to connect two expressions with *inequalities*, signs stating that the two sides of the equation are not equal. Some commonly used symbols for inequalities are shown in Table A-3.

Table A-3. Commonly used inequalities

Sign, abbreviation	Meaning	Examples
≠, <>, NE	Not equal	$a \neq b, a <> b, a$ NE 5
<, LT	Less than	$a < b, a$ LT 5
>, GT	Greater than	$a > b, a$ GT 5
≤, <=, LE	Less than or equal	$a \leq b, a <= b, a$ LE 5
≥, >=, GE	Greater than or equal	$a \geq b, a >= b, a$ GE 5
≈	Approximately equal	$a \approx b, a \approx 5$

The alphabetical abbreviations such as GE and LT in particular are often used to indicate inequalities when writing computer code.

We can evaluate inequalities for their logical or truth value. For instance, if $a = 5$ and $b = 6$, then $a < 6$ and $a < b$ are both true, whereas $a > 5$ and $a > $ b are both false. The following laws govern linear inequalities:

1. If the same number is added or subtracted from both sides of an inequality, the inequality remains in the same direction.

   If $a < b$, then $a + x < b + x$ and $a - x < b - x$. $6 < 10$, so $(6 + 4) < (10 + 4)$ and $(6 - 1) < (10 - 1)$.

2. If the same positive number is used to multiply or divide both sides of an inequality, the inequality remains in the same direction.

   If $a > b$, then $ax > bx$ and $a/x > b/x$. $5 > 3$, so $(5 \times 2) > (3 \times 2)$ and $(5/2) > (3/2)$.

3. If the same negative number is used to multiply or divide both sides of an inequality, the direction of the inequality is reversed.

   If $a < b$, then $a(-x) > b(-x)$. $2 < 4$, so $2(-3) > 4(-3)$ and $2/-3 > 4/-3$, i.e., $-6 > -12$ and $-2/3 > -4/3$.

A linear inequality can be solved using the same steps used to solve linear equations. For instance:

$$4(3x + 2) < 20$$
$$12x + 8 < 20$$
$$12x < 12$$
$$x < 1$$

Fractions

A fraction is simply a way of expressing one number divided by another. The top number is called the numerator, and the bottom number is the denominator, as shown in Figure A-13.

$$\text{fraction} = \frac{\text{numerator}}{\text{denominator}}$$

Figure A-13. The numerator and denominator of a fraction

Figure A-14 presents some basic properties of fractions. (All assume no division by 0.)

$$1.\ \frac{a}{b} = \frac{c}{d} \text{ if and only if } ad = bc$$

$$2.\ \frac{a}{1} = a$$

$$3.\ \frac{a}{a} = 1$$

$$4.\ \frac{a}{b} = \frac{ac}{bc}$$

$$5.\ -\frac{a}{b} = \frac{-a}{b} = \frac{a}{-b}$$

Figure A-14. Properties of fractions

Note that property 4 follows from property 3: anything divided by itself = 1, so multiplying by *c/c* as in this case is simply multiplying by 1 and does not change the value of the fraction. This property also allows us to *simplify* fractions by dividing out common factors, as shown in Figure A-15.

$$\frac{8}{24} = \frac{8 \times 1}{8 \times 3} = \frac{1}{3}$$

$$\frac{4x^3 y^2}{2xy^3} = 2xy^{-1}$$

Figure A-15. Simplifying fractions

Remember from our review of exponents that $y^{-1} = 1/y$.

To *add or subtract fractions*, they need to have a common denominator. You might remember from grade school an exercise called "finding the least common denominator" or "finding the LCD," but for our purposes, any common denominator will do. When you have a common denominator, you just add or subtract the numerators and keep the denominator, as shown in Figure A-16.

$$\frac{a}{c} + \frac{b}{c} = \frac{a+b}{c}$$

Figure A-16. Adding fractions with a common denominator

If the fractions don't have a common denominator, you need to multiply or divide as necessary to get a common denominator, and then do the addition or subtraction and simplify the result by dividing out the common factors—for instance, as in Figure A-17.

$$\frac{5}{6} + \frac{2}{4} = \frac{10}{12} + \frac{6}{12} = \frac{16}{12} = \frac{4}{3} \text{ or } 1\frac{1}{3}$$

Figure A-17. Adding fractions by using a common denominator

1 1/3 is called a *mixed number* because it has both an integer part and a fractional part. 4/3 is called an *improper fraction* because its numerator is larger than its denominator. To convert an improper fraction to a mixed number, remove as many whole units as possible from the fraction, so the final expression is the whole units plus the remainder expressed as a fraction, as in Figure A-18.

$$\frac{4}{3} = \frac{3}{3} + \frac{1}{3} = 1\frac{1}{3}$$

Figure A-18. Converting an improper fraction to a mixed number

To *multiply fractions*, multiply the numerators and denominators separately and simplify the result, as in Figure A-19.

$$\frac{a}{c} \times \frac{b}{d} = \frac{ab}{cd}$$
$$\frac{9}{5} \times \frac{10}{27} = \frac{90}{135} = \frac{2}{3}$$

Figure A-19. Multiplying fractions

To *divide fractions*, invert and multiply. This is possible because dividing by x is the same as multiplying by $1/x$ (that is, dividing is the same as multiplying by the *reciprocal* of the divisor). This is shown in Figure A-20.

$$\frac{a}{b} \div \frac{c}{d} = \frac{a}{b} \times \frac{d}{c} = \frac{ad}{bc}$$

$$\frac{3}{4} \div \frac{1}{2} = \frac{3}{4} \times \frac{2}{1} = \frac{6}{4} = 1\frac{1}{2}$$

Figure A-20. Dividing fractions

Fractions can also be expressed as *decimals* or *percents*. A percent is just a fraction in which the denominator is 100 (*cent* = 100 in Latin). With calculators, it's easy to convert any fraction to a decimal and then convert it to a percent by multiplying by 100; some calculators even have a special key to return divisions automatically as percents. So:

1/4 = 0.25 = 25%, 6/4 = 1.5 = 150%

To take a percent of a number, multiply by the decimal equivalent of that number. For instance, 40% of 30 = 0.4(30) = 12. To calculate an increase over some base number, multiply by 1.0 plus the increase; for instance, calculate a 20% increase by multiplying by 1.2 because multiplying by 1.0 gives you the original number, and multiplying by 0.2 gives you the 20% increase. For this reason, a 100% increase, which is the same as doubling, means multiplying by 2.0 (1.0 for the original number, 1.0 for the increase). To find a decrease from a total, multiply by 1 – the decrease; for instance, to find the number that represents a 10% decrease from 100, multiply 100 by 0.9, so 100(.9) = 90.

Factorials, Permutations, and Combinations

The *factorial* of a number is that number multiplied by all the smaller integers down to 1. The factorial of n is written $n!$ and means $n(n - 1)(n - 2) \ldots (1)$, so:

5! = 5(4)(3)(2)(1) = 120

and:

10! = 10(9)(8)(7)(6)(5)(4)(3)(2)(1) = 3,628,800

Many calculators have a factorial key, usually indicated by *!* or *x!*, as well as permutation and combination keys, often indicated by *nPr* and *nCr*. If your calculator has these keys, experiment with them as you work through this section. Fractions that include factorials can often be simplified by canceling common factors, a useful property because factorials quickly become very large numbers, as we saw in the example of 10!. The utility of canceling common factors should be clear from the example in Figure A-21.

$$\frac{10!}{8!} = \frac{10 \times 9 \times 8 \times 7 \times 6 \times 5 \times 4 \times 3 \times 2 \times 1}{8 \times 7 \times 6 \times 5 \times 4 \times 3 \times 2 \times 1} = 10 \times 9 = 90$$

Figure A-21. Cancelling common factors in a factorial problem

Factorials are useful in problems in which you are concerned with arranging a finite number of objects in order. For instance, how many ways are there to arrange five books on a shelf? You have five choices for the first book, four for the second (because the first book has already been used and can't be chosen again), three for the third, two for the fourth, and one for the fifth. The answer is therefore 5! = 120.

If you are interested in the number of ways to arrange a subset of objects from a finite set of distinct objects (that is, all the objects are different), you can use *permutations* to calculate the answer. In fact, the number of ways to arrange five out of five objects, as in the previous paragraph, is a permutation problem in which the subset is the same as the entire set. However, more typically, a permutation question deals with something like the number of ways to arrange three books from a set of five. There are several conventions in permutation notation, so see Figure A-22, which denotes the number of ways to arrange *r* objects chosen from a set of *n*.

$$P(n, r) = nPr = \frac{n!}{(n-r)!}$$

Figure A-22. Formula for a permutation

The number of ways to arrange three objects selected from five is shown in Figure A-23.

$$5P3 = \frac{5!}{(5-3)!} = 60$$

Figure A-23. Solving a permutation

Note that, by convention, 0! is defined as 1, not 0, to avoid the problem of division by 0.

In a permutation, the order of objects is significant. If we were arranging sets of three from the first five letters of the alphabet, for instance, (a, b, c) would be a different permutation than (a, c, b). If order is not a concern, we are dealing with combinations rather than permutations. In a *combination*, we are interested in the number of distinct sets of *r* objects that can be selected from a set of *n* objects but do not count different orders of the same objects as a different set. When choosing sets of three from the first five letters of the alphabet, (a, b, c) would be considered the same combination as (a, c, b). Like permutations, there is not one standard notation for combinations, and you might see any of the equations in Figure A-24 used to denote the number of combinations of r objects from a set of n.

$$C(n, r) = nCr = \binom{n}{r} = \frac{n!}{r!(n-r)!}$$

Figure A-24. Different ways to write a combination

The number of ways to select three objects from five, when order is not important, is shown in Figure A-25.

$$5C3 = \frac{5!}{3!(5-3)!} = 10$$

Figure A-25. Solving a combination

Exercises

Here's a review of the concepts in this appendix.

Laws of Arithmetic and Real Numbers

You will get a better diagnosis of your current state of mathematical understanding if you do the first seven sections without using a pocket calculator—that is, if you use your knowledge of algebra to solve them by hand. In the case of answers with unresolved variables (such as x or y), just restate them in simplest form.

1. $3 + (-8) =$
2. $6/-3 =$
3. $(-8y)(-6z) =$
4. $2 + 5/10 =$
5. $(2 + 5)/10 =$
6. $6 + 3^2 - 5 =$
7. $(3 + 2)^2 =$
8. $[12(5) - 2(3)] / (3 \times 2) =$
9. $-(3 - 5x) =$
10. $6(4 + 2x) - x(5) =$
11. $3(4/x) =$
12. $5x (4 - 2) =$

13. $(5x + 6)(3) =$

Exponents, Roots, and Logarithms

1. $2^0 =$
2. $(1/4)^2 =$
3. $(-x)^4 =$
4. $(x^3)^2 =$
5. $2^2(2^3) =$
6. $x^5(x^{-2}) =$
7. $(4 \times 2)^2 =$
8. $2^{-1} =$
9. $x^2/x^4 =$
10. $(2/3)^2 =$
11. $(7y^2)^1 =$
12. $(5/9)^{-1} =$
13. $x^5/x^{-2} =$
14. $(27/8)^{-\frac{1}{3}} =$
15. $(4/9)^{\frac{1}{2}} =$
16. $\sqrt{x^4}$
17. $\sqrt[3]{27y^3}$
18. $\sqrt{4 \times 16}$
19. $\sqrt{\dfrac{25}{81}}$
20. $\sqrt[4]{\dfrac{x^4}{y^6}}$

Natural Logarithms

1. $e^0 =$
2. $\ln 1 =$
3. $\log_{10} 100 =$
4. $\log_{10} (5 \times 2) =$
5. $\ln e^3 =$

Solving Equations for x

1. $3x + 7 = 20$
2. $(1/3)x = 6$
3. $3(x + 2) = 2(x + 1)$
4. $4x = 3(x - 2) + 7$

Systems of Linear Equations

1. $3x - 2y = 6$ and $x + 2y = 14$
2. $x + 3y = -1$ and $2x + y = 3$

Linear Equations and Cartesian Coordinates

1. Given a line with the equation $y = 3x + 2$, fill in the following table.

Table A-4. Solving for Cartesian coordinates

x	y
0	
	0
1	
−1	

2. In the equation $y = -x + 5$, what is the slope and what is the y-intercept?
3. Given the equation $y = 6 - 2x$, if x increases by 2, what happens to y?
4. Find the slope for the following pair of points: $(5, 3)$ and $(2, -1)$.
5. Given a line with slope -1 that runs through the point $(2, 4)$, find the y-coordinate for the line when it passes through $x = -3$.

Linear Equalities

1. If $a < b$, what is the relationship of $3a$ to $3b$?
2. If $a < b$, what is the relationship of $-2a$ to $-2b$?
3. Solve down to an inequality for x: $5(2x - 1) > 8$
4. Solve down to an equality for x: $3x(2)$ GE 4

Fractions, Decimals, and Percents

1. $\dfrac{3x^2 y}{1} =$

2. $\dfrac{5xy^3 z^2}{6y^5} =$

3. $\dfrac{8}{10} + \dfrac{3}{15} =$

4. $\dfrac{8y^3}{2y} + \dfrac{9y^2}{3} =$

5. $\dfrac{5}{4} \times \dfrac{7}{3} =$

6. $\dfrac{3x}{7} \times \dfrac{2}{x} =$

7. $\dfrac{7}{5} \div \dfrac{14}{10} =$

8. $\dfrac{x}{3} \div \dfrac{2}{3x} =$

Note: You may use a calculator for the next four questions.

9. What is 20% of 75?

10. What is the decimal equivalent of 7/21?

11. If we sold 500 units last year and sales increased by 10% this year, how many units did we sell this year?

12. If we sold 500 units last year and sales declined by 20% this year, how many units did we sell this year?

Factorials, Permutations, and Combinations

You may use a calculator for this section.

1. $7! =$

2. 6P4 =

3. 8C3 =

4. $\dfrac{x!}{(x-1)!} =$

5. How many ways are there to choose a batting lineup (9 players) from 15 players total (order does count)?

6. How many unique combinations (order does not count) of 5 items can you select from 10 unique items?

Answers

Laws of Arithmetic and Real Numbers

1. $3 + (-8) = -5$
2. $6/-3 = -2$
3. $(-8y)(-6z) = 48yz$
4. $2 + 5/10 = 2.5$ or $2\ 1/2$
5. $(2 + 5)/10 = 7/10$ or 0.7
6. $6 + 3^2 - 5 = 10$
7. $(3 + 2)^2 = 25$
8. $[12(5) - 2(3)] / (3 \times 2) = 9$
9. $-(3 - 5x) = -3 + 5x$
10. $6(4 + 2x) - x(5) = 24 + 12x - 5x = 24 + 7x$
11. $3(4/x) = 12/x$ or $12x^{-1}$
12. $5x(4 - 2) = 10x$
13. $(5x + 6)(3) = 15x + 18$

Exponents, Roots, and Logarithms

1. $2^0 = 1$
2. $(1/4)^2 = 1/16$ or 0.0625
3. $(-x)^4 = x^4$
4. $(x^3)^2 = x^6$
5. $2^2 (2^3) = 2^5 = 32$
6. $x^5 (x^{-2}) = x^3$
7. $(4 \times 2)^2 = 8^2 = 64$
8. $2^{-1} = 1/2$ or 0.5
9. $x^2/x^4 = x^{-2}$ or $1/x^2$
10. $(2/3)^2 = 4/9$ or $0.444...$
11. $(7y^2)^1 = 7y^2$
12. $(5/9)^{-1} = 9/5$ or $1\ 4/5$ or 1.8
13. $x^5/ x^{-2} = x^7$
14. $(27/8)^{-1/3} = 2/3$
15. $(4/9)^{1/2} = 2/3$
16. $\sqrt{x^4} = x^2$
17. $\sqrt[3]{27y^3} = 3y$
18. 8

19. $\sqrt{\dfrac{25}{81}} = \dfrac{5}{9}$

20. $\sqrt[4]{\dfrac{x^4}{y^6}} = \dfrac{x}{y^{\frac{3}{2}}} = xy^{-\frac{3}{2}}$

Natural Logarithms

1. $e^0 = 1$
2. $\ln 1 = 0$
3. $\log_{10} 100 = 2$
4. $\log_{10}(5 \times 2) = 1$
5. $\ln^3 = 3$

Solving Equations for x

1. $3x + 7 = 20$: $x = 13/3$ or $4\,1/3$
2. $(1/3)x = 6$: $x = 18$
3. $3(x + 2) = 2(x + 1)$: $x = -4$
4. $4x = 3(x - 2) + 7$: $x = 1$

Solving Systems of Linear Equations

1. $3x - 2y = 6$ and $x + 2y = 14$: solution = $(5, 4.5)$
2. $x + 3y = -1$ and $2x + y = 3$: solution = $(2, -1)$

Linear Equations and Cartesian Coordinates

1.

x	y
0	2
−2/3	0
1	5
−1	−1

2. Slope = -1, y–intercept = 5
3. y decreases by 4.
4. 4/3
5. $y_1 = 9$

Linear Equalities

1. $3a < 3b$
2. $-2a > -2b$
3. $10x > 13$ or $x > 13/10$
4. x GE $4/6$ or x GE ⅔

Fractions, Decimals, and Percents

1. $\dfrac{3x^2 y}{1} = 3x^2 y$

2. $\dfrac{5xy^3 z^2}{6y^5} = \dfrac{5xz^2}{6y^2}$

3. $\dfrac{8}{10} + \dfrac{3}{15} = \dfrac{24}{30} + \dfrac{6}{30} = 1$

4. $\dfrac{8y^3}{2y} + \dfrac{9y^2}{3} = 7y^2$

5. $\dfrac{5}{4} \times \dfrac{7}{3} = \dfrac{35}{12} = 2\dfrac{11}{12}$

6. $\dfrac{3x}{7} \times \dfrac{2}{x} = \dfrac{6}{7}$

7. $\dfrac{7}{5} \div \dfrac{14}{10} = \dfrac{7}{5} \times \dfrac{10}{14} = 1$

8. $\dfrac{x}{3} \div \dfrac{2}{3x} = \dfrac{x}{3} \times \dfrac{3x}{2} = \dfrac{x^2}{2}$

9. 15
10. 0.333
11. 550
12. 400

Factorials, Permutations, and Combinations

1. $7! = 5040$
2. $6P4 = 360$
3. $8C3 = 56$
4. x
5. $15P9 = 1{,}816{,}214{,}400$
6. $10C5 = 252$

B

Introduction to Statistical Packages

At some point in your statistics career, you will probably need to use statistical software; theoretical understanding and a pocket calculator can take you only so far. Fortunately, we live in an age when many types of software are available to make the task of doing statistics easier. Most statisticians work with one or more of the standard *statistical packages*, such as SAS or SPSS. A statistical package is basically a collection of software routines with a common interface that has been designed to simplify the job of performing statistical analysis and related tasks such as data management. The main thing to remember with regard to statistical packages is that, like any computer software, they are only a means to an end. Each package has its advantages and disadvantages, and at least at the beginning level, you will probably need to use whatever is available at your workplace or at your school. If you then need to learn a new package (say, for a different job) it should not pose great difficulty. If you have a good theoretical understanding of statistics and at least minimal computer aptitude, you can figure out how to use just about any statistical package.

However, starting to work with a new statistical package might seem a daunting task, particularly if your boss or instructor assumes that you are already an expert in it! Printed manuals or online help files might or might not be useful at the very start; a surprising number assume you are already familiar with the software in question when that familiarity is the very thing you lack. Therefore, the purpose of this appendix is to give you a brief overview of several of the most popular packages, with particular emphasis on matters that might be crucial to the new user or are not always clearly stated in the documentation.

Another thing I try to accomplish in this appendix is to provide a sense of the particular strengths and weaknesses of each package and what typical uses are for each. Of course, I can speak only from my experience, and my thoughts are certainly not the last word on the subject. Many reviews of different types of software are available, and if you are ever in the position of needing to choose a package to purchase for your department that will perform specific functions, you might want to begin by searching the Internet, the literature, or both of your profession for phrases such as "comparison of statistical packages."

Minitab

Minitab is a statistical package developed at Pennsylvania State University in the 1980s and now sold by the privately owned company Minitab, Inc. It is commonly used as instructional software in beginning statistics classes and is commonly used for business and quality improvement applications. Although Minitab is a proprietary product, a free 30-day trial copy may be downloaded from the company website (*http://www.minitab.com*).

Minitab is favored in some beginning statistics classes because it is easy to use; according to the company website, it is the most common statistical software used for instruction in colleges and universities worldwide. The standard installation includes an extensive system of help files and demonstrations, which makes it popular with beginners. However, the features that make it easy for beginners to learn, such as reliance on a menu interface and the provision of only a limited number of analytical choices, can make it unsuitable for more advanced applications.

Minitab can import and export files in several formats, including its proprietary Minitab worksheet (identified with the extension **.mtw*) and Minitab project (**.mpj*) formats and Excel (**.xls*) and text (**.txt*) files. Data is stored in rectangular files, as shown in Figure B-1. Rows are numbered, and columns are identified as C1, C2, and so on. Variable names may be added in the shaded row between the column label and data set. Both data and variable names may be typed directly into the Minitab worksheet.

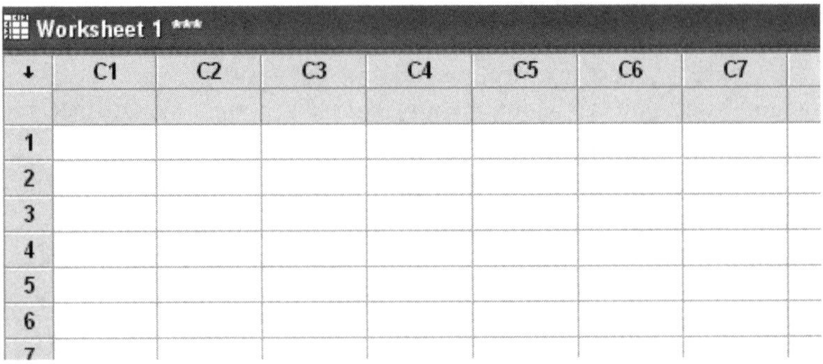

Figure B-1. Minitab worksheet

Commands in Minitab are usually generated through the menu interface; they are recoded in the *session window* along with output that can be expressed as text; an excerpt of a session window for a binary logistic regression analysis is shown in Figure B-2. Each graphical result is written to a separate window (which can make for quite a proliferation of open windows during an analysis!). All results plus the data set for an analysis may be saved as a Minitab project, and data sets and graphs may be saved as separate files in a number of formats.

```
MTB > Blogistic 'CHD' = CHD CAT AGE CHL SMK ECG;
T

Results for: evans

Binary Logistic Regression: CHD versus CAT, AGE, CHL, SMK, ECG

Link Function: Logit


Response Information

Variable  Value  Count
CHD       1         71  (Event)
          0        538
          Total    609


Logistic Regression Table
                                                     Odds      95% CI
Predictor      Coef     SE Coef      Z      P      Ratio  Lower  Upper
Constant   -6.76472    1.13218  -5.97  0.000
CAT         0.776079   0.333091   2.33  0.020   2.17   1.13   4.17
AGE         0.0325374  0.0151541  2.15  0.032   1.03   1.00   1.06
CHL         0.0093670  0.0032332  2.90  0.004   1.01   1.00   1.02
SMK         0.828039   0.304211   2.72  0.006   2.29   1.26   4.15
ECG         0.416540   0.292459   1.42  0.154   1.52   0.85   2.69


Log-Likelihood = -201.337
Test that all slopes are zero: G = 35.884, DF = 5, P-Value = 0.000


Goodness-of-Fit Tests

Method            Chi-Square  DF      P
Pearson              588.700  586  0.461
Deviance             397.129  586  1.000
Hosmer-Lemeshow       16.062    8  0.041


Table of Observed and Expected Frequencies:
(See Hosmer-Lemeshow Test for the Pearson Chi-Square Statistic)

                                  Group
Value   1     2     3     4     5     6     7     8     9    10   Total
1
  Obs   0     2     5     9     6     8     8     4     6    23     71
  Exp  1.8   2.8   3.7   4.4   5.2   6.3   7.3   8.8  11.5  19.2
0
  Obs  60    59    56    52    55    53    57    55    55    38    538
  Exp 58.2  58.2  57.3  56.6  55.8  54.7  53.7  52.2  49.5  41.8
Total 60    61    61    61    61    61    61    61    61    61    609


Measures of Association:
(Between the Response Variable and Predicted Probabilities)

Pairs        Number  Percent  Summary Measures
Concordant    25869    67.7   Somers' D               0.36
Discordant    11933    31.2   Goodman-Kruskal Gamma   0.37
Ties            396     1.0   Kendall's Tau-a         0.08
Total         38198   100.0
```

Figure B-2. Minitab session window

Minitab can perform many basic descriptive statistics, graphical displays, power and sample size calculations, random number generation, and some more advanced statistical analyses such as linear and logistic regression; however, the options available are often surprisingly limited compared to statistical packages such as SPSS or SAS. Therefore, if Minitab is under consideration for purchase, it is wise to run some proposed analyses by using the trial copy to see whether these limitations will be a problem for your proposed uses.

The greatest strength of Minitab might be in quality control and related business applications; it is the world leader in that context, according to the company website. Minitab is often the statistical package taught in conjunction with Six Sigma and similar types of quality improvement training. Specific business and quality control functions are easily produced in Minitab, including DOE (Design of Experiments) analyses, run charts, control charts (Minitab was used to create the control charts for Chapter 14 of this book), time series methods, fishbone diagrams (cause and effect diagrams), Pareto charts, and capability analyses.

Many guides to Minitab are on the market, as well as statistical textbooks that incorporate Minitab; a search on amazon.com or your favorite technical bookstore should turn up many possibilities. In addition, the Minitab home page (*http://www.minitab.com*) includes a number of tutorials and papers to assist Minitab users. A web search will reveal many tutorials and other help sites for Minitab as well.

SPSS

SPSS is a general-purpose statistical computing package that was first released in 1968. It is widely used by social scientists (the name originally meant Statistical Package for the Social Sciences) and is used extensively in other areas, including health research, business, and education. This software package has had several names over its history. A version called SPSS-X was released in the 1980s (this name is reflected in a major listserv devoted to SPSS); from 2009 to 2010, SPSS was called PASW; and since SPSS was bought by IBM in 2010, new releases of the software have been called IBM SPSS (such as IBM SPSS Statistics 19.0, released in August 2010). For the sake of simplicity, we will stick with the name SPSS for all versions of this software.

SPSS can be characterized as offering capabilities somewhere between Minitab and SAS; it is more complex and offers many more analytical possibilities than Minitab but is more limited than SAS. On the other hand, many beginners find SPSS easier to learn than SAS, and many feel SPSS is superior for data formatting and documentation. Particularly since its acquisition by IBM, SPSS has placed strong emphasis on developing applications for predictive analytics and thus might particularly appeal to people working in that field.

SPSS can import and export data in many formats and in nonrectangular configurations; however, the data set is always translated to an SPSS rectangular data file, known as a system file (using the extension **.sav*). *Metadata* (information about the data) such as variable formats, missing values, and variable and value labels are stored with the data set. Two views are offered of the data: the *data view* (Figure B-3) and the *variable view* (Figure B-4), which shows the metadata. You can type directly into either window, so data may be typed into the data view and variable names, labels, and so on typed into the variable view window.

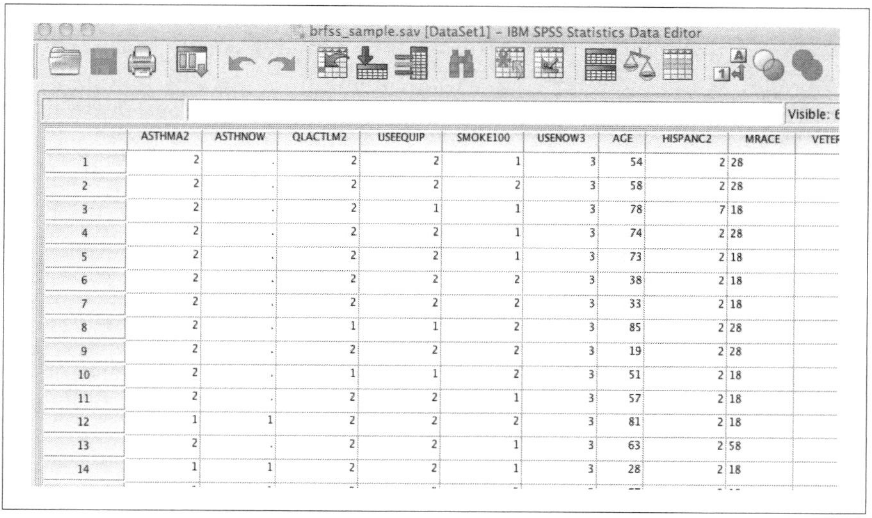

	ASTHMA2	ASTHNOW	QLACTLM2	USEEQUIP	SMOKE100	USENOW3	AGE	HISPANC2	MRACE	VETER
1	2	.	2	2	1	3	54	2	28	
2	2	.	2	2	2	3	58	2	28	
3	2	.	2	1	1	3	78	7	18	
4	2	.	2	2	1	3	74	2	28	
5	2	.	2	2	1	3	73	2	18	
6	2	.	2	2	2	3	38	2	18	
7	2	.	2	2	2	3	33	2	18	
8	2	.	1	1	2	3	85	2	28	
9	2	.	2	2	2	3	19	2	28	
10	2	.	1	1	2	3	51	2	18	
11	2	.	2	2	1	3	57	2	18	
12	1	1	2	2	2	3	81	2	18	
13	2	.	2	2	1	3	63	2	58	
14	1	1	2	2	1	3	28	2	18	

Figure B-3. SPSS data view

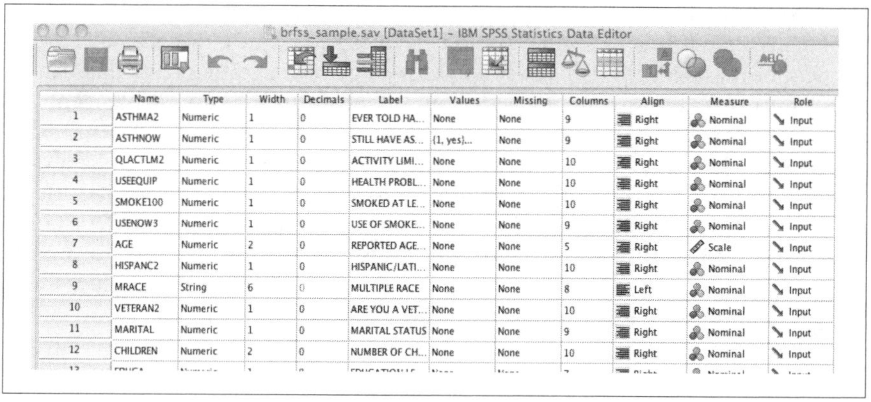

	Name	Type	Width	Decimals	Label	Values	Missing	Columns	Align	Measure	Role
1	ASTHMA2	Numeric	1	0	EVER TOLD HA...	None	None	9	Right	Nominal	Input
2	ASTHNOW	Numeric	1	0	STILL HAVE AS...	{1, yes}...	None	9	Right	Nominal	Input
3	QLACTLM2	Numeric	1	0	ACTIVITY LIMI...	None	None	10	Right	Nominal	Input
4	USEEQUIP	Numeric	1	0	HEALTH PROBL...	None	None	10	Right	Nominal	Input
5	SMOKE100	Numeric	1	0	SMOKED AT LE...	None	None	10	Right	Nominal	Input
6	USENOW3	Numeric	1	0	USE OF SMOKE...	None	None	9	Right	Nominal	Input
7	AGE	Numeric	2	0	REPORTED AGE...	None	None	5	Right	Scale	Input
8	HISPANC2	Numeric	1	0	HISPANIC/LATI...	None	None	10	Right	Nominal	Input
9	MRACE	String	6	0	MULTIPLE RACE	None	None	8	Left	Nominal	Input
10	VETERAN2	Numeric	1	0	ARE YOU A VET...	None	None	10	Right	Nominal	Input
11	MARITAL	Numeric	1	0	MARITAL STATUS	None	None	9	Right	Nominal	Input
12	CHILDREN	Numeric	2	0	NUMBER OF CH...	None	None	10	Right	Nominal	Input

Figure B-4. SPSS variable view

SPSS can be operated entirely through syntax (computer code), which may be typed directly into the syntax window, or written using any text or word processing program and pasted into the syntax window (Figure B-5). SPSS syntax files are stored with the extension **.sps*. SPSS syntax is relatively easy to write and interpret, as should be evident in the code excerpt in Figure B-5. You can probably guess what this code is doing without ever having used SPSS. Here's a hint: lines beginning with * are comments, notes to the programmer rather than executable lines of code. The actual program recodes the continuous variable *exercise* into the dichotomous variable *exerc_cat*, adds labels to the new variable and its values, and creates a cross-tabulation table and frequency table for the two variables.

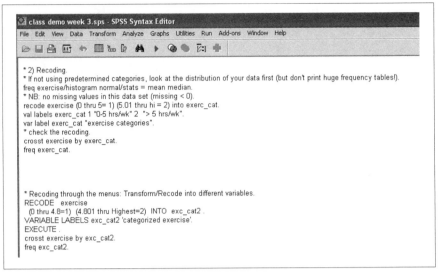

Figure B-5. SPSS syntax window

Some people prefer to use the menu interface, and almost any statistical analysis or data management function in SPSS can be accomplished by either means. I prefer to think of the menu system as an alternative way of generating code that can be saved in a syntax file so I can enjoy the best of both worlds. I can use the menus to write the syntax for an unfamiliar command and then save the syntax as a record of the analysis performed, and I can reuse or alter the syntax for subsequent analyses. The second paragraph of syntax in Figure B-5 was created using the menu interface; the tell-tale sign of menu-generated syntax is the capitalized commands (RECODE, VARIABLE LABELS, and so on). To generate syntax by using the menu system, make all relevant selections in the menu command interface and then select Paste rather than OK as the final step, as shown in Figure B-6. This results in the syntax being saved in a syntax file or appended to an existing syntax file if one is already open. On the other hand, if you simply want to run an analysis and don't care about saving the syntax, click OK instead, and the analysis executes immediately. The statistical results are the same either way.

It would be impossible in this brief space to enumerate all the types of analyses available; an overview of SPSS capabilities can be found on the SPSS web page (*http://www.spss.com*). It is an expensive program, but educational prices are lower, and often universities obtain a site license so they can provide students and employees with access to SPSS free or at a much lower cost.

SAS

SAS is a statistical software package developed at North Carolina State University in the 1960s, and since 1976 it has been a commercial product sold by SAS Institute. SAS is a step up in complexity from SPSS. It is somewhat more difficult to use but offers much more in terms of the types of analyses available and the flexibility

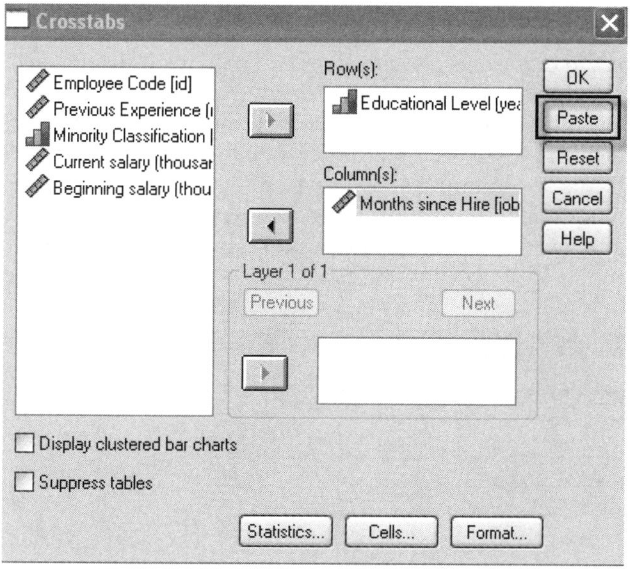

Figure B-6. Using the SPSS menu system to generate syntax

provided for specifying and executing those analyses. The major disadvantage for beginners is that SAS is a syntax-based system, and there are so many choices to be made for even a simple analysis that it can seem overwhelming at first. SAS is also less friendly in terms of managing data files and metadata; for instance, it stores formats in files separate from the data file and requires the format location to be specified in the syntax every time the data file is opened (rather than attaching the format information to the data file, as SPSS does). However, SAS has become the standard language in many professional fields, and there is more assistance for learning and using SAS, both from the SAS web page (*http://www.sas.com*) and help desk and from many published books and websites, than is available for SPSS.

SAS is similar to SPSS in many ways. It is a comprehensive statistical package that can conduct more types of analyses than can possibly be enumerated here and can read and write data sets in many formats. SAS is expensive for an individual to buy but might be affordable if your school or place of business has a site license. One major difference between SAS and SPSS is that SAS is primarily a syntax-based system. Many statisticians prefer to work with syntax anyway, partly because they (like me!) are so old they learned to use computers before graphical interfaces were available and partly because (as mentioned in the SPSS section) syntax may be shared and reused. In addition, writing syntax forces programmers to think through an analysis in a way that can be avoided if the analysis is conducted just by clicking the menus. To someone just starting out in statistics, however, the lack of a menu interface might seem more of a barrier than an advantage. This can be somewhat ameliorated by using the time-tested method of altering someone else's code to fit your needs, and so much annotated SAS code is available on the Internet that you could teach yourself to write SAS programs just by using this method.

SAS has three main windows: the *syntax window*, where you can type your syntax or paste it in from another text or word processing program; the *log window*, which contains a record or log of everything done in a particular session, including warnings and other messages from the SAS system; and the *output window*, where output from statistical procedures is sent by default. (It can be directed to other locations, such as through an HTML or *.rtf* file or through the ODS system.) To use SAS, you open an SAS data set or import another type of data (such as a file stored in Excel or text format), submit commands through the syntax window, and check the output in the output window. The log and syntax windows are illustrated in Figure B-7.

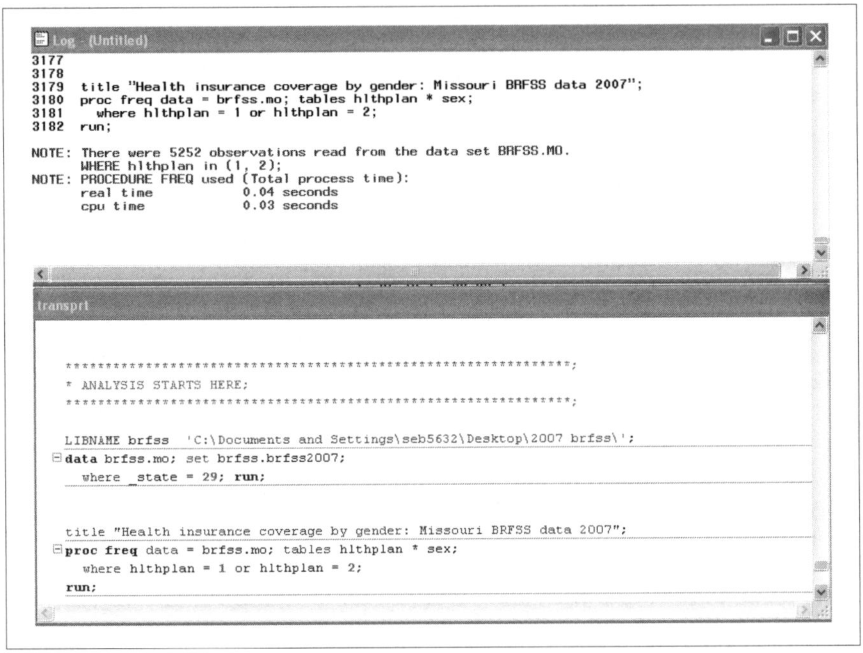

```
Log - (Untitled)
3177
3178
3179   title "Health insurance coverage by gender: Missouri BRFSS data 2007";
3180   proc freq data = brfss.mo; tables hlthplan * sex;
3181     where hlthplan = 1 or hlthplan = 2;
3182   run;

NOTE: There were 5252 observations read from the data set BRFSS.MO.
      WHERE hlthplan in (1, 2);
NOTE: PROCEDURE FREQ used (Total process time):
      real time          0.04 seconds
      cpu time           0.03 seconds
```

```
transprt

*********************************************************;
* ANALYSIS STARTS HERE;
*********************************************************;

LIBNAME brfss   'C:\Documents and Settings\seb5632\Desktop\2007 brfss\';
data brfss.mo; set brfss.brfss2007;
  where _state = 29; run;


title "Health insurance coverage by gender: Missouri BRFSS data 2007";
proc freq data = brfss.mo; tables hlthplan * sex;
  where hlthplan = 1 or hlthplan = 2;
run;
```

Figure B-7. SAS log and syntax windows

The syntax window (Editor – Untitled2*) illustrates three main features of SAS programming. The first is that the location of SAS data files is declared using the *libname* command and the data files themselves referenced with a two-part name: *library.datasetname*. In this case, we declared the library *y* (the actual name is arbitrary, and many people use one-letter *libnames* because they are easier to type) to exist at the physical location:

C:\Documents and Settings\sboslaugh\Desktop\CHQE Projects\
BH Dip Analysis\

and then referenced the *y.sbdip0607* data set that is stored in that location.

SAS programs consist primarily of two types of steps:

1. DATA steps, which open, manipulate, and save data file
2. PROC steps, which perform statistical analyses on the files

The log window (Log - (Untitled)) echoes the syntax submitted and contains messages from the SAS system—for instance, that our *libname* command was successful.

Figure B-8 shows an excerpt from an SAS output window.

```
                         The FREQ Procedure

                Statistics for Table of VENDOR by PERIOD

          Statistic                    DF      Value      Prob

          Chi-Square                    6      3.4970     0.7444
          Likelihood Ratio Chi-Square   6      4.2716     0.6400
          Mantel-Haenszel Chi-Square    1      0.4637     0.4959
          Phi Coefficient                      0.1164
          Contingency Coefficient              0.1156
          Cramer's V                           0.1164

          WARNING: 43% of the cells have expected counts less
                   than 5. Chi-Square may not be a valid test.


                          Fisher's Exact Test

          Table Probability (P)        5.081E-05
          Pr <= P                         0.8013

                     Sample Size = 258
```

Figure B-8. SAS output

SAS has two other windows that may be toggled between by use of the tabs in their lower corners. The *Results* window (Figure B-9) shows an outline of the results produced during a session; clicking any folder causes the next greater level of detail to be displayed. The *Explorer* window (Figures B-10 and B-11) allows access to different SAS libraries. (Any libraries created by the user, such as *y* in this case, must have been declared by a *libname* command during the current SAS session.) Clicking the folders moves the display to the next greater level of detail.

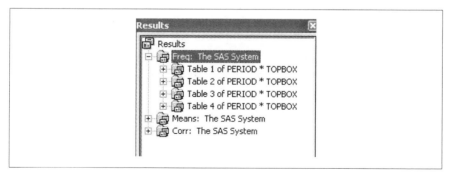

Figure B-9. SAS Results window

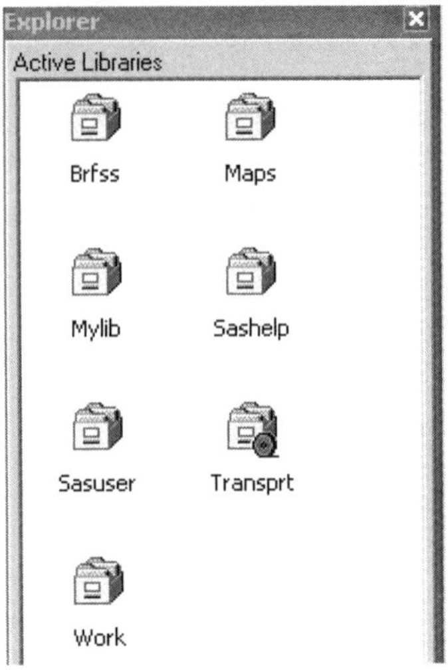

Figure B-10. SAS Explorer window

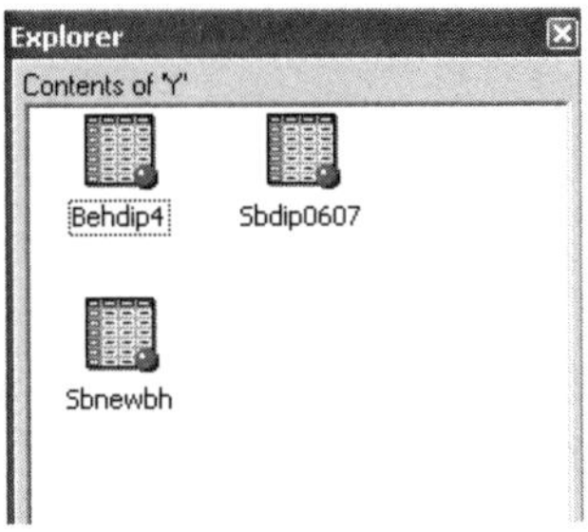

Figure B-11. Contents of a data library (three SAS data files) from the SAS Explorer window

Note that it is possible to open an SAS data set in spreadsheet form (as in Figure B-12), which SAS calls *Viewtable* format, by clicking it in the Explorer window, and that it is possible to enter or edit data directly by this method. Normally, however, in SAS, these procedures are accomplished using syntax.

There are many books to help you learn SAS and many good Internet resources as well, and this community of SAS programmers is a major plus for anyone using this language.

	WILLHLP	RSPREQ	CHLP	COORD	DISRSPCT	MEDS	MDATT
1	3.00	3.00	3.00	3.00	3.00	4.00	4.00
2	5.00	4.00	5.00	5.00	5.00	5.00	5.00
3	5.00	5.00	5.00	5.00	5.00	5.00	5.00
4	4.00	4.00	4.00	4.00	4.00	5.00	4.00
5	5.00	5.00	5.00	5.00	5.00	4.00	5.00
6	5.00	5.00	5.00	5.00	5.00	5.00	5.00
7	3.00	3.00	5.00	5.00	5.00	3.00	4.00
8	4.00	4.00	5.00	5.00	5.00	5.00	4.00
9	4.00	4.00	3.00	3.00	4.00	5.00	5.00
10	3.00	3.00	5.00	3.00	5.00	4.00	5.00

Figure B-12. SAS data set in Viewtable format

R

R is a programming language that also functions as a statistical package because of the many prewritten statistical routines (computer code written to perform a particular task) available for it. It differs from the other packages discussed in this appendix because, rather than being a proprietary product sold or licensed by a corporation, R is freely available for download from the Internet. R is an extremely powerful language, and new routines are constantly being written and made available on the Internet by statisticians and programmers from all over the world.

Free is a tough price point to beat, so you might wonder why everyone isn't already using R to do their statistical work. The answer is that R is harder to use than the other packages discussed in this appendix, particularly at first and for someone who doesn't have a lot of aptitude or experience as a programmer. Using R also requires the programmer to think about what she is doing to a greater extent than when programming in SPSS or SAS. Although this is certainly an educational advantage, people who just want to produce a few simple statistics might feel that the investment of time required to get over the initial hurdles of using R is too great.

On the other hand, if you start out learning R at the same time you learn about statistics, it might be no more difficult to learn than any other package. Several GUI implementations are available, and as R becomes increasingly common, even more user-friendly adaptations might be developed. A sort of natural experiment is currently taking place as R is increasingly being adopted as a teaching language for beginning statistics, so perhaps in 10 years we will be able to answer this question. If you are serious about statistics as a career, you need to become familiar with R because it is the most powerful and flexible language available and might become the *lingua franca* of statistical programming in the near future.

To use R, you must first download it to your computer. The easiest way to do this is to go to the CRAN (Comprehensive R Archive Network) web page (*http://cran.r -project.org*) and follow the instructions. The next step, unless you are very stout of heart (or already an ace programmer), is to find a good instructional text for R; numerous books are available, and many resources are available on the Internet as well, including those available here (*http://www.r-project.org/*).

R is a command-oriented language; you type commands at a command prompt, and the R-interpreter responds interactively, either executing the command or giving you an error message. The commands are quite compact compared to those used in SPSS and SAS and can appear cryptic to the uninitiated; however, after you learn to use R, you might come to appreciate this efficiency. Even more so than with the other languages discussed in this appendix, the best way to get comfortable with R is to get some basic instructional materials and run through some very simple examples on your computer. The logic of the R language is easier to recognize through use and practice than by reading someone else's explanation.

Another thing you should know about R is that it is an *object-oriented language* (as are Java, C++, and Smalltalk, among others); this means that everything you create using R is an object that can be further manipulated by other commands. An object is also a member of a *class*, meaning that it has certain characteristics and internal organization that allow you to perform operations on it.

Microsoft Excel

Microsoft Excel is, properly speaking, not a statistical package at all, although it is sometimes used as one. Excel is a spreadsheet application produced by the Microsoft Corporation that is frequently used for data management because of its ubiquity (it is preloaded on many new computers sold in the United States, for instance), ease of use, and the fact that several major statistical packages have prewritten routines to import and export data in Excel format. Excel also can produce graphs and charts and perform some statistical analyses, although you should know that Excel has some well-known flaws in statistical accuracy (*http://www.daheiser.info/excel/front page.html*), so the advisability of using it for anything beyond the most basic displays and calculations is arguable. On the other hand, Excel might be entirely adequate for your needs, or it might be the software of choice in a class you are taking. Just remember that Excel is a spreadsheet application, not a statistical package, and proceed accordingly.

Excel stores data in individual *spreadsheets*, which it calls *worksheets*; multiple worksheets are collected into *workbooks*. Individual data points are stored in *cells* (the rectangular boxes in the worksheets) identified by column and row. For example, cell A1 is the intersection of column A and row 1. Both individual worksheets and workbooks use the extension *.xls* (or *.xlsx* for newer versions). A spreadsheet looks like a rectangular data set but has many more capabilities, including built-in functions to perform computations on sets of cells such as rows or columns of data. Excel also offers many choices regarding how data is stored, how it appears on the screen, and how it is printed; a given cell, column, or row can be formatted for string or numeric data, to appear in different date formats, and so on.

In Figure B-13, you can see a worksheet (Sheet3, as you can tell from the lower tab) within a workbook that includes three worksheets; you maneuver between worksheets by clicking the tabs at the bottom of the window (labeled Sheet1, Sheet2, and Sheet3 in this example). Rows are horizontal, as in the standard rectangular data set, so we have row 1, row 2, and so on. Columns are vertical, so we have column

A, column B, and so on. Individual cells are defined by row and column, so the cell in the upper left corner is A1, the next to its right is B1, and the next below is A2. The designations A1, A2, and so on are called *cell references*.

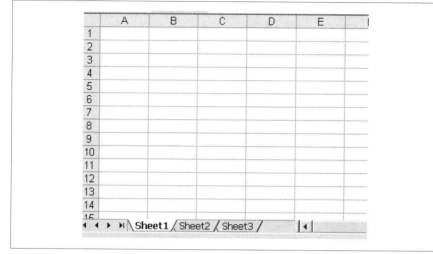

Figure B-13. Microsoft Excel worksheet

Data can be entered simply by typing in the worksheet, in which case Excel applies default formats based on its best guess of the type of data entered. These formats can be changed using the menu commands Format/Format Cells; Figure B-14 shows some of the choices available for date-format data. If you are using Excel to collect data that will be transferred to a different program for analysis, you should be aware that formatting is often lost or garbled in the transfer process. For this reason, particularly when working with time and date variables (which, because of their complexity and the different ways they are stored in the different programs, are frequently mistranslated between programs), some researchers prefer to use text format for all Excel data to be imported and to format it after importation into the program in which it will be analyzed.

Variable names can be added in the first row, and many statistical packages include the option to retain those names when importing data. However, because the row containing the variable names is counted as a data row in Excel but not in programs such as SPSS and SAS, the imported file will have one fewer row than the Excel file. This might cause panic because it appears a case has been lost, but in fact, the discrepancy is just due to differing ways of storing data.

Another trap for the unwary when transferring data between systems is the fact that each system has a different set of rules for variable names. It can be disheartening to spend a lot of time entering meaningful variable names in a spreadsheet only to have them appear as Var1, Var2, and so on when the file is imported into a statistical package. If you are going to import variable names, follow the rules of the target program, so if you are going to import the data into SPSS, follow the SPSS naming conventions when entering the names in your Excel spreadsheet. Another solution

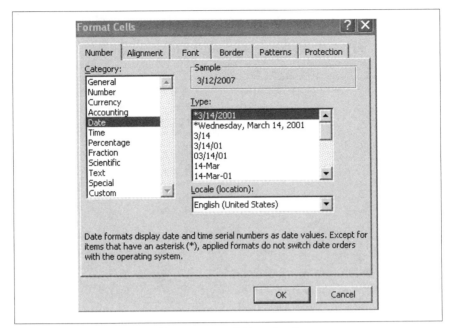

Figure B-14. Some examples of formatting available in Excel

is to use simple names (such as v1, v2, and so on) in Excel and then write code in the target program to add meaningful names to the variables after they are imported.

Excel can create many types of charts and graphs. To create a chart or graph, you insert it into a worksheet, and it can then be saved as a separate object or inserted into other programs such as Microsoft Word.

You can easily do basic arithmetic in Excel, and its spreadsheet capabilities are particularly useful if you need to do calculations on many rows or columns of numbers. Excel also includes a number of built-in functions that allow you to compute basic statistics for any collection of cells, and you can perform arithmetic operations by specifying the equation to be performed. In either case, the function or formula is entered into a cell, which will also be used to store the results of the calculation.

C

References

Preface and General Sources

- Abelson, Robert P. 1995. *Statistics as Principled Argument*. Hillsdale, NJ: Lawrence Erlbaum.

  Abelson, who taught at Yale University for 42 years, provides an excellent discussion of how to think through, and with, statistics.

- Frey, Bruce. 2006. *Statistics Hacks: Tips and Tools for Measuring the World and Beating the Odds*. Sebastopol, CA: O'Reilly.

  Statistics Hacks is a collection of entertaining short essays that use everyday examples to introduce statistical concepts, from testing the randomness or lack thereof in your iPod's "random" shuffle feature to using Benford's law to detect fabricated data.

- Huff, Darryl. 1954. *How to Lie with Statistics*. Repr., New York: W.W. Norton, 1993.

  Originally published in 1954, Huff's work remains a classic introduction to how even the simplest statistical techniques can be used to mislead, confuse, or even outright lie. Readers who can look past the dated examples and (in particular) stereotypical illustrations will find this slim volume an excellent resource and a lot of fun as well.

- Levitt, Steven D., and Stephen J. Dubner. 2005. *Freakonomics: A Rogue Economist Explores the Hidden Side of Everything*. New York: HarperCollins.

  In this *New York Times* bestseller, a University of Chicago economist uses economic theory and statistical analysis to examine questions from the existence of cheating in sumo wrestling to whether legalizing abortion lowered the crime rate. Although written for the public, *Freakonomics* has been adopted as require reading at some universities.

- Salsburg, David. 2001. *The Lady Tasting Tea: How Statistics Revolutionized Science in the Twentieth Century*. New York: W.H. Freeman.

  This popular history examines the application of statistics and probability to scientific problems in the twentieth century, shaping the story around the lives and accomplishments of pioneers such as Ronald Fisher, Karl Pearson, and Jerzy Neyman.

- Tucker, Martha A., and Nancy D. Anderson. 2004. *Guide to Information Sources in Mathematics and Statistics*. Westport, CT: Libraries Unlimited.

  This is a guide to sources of information about mathematics and statistics; the target market is librarians, but researchers will also find it useful. Categories include finding tools, journals, reference books, biographical and historical materials, and math books for science collections (for example, applications of math to other disciplines).

Chapter 1

- Carmines, Edward G., and Richard A. Zeller. 1979. *Reliability and Validity Assessment*. Thousand Oaks, CA: Sage.

  One of the earliest entries in the Sage "little green books" series, this volume introduces classical methods to evaluate reliability and validity assessment, and gives a brief discussion of factor analytic methods.

- Fleming, Thomas R. 2005. "Surrogate endpoints and FDA's accelerated approval process." *Health Affairs* 24 (January/February): 67–78.

  Fleming examines the use of surrogate endpoints in clinical trials intended to provide definitive evidence about the benefits of drugs and other treatments, and describes several situations in which a treatment apparently effective on surrogate endpoints might not be effective with regard to a true clinical endpoint.

- Hand, D.J. 2004. *Measurement Theory and Practice: The World Through Quantification*. London: Arnold.

  Hand provides an excellent discussion of the theory and practice of measurement, including chapters devoted to special problems in the fields of psychology, medicine, the physical sciences, and economics and the social sciences.

- Michiels, Stefan, Aurelie Le Maitre, Marc Buyse, Tomasz Byrzykowski, Emilie Maillard, Jan Bogaerts, et al. 2009. "Surrogate endpoints for overall survival in locally advanced head and neck cancer: Meta-analyses of individual patient data." *The Lancet Oncology* 10 (April): 341–350.

  In an article based on 104 clinical trials, Michiels and colleagues examine the usefulness of two surrogate endpoints in evaluating the success of treating locally advanced head and neck squamous-cell cancer. They conclude that event-free survival correlates more closely than does locoregional control with overall survival (the true clinical endpoint).

- Uebersax, John. "Kappa coefficients." *http://www.john-uebersax.com/stat/ kappa.htm.*

  Uebersax provides a thorough discussion of the strengths and weaknesses of kappa as part of his discussion of agreement statistics in general.

Chapter 2

- Hacking, Ian. 2001. *An Introduction to Probability and Inductive Logic.* Cambridge: Cambridge University Press.

  This volume was written as an introductory text for philosophy students but will be appreciated by anyone who would like a verbal, rather than mathematical, introduction to the basic ideas of statistics.
- Mendenhall, William, et al. 2008. *Introduction to Probability and Statistics.* 13th ed. Pacific Grove, CA: Duxbury Press.

  This is a popular probability and statistics textbook for students who have not taken calculus.
- Packel, Edward W. 2006. *The Mathematics of Games and Gambling.* Washington, D.C.: Mathematical Association of America.

  Packel traces the connections between games and gambling (including backgammon, roulette, and poker) and mathematics and statistics in a manner that assumes only standard high school preparation in mathematics. Many illustrations and exercises are included.
- Ross, Sheldon. 2005. *A First Course in Probability.* 7th ed. Prentice Hall.

  Ross provides a basic introduction to probability theory, illustrated with many examples, for students who have taken elementary calculus.

Chapter 3

- Cohen, J. 1994. "The earth is round (p < .05)." *American Psychologist* 49: 997–1003.

  This is a classic article by one of the most vocal critics of the enshrinement of alpha = 0.05 as absolute indicator of the statistical significance or lack thereof.
- Dorofeev, Sergey, and Peter Grant. 2006. *Statistics for Real-Life Sample Surveys: Non-Simple-Random Samples and Weighted Data.* Cambridge: Cambridge University Press.

  This is a well-written guide to sampling and the analysis of survey data when simple random sampling is not possible (which is most of the time).
- Mosteller, Frederick, and John W. Tukey. 1977. *Data Analysis and Regression: A Second Course in Statistics.* Reading, MA: Addison Wesley.

  This classic textbook in inferential statistics includes a chapter on data transformation.

- National Institute of Standards and Technology. *Engineering Statistics Handbook: Gallery of Distributions. http://www.itl.nist.gov/div898/handbook/eda/section3/eda366.htm.*

  This is a nice presentation of 19 common statistical distributions, including ample illustrations, formulas, and common uses for each.

- Peterson, Ivars. 1997. "Sampling and the census: Improving the decennial count." *Science News* (October 11).

  This clearly written article discusses problems with the data collection efforts of the U.S. census and the controversy over using sampling as part of the process.

- Rice Virtual Lab in Statistics. "Simulations/Demonstrations." *http://onlinestatbook.com/stat_sim/index.html.*

  This Internet site has links to many Java simulations demonstrating statistical concepts, including the central limit theorem, confidence intervals, and data transformations.

Chapter 4

- Cleveland, William S. 1993. *Visualizing Data.* Summit, NJ: Hobart Press.

  This book discusses effective graphical presentation of data with many examples; it also includes a discussion of the visual and psychological principles that lie behind effective graphical presentation of information.

- Erceg-Hurn, David M., and Vikki M. Mirosevich. 2008. "Modern statistical methods: An easy way to maximize the accuracy and power of your research." *American Psychologist* 63: 591–601.

  This discusses robust statistical methods, including trimmed means, and argues for their wider use.

- Robbins, Naomi. 2004. *Creating More Effective Graphs.* Hoboken, NJ: Wiley.

  An easy-to-use guide that shows good and bad examples of graphs presenting the same information, this book always has an eye to using graphical techniques to communicate statistical information more effectively.

- Tufte, Edward R. 2001. *The Visual Display of Quantitative Information.* 2nd ed. Cheshire, CT: Graphics Press.

  This book is a landmark that forever changed the way researchers use graphics to display information. Admirers of Tufte's sometimes contentious approach will want to check out his other works as well, including *Beautiful Evidence* (2006).

- Wand, M.P. 1996. "Data-based choice of histogram bin width." *The American Statistician* 51(1): 59–73.

  Not for the faint of heart or the mathematically underprepared, but this book is a thorough technical investigation of various rules for determining the appropriate number of bins for a histogram.

- Wilkins, Jesse L.M. 2000. "Why divide by N-1?" *Illinois Mathematics Teacher* (Fall): 13–18. *https://scholar.vt.edu/access/content/user/wilkins/Public/IMT.pdf*.

  This is a clear and detailed explanation of a question that invariably arises in statistics classes and proves surprisingly difficult to answer: why, when calculating the sample variance, do we divide by $(n - 1)$ rather than n?

Chapter 5

- Agresti, Alan. 2002. *Categorical Data Analysis*. 2nd ed. Hoboken, NJ: Wiley.

  This is the standard textbook for advanced classes on categorical data analysis. It can be heavy going for the beginner but is clearly written and covers everything from 2×2 tables to linear models.

- Davenport, Ernest C., and Nader A. El-Sanhurry. 1991. "Phi/phimax: Review and synthesis." *Educational and Psychological Measurement* 51(4): 821–828.

  This is a discussion of the range of phi in relation to different data distributions and investigation of a potential solution.

Chapter 6

- Fisher, R.A. 1925. "Applications of 'student's' distribution." *Metron* 5: 90–104.

  This discusses testing for differences between means using the characteristics of the *t* distribution.

- Gosset, William Sealy. 1908. "The probable error of a mean." *Biometrika* 6(1): 1–25.

  This is the original paper describing characteristics of the *t* distribution.

- Senn, S., and W. Richardson. 1994. "The first *t*-test." *Statistics in Medicine* 13(8): 785–803.

  This article is about the first application of the *t*-test in a medical clinical trial.

Chapter 7

- Case, Anne, and Christina Paxson. 2008. "Stature and status: Height, ability, and labor market outcomes." *Journal of Political Economy* 116(3), 499–532.

  This article discusses the positive relationship between height and income, arguing that this observed relationship is due to the positive relationship between height and cognitive ability.

- Holland, Paul W. 1986. "Statistics and causal inference." *Journal of the American Statistical Association* 81(396): 945–960.

  This describes the problematic relationship between the need to determine causal inference and the statistical tools available to analyze certain types of data.

References

- Spearman, C. 1904. "The proof and measurement of association between two things." *American Journal of Psychology* 15: 72–101.

  This is perhaps the most influential paper on measures of association in the history of psychology.
- Stanton, Jeffrey M. 2001. "Galton, Pearson, and the peas. A brief history of linear regression for statistics instructors." *Journal of Statistics Education* 9(3).

  This is a very readable introduction to the development of ideas underlying correlation and regression.

Chapter 8

- Cohen, J., P. Cohen, S.G. West, and L.S. Aiken. 2003. *Applied Multiple Regression/Correlation Analysis for the Behavioral Sciences*. 2nd ed. Hillsdale, NJ: Lawrence Erlbaum Associates.

  This is an excellent textbook introduction to simple and multiple regression.
- Dunteman, George H., and Moon-Ho R. Ho. 2006. *An Introduction to Generalized Linear Models*. Thousand Oaks, CA: SAGE Publications.

  One of the Sage "little green books," this slim (72 pages) volume provides an excellent overview of the general linear model for those who are comfortable with reading mathematical equations.
- Galton, Francis. 1886. "Regression towards mediocrity in hereditary stature." *Journal of the Anthropological Institute* 15: 246–263. *http://galton.org/essays/1880-1889/galton-1886-jaigi-regression-stature.pdf.*

  This is the original paper on regression to the mean.
- Glass, G.V., P.D. Peckham, and J.R. Sanders. 1972. "Consequences of failure to meet assumptions underlying the analysis of variance and covariance." *Review of Educational Research* 42: 237–288.

  This is a technical paper on the assumptions underlying ANOVA and ANCOVA and the consequences for the analysis when they are not met.

Chapter 9

- Fisher, R.A. 1931. "Studies in crop variation. I. An examination of the yield of dressed grain from Broadbalk." *Journal of Agricultural Science* 11: 107–135.

  This covers the original experiments and formulation underlying ANOVA.
- Miler, G.A., and J.P. Chaplin. 2001. "Misunderstanding analysis of covariance." *Journal of Abnormal Psychology* 110(1): 40–48.

  This is a clear discussion of the appropriate use of ANCOVA and what this technique can and can't do for a research project.

Chapter 10

- Achen, Christopher H. 1982. *Interpreting and Using Regression*. Thousand Oaks, CA: Sage Publications.

  A Sage "little green book," this offers an excellent introduction to the correct (and cautious) interpretation of multiple linear regression models.

- Jacard, James, Robert Turrisi, and C.K. Wan. 1990. *Interaction Effects in Multiple Regression*. Thousand Oaks, CA: Sage Publications.

  Another Sage "little green book," this one offers a straightforward synthesis of theory and practice regarding interaction effects in regression models.

- O'Brien, R.M. 2007. "A caution regarding rules of thumb for variance inflation factors." *Quality & Quantity* 41: 673–690.

  O'Brien argues that applying conventional rules of thumb exaggerates the problems caused by multicollinearity and that typical solutions to perceived multicollinearity can cause more problems than they solve.

Chapter 11

- Bates, Douglas M., and Donald G. Watts. 1988. *Nonlinear Regression Analysis and Its Applications*. New York: Wiley.

  This is a very practical textbook introduction to curve fitting and nonlinear modeling.

- Efron, Bradley. 1982. *The Jackknife, the Bootstrap, and Other Resampling Plans*. Philadelphia: Society for Industrial and Applied Mathematics.

  This is a classic textbook on resampling methods.

- Hosmer, David W., and Stanley Lemeshow. 2000. *Applied Logistic Regression*, 2nd ed. New York: Wiley.

  This is a practical presentation of logistic regression and its applications for advanced students and specialists.

Chapter 12

- Gould, Stephen Jay. 1996. *The Mismeasure of Man*. W.W. Norton & Company.

  This excellent book sets out the historical context of intelligence testing and the (mis)use of various multivariate techniques in the understanding of individual differences.

- Hartigan, J.A. 1975. *Clustering Algorithms*. New York: Wiley.

  This book is a modern classic with complete coverage of foundation concepts in clustering, including distance measures, with sufficient detail to implement all the algorithms.

Chapter 13

- Conover, W.J. 1999. *Practical Nonparametric Statistics*. Hoboken, NJ: Wiley.

  This is one book that lives up to its title; it's a great reference for people who need to learn how to do the appropriate nonparametric test for a particular situation and don't want a lengthy theoretical discussion of each statistic. Conover's book includes a handy chart for finding nonparametric equivalents for a parametric test.

- HealthKnowledge. "Parametric and non-parametric tests for comparing two or more groups." *http://www.healthknowledge.org.uk/public-health-textbook/re search-methods/1b-statistical-methods/parametric-nonparametric-tests*.

  This is a series of handy charts to help you locate the appropriate nonparametric statistics for different analytic situations, produced as part of an online public health course by the Department of Health of the United Kingdom.

- Mann, H.B., and D.R. Whitney. 1947. "On a test of whether one of two random variables is stochastically larger than the other." *Annals of Mathematical Statistics* 18: 50–60.

  This paper extends the Wilcoxon Mann Whitney-U test to unequal sample sizes.

- Wilcoxon, F. 1945. "Individual comparisons by ranking methods." *Biometrics Bulletin* 1: 80–83.

  This is the original paper describing the Wilcoxon Mann Whitney-U test for equal sample sizes.

- Wilcoxon, F. 1957. *Some Rapid Approximate Statistical Procedures*. Stamford, CT: American Cyanamid. Revised with R.A. Wilcox, 1964.

  These are the original and revised papers that describe Wilcoxon's Signed Rank Test, including a table of critical values.

Chapter 14

- Clemen, Roger T. 2001. *Making Hard Decisions: An Introduction to Decision Analysis*. Pacific Grove, CA: Duxbury Press.

  This textbook emphasizes the logical and philosophical problems behind decision making while discussing different approaches to decision analysis.

- The Economist Newspaper. 1997. *Numbers Guide: The Essentials of Business Numeracy*. Hoboken, NJ: Wiley.

  This handy pocket guide describes numerical operations useful in business, including index numbers, interest and mortgage problems, forecasting, hypothesis testing, decision theory, and linear programming.

- Gordon, Robert J. 1999. "The Boskin Commission Report and its aftermath." Paper presented at the Conference on the Measurement of Inflation, Cardiff, Wales. *http://faculty-web.at.northwestern.edu/economics/gordon/346.pdf*.

This summarizes criticisms regarding the U.S. Consumer Price Index, including those identified by the 1995 Boskin Commission report, which suggested that the CPI overstated inflation.

- Shumway, Robert, and David S. Stoffer. 2006. *Time Series Analysis and Its Applications: With R Examples*. New York: Springer.

   This popular time series textbook includes code in R (a free computer language) to execute time series analyses.

- Tague, Nancy. 2005. *The Quality Toolbox*. 2nd ed. Milwaukee, WI: American Society for Quality.

   This reference book provides an overview and brief history of Quality Improvement (QI), followed by an alphabetical guide to QI tools, including standard statistical and graphical procedures such as the box plot and hypothesis testing, and more specialized tools such as control charts and fishbone diagrams.

Chapter 15

- Cohen, Jacob. 2002. "A power primer." *Psychological Bulletin* 112 (July).

   This very readable introduction to power concepts is prefaced by research by Cohen and others into the neglect of power considerations in published studies.

- Ahrens, Wolfgang, and Iris Pigeot, Eds. 2004. *Handbook of Epidemiology*. New York: Springer.

   This guide to epidemiology consisting of chapters on specialized topics written by experts in each field. The chapter on sample size calculations and power analysis includes formulas and examples for the most common study designs used in medicine and epidemiology.

- Hennekens, Charles H., and Julie E. Buring. 1987. *Epidemiology in Medicine*. Boston: Little, Brown.

   This is an easy-to-read introduction to epidemiology, from basic concepts through study design and types of analysis.

- Pagano, Marcello, and Kimberlee Gauvreau. 2000. *Principles of Biostatistics*. 2nd ed. Pacific Grove, CA: Duxbury Press.

   This introduction to biostatistics is suitable for an undergraduate course; it's less detailed and easier to use than Rosner's text.

- Rosner, Bernard. *Fundamentals of Biostatistics*. 6th ed. Pacific Grove, CA: Duxbury Press, 2005.

   An excellent introduction to biostatics for graduate students or those who are willing to grapple with more theoretical details than are provided in Pagano and Gauvreau's text.

- Rothman, Kenneth J., et al. 2008. *Modern Epidemiology*. 3rd ed. Philadelphia: Lippincott, Wilkins, and Williams.

   This is a very thorough discussion of epidemiology, including several chapters written by guest authors, for students willing and able to grapple with the subject.

Chapter 16

- Crocker, Linda, and James Algina. 2006. *Introduction to Classical and Modern Test Theory.* Independence, KY: Wadsworth.

  This is an updated version of a standard textbook that is strongest in its descriptions of models based on classical test theory.

- Ebel, R.L. 1965. *Measuring Educational Achievement.* Englewood Cliffs, NJ: Prentice Hall.

  This text is the source of the rules to interpret item discrimination that are cited in Chapter 16.

- Embretson, Susan, and Steven Reise. 2000. *Item Response Theory for Psychologists.* Mahwah, NJ: Erlbaum.

  This is an introductory textbook that takes an intuitive approach to IRT, with many graphical displays and analogies with classic measurement approaches.

- Hambleton, Ronald K., et al. 1991. *Fundamentals of Item Response Theory.* Thousand Oaks, CA: Sage Publications.

  This provides a very clear introduction to item response theory that explains how it overcomes some of the limitations of classic test theory.

- Tanner, David E. 2001. *Assessing Academic Achievement.* Boston: Allyn and Bacon.

  This straightforward text, written for teachers and administrators, covers the major issues in academic testing and evaluation; it discusses contemporary issues (authentic assessment, high-stakes testing, computer-adaptive testing) as well as traditional topics such as classic test theory and norm-referenced versus criterion-referenced assessment.

Chapter 17

- Boslaugh, Sarah. 2004. *An Intermediate Guide to SPSS Programming: Using Syntax for Data Management.* Thousand Oaks, CA: Sage.

  Boslaugh covers the basic aspects of data management for people who will be managing and analyzing data by using SPSS and includes the code to perform many tasks.

- Cody, Ron. 1999. *Cody's Data Cleaning Techniques Using SAS Software.* Cary, NC: SAS Institute.

  Cody presents techniques for checking and cleaning data by using SAS, including many examples of standard procedures and the SAS code to carry them out.

- Hernandez, M.J. 2003. *Database Design for Mere Mortals: A Hands-On Guide to Relational Database Design.* 2nd ed. Upper Saddle River, NJ: Addison Wesley.

  This is a good guide to the theory and practice of setting up databases, discussed in terms of principles applicable to any database rather than instructions in using any particular software product.

- Levesque, Raynald. Raynald's SPSS Pages. *http://www.spsstools.net/*.

  Two websites run by the experienced SPSS programmer Raynald Levesque; both are loaded with tips, tricks, and sample code.

- Little, Roderick J.A., and Donald B. Rubin. 2002. *Statistical Analysis with Missing Data*. 2nd ed. Hoboken, NJ: Wiley.

  Little and Rubin wrote the book on missing data, and this is the standard reference on the subject. However, it's not for the faint of heart and assumes considerable mathematical sophistication on the part of the reader.

Chapter 18

- Christensen, Larry B. 2006. *Experimental Methodology*, 10th ed. Boston: Allyn & Bacon.

  This is a very readable and comprehensive introduction to research and experimental design with a focus on educational and psychological topics.

- Fisher, R.A. 1990. *Statistical Methods, Experimental Design, and Scientific Inference: A Re-issue of Statistical Methods for Research Workers, the Design of Experiments, and Statistical Methods and Scientific Inference*. Oxford: Oxford University Press.

  If you want to read the original rationale for many of the designs and issues described in this chapter, there is no better place than the original source.

- The Framingham Heart Study. *http://www.framinghamheartstudy.org/*.

  This is the official website of one of the largest, longest, and most famous prospective cohort studies in the history of medicine.

- Martin, F., and D. Siddle. 2003. "The interactive effects of alcohol and Temazepam of P300 and reaction time." *Brain and Cognition*, 53(1): 58–65.

  This is the article used as an example of research design in Chapter 18.

- Robinson, W.S. 1950. "Ecological correlations and the behavior of individuals." *American Sociological Review* 15(3): 351–357. Reprinted in the *International Journal of Epidemiology* (2009). *http://ije.oxfordjournals.org/content/early/2009/01/28/ije.dyn357.full.pdf+html*.

  This is a classic paper on the ecological fallacy, demonstrated with data correlating literacy with race and national origin.

- Rosenbaum, Paul R., and Donald B. Rubin. 1983. "The central role of the propensity score in observational studies for causal effects." *Biometrika* 70: 41–55.

  This is the article in which Rosenbaum and Rubin introduced the concept of the propensity score, which is now commonly used in case control studies in medical research.

- Shadish, William R., Thomas D. Cook, and Donald T. Campbell. 2001. *Experimental and Quasi-Experimental Designs for Generalized Causal Inference.* Florence, KY: Wadsworth Publishing.

   This is an updated version of the classic text on research design; not the best book for beginners but essential for anyone who really wants to understand the issues.

- Wolff, Alexander, Albert Chen, and Tim Smith. 2002. "That Old Black Magic." *Sports Illustrated* 96 (January): 50–62.

   This article examines the validity for claims of the "*Sports Illustrated* jinx," a phenomenon often cited as a classic case of regression to the mean.

Chapter 19

- Alley, Michael. 2003. *The Craft of Scientific Presentations.* New York: Springer.

   A book-length consideration of different styles of scientific presentation (for example, informative versus persuasive), this book has many examples as well as general principles for what makes a presentation successful or unsuccessful.

- LaMontaigne, Mario. "Planning a scientific presentation." *http://www.biomech .uottawa.ca/english/teaching/apa6905/lectures/presentation-style.pdf.*

   This is a slide presentation on how to create good slide presentations, with humorous illustrations of some ways to go wrong as well.

- "Slides from NISS/ASA Technical Writing Workshop for Young Researchers . . . and Some Other Stuff." (August 2007). *http://www.public.iastate.edu/~varde man/RTGWritingStuff.html.*

   This collection of slides and other resources about writing scientific articles for professional journals is from a workshop sponsored by the American Statistical Association and the National Independent Statistical Service; the primary target is students and young researchers writing their first article, but there's lots of advice that will be useful to more experienced writers as well.

- Ternes, Reuben. 2011. "Writing with statistics." Purdue Online Writing Lab. *http://owl.english.purdue.edu/owl/resource/672/1/.*

   This is a basic guide to communicating with statistics, intended for undergraduate students. The OWL (Online Writing Lab) has other useful information for scientific writers, including a guide to writing abstracts, guides to the major citation systems, and guides to writing in medicine, nursing, and engineering.

- The OPEN Notebook. *http://www.theopennotebook.com.*

   A website devoted to science journalism, The OPEN Notebook focuses on writing for general audiences and presents a combination of technical advice and behind-the-scenes looks at the process behind the writing of well-known articles and books (such as Rebecca Skloot's *The Immortal Life of Henrietta Lacks*).

- United Nations Economic Commission for Europe. 2009. *Making Data Meaningful.*

    http://www.unece.org/fileadmin/DAM/stats/documents/writing/MDM_Part1_English.pdf
    http://www.unece.org/fileadmin/DAM/stats/documents/writing/MDM_Part2_English.pdf
    http://www.unece.org/fileadmin/DAM/stats/documents/writing/MDM_Part3_English.pdf

    This three-part series is written for managers, public relations officers, statisticians, and others who communicate statistical information to the public and other nontechnical audiences. Part 1 explains how to turn statistical information into a story that will capture the public's imagination and communicate important information, Part 2 discusses how to present statistics (both verbally and graphically), and Part 3 discusses media relations.

Chapter 20

- Good, Phillip I., and James W. Hardin. 2006. *Common Errors in Statistics (and How to Avoid Them)*. Hoboken, NJ: Wiley.

    This is a guide to avoiding common mistakes in statistical methodology and reasoning.

- Alderson, Phil, and Sally Green, eds. 2009. *The Cochrane Collaboration Learning Material for Reviewers. http://www.cochrane-net.org/openlearning/.*

    This includes a clear discussion of publication bias, written to support the efforts of The Cochrane Collaboration, an international organization whose purpose is to support informed decision making in health care.

- Darryl Huff's *How to Lie with Statistics*, cited as a general reference at the start of this appendix, is also highly relevant to this chapter.

Probability Tables for Common Distributions

Probability tables for different distributions are available in many reference books and online; the tables in this section are included for your convenience and include solved examples from the main text. One caution: there is more than one way to display the probability values for any distribution, so it is always wise to spend a few minutes observing how a given table is constructed before you start to use it.

Probability tables are partly a vestige of an era before statistical calculators and software packages were readily available, but they serve a useful purpose even in our electronic age. To use a probability table correctly, you have to think about the distribution in question and how it applies to your research question, so a few minutes working with a probability table is worthwhile even if you expect to do most or all of your statistical calculations with a computer software package.

The tables included in this chapter, except those for the binomial distribution, are taken from the *NIST/SEMATECH e-Handbook of Statistical Methods*, a public domain resource available online (*http://itl.nist.gov/div898/handbook/index.htm*) from the National Institute of Standards and Technology in the United States. The binomial distribution tables, also in the public domain, were created by William Knight, a former professor of computer science and mathematics at the University of New Brunswick, and are available from his website (*http://www.math.unb.ca/~knight/util ity/*).

Note that for continuous distributions such as the normal distribution, we always speak in terms of the probability of an *area* of the distribution (which is equivalent to the probability of all the results included in that area) rather than the probability of a *single point* in the distribution. This means that we can find $P(Z > 2.00)$ or $P(Z < -1.80)$ but not $P(Z = 2.00)$ or $P(Z = -1.80)$. The reason is technical; in a discrete distribution, a point (such as 2.00) has no area and, thus, no probability. This restriction is specific to continuous distributions, and for discrete distributions such as the binomial, we can find the probability of specific values.

The Standard Normal Distribution

Figure D-3 is a table of the area under the normal curve, expressing the probability that $(0 < x < |a|)$, the probability that a value of x lies in the range between 0 and the absolute value of some value a. Suppose $a = 0.5$. The area $(0 < x < 0.5)$ is represented by the shaded area in Figure D-1.

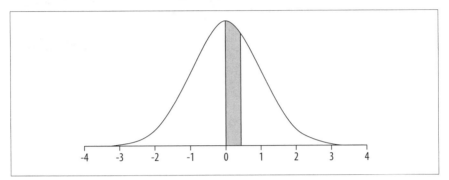

Figure D-1. Area (0 < x < 0.5) of the standard normal distribution

We can see from the normal table presented in Figure D-3 that the area of this region, which is the same as the probability of a value in the range $(0, 0.5)$, is 0.19146. (Remember, the total area under the normal curve is 1.0.) We find this value by going down the column labeled x until we come to the row for 0.5 and then going across that row to the 0.00 column. The value in the cell where the column and row intersect is the probability of a value in the range between 0 and the absolute value of a (in this case, between 0 and 0.5). This value, 0.19146, is both the area under the normal curve between 0 and 0.5 and the probability, in a standard normal distribution, of a value between 0 and 0.5.

Because the standard normal distribution is symmetric, only positive values are given in this table, but the area for a value of a less than 0 can be found easily. For instance, $P(0 < x < 0.5) = P(0 > x > -0.5) = P(-0.5 < x < 0)$. The shaded area in Figure D-2 represents the area $(-0.5 < x < 0)$, and its area and probability are 0.19146.

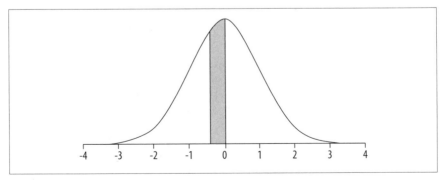

Figure D-2. Area (-0.5 < x < 0) of the standard normal distribution

x	0.00	0.01	0.02	0.03	0.04	0.05	0.06	0.07	0.08	0.09
0.0	0.00000	0.00399	0.00798	0.01197	0.01595	0.01994	0.02392	0.02790	0.03188	0.03586
0.1	0.03983	0.04380	0.04776	0.05172	0.05567	0.05962	0.06356	0.06749	0.07142	0.07535
0.2	0.07926	0.08317	0.08706	0.09095	0.09483	0.09871	0.10257	0.10642	0.11026	0.11409
0.3	0.11791	0.12172	0.12552	0.12930	0.13307	0.13683	0.14058	0.14431	0.14803	0.15173
0.4	0.15542	0.15910	0.16276	0.16640	0.17003	0.17364	0.17724	0.18082	0.18439	0.18793
0.5	0.19146	0.19497	0.19847	0.20194	0.20540	0.20884	0.21226	0.21566	0.21904	0.22240
0.6	0.22575	0.22907	0.23237	0.23565	0.23891	0.24215	0.24537	0.24857	0.25175	0.25490
0.7	0.25804	0.26115	0.26424	0.26730	0.27035	0.27337	0.27637	0.27935	0.28230	0.28524
0.8	0.28814	0.29103	0.29389	0.29673	0.29955	0.30234	0.30511	0.30785	0.31057	0.31327
0.9	0.31594	0.31859	0.32121	0.32381	0.32639	0.32894	0.33147	0.33398	0.33646	0.33891
1.0	0.34134	0.34375	0.34614	0.34849	0.35083	0.35314	0.35543	0.35769	0.35993	0.36214
1.1	0.36433	0.36650	0.36864	0.37076	0.37286	0.37493	0.37698	0.37900	0.38100	0.38298
1.2	0.38493	0.38686	0.38877	0.39065	0.39251	0.39435	0.39617	0.39796	0.39973	0.40147
1.3	0.40320	0.40490	0.40658	0.40824	0.40988	0.41149	0.41308	0.41466	0.41621	0.41774
1.4	0.41924	0.42073	0.42220	0.42364	0.42507	0.42647	0.42785	0.42922	0.43056	0.43189
1.5	0.43319	0.43448	0.43574	0.43699	0.43822	0.43943	0.44062	0.44179	0.44295	0.44408
1.6	0.44520	0.44630	0.44738	0.44845	0.44950	0.45053	0.45154	0.45254	0.45352	0.45449
1.7	0.45543	0.45637	0.45728	0.45818	0.45907	0.45994	0.46080	0.46164	0.46246	0.46327
1.8	0.46407	0.46485	0.46562	0.46638	0.46712	0.46784	0.46856	0.46926	0.46995	0.47062
1.9	0.47128	0.47193	0.47257	0.47320	0.47381	0.47441	0.47500	0.47558	0.47615	0.47670
2.0	0.47725	0.47778	0.47831	0.47882	0.47932	0.47982	0.48030	0.48077	0.48124	0.48169
2.1	0.48214	0.48257	0.48300	0.48341	0.48382	0.48422	0.48461	0.48500	0.48537	0.48574
2.2	0.48610	0.48645	0.48679	0.48713	0.48745	0.48778	0.48809	0.48840	0.48870	0.48899
2.3	0.48928	0.48956	0.48983	0.49010	0.49036	0.49061	0.49086	0.49111	0.49134	0.49158
2.4	0.49180	0.49202	0.49224	0.49245	0.49266	0.49286	0.49305	0.49324	0.49343	0.49361
2.5	0.49379	0.49396	0.49413	0.49430	0.49446	0.49461	0.49477	0.49492	0.49506	0.49520
2.6	0.49534	0.49547	0.49560	0.49573	0.49585	0.49598	0.49609	0.49621	0.49632	0.49643
2.7	0.49653	0.49664	0.49674	0.49683	0.49693	0.49702	0.49711	0.49720	0.49728	0.49736
2.8	0.49744	0.49752	0.49760	0.49767	0.49774	0.49781	0.49788	0.49795	0.49801	0.49807
2.9	0.49813	0.49819	0.49825	0.49831	0.49836	0.49841	0.49846	0.49851	0.49856	0.49861
3.0	0.49865	0.49869	0.49874	0.49878	0.49882	0.49886	0.49889	0.49893	0.49896	0.49900
3.1	0.49903	0.49906	0.49910	0.49913	0.49916	0.49918	0.49921	0.49924	0.49926	0.49929
3.2	0.49931	0.49934	0.49936	0.49938	0.49940	0.49942	0.49944	0.49946	0.49948	0.49950
3.3	0.49952	0.49953	0.49955	0.49957	0.49958	0.49960	0.49961	0.49962	0.49964	0.49965
3.4	0.49966	0.49968	0.49969	0.49970	0.49971	0.49972	0.49973	0.49974	0.49975	0.49976
3.5	0.49977	0.49978	0.49978	0.49979	0.49980	0.49981	0.49981	0.49982	0.49983	0.49983
3.6	0.49984	0.49985	0.49985	0.49986	0.49986	0.49987	0.49987	0.49988	0.49988	0.49989
3.7	0.49989	0.49990	0.49990	0.49990	0.49991	0.49991	0.49992	0.49992	0.49992	0.49992
3.8	0.49993	0.49993	0.49993	0.49994	0.49994	0.49994	0.49994	0.49995	0.49995	0.49995
3.9	0.49995	0.49995	0.49996	0.49996	0.49996	0.49996	0.49996	0.49996	0.49997	0.49997
4.0	0.49997	0.49997	0.49997	0.49997	0.49997	0.49997	0.49998	0.49998	0.49998	0.49998

Figure D-3. Probability table for the standard normal distribution

One way to orient yourself with respect to an unfamiliar distribution table is to find a few familiar values. For instance, the Z-value of 1.96 should be familiar to you due to its association with 95% two-tailed confidence intervals. If you find the cell where 1.9 (row) and .06 (column) intersect, you will see the cell value of 0.47500. This is $P(0 < x < 1.96)$, and if you double it, adding in the probability for $P(-1.96 < x < 0)$, you get 0.95 or 95%. To look at it another way, 0.025 of the area in the standard normal distribution lies above the value of 1.96 and 0.025 below the value of –1.96, so only 5% of the values in a standard normal distribution lie outside the range (–1.96, 1.96). This is why, when we use the alpha value of 0.05 for a statistical test based on the normal distribution, a result that translates to a standard normal score outside the range (–1.96, 1.96) is considered significant; a value this extreme would occur less than 5% of the time if the null hypothesis were true.

Working a few examples might make this table easier to understand. Often, finding the probability of a result involves adding probabilities from both sides of 0. For instance, in Chapter 3 (Figure 3-4) we found that the Z-value for a score of 105, from a population distributed $x \sim N(100, 5)$, was 1.00. To find the probability of a value at least as high as 105, given a population distributed $x \sim N(100, 5)$, is the same thing as finding the probability of a Z-score at least as high as 1.00 in the standard normal distribution. The easiest way to find $P(Z > 1.0)$ is to find the area below $Z = 1.00$ and then subtract that area from 1.0 (the total area under the normal curve). The area below $Z = 1.00$ includes the area from negative infinity to zero and the area from 0 to 1.00. We know the former is 0.5 (because half the area in the standard normal distribution lies below 0 and half above 0), and using the table, we find the latter is 0.34134. Therefore, if we denote the score as X:

$P(X < 105) = P(Z < =1.00) = 0.50000 + 0.34134 = 0.84134$, $P(X > 105) = P(Z > 1.00) = 1 - 0.84134$ or 0.15866.

Returning to the original problem, this result tells us that there is about a 15.9% chance of a score greater than 105, if the scores in a population are distributed $\sim N(100, 5)$. To put it another way, a score of 105 from such a population ranks in about the top 16% of this population. The shaded area in Figure D-4 represents this area.

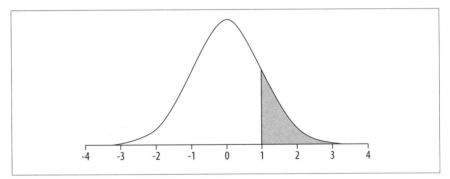

Figure D-4. Area of the standard normal distribution for P (Z > 1.00)

We also calculate in Chapter 3 (Figure 3-5) that a value of 95, from a population distributed $N \sim (100, 5)$, translates to a Z-score of –1.00. Suppose we want to know what proportion of values lie below this score. To do this, we use two facts:

- By the definition of the standard normal distribution, the area below 0 (between negative infinity and 0) is 0.5000.
- The area between –1.00 and 0 is 0.34134 (the same as the area between 0 and 1.00).

Therefore, the area below –1.00 is 0.5000 – .34134, or 0.15866. Note that this is the same area as that above $Z = 1.00$, not a surprising result given the symmetry of the standard normal distribution. The shaded area in Figure D-5 represents this area.

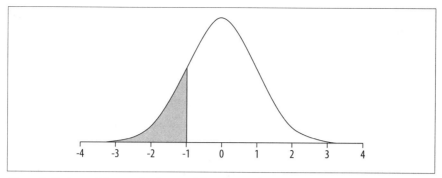

Figure D-5. Area of the standard normal distribution for P(Z < –1.00)

It is easy to get confused about which area you are calculating when working with a probability table, so it might help to draw a sketch to indicate exactly what you need to add or subtract to come up with the answer.

In Chapter 3 (Figure 3-6), we calculate that a value of 35, from a population ~N(50, 10), translates to a Z-score of –1.50. To calculate the probability of a score *higher* than 35 from this population, we note that the answer will include the area between –1.5 and 0, and the area above 0 (between 0 and positive infinity).

$$P(Z > –1.5) = .43319 + 0.50000 = 0.93319.$$

Therefore, the probability of a score above 35 in this population is 93.3%. This corresponds to the shaded area in Figure D-6.

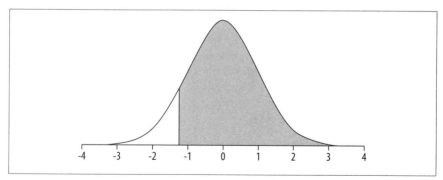

Figure D-6. Area of the standard normal distribution for P(Z > –1.50)

For the purpose of finding probabilities using a standard normal table, it doesn't matter whether a Z-value represents a single score or a sample mean. In Chapter 3 (Figure 3-21), we calculated the Z-statistic for a sample mean of 52 from a sample of 30 drawn from a population with a mean of 50 and a standard deviation of 10. This sample mean corresponds to a Z-statistic of 1.10. To find the probability of a Z-statistic at least this high, we calculate as follows:

$$P(Z > 1.10) = 1 – P(Z < 1.10) = 1 – (0.5000 + 0.36433) = 0.13567$$

If instead we were interested in the probability of a score at least this low, we would calculate:

$$P(Z < 1.10) = (0.5000 + 0.36433) = 0.86433$$

As previously, the 0.5000 represents the probability from negative infinity to 0, whereas 0.36433 represents the probability from 0 to 1.10.

Here are the probabilities for the other Z-distribution examples from Chapter 3.

Figure 3-22: $Z = 1.55$

$$P(Z > 1.55) = 1 - P(Z < 1.55) = 1 - (0.50000 + 0.43943) = 0.06057\ P(Z < 1.55)$$
$$= (0.50000 + 0.43943) = 0.93943$$

Figure 3-23: $Z = 2.00$

$$P(Z > 2.00) = 1 - P(Z < 2.00) = 1 - (0.50000 + 0.47725) = 0.02275\ P(Z < 2.00)$$
$$= (0.50000 + 0.47725) = 0.97725$$

The t-Distribution

Because the t-statistic has a different distribution for every degree of freedom, t-tables are usually abbreviated to show only certain critical values (otherwise, the tables would be immense). In the table presented in Figure D-7, the column labeled v represents degrees of freedom, whereas the columns 0.10, 0.05, and so on present the probability of exceeding the critical value for a t-distribution with the degrees of freedom indicated by v. These are one-tailed values and, because the t-distribution is symmetric, to get the two-tailed probability, you choose the probability column whose value is half of the α-value you want.

Suppose you want to find the critical value for a two-tailed t-test with 20 degrees of freedom and $\alpha = 0.05$. You choose the column for 0.025 (because $0.05/2 = 0.025$), read down to the row $v = 20$, and find the critical value of 2.086. This means that the t-statistic from your experiment must be greater than 2.086, or less than –2.086, for you to reject the null hypothesis.

If we were performing a one-tailed test with $v = 20$ and $\alpha = 0.05$, we would use the column labeled 0.05, go down to the row $v = 20$ as before, and find that our critical value is 1.725.

Upper critical values of Student's t distribution with ν degrees of freedom

Probability of exceeding the critical value

ν	0.10	0.05	0.025	0.01	0.005	0.001
1.	3.078	6.314	12.706	31.821	63.657	318.313
2.	1.886	2.920	4.303	6.965	9.925	22.327
3.	1.638	2.353	3.182	4.541	5.841	10.215
4.	1.533	2.132	2.776	3.747	4.604	7.173
5.	1.476	2.015	2.571	3.365	4.032	5.893
6.	1.440	1.943	2.447	3.143	3.707	5.208
7.	1.415	1.895	2.365	2.998	3.499	4.782
8.	1.397	1.860	2.306	2.896	3.355	4.499
9.	1.383	1.833	2.262	2.821	3.250	4.296
10.	1.372	1.812	2.228	2.764	3.169	4.143
11.	1.363	1.796	2.201	2.718	3.106	4.024
12.	1.356	1.782	2.179	2.681	3.055	3.929
13.	1.350	1.771	2.160	2.650	3.012	3.852
14.	1.345	1.761	2.145	2.624	2.977	3.787
15.	1.341	1.753	2.131	2.602	2.947	3.733
16.	1.337	1.746	2.120	2.583	2.921	3.686
17.	1.333	1.740	2.110	2.567	2.898	3.646
18.	1.330	1.734	2.101	2.552	2.878	3.610
19.	1.328	1.729	2.093	2.539	2.861	3.579
20.	1.325	1.725	2.086	2.528	2.845	3.552
21.	1.323	1.721	2.080	2.518	2.831	3.527
22.	1.321	1.717	2.074	2.508	2.819	3.505
23.	1.319	1.714	2.069	2.500	2.807	3.485
24.	1.318	1.711	2.064	2.492	2.797	3.467
25.	1.316	1.708	2.060	2.485	2.787	3.450
26.	1.315	1.706	2.056	2.479	2.779	3.435
27.	1.314	1.703	2.052	2.473	2.771	3.421
28.	1.313	1.701	2.048	2.467	2.763	3.408
29.	1.311	1.699	2.045	2.462	2.756	3.396
30.	1.310	1.697	2.042	2.457	2.750	3.385
31.	1.309	1.696	2.040	2.453	2.744	3.375
32.	1.309	1.694	2.037	2.449	2.738	3.365
33.	1.308	1.692	2.035	2.445	2.733	3.356
34.	1.307	1.691	2.032	2.441	2.728	3.348
35.	1.306	1.690	2.030	2.438	2.724	3.340
36.	1.306	1.688	2.028	2.434	2.719	3.333
37.	1.305	1.687	2.026	2.431	2.715	3.326
38.	1.304	1.686	2.024	2.429	2.712	3.319
39.	1.304	1.685	2.023	2.426	2.708	3.313
40.	1.303	1.684	2.021	2.423	2.704	3.307
41.	1.303	1.683	2.020	2.421	2.701	3.301
42.	1.302	1.682	2.018	2.418	2.698	3.296
43.	1.302	1.681	2.017	2.416	2.695	3.291
44.	1.301	1.680	2.015	2.414	2.692	3.286
45.	1.301	1.679	2.014	2.412	2.690	3.281
46.	1.300	1.679	2.013	2.410	2.687	3.277
47.	1.300	1.678	2.012	2.408	2.685	3.273
48.	1.299	1.677	2.011	2.407	2.682	3.269
49.	1.299	1.677	2.010	2.405	2.680	3.265
50.	1.299	1.676	2.009	2.403	2.678	3.261
51.	1.298	1.675	2.008	2.402	2.676	3.258
52.	1.298	1.675	2.007	2.400	2.674	3.255
53.	1.298	1.674	2.006	2.399	2.672	3.251
54.	1.297	1.674	2.005	2.397	2.670	3.248
55.	1.297	1.673	2.004	2.396	2.668	3.245
56.	1.297	1.673	2.003	2.395	2.667	3.242
57.	1.297	1.672	2.002	2.394	2.665	3.239
58.	1.296	1.672	2.002	2.392	2.663	3.237
59.	1.296	1.671	2.001	2.391	2.662	3.234
60.	1.296	1.671	2.000	2.390	2.660	3.232
61.	1.296	1.670	2.000	2.389	2.659	3.229
62.	1.295	1.670	1.999	2.388	2.657	3.227
63.	1.295	1.669	1.998	2.387	2.656	3.225
64.	1.295	1.669	1.998	2.386	2.655	3.223
65.	1.295	1.669	1.997	2.385	2.654	3.220
66.	1.295	1.668	1.997	2.384	2.652	3.218
67.	1.294	1.668	1.996	2.383	2.651	3.216
68.	1.294	1.668	1.995	2.382	2.650	3.214
69.	1.294	1.667	1.995	2.382	2.649	3.213
70.	1.294	1.667	1.994	2.381	2.648	3.211
71.	1.294	1.667	1.994	2.380	2.647	3.209
72.	1.293	1.666	1.993	2.379	2.646	3.207
73.	1.293	1.666	1.993	2.379	2.645	3.206
74.	1.293	1.666	1.993	2.378	2.644	3.204
75.	1.293	1.665	1.992	2.377	2.643	3.202
76.	1.293	1.665	1.992	2.376	2.642	3.201
77.	1.293	1.665	1.991	2.376	2.641	3.199
78.	1.292	1.665	1.991	2.375	2.640	3.198
79.	1.292	1.664	1.990	2.374	2.640	3.197
80.	1.292	1.664	1.990	2.374	2.639	3.195
81.	1.292	1.664	1.990	2.373	2.638	3.194
82.	1.292	1.664	1.989	2.373	2.637	3.193
83.	1.292	1.663	1.989	2.372	2.636	3.191
84.	1.292	1.663	1.989	2.372	2.636	3.190
85.	1.292	1.663	1.988	2.371	2.635	3.189
86.	1.291	1.663	1.988	2.370	2.634	3.188
87.	1.291	1.663	1.988	2.370	2.634	3.187
88.	1.291	1.662	1.987	2.369	2.633	3.185
89.	1.291	1.662	1.987	2.369	2.632	3.184
90.	1.291	1.662	1.987	2.368	2.632	3.183
91.	1.291	1.662	1.986	2.368	2.631	3.182
92.	1.291	1.662	1.986	2.368	2.630	3.181
93.	1.291	1.661	1.986	2.367	2.630	3.180
94.	1.291	1.661	1.986	2.367	2.629	3.179
95.	1.291	1.661	1.985	2.366	2.629	3.178
96.	1.290	1.661	1.985	2.366	2.628	3.177
97.	1.290	1.661	1.985	2.365	2.627	3.176
98.	1.290	1.661	1.984	2.365	2.627	3.175
99.	1.290	1.660	1.984	2.365	2.626	3.175
100.	1.290	1.660	1.984	2.364	2.626	3.174
∞	1.282	1.645	1.960	2.326	2.576	3.090

Figure D-7. Selected critical values for the t-distribution

Although this *t*-table will not give you the precise probability of every *t*-value, you can use it for hypothesis testing. For instance, in Chapter 6, Figure 6-7, we calculated a *t*-statistic of –3.87 for a one-sample *t*-test with 14 degrees of freedom. For a two-tailed test with an alpha value of 0.05, the critical value is 2.145 (in Figure D-7, the row for 14 *df* and a probability of 0.25). Our value as calculated from the data is more extreme than this, so we reject the null hypothesis.

In Chapter 6, Figure 6-12, we calculated the *t*-statistic of 1.01 for a two-tailed *t*-test for two independent groups with 18 degrees of freedom and alpha = 0.05. Looking at Figure D-7, we see that the critical value in this case is 2.101. The *t*-statistic calculated from our data is less extreme (closer to 0) than the critical value, so for this data, we fail to reject the null hypothesis.

The Binomial Distribution

Because the binomial has a different distribution for every combination of *n* and *p*, binomial tables can be quite large. Fortunately, the normal distribution can be used to approximate the binomial when *np* and *n(1 – p)* are both greater than or equal to 5, hence there is less need for tables with large values for *n*. We include an excerpt here from binomial probability and cumulative binomial probability tables generated by University of New Brunswick professor William Knight; the complete tables are available from his website (*http://www.math.unb.ca/~knight/utility/*). Figure D-8 presents binomial probabilities for *n* = 3 – 10, whereas Figure D-9 presents cumulative binomial probabilities for *n* = 3 – 10.

To find a binomial probability, you need to know *n* (the number of trials), *k* (the number of successes), and *p* (the probability of success on any trial). First find the table for *n*, and then find where the row for *k* and the column for *p* intersect; that cell holds the binomial probability (in Figure D-8) or cumulative binomial probability (in Figure D-9) for the specified result. For example, in Chapter 3 (Figure 3-10), we calculated *b*(1;5, 0.5), the probability of exactly 1 success in 5 trials, with *p* = 0.5, to be 0.16. To confirm this probability in Figure D-8, we find the table for *n* = 5 and then the intersection of the row for *k* = 1 and the column for *p* = 0.5. The probability of this result is 0.15625, which, after rounding, is the same as the value we calculated. If we want to know the probability of *no more than* 1 success in 5 trials—the probability of 0 or 1 successes—we would use the cumulative probability table (Figure D-9). Following the same procedure, we find that the cumulative probability for 0 or 1 successes in 5 trials at *p* = 0.5 is 0.18750. We could have gotten the same value by adding the probabilities of 0 and 1 success from Figure D-8: 0.03125 + 0.156250 = 0.18750.

Suppose we want to answer a different question: what is the probability of *at least* 1 success in 5 trials, with *p* = 0.5? The easiest way to answer this is to calculate the probability of 0 successes and then subtract that probability from 1. Because *b*(0;5, 0.5) = 0.03125 (the probability of 0 successes), the probability of more than 0 successes, that is, 1 or more successes, is 1 – 0.03125, or 0.96875.

```
N = 3
K \ P=.1    .2      .3      .4      .5      .6      .7      .8      .9
-----------------------------------------------------------------------
0 | 0.729  0.512  0.343  0.216  0.125  0.064  0.027  0.008  0.001
1 | 0.243  0.384  0.441  0.432  0.375  0.288  0.189  0.096  0.027
2 | 0.027  0.096  0.189  0.288  0.375  0.432  0.441  0.384  0.243
3 | 0.001  0.008  0.027  0.064  0.125  0.216  0.343  0.512  0.729

N = 4
K \ P=.1    .2      .3      .4      .5      .6      .7      .8      .9
-----------------------------------------------------------------------
0 | 0.6561 0.4096 0.2401 0.1296 0.0625 0.0256 0.0081 0.0016 0.0001
1 | 0.2916 0.4096 0.4116 0.3456 0.2500 0.1536 0.0756 0.0256 0.0036
2 | 0.0486 0.1536 0.2646 0.3456 0.3750 0.3456 0.2646 0.1536 0.0486
3 | 0.0036 0.0256 0.0756 0.1536 0.2500 0.3456 0.4116 0.4096 0.2916
4 | 0.0001 0.0016 0.0081 0.0256 0.0625 0.1296 0.2401 0.4096 0.6561

N = 5
K \ P=.1     .2       .3       .4       .5       .6       .7       .8       .9
-----------------------------------------------------------------------------------
0 | 0.59049 0.32768 0.16807 0.07776 0.03125 0.01024 0.00243 0.00032 0.00001
1 | 0.32805 0.40960 0.36015 0.25920 0.15625 0.07680 0.02835 0.00640 0.00045
2 | 0.07290 0.20480 0.30870 0.34560 0.31250 0.23040 0.13230 0.05120 0.00810
3 | 0.00810 0.05120 0.13230 0.23040 0.31250 0.34560 0.30870 0.20480 0.07290
4 | 0.00045 0.00640 0.02835 0.07680 0.15625 0.25920 0.36015 0.40960 0.32805
5 | 0.00001 0.00032 0.00243 0.01024 0.03125 0.07776 0.16807 0.32768 0.59049

N = 6
K \ P=.1     .2       .3       .4       .5       .6       .7       .8       .9
-----------------------------------------------------------------------------------
0 | 0.53144 0.26214 0.11765 0.04666 0.01562 0.00410 0.00073 0.00006 0.00000
1 | 0.35429 0.39322 0.30253 0.18662 0.09375 0.03686 0.01021 0.00154 0.00005
2 | 0.09842 0.24576 0.32414 0.31104 0.23438 0.13824 0.05954 0.01536 0.00122
3 | 0.01458 0.08192 0.18522 0.27648 0.31250 0.27648 0.18522 0.08192 0.01458
4 | 0.00122 0.01536 0.05954 0.13824 0.23438 0.31104 0.32414 0.24576 0.09842
5 | 0.00005 0.00154 0.01021 0.03686 0.09375 0.18662 0.30253 0.39322 0.35429
6 | 0.00000 0.00006 0.00073 0.00410 0.01562 0.04666 0.11765 0.26214 0.53144

N = 7
K \ P=.1     .2       .3       .4       .5       .6       .7       .8       .9
-----------------------------------------------------------------------------------
0 | 0.47830 0.20972 0.08235 0.02799 0.00781 0.00164 0.00022 0.00001 0.00000
1 | 0.37201 0.36700 0.24706 0.13064 0.05469 0.01720 0.00357 0.00036 0.00001
2 | 0.12400 0.27525 0.31765 0.26127 0.16406 0.07741 0.02500 0.00430 0.00017
3 | 0.02296 0.11469 0.22689 0.29030 0.27344 0.19354 0.09724 0.02867 0.00255
4 | 0.00255 0.02867 0.09724 0.19354 0.27344 0.29030 0.22689 0.11469 0.02296
5 | 0.00017 0.00430 0.02500 0.07741 0.16406 0.26127 0.31765 0.27525 0.12400
6 | 0.00001 0.00036 0.00357 0.01720 0.05469 0.13064 0.24706 0.36700 0.37201
7 | 0.00000 0.00001 0.00022 0.00164 0.00781 0.02799 0.08235 0.20972 0.47830

N = 8
K \ P=.1     .2       .3       .4       .5       .6       .7       .8       .9
-----------------------------------------------------------------------------------
0 | 0.43047 0.16777 0.05765 0.01680 0.00391 0.00066 0.00007 0.00000 0.00000
1 | 0.38264 0.33554 0.19765 0.08958 0.03125 0.00786 0.00122 0.00008 0.00000
2 | 0.14880 0.29360 0.29648 0.20902 0.10938 0.04129 0.01000 0.00115 0.00002
3 | 0.03307 0.14680 0.25412 0.27869 0.21875 0.12386 0.04668 0.00918 0.00041
4 | 0.00459 0.04588 0.13614 0.23224 0.27344 0.23224 0.13614 0.04588 0.00459
5 | 0.00041 0.00918 0.04668 0.12386 0.21875 0.27869 0.25412 0.14680 0.03307
6 | 0.00002 0.00115 0.01000 0.04129 0.10938 0.20902 0.29648 0.29360 0.14880
7 | 0.00000 0.00008 0.00122 0.00786 0.03125 0.08958 0.19765 0.33554 0.38264
8 | 0.00000 0.00000 0.00007 0.00066 0.00391 0.01680 0.05765 0.16777 0.43047

N = 9
K \ P=.1     .2       .3       .4       .5       .6       .7       .8       .9
-----------------------------------------------------------------------------------
0 | 0.38742 0.13422 0.04035 0.01008 0.00195 0.00026 0.00002 0.00000 0.00000
1 | 0.38742 0.30199 0.15565 0.06047 0.01758 0.00354 0.00041 0.00002 0.00000
2 | 0.17219 0.30199 0.26683 0.16124 0.07031 0.02123 0.00386 0.00029 0.00000
3 | 0.04464 0.17616 0.26683 0.25082 0.16406 0.07432 0.02100 0.00275 0.00006
4 | 0.00744 0.06606 0.17153 0.25082 0.24609 0.16722 0.07351 0.01652 0.00083
5 | 0.00083 0.01652 0.07351 0.16722 0.24609 0.25082 0.17153 0.06606 0.00744
6 | 0.00006 0.00275 0.02100 0.07432 0.16406 0.25082 0.26683 0.17616 0.04464
7 | 0.00000 0.00029 0.00386 0.02123 0.07031 0.16124 0.26683 0.30199 0.17219
8 | 0.00000 0.00002 0.00041 0.00354 0.01758 0.06047 0.15565 0.30199 0.38742
9 | 0.00000 0.00000 0.00002 0.00026 0.00195 0.01008 0.04035 0.13422 0.38742

N =10
K \ P= .1    .2       .3       .4       .5       .6       .7       .8       .9
-----------------------------------------------------------------------------------
0  | 0.34868 0.10737 0.02825 0.00605 0.00098 0.00010 0.00001 0.00000 0.00000
1  | 0.38742 0.26844 0.12106 0.04031 0.00977 0.00157 0.00014 0.00000 0.00000
2  | 0.19371 0.30199 0.23347 0.12093 0.04395 0.01062 0.00145 0.00007 0.00000
3  | 0.05740 0.20133 0.26683 0.21499 0.11719 0.04247 0.00900 0.00079 0.00001
4  | 0.01116 0.08808 0.20012 0.25082 0.20508 0.11148 0.03676 0.00551 0.00014
5  | 0.00149 0.02642 0.10292 0.20066 0.24609 0.20066 0.10292 0.02642 0.00149
6  | 0.00014 0.00551 0.03676 0.11148 0.20508 0.25082 0.20012 0.08808 0.01116
7  | 0.00001 0.00079 0.00900 0.04247 0.11719 0.21499 0.26683 0.20133 0.05740
8  | 0.00000 0.00007 0.00145 0.01062 0.04395 0.12093 0.23347 0.30199 0.19371
9  | 0.00000 0.00000 0.00014 0.00157 0.00977 0.04031 0.12106 0.26844 0.38742
10 | 0.00000 0.00000 0.00001 0.00010 0.00098 0.00605 0.02825 0.10737 0.34868
```

Figure D-8. Binomial probabilities for n = 3 – 10

```
N = 2
K \ P=.1    .2    .3    .4    .5    .6    .7    .8    .9
-----------------------------------------------------------
0 | 0.81 0.64 0.49 0.36 0.25 0.16 0.09 0.04 0.01
1 | 0.99 0.96 0.91 0.84 0.75 0.64 0.51 0.36 0.19
2 | 1.00 1.00 1.00 1.00 1.00 1.00 1.00 1.00 1.00


N = 3
K \ P=.1     .2     .3     .4     .5     .6     .7     .8     .9
-----------------------------------------------------------------
0 | 0.729 0.512 0.343 0.216 0.125 0.064 0.027 0.008 0.001
1 | 0.972 0.896 0.784 0.648 0.500 0.352 0.216 0.104 0.028
2 | 0.999 0.992 0.973 0.936 0.875 0.784 0.657 0.488 0.271
3 | 1.000 1.000 1.000 1.000 1.000 1.000 1.000 1.000 1.000


N = 4
K \ P=.1      .2      .3      .4      .5      .6      .7      .8      .9
-------------------------------------------------------------------------
0 | 0.6561 0.4096 0.2401 0.1296 0.0625 0.0256 0.0081 0.0016 0.0001
1 | 0.9477 0.8192 0.6517 0.4752 0.3125 0.1792 0.0837 0.0272 0.0037
2 | 0.9963 0.9728 0.9163 0.8208 0.6875 0.5248 0.3483 0.1808 0.0523
3 | 0.9999 0.9984 0.9919 0.9744 0.9375 0.8704 0.7599 0.5904 0.3439
4 | 1.0000 1.0000 1.0000 1.0000 1.0000 1.0000 1.0000 1.0000 1.0000


N = 5
K \ P=.1       .2       .3       .4       .5       .6       .7       .8       .9
-----------------------------------------------------------------------------------
0 | 0.59049 0.32768 0.16807 0.07776 0.03125 0.01024 0.00243 0.00032 0.00001
1 | 0.91854 0.73728 0.52822 0.33696 0.18750 0.08704 0.03078 0.00672 0.00046
2 | 0.99144 0.94208 0.83692 0.68256 0.50000 0.31744 0.16308 0.05792 0.00856
3 | 0.99954 0.99328 0.96922 0.91296 0.81250 0.66304 0.47178 0.26272 0.08146
4 | 0.99999 0.99968 0.99757 0.98976 0.96875 0.92224 0.83193 0.67232 0.40951
5 | 1.00000 1.00000 1.00000 1.00000 1.00000 1.00000 1.00000 1.00000 1.00000


N = 6
K \ P=.1       .2       .3       .4       .5       .6       .7       .8       .9
-----------------------------------------------------------------------------------
0 | 0.53144 0.26214 0.11765 0.04666 0.01562 0.00410 0.00073 0.00006 0.00000
1 | 0.88574 0.65536 0.42018 0.23328 0.10938 0.04096 0.01094 0.00160 0.00006
2 | 0.98415 0.90112 0.74431 0.54432 0.34375 0.17920 0.07047 0.01696 0.00127
3 | 0.99873 0.98304 0.92953 0.82080 0.65625 0.45568 0.25569 0.09888 0.01585
4 | 0.99994 0.99840 0.98906 0.95904 0.89062 0.76672 0.57982 0.34464 0.11426
5 | 1.00000 0.99994 0.99927 0.99590 0.98438 0.95334 0.88235 0.73786 0.46856
6 | 1.00000 1.00000 1.00000 1.00000 1.00000 1.00000 1.00000 1.00000 1.00000


N = 7
K \ P=.1       .2       .3       .4       .5       .6       .7       .8       .9
-----------------------------------------------------------------------------------
0 | 0.47830 0.20972 0.08235 0.02799 0.00781 0.00164 0.00022 0.00001 0.00000
1 | 0.85031 0.57672 0.32942 0.15863 0.06250 0.01884 0.00379 0.00037 0.00001
2 | 0.97431 0.85197 0.64707 0.41990 0.22656 0.09626 0.02880 0.00467 0.00018
3 | 0.99727 0.96666 0.87396 0.71021 0.50000 0.28979 0.12604 0.03334 0.00273
4 | 0.99982 0.99533 0.97120 0.90374 0.77344 0.58010 0.35293 0.14803 0.02569
5 | 0.99999 0.99963 0.99621 0.98116 0.93750 0.84137 0.67058 0.42328 0.14969
6 | 1.00000 0.99999 0.99978 0.99836 0.99219 0.97201 0.91765 0.79028 0.52170
7 | 1.00000 1.00000 1.00000 1.00000 1.00000 1.00000 1.00000 1.00000 1.00000


N = 8
K \ P=.1       .2       .3       .4       .5       .6       .7       .8       .9
-----------------------------------------------------------------------------------
0 | 0.43047 0.16777 0.05765 0.01680 0.00391 0.00066 0.00007 0.00000 0.00000
1 | 0.81310 0.50332 0.25530 0.10638 0.03516 0.00852 0.00129 0.00008 0.00000
2 | 0.96191 0.79692 0.55177 0.31539 0.14453 0.04981 0.01129 0.00123 0.00002
3 | 0.99498 0.94372 0.80590 0.59409 0.36328 0.17367 0.05797 0.01041 0.00043
4 | 0.99957 0.98959 0.94203 0.82633 0.63672 0.40591 0.19410 0.05628 0.00502
5 | 0.99998 0.99877 0.98871 0.95019 0.85547 0.68461 0.44823 0.20308 0.03809
6 | 1.00000 0.99992 0.99871 0.99148 0.96484 0.89362 0.74470 0.49668 0.18690
7 | 1.00000 1.00000 0.99993 0.99934 0.99609 0.98320 0.94235 0.83223 0.56953
8 | 1.00000 1.00000 1.00000 1.00000 1.00000 1.00000 1.00000 1.00000 1.00000


N = 9
K \ P=.1       .2       .3       .4       .5       .6       .7       .8       .9
-----------------------------------------------------------------------------------
0 | 0.38742 0.13422 0.04035 0.01008 0.00195 0.00026 0.00002 0.00000 0.00000
1 | 0.77484 0.43621 0.19600 0.07054 0.01953 0.00380 0.00043 0.00002 0.00000
2 | 0.94703 0.73820 0.46283 0.23179 0.08984 0.02503 0.00429 0.00031 0.00000
3 | 0.99167 0.91436 0.72966 0.48261 0.25391 0.09935 0.02529 0.00307 0.00006
4 | 0.99911 0.98042 0.90119 0.73343 0.50000 0.26657 0.09881 0.01958 0.00089
5 | 0.99994 0.99693 0.97471 0.90065 0.74609 0.51739 0.27034 0.08564 0.00833
6 | 1.00000 0.99969 0.99571 0.97497 0.91016 0.76821 0.53717 0.26180 0.05297
7 | 1.00000 0.99998 0.99957 0.99620 0.98047 0.92946 0.80400 0.56379 0.22516
8 | 1.00000 1.00000 0.99998 0.99974 0.99805 0.98992 0.95965 0.86578 0.61258
9 | 1.00000 1.00000 1.00000 1.00000 1.00000 1.00000 1.00000 1.00000 1.00000


N =10
K \ P= .1      .2       .3       .4       .5       .6       .7       .8       .9
-----------------------------------------------------------------------------------
0 | 0.34868 0.10737 0.02825 0.00605 0.00098 0.00010 0.00001 0.00000 0.00000
1 | 0.73610 0.37581 0.14931 0.04636 0.01074 0.00168 0.00014 0.00000 0.00000
2 | 0.92981 0.67780 0.38278 0.16729 0.05469 0.01229 0.00159 0.00008 0.00000
3 | 0.98720 0.87913 0.64961 0.38228 0.17188 0.05476 0.01059 0.00086 0.00001
4 | 0.99837 0.96721 0.84973 0.63310 0.37695 0.16624 0.04735 0.00637 0.00015
5 | 0.99985 0.99363 0.95265 0.83376 0.62305 0.36690 0.15027 0.03279 0.00163
6 | 0.99999 0.99914 0.98941 0.94524 0.82812 0.61772 0.35039 0.12087 0.01280
7 | 1.00000 0.99992 0.99841 0.98771 0.94531 0.83271 0.61722 0.32220 0.07019
8 | 1.00000 1.00000 0.99986 0.99832 0.98926 0.95364 0.85069 0.62419 0.26390
9 | 1.00000 1.00000 0.99999 0.99990 0.99902 0.99395 0.97175 0.89263 0.65132
10 | 1.00000 1.00000 1.00000 1.00000 1.00000 1.00000 1.00000 1.00000 1.00000
```

Figure D-9. Cumulative binomial probabilities for n = 3 – 10

The Chi-Square Distribution

The chi-square distribution is not symmetrical, as can be seen from Figure D-10, and for this reason, the upper and lower critical values differ. In practice, the upper critical values are used far more commonly, and therefore only the upper critical values are included in this appendix. The shape of the chi-square distribution varies according to the degrees of freedom, and each distribution has a separate set of critical values. In the interest of space, we have included a chi-square table for up to 40 degrees of freedom; a table for up to 100 degrees of freedom, as well as a table of lower-tail critical values, is available online in the *NIST/SEMATECH e-Handbook of Statistical Methods (http://itl.nist.gov/div898/handbook/eda/section3/eda3674.htm)*.

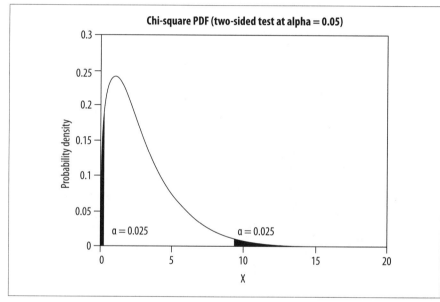

Figure D-10. Chi-square distribution, two-tailed test

To use the chi-square table in Figure D-11, find the row corresponding to the degrees of freedom (labeled v) and then go across to the column for the correct upper-tail probability (assuming a one-tailed test). For a chi-square test with 1 degree of freedom and $\alpha = 0.05$, the critical value is 3.841. This is the value that the test statistic must exceed to reject the null hypothesis. To put it another way, if the null hypothesis is true, there is only a 5% chance of an experiment with one degree of freedom returning a chi-square statistic of 3.841 or greater. For a chi-square test with 5 degrees of freedom and $\alpha = 0.01$, the critical value is 15.086.

Consider the example from Table 5-7 in Chapter 5. The experiment returned a chi-square value of 21.8 with 3 degrees of freedom. We can see from the chi-square table in Figure D-11 that the critical value for alpha = 0.5 and 3 degrees of freedom is 7.815. Our value is larger than that, so we will reject the null hypothesis.

Upper critical values of chi-square distribution with ν degrees of freedom

ν	Probability of exceeding the critical value 0.10	0.05	0.025	0.01	0.001
1	2.706	3.841	5.024	6.635	10.828
2	4.605	5.991	7.378	9.210	13.816
3	6.251	7.815	9.348	11.345	16.266
4	7.779	9.488	11.143	13.277	18.467
5	9.236	11.070	12.833	15.086	20.515
6	10.645	12.592	14.449	16.812	22.458
7	12.017	14.067	16.013	18.475	24.322
8	13.362	15.507	17.535	20.090	26.125
9	14.684	16.919	19.023	21.666	27.877
10	15.987	18.307	20.483	23.209	29.588
11	17.275	19.675	21.920	24.725	31.264
12	18.549	21.026	23.337	26.217	32.910
13	19.812	22.362	24.736	27.688	34.528
14	21.064	23.685	26.119	29.141	36.123
15	22.307	24.996	27.488	30.578	37.697
16	23.542	26.296	28.845	32.000	39.252
17	24.769	27.587	30.191	33.409	40.790
18	25.989	28.869	31.526	34.805	42.312
19	27.204	30.144	32.852	36.191	43.820
20	28.412	31.410	34.170	37.566	45.315
21	29.615	32.671	35.479	38.932	46.797
22	30.813	33.924	36.781	40.289	48.268
23	32.007	35.172	38.076	41.638	49.728
24	33.196	36.415	39.364	42.980	51.179
25	34.382	37.652	40.646	44.314	52.620
26	35.563	38.885	41.923	45.642	54.052
27	36.741	40.113	43.195	46.963	55.476
28	37.916	41.337	44.461	48.278	56.892
29	39.087	42.557	45.722	49.588	58.301
30	40.256	43.773	46.979	50.892	59.703
31	41.422	44.985	48.232	52.191	61.098
32	42.585	46.194	49.480	53.486	62.487
33	43.745	47.400	50.725	54.776	63.870
34	44.903	48.602	51.966	56.061	65.247
35	46.059	49.802	53.203	57.342	66.619
36	47.212	50.998	54.437	58.619	67.985
37	48.363	52.192	55.668	59.893	69.347
38	49.513	53.384	56.896	61.162	70.703
39	50.660	54.572	58.120	62.428	72.055
40	51.805	55.758	59.342	63.691	73.402

Figure D-11. Upper-tail chi-square critical values

E

Online Resources

There are many statistical resources available on the Internet, and no published list could possibly be complete, nor would it want to be; too much information can be as bad as too little. As is true of the Internet in general, not every resource online is accurate or reliable, so it's up to the user to decide whether a particular resource is appropriate to his use. The web pages listed here are all maintained by reputable sources, including the federal government, university departments of statistics, professional statisticians, and companies that produce widely used statistical products.

General Resources

- The Statistics Online Computational Resource (*http://socr.ucla.edu/SOCR .html*)

  Many resources, including interactive tools and course materials, from the UCLA Statistics Online Computational Resource.

- Rice Virtual Lab in Statistics (*http://onlinestatbook.com/rvls.html*)

  A collection of resources, including an online textbook, simulations and demonstrations, cases studies, and statistical analysis tools.

- Web Pages that Perform Statistical Calculations (*http://statpages.org/index .html*)

  Links to many tools, including statistical decision trees, free statistical software, online calculators, and graphing programs, maintained by John C. Pezzullo, a retired professor of biostatistics and pharmacology.

- Wolfram Demonstrations Project: Statistics (*http://demonstrations.wolfram .com/topic.html?topic=Statistics&limit=20*)

  A collection of interactive tools related to statistical topics from the Wolfram Demonstrations Project; none require Mathematica, and all are open-code and designed to run on any standard computer running Windows, Macintosh, or Linux.

- StatLib: Data, Software and News from the Statistics Community (*http://lib.stat.cmu.edu/index.php*)

  A website dedicated to distributing statistical software, data sets, and information, maintained on the website of Carnegie Mellon University.

- CAUSEweb (*http://www.causeweb.org/resources/links.php*)

  A large collection of links relevant to statistics education, compiled by Juha Puranen of the University of Helsinki, and maintained on the website of the Consortium for the Advancement of Undergraduate Statistics Education (CAUSE). Categories include course materials, data sets, demonstrations, statistical software, and texts; many of the links will be useful to practicing statisticians as well as educators.

- "Ask Dr. Math." (*http://mathforum.org/dr.math/*)

  Searchable archive of answers to mathematical and statistical questions, ranging in difficulty from elementary school through college.

- Mathematics Review Manual, Department of Mathematics and Statistics, McMaster University (*http://www.math.mcmaster.ca/lovric/rm/MathReviewManual.pdf*)

  Provides a review of mathematical concepts from basic algebra through calculus, along with advice about how to learn and understand mathematics. There are solved problems and a quiz for each topic covered.

- The World Wide Web Virtual Library: Statistics (*http://www.stat.ufl.edu/vlib/statistics.html*)

  A collection of links compiled by the University of Florida Department of Statistics; categories include data sources, educational institutions, professional organizations, software venders, mailings lists, and news groups.

- College Board: AP Statistics Course Home Page (*http://apcentral.collegeboard.com/apc/public/courses/teachers_corner/2151.html*)

  A collection of links to materials relevant to the AP Statistics course (taught in American high schools), compiled by the College Board (the company that develops and administers the test). Categories include information about the test itself (including practice tests), teaching materials, and short articles on statistical topics relevant to the course.

Glossaries

- StatSoft Statistics Glossary (*http://www.statsoft.com/textbook/statistics-glossary/*)

  A detailed glossary maintained by the company that produces Statistica software.

- EXCITE! Glossary of Epidemiology Terms (*http://www.cdc.gov/excite/library/glossary.htm*)

  A glossary of epidemiological terms maintained by the U.S. Centers for Disease Control and Prevention (CDC); definitions are drawn from *Principles of*

Epidemiology in Public Health Practice, 3rd edition, a self-study course developed by the CDC for health care professionals.

- Pocket Dictionary of Statistics (*http://www.mhhe.com/business/opsci/bstat/key term.mhtml*)

  A glossary of terms used in business statistics, written by Hardeo Sahai and Anwer Khurshid and maintained on the website of the Higher Education division of the publisher, McGraw-Hill.

- Six Sigma Glossary (*http://www.micquality.com/six_sigma_glossary/index.htm*)

  A glossary of terms used in the Six Sigma quality control program, maintained on the website of MiC Quality, a company that provides Six Sigma training courses and educational materials.

- A Glossary for Multilevel Analysis (*http://www.paho.org/English/DD/AIS/be _v24n3-multilevel.htm*)

  A glossary of terms relevant to multilevel analysis, written by Dr. Ana V. Diez Roux, a Columbia University professor, and maintained on the website of the Pan American Health Organization.

Probability Tables

- Tables for Probability Distributions (*http://itl.nist.gov/div898/handbook/eda/sec tion3/eda367.htm*)

  Public domain tables for the standard normal distribution, *t*-distribution, *F*-distribution, and chi-square distribution, from the National Institute of Standards and Technology.

- William Knight: Public Domain Tables (*http://www.math.unb.ca/~knight/util ity/*)

  Public domain tables for confidence intervals for the median, the U-test, the sign test, binomial coefficients, binomial probabilities, the standard normal distribution (including a short table for teaching), the *t*-distribution, the chi-square distribution, the *F*-distribution, and square roots, from Professor William Knight of the University of New Brunswick.

Online Calculators

- QuickCalcs: Online Calculators for Scientists (*http://graphpad.com/quickcalcs/ index.cfm*)

  Page containing a variety of online statistical calculators, maintained by the scientific software company, GraphPad.

- Applets for the Cybergnostics Project (*http://www.stat.tamu.edu/~west/applets/*)

  A collection of statistical calculators and graphical demonstrations of statistical concepts, written by R. Webster West, a professor of statistics at Texas A&M; this collection is particularly useful in teaching because it allows students to perform simulations and to change the parameters of different distributions and observe how the shape of the distributions change.

- Power and Sample Size Programs (*http://www.epibiostat.ucsf.edu/biostat/samp size.html*)

  Links to many power and sample size calculators, and related information and software, on a site maintained by Steve Shiboski, a professor in the UCSF Dept. of Epidemiology and Biostatistics.

- Java Applets for Power and Sample Size (*http://homepage.stat.uiowa.edu/ ~rlenth/Power/*)

  Graphical interface to address many common power and sample size questions, maintained by Russell V. Lenth, a professor of statistics and actuarial science at the University of Iowa; the software may be run on the site or downloaded to the user's PC.

Online Textbooks

- General Statistics Curriculum E-Book (*http://wiki.stat.ucla.edu/socr/index.php/ EBook*)

  Online textbook for the AP statistics course, from the UCLA Statistics Online Computational Resource.

- Statistics at Square One (*http://www.bmj.com/about-bmj/resources-readers/pub lications/statistics-square-one*)

  Statistics at Square One, now in its ninth edition, is an introductory statistics textbook particularly useful for medical personnel.

- Research Methods Knowledge Base (*http://www.socialresearchmethods.net/kb/*)

  The Research Methods Knowledge Base, a web-based textbook created by William M.K. Trochim, a professor of policy analysis and management at Cornell University; it covers topics usually taught in a social science research methods course, including research design, sampling, analytical techniques, and writing the results.

- StatSoft Electronic Statistics Textbook (*http://www.statsoft.com/textbook/*)

  The StatSoft Electronic Statistics Textbook, from the company that created Statistica; it includes information about many advanced techniques, including CHAID analysis, data mining techniques, and structural equation modeling.

F

Glossary of Statistical Terms

One challenge in any profession is learning the specific vocabulary required. This appendix provides a quick guide to key terms and notation used in this book; statistical vocabulary is covered in far greater detail in reference books such as *The Cambridge Dictionary of Statistics* (Cambridge University Press, 2010) and *The Concise Encyclopedia of Statistics* (Springer, 2008).

Table F-1. The Greek alphabet

Capital letter	Lowercase letter	Greek name	Capital letter	Lowercase letter	Greek name
A	α	Alpha	N	ν	Nu
B	β	Beta	Ξ	ξ	Xi
Γ	γ	Gamma	O	o	Omicron
Δ	δ	Delta	Π	δ	Pi
E	ε	Epsilon	P	ρ	Rho
Z	ζ	Zeta	Σ	σ	Sigma
H	η	Eta	T	τ	Tau
Θ	θ	Theta	Y	υ	Upsilon
I	ι	Iota	Φ	φ	Phi
K	κ	Kappa	X	χ	Chi
Λ	λ	Lambda	Ψ	ψ	Psi
M	μ	Mu	Ω	ω	Omega

Table F-2. Statistical notation

Symbol	Meaning
S	Sample space (probability theory)
E	Event (probability theory)
∪	Union of sets

Symbol	Meaning
∩	Intersection of sets
$P(A)$	Probability of event A
$P(A|B)$	Probability of event A, given event B
$P(\sim A)$	Probability of A complement (probability of *not-A*)
e	Euler's constant, the irrational number 2.718 . . .
ln	Natural logarithm (logarithm to base e)
$\log_x$	Logarithm to base x
x_i	The ith member of sample x
$\overline{x}$	Sample mean
μ	Population mean
s	Sample standard deviation
σ	Population standard deviation
s^2	Sample variance
s^2_p	Pooled sample variance
σ^2	Population variance
n	Sample size
N	Population size
r	Sample correlation
r_{pb}	Point-biserial correlation
r_s	Spearman rank-order correlation (Spearman's rho)
γ	Goodman and Kruskal's gamma
τ_a, τ_B, τ_C	Kendall's tau-A, tau-B, tau-C
ϕ	Phi (measure of association between two binary variables)
P	Concordant pairs (ordinal measures of association)
Q	Discordant pairs (ordinal measures of association)
ρ	Population correlation
χ^2	Chi-square
O	Observed value (chi-square)
E	Expected value (chi-square)
$R \times C$	A table with R rows and C columns
E	Expected value (for calculating chi-square)
H_0	Null hypothesis
H_A, H_1	Alternate hypothesis
a	Alpha, the probability of Type I error
β	Beta, the probability of Type II error
Σ	Summation
nPk	Permutation

Symbol	Meaning
$nCk \binom{n}{k}$	Combination
$n!$	n factorial, that is, $n \times (n-1) \times (n-2) \times \dots 1$
t	Student's t
df	Degrees of freedom
Z	Standard normal score/distribution
SS	Sums of squares
MS	Mean square
I_t	Index for time t (business statistics)
Qit	Quantity of product i for time t (business statistics)
Pit	Price of product i for time t (business statistics)
Tt	Secular trend (time series)
Ct	Cyclical effect (time series)
St	Seasonal effect (time series)
Rt	Residual or error effect trend (time series)
RR	Risk ratio (relative risk)
OR	Odds ratio
OR_{MH}	Mantel-Haenszel odds ratio
$D+, D-$	Disease, no disease (epidemiology)
$E+, E-$	Exposure, no exposure (epidemiology)
δ	Delta, effect size (sample size and power calculations)
T	True score, true component (measurement theory)
E	Error component (measurement theory)
X	Observed score (measurement theory)
$\hat{P}_{xx'}$	Estimated reliability (Spearman-Brown prophecy formula, measurement theory)
$\bar{\alpha}$	Coefficient alpha (measurement theory)
KR_{20}, KR_{21}	Kuder-Richardson 20, Kuder-Richardson 21 (measurement theory)

Absolute value

The numerical value of a number, disregarding its sign; the absolute values of -4 and 4 are both 4. Using notation: $|-4| = |4| = 4$.

Alpha (α)

In experimental design, the probability of a Type I error, that is, the probability of rejecting the null hypothesis when it is true.

A priori hypothesis

A hypothesis specified before any testing is done.

Beta (β)

In experimental design, the probability of a Type II error, that is, the probability of failing to reject the null hypothesis when it is false. $1 - \beta$ = power.

Bias

Error that is systematic and might lead to incorrect interpretation of results.

Binary variables

Variables that can take only two values; also called *dichotomous variables*.

Blinding

In experimental design, keeping the people involved in a study ignorant of important aspects of the study, for instance, which participants received an experimental treatment and which received a placebo. In a single-blind trial, information is withheld from the participants. In a double-blind trial, information is withheld from both the subjects and the researchers administering the treatment. In a triple-blind trial, information is withheld from the participants, the researchers administering the treatment, and the researchers evaluating the data.

Categorical data

See *nominal data*.

Cohort

A group of people having a time-related factor in common (for instance, being born in 1950 or entering college in 2000).

Confounding variable

In research design, a variable that correlates with both the independent and dependent variables and is not in the causal pathway between them.

Construct validity

The extent to which a measurement, or series of measurements, adequately measures a construct (such as intelligence).

Content validity

The extent to which an instrument (such as a test) adequately reflects the content domain it is intended to represent.

Continuous data

Data that might take any value or any value within a range.

Criterion validity

The extent to which a measurement correlates with something else, for instance, how well scores on an IQ test correlate with grades in school.

Control variables

Variables included in a study design not because they are the focus of interest but because they are believed to influence the variables of interest and the researcher wants to control for their effect.

Cross-sectional study

A study in which data is collected at a single point in time.

Degrees of freedom

The number of values free to vary in an equation or statistic.

Dependent variables

In research design, variables that are assumed to be influenced by other, independent variables included in the design.

Detection bias
> Bias because some qualities might be more likely to be detected in some people than in others.

Dichotomous variables
> See *binary variables*.

Discrete data
> Data that might take only specified values.

Double blind
> See *blinding*.

Error score
> In measurement theory, the error component of an observed score.

Factorial design
> A design including two or more categorical variables and their interactions; in a *full factorial design*, all possible combinations of the variables are included in the study.

Incidence
> In medicine and epidemiology, the number of new cases of a disease or condition in a population at risk over some period.

Independent variables
> In research design, variables that are believed to exert influence on other, dependent variables included in the design.

Index numbers
> In business and economic statistics, a number used to measure the changes in quantity and/or price over time for a good or combination of goods; a well-known example is the Consumer Price Index (CPI).

Information bias
> Bias due to the way data is collected and recorded.

Interaction variable
> A variable for which the relationships between two other variables are different, depending on the level of the interaction variable.

Internal consistency
> In test theory, the extent to which the items on an instrument (for instance, a test) measure the same thing.

Interquartile range
> The range of numbers that contains the central 50% of the values of a variable.

Interval data
> Data that may be ordered, and in which equal intervals between consecutive values may be assumed; also known as *equal interval data*.

Likert scale
> A type of ordinal rating scale developed by the psychologist Rensis Likert; a Likert scale presents a statement and asks people to indicate their agreement or disagreement using an ordered scale.

Maximax

A method of decision making under uncertainty, with the goal of maximizing the largest anticipated gain.

Maximin

A method of decision making under uncertainty, with the goal of maximizing the smallest anticipated gain.

Mean

The arithmetic average of a set of numbers.

Median

The central value of a set of numbers when they are ranked by value.

Minimax

A method of decision making under uncertainty, with the goal of minimizing opportunity loss.

Mode

The most common value of a variable.

Nominal data

Data that do not have numeric meaning and for which numeric values serve only as labels (such as gender or color). Also called *categorical data.*

Nonparametric statistics

Statistics not based on assumptions about the distribution of the population(s) from which the study data have been drawn or which make less stringent assumptions than parametric statistics.

Nonprobability sampling

Sampling in which the probability of selection for any unit or combination of units is unknown; examples include convenience sampling and quota sampling.

Nonresponse bias

Bias due to some members of a sample declining to participate in a study or declining to supply some requested information.

Observed score

In measurement theory, the value of something as observed, including the error of measurement.

Operationalization

In research, the process of specifying how a concept will be defined and measured.

Ordinal data

Data that may be ordered, that is, ranked in size but without the assumption of equal intervals between consecutive values.

Parametric statistics

Statistics based on assumptions about the distribution of the population(s) from which the study data have been drawn.

Placebo

In experimental design, a treatment that is expected to have no effect on the outcome.

Post hoc test

Tests conducted after some other test; for instance, post hoc tests can be conducted after the overall *F*-test for an ANOVA to see which groups differ from other groups.

Power

In study design, the probability of rejecting the null hypothesis when it is false. Power = $1 - \beta$, that is, $1 - P(\text{Type II error})$.

Prevalence

In medicine and epidemiology, the number of cases of a disease or condition at some point in time; prevalence includes both new and existing cases.

Probability sampling

Sampling methods in which all combinations of members of the population have a known probability of selection; examples include simple random sampling and stratified sampling.

Proportion

A ratio in which all the cases in the numerator are also included in the denominator, for instance, the proportion of females with cancer in the United States (the denominator is all people, both male and female, with cancer in the United States).

Prospective study

A study in which individuals are followed (and data collected) moving forward in time.

Proxy measurement

Substituting one measurement for another.

Random error

Error that is due to chance. Random error makes measurement less precise but does not introduce bias.

Range

The difference between the highest and lower values of a variable.

Rate

A proportion expressed to include a unit of time, such as injuries per year on a job site.

Ratio

A method of expressing the relationship between the magnitude of two numbers; the numbers do not need to share a common unit (for instance, hospital beds per 1,000 population).

Ratio data

Data that may be ordered, is equal interval, and has a natural zero value.

Recall bias

Bias because life experiences can make some people better able to recall events.

Reliability

How consistent or repeatable measurements are over time.

Retrospective study
> A study of events that have already taken place.

Selection bias
> Bias due to the way a sample is selected.

Sensitivity
> In medicine and epidemiology, the probability that a person who has a disease will test positive for it.

Single blind
> See *blinding*.

Social desirability bias
> Bias due to the human tendency to present oneself in the best possible light.

Specificity
> In medicine and epidemiology, the probability that a person without a disease will test negative for it.

Standard deviation
> The square root of variance; for a set of numbers, the square root of the mean squared deviations from the mean.

Standard error
> The standard deviation of the sampling distribution of the sample mean.

Statistical significance
> A result that is unlikely to be due to chance.

Systematic error
> Error due to some cause other than chance; systematic error can make observed values consistently higher or lower than true values and thus introduce bias.

Triple blind
> See *blinding*.

True score
> In measurement theory, the value of something when measured without error.

Type I error
> In experimental design, rejecting the null hypothesis when it is true.

Type II error
> In experimental design, failing to reject the null hypothesis when it is false.

Unique identifier
> A code or variable used to identify all the records belonging to a single unit of analysis (for instance, a patient ID to identify all the hospital services provided to a single patient).

Validity
> How closely a measurement actually measures what it is intended to measure.

Variance

A measure of the variability of a range of numbers, calculated as the mean squared difference from the mean.

Volunteer bias

A type of selection bias resulting from collecting data from a sample of volunteers.

Glossary

Index

We'd like to hear your suggestions for improving our indexes. Send email to *index@oreilly.com*.

Cartesian coordinates (rectangular coordinates), 482
case control design, 429
case-control studies, 367
categorical data, 3, 121–146
  (see also nominal data)
  about, 121–122
  chi-square distributions, 125
  chi-square test
    about, 127
    for equality of proportions, 130
    for independence, 127–130, 131
    McNemar's test, 134–136
    of goodness of fit, 130–131
  correlation statistics for, 138–145
  Fisher's Exact Test, 132–134
  Likert scale, 145
  measures of agreement, 123–125
  nominal data and, 3
  proportions, 136–138
  R×C table, 122
  semantic differential scale, 146
category-specific rates, 358–362
Cattell, James, 292
ceiling effect, 108
Census (U.S.), samples and, 55
central limit theorem, 59–62
Central Moving Average (CMA), 333
central tendency, measures of
  about, 84
  critiquing choice in article of, 466–467
  in descriptive statistics
    mean, 84–87
  mean
    in descriptive statistics, 84–87
    in inferential statistics, 45
  median, 87–88–90
  mode in, 88–90
checklist for statistics based investigations, 461–462
chi-square distributions, 125, 537
chi-square test
  about, 127
  for equality of proportions, 130
  for independence
    about, 127–130

Yates's correction for continuity, in chi-square test, 131
  McNemar's test, 134–136
  of goodness of fit, 130–131
CI (Cumulative Incidence), 355
classic experimental design, 426
classical test theory, 393–394
Cluster analysis, 299–302
cluster samples, 58
CMA (Central Moving Average), 333
codebooks, data management, 413–415
coefficient
  about term, 194
  of determination, 187
  of equivalence, 11, 395, 397
  of precision, 398
  of stability, 11, 395
coefficient alpha
  Cronbach's alpha, 12, 398–399
  Kuder-Richardson formulas, 398, 399
Coefficient of Variation (CV), 95–96
Cohen's kappa, 123–125
cohort, 426, 546
coins, 37
combinations, 491
combinations of elements, 28–29
common causes of variation, 340
communicating with statistics, 449–456
complement of event, 25–26
complex random samples, 58
composite indices, 327
composite test
  reliability of, 394–395
  scores, 391
compound events, 24
conclusions and results, critiquing in articles, 458
concurrent validity, 13
conditional probabilities, 31–32
confidence coefficient, 68
confidence intervals
  about, 67–68
  calculating for risk ratio, 364
  critiquing in articles, 467
  for independent samples (two-sample) t-test, 163
  for one-sample t-test, 159–160
  for proportions, 378

E

EBCDIC (Extended Binary Coded Decimal Interchange Code), 422
ecological
  fallacy, 427
  studies, 427
  validity, 430
educational and psychological statistics
  about, 385–386
  classical test theory, 393–394
  item analysis, 400–403
  item response theory, 403–408
  measures of internal consistency, 395–399
  percentiles, 386–388
  reliability of composite test, 394–395
  standardized scores, 388–390
  test construction, 390–393
80–20 rule, 105
EMA (Exponential Moving Average), 334
email addresses and web page, for book, xviii
Engineering Statistics Handbook (National Institute of Standards and Technology), 167
epidemiological and medical statistics
  about, 351
  category-specific rates, 358–362
  confounding, 370–373
  crude rate, 357–358
  incidence, 355–357
  Mantel-Haenszel (MH) common odds ratio, 373–375
  measures of disease frequency, 351
  odds ratio, 367–369
  power analysis, 375–377
  prevalence, 354–355, 357
  ratio, proportion, and rate, 354–355
  risk ratio, 362–367
  standardization, 358–362
  stratified analysis, 372–373
equations
  graphing, 482–485
  linear, 482
  linear equations, 194
  linear inequalities, 486
  solving, 479–481
  systems of, 480–482
error scores
  definition of, 547
  true scores and, 8–10
etiologic fraction, 366
Euclidean distance, 300
events, definition of, 24
Excel
  bar charts in, 100
  for data management, 418
  graphing in, 97
  rectangular data file in, 416
  using for statistical package, 510–512
expected values, 127
experimental design
  about, 425
  blinding, 444
  blocking and Latin square, 445
  classification of studies, 426
  communicating with statistics, 449–456
  data types, 427
  example of, 447–448
  experimental studies, 436
  factor in, 426
  factorial design, 426
  gathering experimental data, 437–447
  hypothesis testing vs. data mining, 443
  ingredients of good design, 436–437
  issues in, 463–465
  observational studies, 428–430
  physical vs. social sciences definition of treatments, 442
  quasi-experimental studies, 431–434
  retrospective adjustment, 445
  style of notation, 427
  types of, 426
  unit of analysis in study, 427
experimental units
  about, 436
  identifying, 438–440
Exponential Moving Average (EMA), 334
exponents, 476–477
Extended Binary Coded Decimal Interchange Code (EBCDIC), 422

incorrect use of tests in inferential
statistics, 470–471
issues in research design, 463–465
peer review process, 453–454
writing for, 451–453, 451–454

K

Kaiser normalization, 295
kappa (kappa coefficient), 123–125
Kendall, Maurice, 144–145
Kendall's tau-a, 144, 145
Kendall's tau-b, 144–145
Kendall's tau-c, 145
Knight, William, binomial distribution
probability tables, 534
Kolmogorov–Smirnov test, 72–74
Kruskal-Wallis H test, 314–316
Kuder-Richardson formulas, 398, 399

L

lag, 331
large-sample Z test for proportions, 136–
138
Laspeyres index, 327–328
Latin square, in experimental design, 446
LDFs (Linear Discriminant Functions),
302
Levene's test, 167
Likert scale, 18, 145, 547
Likert, Rensis, 147
line graphs, 114–116
linear algebra, 176
Linear Discriminant Functions (LDFs),
302
linear equations, 194, 482
linear inequalities, 486
linear regression
about, 195–197
arbitrary curve-fitting, 285–287
assumptions, 198–205
calculating by hand, 212–214
cubic regression model, 282–285
logistic regression, 273–279
logit outcome variable, 273
multinomial logistic regression, 279–
281
multiple

about, 243–244
adding interaction term, 253–255
assumptions, 245
creating a correlation matrix, 249–
251
dummy variables, 258–260
methods for building regression
models, 260–267
modeling principles, 244–245
regression equation for data, 257–
258
results for individual predictors,
251
standardized coefficients, 251
variables in model, 245–249
polynomial regression, 282–285
quadratic, 282–285
violations of assumptions of, 471
literature review
critiquing in articles, 458
writing, 452
Little, Donald B., 422–424
local independence assumption, 405
logarithms (log)
about, 478–479
in solving equations, 480
logistic regression, 273–279
logit outcome variable, 273, 279

M

Mahalanobis distance, 300
main effects, 426
Manhattan distance, 300
Mann-Whitney U test, 312
Mantel-Haenszel (MH) common odds
ratio, 373–375
marginal frequencies, 362
marginals, 128
matching, 371
The Mathematics of Games and Gambling
(Packel), 43
maturation bias, 446
maximax decision making procedure,
336–337, 548
maximin decision making procedure,
336–337, 548
McNemar's chi-square test, 134–136
mean

multiple-occasions reliability (test-retest reliability), 11, 395
multivariate, 111
mutual exclusive events, 26

N

Naperian logarithms, 478
National Institute of Standards and Technology Engineering Statistics Handbook, 167
National Institute of Standards and Technology (U.S. (NIST/SEMATECH e-Handbook of Statistical Methods)), 527
natural logarithms, 478
negative discrimination, 401
Nelson quality control rules, 345
Nelson, Lloyd S., 345
Nightingale, Florence, 103
NIST/SEMATECH e-Handbook of Statistical Methods (National Institute of Standards and Technology (U.S.)), 527
NNT (Number Needed to Treat), 366
nominal data
    about, 2–3
    definition of, 548
nonparametric statistics
    about, 307
    definition of, 548
    parametric statistics and, 72, 307
nonprobability sampling, 55–56, 548
nonresponse bias, 15, 548
norm group, 386
norm-referenced
    scoring, 386
    tests, 391
normal distribution, 47–50, 109
normal distribution, standard, 528–532
normal score, 48–50, 386
normalized scores, 48–50
null hypothesis, 65
number line, 472, 474
Number Needed to Treat (NNT), 366
numeric and string data, 422

O

observational studies, 428–430
observed score, 548
observed values, 127
odds ratio, 367–369
odds, calculating, 369
OLS (Ordinary Least Squares) regression equation, 54
omnibus F-test, 210
one-group pretest-posttest design, 432
one-sample t-test, 157–160
one-way ANOVA
    about, 206–210
    t-test and, 155
online resources, 539–542
operationalization, 5, 548
opportunity loss table, 336
ordinal data
    about, 3–4, 122
    definition of, 548
    mean rank, 308
    measures of agreement, 123–125
    rank sum, 308
    R×C table, 122
ordinal variables, correlation statistics for
    gamma, 143–144
    Kendall's tau-a, 144, 145
    Kendall's tau-b, 144–145
    Kendall's tau-c, 145
    Somers's d, 145
    Spearman's rank-order coefficient, 142
Ordinary Least Squares (OLS) regression equation, 54
orthogonality, in research design structure, 437
outliers, 96–97
overfitting, 285–287

P

p-values
    about, 68–69
    of Z value, 71
Paasche index, 328–330
Packel, Edward, The Mathematics of Games and Gambling, 43

definition of, 550
in data management, 420
unit of analysis, 417, 427
univariate, 111

V

validity
about, 12
definition of, 550
reliability and, 10–14
variable names, in transfer process to
software, 420
variance
definition of, 551
formula for, 93
in inferential statistics, 46
standard deviation and, 92–96
variation, in research design, 464
Velicer partial correlation procedure, 293
Venn diagrams, 24
Venn, John, Symbolic Logic, 24
The Visual Display of Quantitative
Information (Tufte), 97
volunteer bias, 15, 551
volunteer samples, 56

W

Watters, P.A, "Caffeine and cortical
arousal", 282–283
web page and email addresses, for book,
xviii
weighted composite index, 327
Weighted Moving Average (WMA), 334
Welch's t-test, 167–168
Western Electric quality control rules,
345
Wilcoxon rank sum test, 308–312
Wilcoxon signed ranks test, 317–319
Wilks's lambda, 303
Winsorized mean, 86–87
within-subjects designs, tests for, 317–
321, 446
workplace, writing for, 455–456, 459–
460
writing about statistics
articles, 452–453

checklist for statistics based
investigations, 461–462
common problems with, 459–460
critiquing descriptive statistics, 466–
470
for general public, 454–455
for workplace, 455–456
incorrect use of tests in inferential
statistics, 470–471
issues in research design, 463–465
scientific papers, 452–453

Y

Yates's correction for continuity, in chi-
square test for independence,
131

Z

Z distribution, 47
Z-scores, 48–50, 386–388
Z-statistic, 70–72

About the Author

Sarah Boslaugh holds a PhD in Research and Evaluation from the City University of New York and has been working as a statistical analyst for 20 years in a variety of professional settings, including the New York City Board of Education, the Institutional Research Office of the City University of New York, Montefiore Medical Center, the Virginia Department of Social Services, Magellan Health Services, Washington University School of Medicine, and BJC HealthCare. She has taught statistics in several different contexts and is currently a grant and proposal writer at Kennesaw State University. She has published two previous books: *An Intermediate Guide to SPSS Programming: Using Syntax for Data Management* (SAGE Publications, 2004) and *Secondary Data Sources for Public Health* (Cambridge University Press, 2007) and edited the *Encyclopedia of Epidemiology for SAGE Publications* (2007). Her latest book, *Healthcare Systems Around the World: A Comparative Guide* will be published by SAGE in 2013.

Colophon

The animal on the cover of *Statistics in a Nutshell* is a thornback crab, also known as a spiny spider crab (*Maja squinado, Maja brachydactyla*). Found in the northeast Atlantic Ocean and the Mediterranean Sea, the thornback crab is the largest of the European crabs, with a carapace diameter of two to seven inches. It is easily identifiable by the two hornlike spikes between its eyes, and the six or so smaller spikes that extend from each side of its shell. The thornback's body is reddish with pink, brown, or yellow markings, and its surface is also covered with small spikes, as the crab's name implies.

Thornback crabs are occasionally found on the shore, but they prefer depths of 90 to 600 feet. They are solitary animals except during mating season, when they form large breeding mounds. In years when their numbers are particularly abundant, they can be a source of frustration for lobster fisherman as they infest the lobster pots. Thornbacks are themselves fished for their delicious claw meat.

Male thornbacks are effective predators; their delicate-looking claws are actually quite powerful and can open small mussels to feed on them. Their claws are also double-jointed, so although it is generally safe for a person to hold crustaceans by each side of their shells, thornbacks are able to reach over their backs to pinch the offender. Females have smaller, less flexible claws and are thus more vulnerable to attack. To defend against their predators—which include lobsters, wrasses, and cuttlefish—many species of spider crabs decorate their spiny shells with seaweed, sponges, or aquatic debris to better blend in against the seabed.

The cover image is from *Lydekker's Library of Natural History*. The cover font is Adobe ITC Garamond. The text font is Linotype Birka; the heading font is Adobe Myriad Condensed; and the code font is LucasFont's TheSansMonoCondensed.

Have it your way.

Get even more for your money.

Join the O'Reilly Community, and register the O'Reilly books you own. It's free, and you'll get:

- $4.99 ebook upgrade offer
- 40% upgrade offer on O'Reilly print books
- Membership discounts on books and events
- Free lifetime updates to ebooks and videos
- Multiple ebook formats, DRM FREE
- Participation in the O'Reilly community
- Newsletters
- Account management
- 100% Satisfaction Guarantee

Signing up is easy:

1. Go to: oreilly.com/go/register
2. Create an O'Reilly login.
3. Provide your address.
4. Register your books.

Note: English-language books only

To order books online:
oreilly.com/store

For questions about products or an order:
orders@oreilly.com

To sign up to get topic-specific email announcements and/or news about upcoming books, conferences, special offers, and new technologies:
elists@oreilly.com

For technical questions about book content:
booktech@oreilly.com

To submit new book proposals to our editors:
proposals@oreilly.com

O'Reilly books are available in multiple DRM-free ebook formats. For more information:
oreilly.com/ebooks

Spreading the knowledge of innovators   oreilly.com